Giant Resonances in Atoms, Molecules, and Solids

NATO ASI Series

Advanced Science Institutes Series

A series presenting the results of activities sponsored by the NATO Science Committee, which aims at the dissemination of advanced scientific and technological knowledge, with a view to strengthening links between scientific communities.

The series is published by an international board of publishers in conjunction with the NATO Scientific Affairs Division

A	Life Sciences	Plenum Publishing Corporation
B	Physics	New York and London
C	Mathematical and Physical Sciences	D. Reidel Publishing Company Dordrecht, Boston, and Lancaster
D	Behavioral and Social Sciences	Martinus Nijhoff Publishers
E	Engineering and Materials Sciences	The Hague, Boston, Dordrecht, and Lancaster
F	Computer and Systems Sciences	Springer-Verlag
G	Ecological Sciences	Berlin, Heidelberg, New York, London,
H	Cell Biology	Paris, and Tokyo

Recent Volumes in this Series

Volume 150—Particle Physics: *Cargèse 1985*
   edited by Maurice Lévy, Jean-Louis Basdevant, Maurice Jacob,
   David Speiser, Jacques Weyers, and Raymond Gastmans

Volume 151—Giant Resonances in Atoms, Molecules, and Solids
   edited by J. P. Connerade, J. M. Esteva, and
   R. C. Karnatak

Volume 152—Optical Properties of Narrow-Gap Low-Dimensional Structures
   edited by C. M. Sotomayor Torres, J. C. Portal, J. C. Mann,
   and R. A. Stradling

Volume 153—Physics of Strong Fields
   edited by W. Greiner

Volume 154—Strongly Coupled Plasma Physics
   edited by Forrest J. Rogers and Hugh E. Dewitt

Volume 155—Low-Dimensional Conductors and Superconductors
   edited by D. Jérome and L. G. Caron

Volume 156—Gravitation in Astrophysics: *Cargèse 1986*
   edited by B. Carter and J. B. Hartle

Series B: Physics

Giant Resonances in Atoms, Molecules, and Solids

Edited by
J. P. Connerade
Imperial College
London, England

and
J. M. Esteva and
R. C. Karnatak
University of Paris
Orsay, France

Plenum Press
New York and London
Published in cooperation with NATO Scientific Affairs Division

Proceedings of a NATO Advanced Study Institute on
Giant Resonances in Atoms, Molecules, and Solids,
held June 16–26, 1986,
in Les Houches, France

Library of Congress Cataloging in Publication Data

NATO Advanced Study Institute on Giant Resonances in Atoms, Molecules, and Solids (1986: Les Houches, Haute-Savoie, France)
  Giant resonances in atoms, molecules, and solids.

  (NATO ASI series. Series B, Physics; v. 151)
  "Proceedings of a NATO Advanced Study Institute on Giant Resonances in Atoms, Molecules, and Solids, held June 16–26, 1986, in Les Houches, France"—T.p. verso.
  "Published in cooperation with NATO Scientific Affairs Division."
  Includes bibliographical references and index.
  1. Nuclear magnetic resonance, Giant—Congresses. 2. Atomic theory—Congresses. 3. Molecular spectra—Congresses. 4. Solids—Spectra—Congresses. 5. Atomic spectra—Congresses. I. Connerade, J. P. II. Esteva, J. M. III. Karnatak, R. C. IV. Title. V. Series.
  QC762.N35 1986   538'.362   87-7950
  ISBN 0-306-42564-5

© 1987 Plenum Press, New York
A Division of Plenum Publishing Corporation
233 Spring Street, New York, N.Y. 10013

All rights reserved. No part of this book may be reproduced, stored in a retrieval system, or transmitted in any form or by any means, electronic, mechanical, photocopying, microfilming, recording, or otherwise, without written permission from the Publisher

Printed in the United States of America

FOREWORD

Often, a new area of science grows at the confines between recognised subject divisions, drawing upon techniques and intellectual perspectives from a diversity of fields. Such growth can remain unnoticed at first, until a characteristic family of effects, described by appropriate key words, has developed, at which point a distinct subject is born.

Such is very much the case with atomic 'giant resonances'. For a start, their name itself was borrowed from the field of nuclear collective resonances. The energy range in which they occur, at the juncture of the extreme UV and the soft X-rays, remains to this day a meeting point of two different experimental techniques: the grating and the crystal spectrometer. The impetus of synchrotron spectroscopy also played a large part in developing novel methods, described by many acronyms, which are used to study 'giant resonances' today. Finally, although we have described them as 'atomic' to differentiate them from their counterparts in Nuclear Physics, their occurrence on atomic sites does not inhibit their existence in molecules and solids. In fact, 'giant resonances' provide a new unifying theme, cutting accross some of the traditional scientific boundaries. After much separate development, the spectroscopies of the atom in various environments can meet afresh around this theme of common interest.

Centrifugal barrier effects and 'giant resonances' proper emerged almost simultaneously in the late 1960's from two widely separated areas of physics, namely the study of free atoms and of condensed matter. It soon became apparent that the features later to be described as 'giant resonances' form a unique class of excitations which preserve their character under many changes of environment from the free atom right through to the solid. The effects of the molecular field on the 'giant resonances' were then explored. Thus, the interdisciplinary nature of the subject was actually exploited from the outset as one means of characterising this class of resonances.

A further thread winds through the theory of atomic centrifugal effects. Within the independent particle model, the connection between 'giant resonances' and the lanthanide contraction became apparent, uncovering a rich complexity of related phenomena. Thanks to this connection, the subject of 4f electrons, their occupancy and their remarkable hesitations of valence in condensed matter were perceived in a new light by atomic physicists. Indeed, a quasi-periodic table, obtained semi-empirically from studies of conductivity and magnetism in solids, charts a sophisticated relation between localisation and the nature of the atom, the precise details of which remain to be accounted for. This table is perhaps a central piece of the new subject area.

From a theoretical standpoint, many fundamental issues have arisen in this general area of science. Taking atomic 'giant resonances' first, the inadequacy of the simple independent particle approximations led to the furtherance of many-body theories (the random phase approximation with exchange and many body perturbation theory), to the development of density functional approaches for the calculation of atomic spectra and, indeed, to the refinement of the independent particle pictures themselves, which now include term-dependent Hartree-Fock potentials for the continuum states, incorporating some essential aspects of the more complete theories.

An even more fundamental advance exploits a static mean central field optimised according to the prescription of relativistic quantum field theory for N-electron systems - the so-called g-Hartree equation, which has proved remarkably successful in treating wavefunction collapse. There are few problems as fundamental in low energy physics as the study of many-body phenomena in a quantum system with few particles (the atom). Thus, basic issues of wide significance have been raised.

In molecular physics, several kinds of very broad resonances occur. The simplest to characterise is the atomic 'giant resonance' slightly modified by molecular bonds (as in e.g. LaF_3). Next come 'giant resonances' which closely mimick those in a corresponding united atom (as e.g. in HI). Finally, there are resonances of similar type in large molecules, which exist only by virtue of the molecular field and have no direct analogue in atoms. All these types of resonances are being actively explored, and their nature is currently a subject of discussion. Passing from molecules to chemistry, photon stimulated desorption from surfaces can be resonantly enhanced by tuning to 'giant resonances'. This provides a new means of probing adsorption, as well as an intriguing class of new effects. The subject of centrifugal barriers and its connection with the quasi-periodic table mentioned above could provide new insights into rare-earth catalysis.

In condensed matter, there is the challenge of wedding Anderson-type models with atomic structure theory: the former are highly successful in describing 4f occupancies and interrelating different spectroscopies, while the latter is necessary to account for the persistence of multiplet structure in solids. A unified theory would clearly be an important achievement.

There are also connections between 'giant resonances' and laser-based research which were explored in les Houches. First, the 'giant resonances' are closely related to certain unresolved transition arrays in the spectra of laser-produced rare-earth plasmas. Secondly, collective effects can actually be driven directly by an intense laser field, and this is a novel area of research in its own right.

Against this backcloth of activity in the study of resonant wavefunction localisation, it seemed opportune to hold an interdisciplinary NATO Advanced Study Institute on 'Giant Resonances in Atoms Molecules and Solids'. Most researchers in the field had not had the opportunity to pursue extensive contacts with colleagues from quite different, though related, disciplines, and the subject was in some danger of becoming fragmented. There are few areas of scientific research which combine such breadth and richness of practical applications with a truly fundamental nature.

We have studiously avoided, in this preface, any attempt at defining 'giant resonances' and how one may or may not distinguish them from shape resonances. Authors differ on this issue, and several attempts at answering this question will be found in the present volume. Of course, we have our opinions on the matter, and the reader is encouraged to form his own judgement. Suffice it to say that this was a matter of hot debate between proponents and adversaries of the many-body approaches.

The present Proceedings provide, we believe, a comprehensive pedagogical account of the subject of 'giant resonances' in which the reader will find detailed discussion of all the issues raised above. Since the area is evolving rapidly, the articles collected here may require updating fairly soon in their more technical aspects. However, we hope their pedagogical nature will be durable, and will make the present volume a valuable introduction to this new field for some time to come.

J.P. Connerade
J.M. Esteva
R.C. Karnatak

CONTENTS

I. ATOMIC THEORY AND EXPERIMENTS

A1. Controlled Collapse and the Profiles of "Giant Resonances"
J.P. Connerade ... 3

A2. Mayer-Fermi Theory and the Long Sequences in the Periodic Table
D.C. Griffin, Robert D Cowan, M.S. Pindzola ... 25

A3. Relativistic Correlations in Atoms
K. Dietz ... 49

A4. Atomic Many Body Theory of Giant Resonances
Hugh P. Kelly and Zikri Altun ... 71

A5. Giant Resonances in Atoms and in Fluorine Cage Molecules
M.W.D. Mansfield ... 91

A6. Giant Resonances in the 4d Subshell Photoabsorption Spectra of Ba, Ba+, and Ba++
K. Nuroh, E. Zaremba and M.J. Stott ... 115

A7. Giant Resonances in the Transition Regions of the Periodic Table
C.W. Clark and T.B. Lucatorto ... 137

A8. Giant Resonances in Heavy Elements: Open-shell Effects, Multiplet Structure and Spin - Orbit Interaction
Françoise Combet-Farnoux ... 153

A9. Giant Resoanaces as a Probe of Collective Excitations in Atoms and Solids
G. Wendin ... 171

A10. The State-Specific Theory of Atomic Structure and Aspects of the Dynamics of Photoabsorption
C.A. Nicolaides ... 213

A11. Influence of Centrifugal Barrier Effects on the Autoionising Resonances
M.A. Baig, K. Sommer, J.P. Connerade, J. Hormes ... 225

A12.	Giant Resonance Phenomena in the Electron Impact Ionization of Heavy Atoms and Ions S.M. Younger	237
A13.	The g-Thomas-Fermi Method Th. Millack and G. Weymans	243
A14.	The Giant Resonances in the 3d Transition Elements: Single and Double Photoionization in the Region of the 3p Excitation M. Schmidt and P. Zimmermann	247
A15.	Decay Channels of Core Excitation Resonances in Atomic Lanthanides and 3rd Transition Metals M. Meyer, Th. Prescher, E.v. Raven, M. Richter, E. Schmidt, B. Sonntag and H.E. Wetzel	251

II. MOLECULES - THEORY AND EXPERIMENTS

M1.	Inner Shell Resonances in Molecular Photoionization I. Nenner	259
M2.	Resonances in the K-Shell Spectra of Fluorinated Organic Molecules A.P. Hitchcock, P. Fischer and R. McLaren	281
M3.	Atomic Autoionization Observed in nd O* Molecular Resonances P. Morin	291
M4.	Are the d-f shape Resonances in Diatomic Molecules "Giant Resonances"? H. Lefebvre-Brion	301

III. SOLIDS - THEORY AND EXPERIMENTS

S1.	The Quasi-Periodic Table and Anomalous Metallic Properties A.M. Boring and J.L. Smith	311
S2.	A Condensed Matter View of Giant Resonance Phenomena Andrew Zangwill	321
S3.	Photon and Electron - Stimulated Desorption from Rare Earth Oxides G.M. Loubriel	339
S4.	Applications of Electron Energy Loss Spectroscopy to Giant Resonances in Rare Earth and Transition Metal Systems J.A.D. Matthew	353
S5.	High Resolution Soft X-Ray Spectra of d-f Transitions: New Specific Features in the 3d Resonances of CeO_2, PrO_2 and TbO_2 J-M Esteva and R.C. Karnatak	361

S6.	The Effects of 4f Level Occupancy, Coulomb Interactions, and Hybridization on Core Level Spectra of Lanthanides J.C. Fuggle	381
S7.	Intermediate Valence Spectroscopy O. Gunnarsson and K. Schonhammer	405
S8.	Resonant Photoemission Using the d-f Giant Resonances S.J. Oh	431
S9.	Mixed Electronic States and Magnetic Moments Studied with near Edge X-Ray Absorption Spectroscopy G. van der Laan	447
S10.	Photoelectron Spectroscopic Observation of the Density of Low-Lying Excitations in Ce-Solids W.D. Schneider, F. Patthey, Y. Baer, B. Delley	463
S11.	The 4d - 4f Giant Resonances from Barium through the Rare-Earths U. Becker	473
S12.	Appearance-Potential Study of Resonant Transitions in Lanthanides D.R. Chopra	483
S13.	On the Excitation and Relaxation of 5p Electrons near Threshold in Eu and Yb Metals G. Rossi	491

IV. LASERS AND PLASMAS

L1.	The Spectra of Laser Produced Plasmas with Lanthanide Targets G. O'Sullivan	505
L2.	Ordered Many-Electron Motions in Atoms and X-Ray Lasers C.K. Rhodes	533
	Useful Acronyms	557
	Participants	561
	Author Index	565
	Subject Index	567

I. ATOMIC THEORY AND EXPERIMENTS

CONTROLLED COLLAPSE AND THE PROFILES OF 'GIANT RESONANCES'

J.P. Connerade

Blackett Laboratory
Imperial College
London SW7 2AZ

INTRODUCTION

The purpose of the present course is to provide a pedagogically simple introduction to the idea of quasi-atomic 'giant-resonances' and to some of their superficially paradoxical properties. In fact, they are in many ways far simpler than the more familiar autoionising resonances of atomic physics, with which it is constantly useful to contrast and compare them.

EXPERIMENTALLY DETERMINED PROPERTIES

Let me begin by describing some experimental features which allow us to recognise a 'giant resonance' in an atomic spectrum.

(i) The 'giant resonance' dominates the cross section. It is the strongest feature in the excitation channel, and its effects are so large that they usually affect all the other channels as well near the resonance energy.

(ii) When a 'giant resonance' appears, it takes up so much of the oscillator strength available within a channel that the Rydberg series one might otherwise expect to find is wiped out. There is a definite 'non-Rydberg' character of any channel which contains a 'giant resonance'.

(iii) A 'giant resonance' possesses a very distinctive profile (to be discussed below). It is always asymmetric, with a minimum attaching to the associated threshold and a 'tail' towards higher energies. It becomes broader the further it is removed from the threshold, and is generally a very broad feature of the spectrum, implying lifetimes two or three orders of magnitude shorter than typical autoionising states.

(iv) If a 'giant resonance' is present in an atom, then we expect to find it, with perhaps slight variations of the profile shape, in molecules containing this atom, or indeed in the condensed phase. This persistence of the resonance is unique amonst excited atomic states, and contrasts sharply with Rydberg behaviour. It provides experimental confirmation that the effect originates deep inside the atom, and was indeed (together with (i) above) the first clearly recognised property. What subsequently came came to be known as 'giant-resonances' were actually first observed in the spectra of solids by Zimkina and coworkers (1), who understood their atomic origin.

(v) There is a systematic evolution of 'giant resonances' as a function of atomic number. This property shows that the effect originates from the radial equation rather than from purely angular terms. However, one must add that 'giant resonances' only occur for high angular momenta, i.e. that the centrifugal term in the radial equation plays a crucial role.

(vi) To this list of properties, one should add the possibility of specifically molecular 'giant resonances' i.e. those which originate in a molecular rather than an atomic field. They possess most of the properies above, except that their persistence in different environment is obviously subject to what changes are wrought in the molecular structure. Usually, molecular 'giant resonances' are more subject to external perturbation.

THE ORIGIN OF THE NAME 'GIANT RESONANCE'

Let me now turn to the origin of the designation 'giant resonance' in Atomic Physics. In fact, the name comes from Nuclear Physics, and it is worth noting the similarities which led to its adoption in the present context. First, one should note that nuclei are held together by short-range forces, and that nuclear spectra are therefore definitely non-Rydberg in character. Giant resonances in nuclei have been known for the past 30 years, and were originally observed by Baldwin and Klaiber (2,3) in photoabsorption experiments using monochromatic 17 MeV photons to study induced radioactivity. The cross sections in 'giant resonances' were about two orders of magnitude higher than expected. The resonances are attributed to collective oscillations of the protons against the neutrons (4). There are different modes of the oscillation, of which the most important ones are the giant electric dipole deformation, the giant quadrupole surface oscillation and the giant monopole breathing mode. One also distinguishes between protons and neutrons vibrating together (isoscalar) or in opposition (isovector). Absorption experiments involving photoexcitation usually detect the giant dipole resonance, which is the purest case. A typical example is shown in Fig. 1. Some of the properties of nuclear giant resonances are close to those listed above. For example, nuclear giant resonances exhaust a large part of the energy-weighted sum rule; their energy changes systematically, as do their form and width, with the particle number in the nucleus.

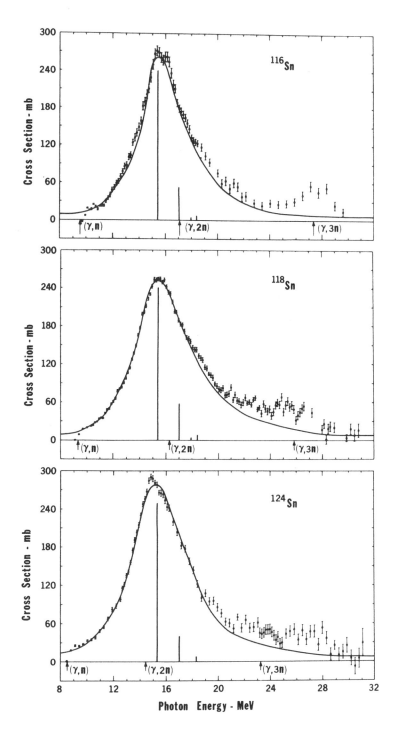

Fig. 1: Showing total photoneutron cross sections for different isotopes of tin, together with theoretical fits based on a dynamical collective model. (After reference 5)

The name 'giant resonance' was originally introduced into the atomic context by many-body theorists (6), in connection with a picture in which a full atomic subshell vibrates collectively. This picture has been much discussed and has occasioned some controversy. It must be moderated by the fact that we have not, in Atomic Physics, observed higher modes of oscillation. This is consistent with the fact that, within a harmonic oscillator model, the 'giant resonances' of atomic physics are strongly damped, i.e. decay within one period of oscillation as measured by their lifetime broadening. Nevertheless, there is strong evidence that the many-body terms are important in calculating the properties of 'giant resonances', and the name is a useful reminder both of this fact and of their other most salient property, which is the non-Rydberg character of the excitation.

THE QUANTUM SCATTERING MODEL OF 'GIANT RESONANCES'

The basis of a quantitative understanding of atomic 'giant resonances' is ab initio atomic theory, which even requires (as implied above) extensions beyond a conventional Hartree-Fock approach to give good agreement with experiment. However, these calculations are very elabrate, and it is therefore helpful to set up a very simple 'toy model' so as to achieve a good understanding of the underlying physics.

We begin with some assumptions: (i) that the 'giant resonance' is entirely a final state effect, ie that we can attribute it to the density of states in the continuum, (ii) that it originates from a short-range potential within the atom and (iii) that it occurs in a continuum of high angular momentum (typically $l = 2$ or 3).

Now, these are clearly simplifications: we are fogetting about inter-channel interactions, about the coulombic behaviour of the potential at large r, etc. Nevertheless, they allow us to define what we mean by a 'giant resonance' in a rather clear and simple way, and to distinguish such resonances from extraneous effects which are generally superposed.

I will leave the question of why a short range inner well develops in high angular momentum states of certain atoms to Professor Griffin's lecture on Mayer-Fermi theory. Let me just say that, if a short range well is present, then a convenient theorem due to Schwinger (1947) tells us that the bound state and low energy scattering spectrum of such a well are determined by the strength of binding alone, and not by the detailed shape of the potential. This theorem has the unpleasant consequence that one cannot deduce the detailed shape of the potential from observations of the low-energy scattering spectrum, but it allows us to substitute any simple short range potential for the true one in order to construct a simple model.

Given this freedom, we can select a potential which provides an analytic solution to the problem. Now, it turns out that the spherical square well with angular momentum has a particularly simple continuum solution, expressible in terms of the phase shift (the derivative of the phase shift with respect to wave vector is of course just the density of states) using spherical Bessel functions thus (7,8):

$$X = \tan \delta = \frac{z j_3(z') j_2(z) - z' j_3(z) j_2(z')}{z j_3(z') j_{-3}(z) + z' j_{-4}(z) j_2(z')}$$

in the specific case $l = 3$. The profile form is of course given by the standard relation:

$$\sigma(k) = \sin^2(\delta)/k^2 = X^2/k^2(1+X^2)$$

where k, k' are the wavevectors outside and inside the well respectively, $z = ka$, $z' = k'a$ and a is the radius of the well.

Also, we know from elementary quantum mechanics that a square well with an infinite repulsive wall on one side will only hold bound states above a certain critical strength (measured by the depth times the width squared). So it makes good sense to search in the continuum of such a spherical square well, just below the critical strength, and look at what happens to the low energy scattering spectrum as the strength of binding is increased.

The result is shown in Fig. 2. A resonance appears, which sweeps down in energy through the continuum, becoming sharper as it does so, and which turns into a bound state at the point of critical binding.

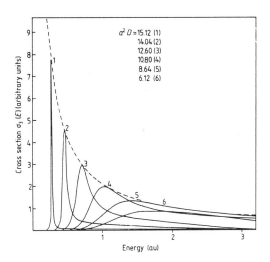

Fig. 2: A plot of the evolution of 'giant resonance profiles as a function of the strength of the scattering potential for a spherical square well with angular momentum. The unitary limit is shown as a dashed curve.

To see the relevance of this picture, consider the result of Hartree-Fock calculations of the d -> f 'giant resonances' in the elements approaching lanthanum, shown in Fig. 3. The impression is clearly very similar. The quantum scattering model implies the existence of the so-called 'unitary limit', which falls off as $1/k^2$: even this feature is quite well reproduced in the ab initio calculations, in spite of a slight overshoot.

Now, this simple model shows that the continuum is not segregated from the lowest energy bound states in the presence of a short range well in the same way as it would be were the potential coulombic. In effect, there is now some communication between the continuum and the inner part of the atom. One way of looking at this connection is via the Uncertainty Principle. We may ask: how short-lived would a state in the continuum have to be in order to be resonant in the well?

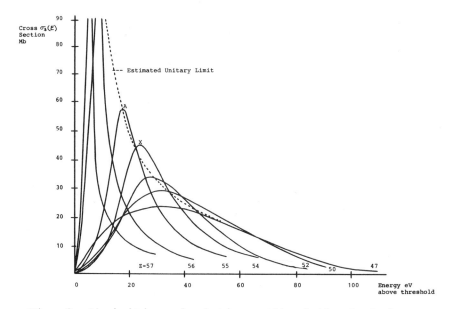

Fig. 3: Ab initio calculations (9) of the d -> f 'giant resonances' in the atoms approaching lanthanum. The calculated curves of ref (9) have been re-plotted to bring the thresholds in line with each other for comparison with Fig. 2. The present curves are of course based on a sophisticated ab initio model of the atom, rather than on the simple short range well used for the curves of Fig. 2, so the comparison tests the validity of Schwinger's theorem. In fact, the unitary limit is well respected by the ab initio calculations: there is only a slight overshoot at X if one fits a unitary curve through A in the Figure, and the extent of the overshoot is well within the expected accuracy of the ab initio calculations.

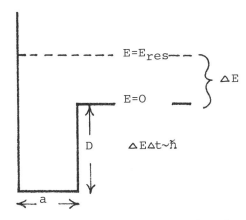

Fig. 4: A simple picture expressing how a the 'giant resonance' can be viewed as a virtual state in the continuum of the atom. If the lifetime of the state is very short, the uncertainty in energy encompasses the energy difference between the threshold energy ($E = 0$) and the resonance energy (E_{res}). If the well is just below the critical strength which would allow it to hold a bound state, then the electron can be trapped for a very short time. The Uncertainty Principle therefore implies that the breadth of the resonance should be proportional to E_{res}.

The answer is expressed in Fig. 4: the lifetime broadening of the resonance should be proportional to the resonance energy. Again, this is a simple result, which we can test against experiment, at least for free atoms, where reasonable estimates for threshold energies can be made. In Fig. 5, we see that, indeed, this prediction is quite correct over a wide range of energies. A practical implication of this result is that, provided the 'giant resonance' profile is reasonably free of multiplet smearing, one can use this general behaviour to infer the energy of the threshold in cases where it is not accessible experimentally.

Before we leave this simple model, I would like to stress that Schwinger's theorem is really remarkably powerful: although the square well potential is an otherwise unphysical choice, the precise form of the potential is actually unimportant provided it is of short range. We can thus obtain excellent agreement with experiment, and the profile formula can be used for parametric fits of 'giant resonances' to a surprising accuracy.

DISTINCTION BETWEEN 'GIANT' AND AUTOIONISING RESONANCES

A specific feature of 'giant resonances' now emerges rather clearly. The minimum in the cross section always occurs near the threshold. This is in contrast to the behaviour of the

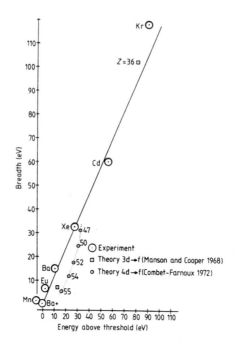

Fig. 5: A plot of breadth against energy based on experimental data (open circles) for atomic 'giant resonances', illustrating the linear relation described in the text. Some theoretical points are also plotted. It is interesting that the experimental data actually conform more closely to a linear trend than does ab initio theory.

minimum in the cross section of a Beutler-Fano resonance, which is tied to the resonance energy. The reason is that a 'giant resonance', within the description just set out, is a single-channel effect, whereas the Beutler-Fano resonance appears as the result of interchannel coupling. In the case of a 'giant resonance', the continuum always contains a modulation, whereas the continuum is assumed to be 'flat' over the width of an Beutler-Fano resonance. They are, therefore, essentially distinct: in the latter case, we can 'deperturb' the problem to produce a discrete state in the continuum, whereas, in the former, there is no perturbation to be removed.

An illustration of how well the profile formula works is provided in Fig. 6, which shows experimental results obtained in Bonn on the 'giant resonance' of gadolinium, analysed by the quantum scattering approach (10).

What more does our simple picture tell us? One of the most important points is that the profile of a 'giant continuum resonance' depends mainly on the strength of binding of the inner well, and is comparatively insensitive to what is going on in the outer reaches of the atom. Obviously, our model would be totally inadequate at large radius. Indeed, as I have stressed, it only works at small r by virtue of Schwinger's theorem. Localisation in a short range well explains why 'giant resonances' are essentially

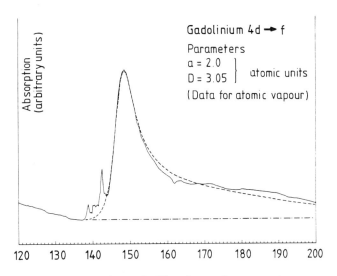

Fig. 6: An example of fitting the quantum scattering formula given in the text to an experimental 'giant resonance' (the d -> f resonance of Gadolinium). The parameters are a, the radius of the well and D its depth. The data were obtained at the 500 MeV synchrotron in Bonn (see ref. 10)

unaffected by environment, provided they stand well clear of the threshold. They survive in molecules and solids, which is a good experimental test of their nature. In a sense, they then appear as a continuation of the electron scattering spectra of EXAFS and XANES into the inner reaches of the atomic potential. Notice that I have suddenly introduced here the notion of 'continuum resonance'. There is another kind of 'giant resonance' with somewhat different properties, as we shall see below. This other kind (giant resonances amongst discrete states) cannot be accounted for within the simple quantum scattering model used above.

One should also note in passing that Beutler-Fano autoionising resonances normally form Rydberg series which do not survive into the condensed phase. Since the 'giant continuum resonance' is formed in the non-coulombic part of the potential, it is definitely a non-Rydberg feature of the spectrum.

I have presented the 'giant resonance' as an essentially single channel phenomenon, because this provides the simplest picture. However, a pure single channel problem does not occur at high energies. A compact (small radius) atom necessarily exhibits strong many-body interactions. So the short-range potential hides much of the true complexity of the problem. Also, and for the same reason, 'giant resonances' will always encourage interchannel coupling, and therefore a real resonance, as will unfold in the course of the present Advanced Study Institute, is truly a rather complicated affair. I must emphasise that this many-body character and complexity are the real motivation for much of the theoretical interest in 'giant resonances'.

CONTROLLED ORBITAL COLLAPSE

We can now ask some slightly more subtle questions, which cannot actually be answered within the simple model I have described. The notion of tuning the 'giant resonance' so that it sweeps down in energy towards the threshold is called the 'controlled orbital collapse' problem, and was originally suggested (11,12) purely on the basis of self-consistent field atomic structure calculations. In principle, it is accessible to experiment, either through changes of environment, as in the experiments of Elango and co-workers (13,14), or by controlled ionisation, as described by Dr. Charles Clark of N.B.S. in the present volume. As the 'giant resonance' sweeps down in energy in a real atom, it eventually becomes degenerate with the bound states of the coulombic outer well before finally dropping deep into the inner well of the atom.

Thus, we necessarily encounter a new problem when dealing with controlled collapse: the 'giant resonance' will interact with the coulombic manifold of states before it turns into a fully localised bound state of the inner well. Since orbital collapse is (at least for f-electrons) a rather sensitive balance, atoms tend to possess either collapsed or diffuse orbitals in a double-well potential, and it is very unlikely that they will be poised on the 'knife-edge', with half the wavefunction in the inner and half in the outer well. Nevertheless, nature has been kind to us in providing just one example of this highly unusual situation: the 5d -> nf spectrum of Ba^+.

It has been known for a very long time that something unusual occurs in the nf series of Ba^+. The anomaly was originally pointed out in 1934 by Saunders Schneider and Buckingham (15) who remarked that '... the perturbation ... appears to be of a novel type'. This remark shows that they possessed a profound understanding of series perturbations. However, they were unable at that date to advance any explanation for whatever prompted their observation. Most probably, they were referring to the behaviour of the quantum defects in the series, illustrated in Fig. 7, which shows that the complete f series is perturbed and that the quantum defects change by more than unity as a result of the interaction. Notice that the curvature of the nf plot cannot be due to an error in determining the threshold energies, since the np series in the same plot converge to the same ionisation potentials. Now, the remarkable thing, which must surely have puzzled the authors just cited, is that there is no level anywhere on the horizon which could account for the enormous perturbation.

The explanation for this anomaly came in 1975, when it was realised (16) that Ba^+, possesses a double valley potential with a 'knife edge' equilibrium between the inner and the outer well, and that the nf-eigenfunctions are in the critically poised condition described above. Under such circumstances, there ceases to be a recapitulation of atomic orbitals. Recapitulation means that, as n is increased, the nodes of atomic orbitals of high enough n return at constant r, and that the inner part of the radial wavefunction has a form which depends only weakly on energy. Recapitulation

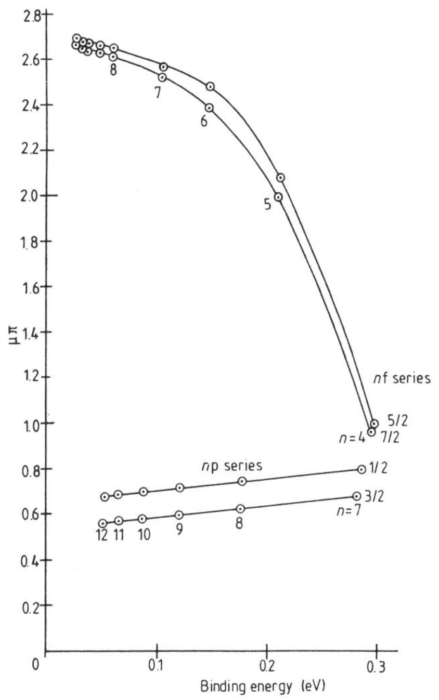

Fig. 7: Showing the variation of the experimental quantum defect for the np and nf series of Ba^+. Note the large variation for the nf series in spite of the absence of any perturber. Both the np and nf series converge to the same limits.

is the cornerstone of quantum defect theory. If it breaks down, the very foundations of the quantum defect approach are undermined. This is yet another manifestation of non-Rydberg behaviour, which we can describe as a centrifugal barrier perturbation. Again, as in the case of the continuum resonance, it is a single channel effect. In other words, there is no external perturber, but simply a varying degree of penetration of coulombic states into the inner well, which results in n-dependent phase shifts of all the coulombic functions.

Now, quantum scattering methods can again come to the rescue: effective range theory gives us the form of the phase shift near a resonance in the continuum as $\tan\delta = \Gamma_o/2(E_o-E)$ and Levinson's theorem tells us how to relate the phase shift to the quantum defect in the presence of resonances which can drop into the inner well to become bound states: we write $\delta = (\mu - m)\pi$, where m is an index which counts the number of shape resonances which have dropped below threshold. This slight generalisation of Seaton's theorem, with an analytic continuation from the continuum into the Rydberg states, yields the relation

$$\coth(m - \mu(E))\pi = 2(E_o - E)/\Gamma_o$$

which provides a more appropriate function than the quantum defect u in the presence of a double well potential, because it yields a straight line when plotted against energy. In the case of Ba^+, m = 1, and the 'straightening out' of the quantum defect plot is illustrated in Fig. 8. Notice that both the nf and np plots now yield straight lines, using the same ionisation potentials as in Figure 7.

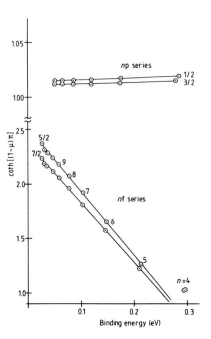

Fig. 8: The function coth$(1 - \mu)\pi$ plotted against energy for the same experimental data as in Fig. 7. Notice how the nf series and np series both yield straight line plots. The point corresponding to 4f lies below the 'giant resonance' energy.

The slope of the straight line in Fig. 8 yields the width Γ_o, which would be the width of the 'giant resonance' above threshold. In the regime of strong hybridisation, in which the 'giant resonance' is mixed into the coulombic bound states, this quantity becomes an interaction strength, which measures the extent of the series perturbation. The value plotted for Ba$^+$ in Fig. 5 was in fact obtained by analysing the bound states in this way, and is consistent with widths deduced directly above threshold.

The intercepts of the straight line in Fig. 8 determine E_o, and tell us that the giant resonance lies below 5f. However, since the 4f point lies off the straight line, we cannot deduce E_o exactly in this way. A useful trick (17) is to fit a Morse potential to the ab initio effective radial potential of Ba$^+$, which can be done very accurately in the inner well. It turns out that the eigenstate of this Morse potential lies between the energies of the 4f and 5f hybrids, which is probably the best estimate of E_o.

ACCORDION RESONANCES

In a recent text on ab initio calculations of autoionisation resonances, a distinction has been drawn between 'accordion resonances' in which a phase shift occurs at resonance (see Fig. 9), and 'rope resonances', in which the wavefunction changes sign at the origin, regarded by the authors of the text as the only genuine examples of Feshbach-type autoionising resonances. Of course, wavefunctions and potentials are not measurable quantities, so the distinction has only a limited practical value. Nevertheless, it is interesting to consider 'giant resonances' from this point of view.

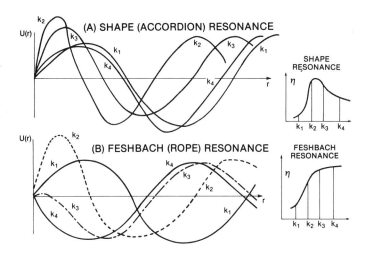

Fig. 9: Typical behaviour of an accordion resonance (schematic, after ref. 18).

Ab initio calculations confirm that 'giant resonances' are excellent examples of accordion resonances. This point is illustrated in Fig. 10, which shows Hartree-Fock wavefunctions for the nf states of Ba^+. From this figure, we can also deduce that the 'giant resonance' lies between n = 4 and n = 5, and is somewhat closer to the latter. We see, by considering the 'phase shifts' of the coulombic parts of the nf eigenfunctions, how it is that a 'giant resonance' can actually propagate downwards in energy accross the threshold into the bound states.

Before leaving the Ba^+ problem, I would like to point out that a quantitative understanding of the nf states requires extensions beyond ordinary Hartree-Fock theory, a matter which will be discussed in detail by Professor Klaus Dietz, in the present volume.

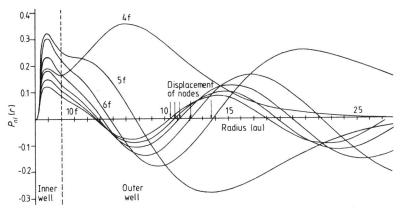

Fig. 10: Hartree-Fock wavefunctions for the nf hybrid states of Ba^+ (16), illustrating the 'phase shifts' of the coulombic parts of the eigenfunctions and the 'accordion' nature of the 'giant resonance'.

INTERCHANNEL INTERACTIONS

Another reason for which 'giant resonances' deserve their name is of course the enormous concentration of oscillator strength into a single feature of the continuum which is apparent when they are produced by inner-shell excitation. Again, we can achieve a qualitative understanding of the excitation mechanism within an independent electron model with an effective potential. We now assume that the initial state involves a compact d wavefunction, whose principal maximum lies around the radius of the inner well. The spatial overlap between the initial d and final f state then depends entirely on the penetration of the excited state wavefunctions into the inner well. Thus, coulombic states whose wavefunctions are dominantly formed from eigenfunctions or continuum functions of the outer reaches are suppressed, whereas localised continuum states or eigenstates of the inner well, which possess a high amplitude in the inner region of the atom, are enormously enhanced. In the case of excited f-states in Ba or the lanthanides, the inner well can only support one resonance or one bound state. There is, in fact, a term dependence of the well-strength, so that bound states appear earlier in the triplet channel and later in the singlet channel as the atomic number is increased. Thus, in a case such as Ba, there *is* a spin-dependent coexistence of 4f bound states and $\overline{4,\epsilon}$f resonantly localised continuum states within the same excitation spectrum.

Either way, it is clear that the excitation of Rydberg states within the channel is suppressed. The sum rules for the conservation of oscillator strength then operate in such a fashion that the whole of the available oscillator strength becomes concentrated into a single feature, which dominates the photoabsorption spectrum, <u>provided interchannel interactions do not spoil the picture</u>. So, we can say that 'giant resonances' yield the simplest spectrum imaginable, much simpler even than hydrogen, with each excitation channel dominated by a single feature. As an example of this simplicity, Fig. 11 shows the photoabsorption spectrum of Th I (17)

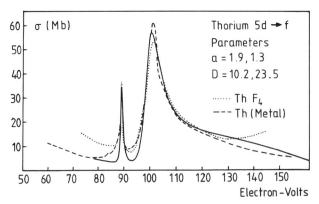

Fig. 11: The 5d -> f 'giant resonance' in the spectrum of thorium, together with a fitted curve (full curve) based on two spherical square wells of l = 3, one for singlet and one for triplet states.

Now, interchannel interactions feed on oscillator strength from the single excitation channels, which they borrow for their own purposes. 'Giant resonances' are potentially the largest available supply of their most essential ingredient. Also, we must remember that all the independent particle descriptions I am using to keep this introductory discussion simple are deceptive, because we are dealing with a very compact state of the atom, in which many-body interactions will come into play at the slightest provocation. Multichannel quantum defect theory (18) is the usual framework for the discussion of interchannel interactions in atoms. This theory partitions the space of the atom into a large volume (greater than r_o), where the wavefunctions are essentially coulombic, and therefore possess independent particle character, and a small volume (smaller than r_o), dominated by many-body forces, where the interchannel mixing actually takes place. Not only do 'giant resonances' possess a huge supply of oscillator strength, they are also packed into the small volume in which interchannel interactions are determined. If we open this can of worms, the simplest spectrum in the book may ultimately turn into the most complicated object we can think of.

Before we get into this aspect, it is worth pointing out the effect of a 'giant resonance' on other channels when the latter are essentially featureless at the energy of interest. This is obtained from measurements of partial cross sections within a 'giant resonance', as shown in Fig. 12: we see that partial cross sections of all the channels are enhanced near the energy of the 'giant resonance'. This figure emphasises that measuring the total photoabsorption cross section is not a good way to detect subtle many-body effects: within an apparently simple spectrum are hidden many essentially many-body inter-channel couplings.

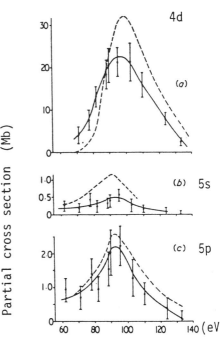

Fig. 12: Showing the interchannel couplings which lead to enhancement of partial cross sections near the 4d -> f 'giant resonance' of Xe I (19). The dashed curves show calculations by the RPAE many-body theory (20).

In short, the tendency to suppress Rydberg excitation within the channel can be offset by the tendency to enhance any other available channel which just happens to lie close enough in energy to couple with the 'giant resonance'. Indeed, this enhancement can even result in a multiplicity of supernumary Rydberg states, as we shall see below.

Interchannel coupling involving a Rydberg series and a broad resonance acting as a perturber is a subject of considerable interest at the present time (21), and leads to a variety of fascinating effects, such as 'q-reversals' (22) and vanishing widths (23) as well as perturbations of intensity and quantum defect. The examples cited do not yet involve interactions with 'giant resonances', because really high resolution data on such Rydberg series are scant. Where available, they reveal a picture of great complexity, but the principles are the same (24). A clear case was reported very recently by Sandner et al (25), who have studied angular distributions of autoionising electrons from the $6p_{3/2}ns$ J=1 Rydberg states. The strong energy dependence of the asymmetry parameter could only be explained as resulting from a shape resonance in the 5d f channel, and this inference was verified both by ab initio Hartree-Fock calculations and by using the general scaling laws for resonance energy and width described above (see Fig. 5).

Another case in which the excited state wavefunction is not so strongly localised, but nevertheless possesses the essential features of a shape resonance involves the 5d wavefunction of core-excited Ba I, which is similar to the 5d function of La I, and exemplifies l = 2 orbital collapse. Before experimental data became available, it was predicted by RPAE calculations that a giant collective dipole resonance would appear in the $(5p;5d)^1P$ channel below the O_{II} and O_{III} thresholds (treated as degenerate). When it was searched for, the data revealed, instead of a 'giant resonance' dominating the discrete structure as had been expected, a bewildering profusion of Rydberg series (far more than allowed by the independent particle excitation scheme). As subsequent calculations confirmed, interchannel interactions, turned on by the proximity in energy of 5d and 6s, effectively kill the 'giant resonance' and shatter the $6s^2$ subshell. Here, correlations conspire to quench a resonance, which then breaks up into a large number of fragments, each one carrying a different Rydberg signature.

The analysis of the 5p spectrum is very complex, and appeals to the most sophisticated multiconfigurational Dirac-Fock procedures (26). Indeed, the most recent high resolution data (27) reveal that the structure is even more complex than had initially been suspected.

Perhaps the most interesting postscript to these detailed studies is provided by experimental data on 5p-excitation in the sequence of elements following barium. As the 5d and 6s subshells move apart in energy, the interchannel interactions fade away. Suddenly, around Sm, the $(5p;5d)^1P$ 'giant resonance' emerges as the dominant feature of the spectrum, with the characteristic profile expected from quantum scattering theory (see Fig. 13).

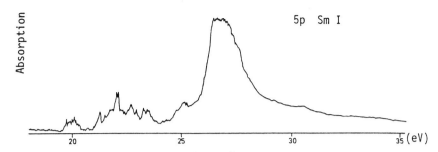

Fig. 13: Showing the $(5p;5d)^1P$ 'giant resonance' in Sm, which emerges to dominate the spectrum as mixing between 5d and 6s subsides.

COLLECTIVE VERSUS INDEPENDENT PARTICLE DESCRIPTIONS

One aspect, however, stands out, and seems to be a quite general conclusion for atomic, as opposed to nuclear 'giant resonances'. In atoms, it is generally found that the predictions of good independent particle models overestimate the sharpness and intensity of 'giant resonances'. As more and more many-body interactions are turned on, the effect is to shift and dampen rather than enhance the resonance. The same is true in nuclear RPA models, but damping effects are stronger in the atomic case. Thus, it is correct to say that 'giant resonances' exhibit a strong collective nature, but the oscillations are not collective for atomic shells in quite the same sense as in nuclear shell models.

In fact, we can go one step further by examining the characteristic lifetime of a typical 'giant resonance': it has been argued (28) that the 4d -> f 'giant resonance' of Ba can be considered as a harmonic oscillator representing the collective motions of the 4d subshell, centred at $\hbar\omega$ =105 eV. If one pursues this line of thought, a damping constant must be introduced to account for the observed width. Comparison with the data indicates (29) that the damping constant corresponds to a lifetime of 3×10^{-17} s, which is actually shorter than the period of the proposed oscillation (4×10^{-17} s). So the notion of collective 'oscillations' is a little misleading: they are strongly quenched and, in fact, no harmonics are observed. A similar conclusion for rare gas atoms was reached by a purely theoretical many-body approach (30).

The very short lifetime is a rather interesting point, since in 3×10^{-17} s, a photon travels only about 90 Å, an indication of how far in space the atomic mean field is defined during the excitation of a 'giant resonance'.

Now, returning again to the comparison with 'giant resonances' in Nuclear Physics, the main disadvantage of Atomic studies is the absence of data on 'giant resonances' of higher multipolarity. In Nuclear studies, the 'giant dipole' resonances are the most firmly characterised, since they can be isolated from all the others in photoabsorption experiments. However, hadronic scattering experiments establish the existence of e.g. quadrupole 'giant resonances' as well as the occurrence of harmonics. Even though the scattering experiments yield information of lesser purity, they are very important in establishing the collective nature of the oscillations.

In atoms, it seems fair to summarise the situation by remarking that there is a strong collective shift and damping of the independent particle prediction.

APPLICATIONS OUTSIDE ATOMIC PHYSICS

The present Institute is interdisciplinary, and this introduction is an excellent opportunity to emphasise the wider relevance of atomic 'giant resonances', emphasising some aspects in which an atomic approach may well shed further light.

(i) The 'quasi-Periodic' Table. This Table, described by Dr. James Smith in the present volume, summarises a vast amount of experimental information on conductivity and magnetism. The ridge of pressure-sensitive elements described by Dr. Smith as separating itinerant (conductive) from localised (magnetic) behaviour, is immediately recognisable to atomic physicists from a different standpoint. We see it as the locus of 'giant resonances', and more specifically as where they lie closest to their associated thresholds.

(ii) Photon-stimulated desorption. The effects described by Dr. Guillermo Loubriel and co-workers (31), in connection with the photon-stimulated desorption of hydrogen from cerium oxide, show a clear persistence of 'giant resonances' not only in a solid state environment, but also in the chemistry of surfaces. The experiments signals clearly that the double-well is important at least in modelling the ejection process, and possibly also in the adsorption mechanism.

(iii) Valence fluctuations and intermediate valence. There is, here, a whole area of study of considerable importance in solid state physics. A good understanding of the double-well problem, with the possible coexistence of collapsed and diffuse solutions (32, 33), potential violations of Koopman's theorem, and also the calculation of the energy required to drive an atom from the collapsed to the diffuse condition, are worthwhile input to current discussions in this area.

(iv) Catalysis. The word is of course a dangerous one, since the experts disagree on its definition, but it is certainly true that a 'soft' radial potential must assist in a greater chemical adaptability, and that the critical double valley structure may possess quite significant chemical properties. Indeed, it comes as no surprise that the ridge in the quasi-Periodic table mentioned above also contains some good catalysts. The challenge is to take this further.

In several of (i) to (iv), a basic question one may ask is how much external perturbation needs to be applied in order to transfer the atom from a condition in which the f-wavefunction is predominantly diffuse to one in which it is predominantly collapsed. Atomic Physics can give us an order of magnitude answer to this question: if one takes as a working medium a 4d-excited Cs atom in the configuration average approximation, the excitation of the outermost electron to successively higher Rydberg orbitals precipitates a collapse of the 4f-functions, as shown in Fig. 14. Now, the Hartree-Fock calculation provides us with (i) an energy difference W and (ii) with a difference in volume ΔV (from the average radii of the outer electrons). Thus, the elementary relation $W = P \Delta V$ allows us to translate this information into an effective pressure, which turns out to be about one kilobar. Ideed, many of the pressure sensitivities of interest to solid state physicists dealing with the Quasi-Periodic Table are of this order, which is a very interesting fact.

CONCLUSION

The present report is intended as a brief introduction to a collection of specialist lectures by experts in a wide variety of disciplines, brought together by an interest in 'giant resonances', potential barriers and related effects. Most of the topics and questions raised are covered in greater depth in the other contributions in the present volume.

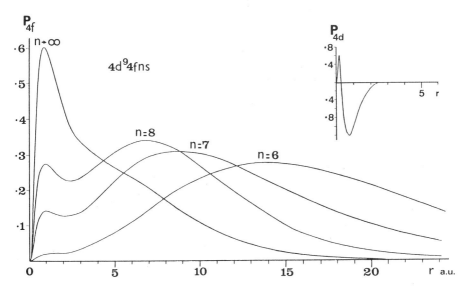

Fig. 14: Showing the controlled collapse of $4f_{av}$ atomiv wavefunctions in 4d-excited Cs I as the external electron is progessively excited to higher and higher Rydberg orbitals. This effect (controlled collapse) allows one to estimate the pressure required to induce localisation for e.g. an atom with a double-well potential in a solid.

REFERENCES

1. A.P. Lukirskii and T.M. Zimkina, Bull. Acad. Sci. USSR Phys. Ser. 27, 808 (1963)
2. G.C. Baldwin and G.S. Klaiber, Phys. Rev. 71: 3 (1947)
3. G.C. Baldwin and G.S. Klaiber, Phys. Rev. 73: 1156 (1948)
4. M. Goldhaber and E. Teller, Phys. Rev. 74: 1046 (1948)
5. B.L. Berman and S.C. Fultz, Rev. Mod. Phys. 47: 713 (1975)
6. G. Wendin, Phys. Letters 46A: 119 (1973)
7. N.F. Mott and H.S.W. Massey, in "The Theory of Atomic Collisions" 3rd Ed, (Oxford:Clarendon Press 1965)
8. J.P. Connerade, J. Phys. B. At. Mol. Phys. 17: L165 (1984)
9. F. Combet-Farnoux, Proc. Int. Conf. on Inner Shell Ionization Phenomena (Atlanta) Vol. II: 1130 (1972)
10. J.P. Connerade and M. Pantelouris, J. Phys. B At. Mol. Phys. 17: L173 (1984)
11. J.P. Connerade, J. Phys. B At. Mol. Phys. 13: L381 (1978)
12. J.P. Connerade, J. Phys. B At. Mol. Phys. 13: L409 (1978)
13. M. Elango A. Maiste and R. Ruus, Phys. Letters A72: 16 (1979)
14. R.I. Karaziya, Sov. Phys. Usp. 24: 775 (1981)
15. F.A. Saunders E.G. Schneider and E. Buckingham, Proc. Natn. Acad. Sci, (Washington) 20: 291 (1934)
16. J.P. Connerade and M.W.D. Mansfield, Proc. R. Soc. Lond. A 346, 565 (1975)
17. J.P. Connerade, Course 6 "The Physics of Non-Rydberg States" in les Houches, Session XXXVIII New Trends in Atomic Physics (Grynberg and Stora Eds), Elzevier Science Publishers (1984)
18. A. Temkin in "Autoionization - New Developments and Applications", Plenum 1985
19. J.B. West P.R. Woodruff K. Codling and R.G. Houlgate, J. Phys. B At. Mol. Phys. 9: 407 (1976)
20. M.Ya. Amusia V.K. Ivanov N.A. Cherepkov and L.V. Chernysheva, Sov. Phys. J.E.T.P. 66:1537 (1974)
21. J.P. Connerade A.M. Lane and M.A. Baig, J. Phys. B At. Mol. Phys. 18: 3507 (1985)
22. J.P. Connerade, Proc. R. Soc. London A 362: 361 (1978)
23. J.P. Connerade and A.M. Lane, J. Phys. B At. Mol. Phys. 18: L605 (1985)
24. J.P. Connerade and A.M. Lane, to be published
25. W. Sandner U. Eichmann V. Lange and M. Volkel, J. Phys. B At. Mol. Phys. 19: 51 (1986)
26. S.J. Rose I.P. Grant and J.P. Connerade, Phi. Trans. Roy. Soc. 296: 527 (1980)
27. M.A. Baig J.P. Connerade C. Mayhew and K. Sommer, J. Phys. B At. Mol. Phys. 17: 371 (1984)
28. K. Nuroh M.J. Scott and E. Zaremba, Phys. Rev. Lett. 49:862 (1982)
29. J.P. Connerade, J. Phys. B At. Mol. Phys. 16: L257 (1983)

30. M.Ya. Amusia N.A. Cherepkov R.K. Janev and D.J. Zivanovic in "The Physics of Electronic and Atomic Collisions (Ed B. Cobic and M. Kurepa. Beograd Institute of Physics) Vol. I: 242 (1973)
31. B.E. Koel G.M. Loubriel M.L. Knotek R.H. Stulen R.A. Rosenberg and C.C. Parks, Phys. Rev. B $\underline{25}$: 5551 (1982)
32. I.M. Band and V.I. Fomichev, Physics Letters $\underline{75A}$: 178 (1980)
33. J.P. Connerade, Journal of the Less Common Metals $\underline{93}$: 171 (1983), and to be published.

MAYER-FERMI THEORY AND THE LONG SEQUENCES IN THE PERIODIC TABLE

D. C. Griffin

Department of Physics, Rollins College
Winter Park, Florida 32789
and Physics Division, Oak Ridge National Laboratory
Oak Ridge, Tennessee 37831

Robert D. Cowan

Los Alamos National Laboratory
Los Alamos, New Mexico 87545

M. S. Pindzola

Department of Physics, Auburn University
Auburn, Alabama 36849
and Physics Division, Oak Ridge National Laboratory
Oak Ridge, Tennessee 37831

INTRODUCTION

58 years ago, Fermi[1] employed the Thomas-Fermi (TF) statistical model of the atom to predict a collapse of the 4f orbital for atomic numbers between 55 and 60, and thereby explained the formation of the first rare-earth series of elements. In 1941, Maria Mayer[2] used the same model to predict the onset of a second rare-earth series with the orbital collapse of the 5f electron for atomic numbers between 86 and 91. She also showed that the abrupt changes in the character of these electrons can be explained in terms of the f-electron effective potential which, in these regions of the periodic table, consists of two wells separated by a positive centrifugal barrier.

Since Mayer's classical paper, the results of a number of more complete calculations using the Thomas-Fermi-Dirac (TFD) model,[3] as well as the self-consistent Hartree-Fock-Slater,[4] Hartree plus statistical exchange scheme (HX)[5] and Hartree-Fock (HF)[6] methods have been employed to study the variation of d-, f- and even g-electron wave functions within the excited states of neutral atoms. In the next section of this paper, we employ the results of the more recent self-consistent-field calculations to consider the now well known changes in radial wave functions for d electrons which occur preceding the onset of the transition series of elements, and for f electrons preceding the onset of the lanthanide and actinide series. The sensitivity of the radial wave functions to variations in the effective potential is discussed, and the large variation of the radial wave functions between the LS terms of certain types of excited configurations in these regions of the periodic system is analyzed.

The changes in the d- and f-radial wave functions along isoelectronic and isonuclear sequences are much more gradual than the sudden contractions of the wave functions that occur within neutral atomic species. Nevertheless, such variations may lead to large term-dependent effects in certain excited configurations which are important in electron-impact excitation, as well as photoabsorption. The excitation of inner-shell electrons followed by autoionization can in some cases dominate electron-impact ionization cross sections.[7,8,9] This is especially true in species with intermediate to heavy atomic mass where term-dependent effects may also be important. In the third section of this paper, several examples of electron-impact ionization, where the indirect mechanism is dominant and term dependence significantly affects the excitation of the inner-shell electron, are explained by analyzing the effective potentials for the excited electrons in the intermediate autoionizing states.

Finally, potential barriers can lead to significant term-dependent effects in the continuum which play an important role, not only in photoionization, but also direct electron-impact ionization. In the fourth section, we consider several examples where such effects are essential to the accurate determination of both single and double electron-impact ionization cross sections.

THE DOUBLE-WELL EFFECTIVE POTENTIAL AND WAVE FUNCTION COLLAPSE

Near the beginning of the rare-earth series of elements the excited f electrons undergo abrupt changes in binding energy. This is shown graphically in Fig. 1 in terms of the effective quantum number n^*, which

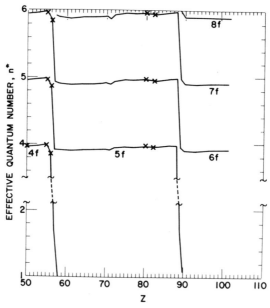

Fig. 1. Effective quantum numbers for f electrons as a function of atomic number from Ref. 5. Curves, theoretical values calculated using HX binding energies; x, experimental values.

for neutral atoms is defined by the equation:

$$n^* = E_B^{-1/2}, \qquad (1)$$

where E_B is the binding energy of the electron in rydbergs. These sudden changes are accompanied by a corresponding change in the nature of the f electrons consisting of contractions of the wave functions to smaller radii. In order to understand these effects it is useful to introduce an effective central potential, so that the radial differential equation can be written as:

$$\left[-\frac{d^2}{dr^2} + V_{eff}(r) \right] P_{n\ell}(r) = E_{n\ell} P_{n\ell}(r), \qquad (2)$$

where the effective potential, $V_{eff}(r)$, and the eigenvalue, $E_{n\ell}$, are in rydberg units. If one employs the Hartree-Fock method, $V_{eff}(r)$ is defined by the equation:

$$V_{eff}(r) = \frac{\ell(\ell+1)}{r^2} - \frac{2Z}{r} + \frac{2Y(n\ell;r)}{r} + \frac{2X(n\ell;r)}{rP_{n\ell}(r)} + \sum_{n'} \frac{E_{n\ell,n'\ell} P_{n'\ell}(r)}{P_{n\ell}(r)}, \qquad (3)$$

where we use a notation similar to Fischer's.[10] In Eq. (3), Z is the atomic number; $2Y(n\ell;r)/r$ is simply the direct potential when the subshell is singly occupied, as it is for the cases of interest here; $2X(n\ell;r)/r$ is the exchange function; and $E_{n\ell,n'\ell}$ is a Lagrangian multiplier used to force orthogonality between radial wave functions with equal values of ℓ but different n. This function is well defined for nodeless radial wave functions, such as $P_{4f}(r)$, and in other cases a simple interpolation may be used to prevent singularities in a plot of $V_{eff}(r)$.

For electrons in an open subshell, $V_{eff}(r)$ will be different for the different LS terms, resulting in term-dependent radial wave functions. We will shortly consider cases for which such variation among the terms is quite sizable; however, for many other cases, $V_{eff}(r)$ and $P_{n\ell}(r)$ are nearly the same for all terms of a configuration, and we may then employ a configuration-average Hartree-Fock (CAHF) method for which the direct potential and the exchange function are the same for all terms. When this is the case, it is possible to employ various approximations to the CAHF method, in which the Lagrangian multipliers, $E_{n\ell,n'\ell}$, are set to zero and the exchange function is replaced by some approximate, local exchange potential. The best known of these are the Hartree-Fock-Slater method,[11] in which the exchange function is replaced by Slater's free-electron gas exchange potential, and the modified HFS potential by Herman and Skillman.[12] In addition, there is the Hartree plus statistical exchange scheme[13] which incorporates an improved free-electron gas exchange potential and yields one-electron energies which agree quite well with those obtained from the CAHF method.

With today's high speed computers and improved numerical techniques, the HF or CAHF problem can be solved routinely and the various local approximations to the Hartree-Fock method are no longer widely used. However, most of the work done on wave function collapse in the neutral atoms was performed with one of these methods, and therefore they are included here. Furthermore, it is important to note that the HF equations are not solved in the form given in Eq. (2), but rather in their proper inhomogeneous form. The HF form of $V_{eff}(r)$ is used only as a descriptive device in explaining the nature of wave function collapse.

In hydrogen, the effective potential reduces to the form:

$$V_{eff}(r) = \frac{\ell(\ell+1)}{r^2} - \frac{2}{r} . \qquad (4)$$

With $\ell=3$, for f electrons, this is positive for r<6 (Bohr units), and beyond this there exists a shallow potential well with a mimimum at r=12. In multi-electron atoms, the core-electron wave functions do not extend appreciably into this outer region, and therefore, the potential for r>6 is nearly hydrogenic. However, for large Z, a second inner well develops at small radii where the effective nuclear charge is quite large. These two wells are separated by a positive centrifugal barrier, which if high enough, results in the existence of two quasi-independent sets of negative energy levels. Since the inner well is very narrow as compared to the outer hydrogenic region, the energy levels associated with this well are very widely spaced, and until Z becomes quite large, the lowest energy eigenvalue for the system is that of the outer well. Thus the binding energy for an f electron will be nearly equal to the hydrogenic value and $n^* \cong n$. As Z continues to increase, we reach a point where the inner well is sufficiently deep and wide that the inner-well level becomes the lowest one, and we then see an abrupt increase in the binding energy and an associated decrease in n^*.

The f-electron wave function has appreciable amplitude in only one of the two well regions, depending on whether the eigenvalue corresponds to an energy level of the inner or outer well. These two alternatives are shown in Fig. 2 for the 4f wave function in neutral $_{56}$Ba and $_{57}$La. The 4f wave function in Ba lies almost entirely in the outer-well hydrogenic region, the eigenvalue is nearly equal to that of hydrogen and $n^*=3.8$. However, in La the inner well is sufficiently deep and wide that the 4f wave function has collapsed completely into the

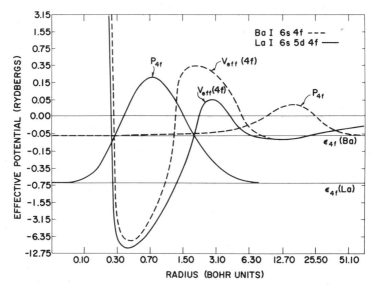

Fig. 2 Plot of the effective potentials and radial wave functions for the 4f electron in neutral Ba and La calculated using the HX approximation (Ref. 5). Nonlinear scales are used for both the effective potential and the radius. The radial wave functions are plotted on a linear scale the zero of which lies at the corresponding eigenvalue, noted in the figure by ε_{4f}.

inner-well region with a corresponding abrupt decrease in n*.

At the same point where the 4f wave function collapses into the inner well, the first antinode for the 5f wave function moves abruptly from the outer region into the inner-well and barrier regions and becomes quite small in magnitude. This is illustrated in Fig. 3 where, in La, only the second antinode of the 5f wave function has any appreciable amplitude, and it lies in the outer-hydrogenic region where its shape resembles that of the 4f wave function in Ba. Thus the quantum defect (n-n*) changes abruptly from a value close to zero to a value of nearly one in going from Z=56 to Z=57. More generally, if for a given atomic number the wave functions for the m lowest-lying nf subshells are in the inner well, then all excited nf wave functions have m antinodes with relatively small amplitudes in the inner-well or potential-barrier region, and therefore n-m-3 antinodes with much larger amplitudes in the outer-well, hydrogenic region. This implies an effective quantum number $n^* \cong n-m$, in agreement with the trends for f electrons shown in Fig. 1.

It is possible that an inner- and outer-well level might have equal energies. In such a case, the wave function would have large amplitudes in both wells and a minimum in the barrier region. For a fixed potential such as TF or TFD, such a situation will exist at some, in general non-integral, value of Z. However, for neutral atoms with an excited electron for which $\ell \geq 3$, where the barrier height is quite large, such a situation will be unstable when one uses a self-consistent-field model. If a small portion of charge is transferred from the outer to the inner well during the iterative process, the other electrons become more completely shielded from the nucleus and move to larger r; this in turn causes the excited electron to be less well shielded and it moves further toward the inner well; the inner well becomes deeper and wider and its level finally drops below the outer well level causing the excited electron to collapse completely into the inner region. This argument, which explains the

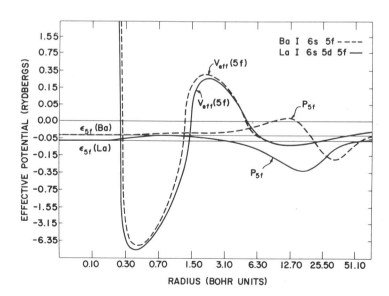

Fig. 3. Plot of the effective potentials and radial wave functions for the 5f electron in neutral Ba and La calculated using the HX approximation (Ref. 5). Further description is given under Fig. 2.

abrupt nature of the contraction for f electrons in neutral species, is valid only when the barrier height is large and we can correctly discuss two quasi-independent sets of energy levels. As we shall see, when the barrier height is small, there are cases where even a self-consistent procedure will yield appreciable amplitudes in each of the two regions.

Now let us consider the situation for d electrons. On the basis of the TF potential, Mayer[2] noted that double wells with positive potential barriers cannot occur for d electrons, and thus concluded that abrupt wave function and binding-energy changes should not occur for d electrons. However, Latter's subsequent calculations,[3] which employed both the TF and TFD potentials, corrected to have the proper $-2/r$ long range behavior, showed changes in energy similar to those for f electrons, although somewhat more gradual. Furthermore, Rau and Fano[4] pointed out that potential barriers do exist for d electrons when one uses a reasonably accurate potential, such as HFS with a tail cut-off at $-2/r$.[12]

The plot of effective quantum numbers for d electrons calculated using the HX potential is shown in Fig. 4. As can be seen, the changes in n^* with Z are still pronounced, although somewhat less abrupt than in the case of f electrons. The potential barriers for d electrons, when they do exist, are much more tenuous. For $\ell=2$, the hydrogenic well begins at r=3 and has its minimum at r=6. Thus, in a multi-electron atom, the core-electron wave functions will tend to overlap into the hydrogenic region, reducing the possibility of barrier formation. The barrier will be the most pronounced when the core electrons are the most compact, such as in the case of a closed-shell core like $K(3p^63d)$ where $\langle r \rangle_{3p}=1.43$.

The effective potentials and radial wave functions for the 3d electron in K and Ca are shown in Fig. 5. The effective potential for K does indeed have a well defined barrier region, although its maximum is actually below zero. The 3d wave function lies predominately in the

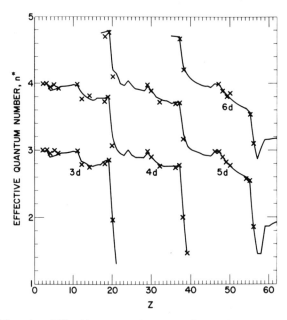

Fig. 4. Effective quantum numbers for d electrons as a function of atomic number from Ref. 5. The notation is the same as in Fig. 1.

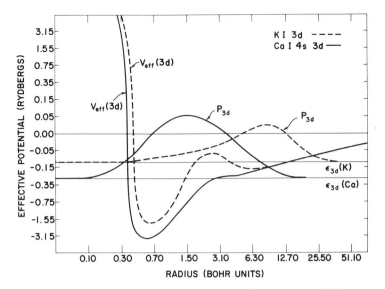

Fig. 5. Plot of the effective potentials and radial wave functions for the 3d electron in neutral K and Ca calculated using the HX approximation (Ref 5). Further description is given under Fig. 2.

outer well; however, it does have a noticeable overlap with the core region. In the next element Ca($3p^64s3d$), the 4s electron extends far into the outer-well region and provides very poor shielding for the added unit of nuclear charge. As a result, the barrier is wiped out, and the 3d wave function moves much further into the core region with the accompanying large change in n^* displayed in Fig. 4. The situation with the excited d electrons for which $n \geq 4$, is similar to that for the nf electrons already discussed, although again less dramatic.

Since the potential barriers for d-electron effective potentials are low in height, or not even present, the outer well regions are much less hydrogenic than they are for f electrons. This is the reason why, in Fig. 4, the n^* curves are not nearly as flat between drops, the values of n^* are not as close to integer values and the changes in n^* are not as close to unity as in the case of the f electrons shown in Fig. 1. This is especially true for high Z where the core-electron wave functions are less compact, and do not tend to provide the shielding necessary for the formation of well defined barriers. In general, the sudden variations in d-electron wave functions are much more a function of the detailed nature of the shell structure of the atom, and therefore much more sensitive to the core-electron configuration. Furthermore, predicted locations of d-wave function collapse within the periodic system are much more dependent on the central-field model employed.

All of the above discussion pertains only to neutral atoms; however, some of the more interesting physics associated with these centrifugal barrier effects occurs in ionized species. The effective potential for a multiply-ionized atom, can be written in the form:

$$V_{eff}(r) = \frac{\ell(\ell+1)}{r^2} - \frac{2Z_{eff}}{r} \qquad (5)$$

where Z_{eff} is the effective nuclear charge. For an excited electron which remains outside the core electrons in a Z-N ionized atom, Z_{eff}

reduces to the hydrogenic-like value Z-N+1, where N is the number of electrons. We see from Eq. (5) that the radius where the outer, hydrogenic-like well begins is inversely proportional to the value of the effective nuclear charge, and thus it decreases with ionization stage. Therefore, the core-electron wave functions, which have about the same radii in ions as in neutrals, have much larger overlaps with the outer well, even for $\ell \geq 3$. We then would expect much smaller barrier heights in ions, and these should decrease and become less important as a function of ionization stage.

In Fig. 6, we show the HX effective potential and radial wave function for the 4f electron in Ba$^+$(5p^{6}4f).[14] As can be seen, the effective potential is much less pronounced than in the case of neutral Ba shown in Fig. 2, and in fact, has a top which is negative. As a result, the 4f wave function is bimodal in shape with an appreciable amplitude in both the inner- and outer-well regions. Nevertheless, the largest portion of the amplitude of the wave function remains outside, and the quantum defect is relatively small ($\cong .32$).

As n increases, the barrier has gradually less effect on the nf wave functions, and their first antinodes penetrate deeper into the inner-well region.[15] This leads to some rather unusual properties in this nf Rydberg series. For example, the value of the quantum defect increases toward unity with increasing n,[5] more rapidly at first (for n=5, n-n*=.65), and then very gradually, reaching a value of .85 by n=10. Such variations are so unusual that, for a time, they were suspected to be the result of misinterpretation of the observed spectra.[14,16] The behavior of these nf wave functions also leads to some unusual features in the absorption spectrum of Ba$^+$ from the long lived 5d ^{2}D states,[17] as discussed by Connerade.[15]

As we would expect from the above considerations, the d-electron

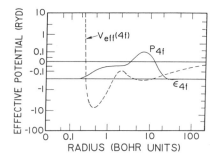

Fig. 6. Plot of the effective potential and radial wave function for the 4f electron in Ba$^+$ calculated using the HX approximation (Ref. 14). Further description is given under Fig. 2.

and f-electron orbitals collapse gradually as a function of ionization stage as compared to their variation as a function of Z for the neutral atoms. Nevertheless, these contractions can have important effects on the atomic structure of excited states involving d and f electrons. A discussion of the effects of such wave function variations for d electrons along isoelectronic sequences is given, for example, in Ref. 18. The contraction of f electrons as a function of ionization stage will be discussed in detail in the next section.

We now consider cases in which there is significant term dependence, or variation of the radial orbitals among the terms of a configuration. When this occurs, the CAHF method or the various local, configuration-average approximations mentioned in the last section no longer provide an accurate description for the system. Such term-dependent effects were first studied by the Hartrees[19] in a calculation of the 2s2p configuration, and a discussion of the results for the 2s2p ^{3}P and ^{1}P terms in Be is given by Fischer.[10] In this example, the 2s-2p exchange interaction in Be causes the 2p ^{3}P orbital to be slightly contracted ($\langle r \rangle$=2.9), while the 2p ^{1}P orbital is repelled to much larger radii ($\langle r \rangle$=5.0).[10]

These term-dependent effects are even more pronounced in configurations of the type p^{5}d near the beginning of the transition series, and d^{9}f near the beginning of the rare-earth series of elements. In order to understand this, we consider the energy expressions for the ^{1}P terms of these two configurations:

$$E(p^5d\ ^1P) = E_{ave}(p^5d) - 0.200F^2(pd) + 1.267G^1(pd) - 0.043G^3(pd) \quad (6)$$

$$E(d^9f\ ^1P) = E_{ave}(d^9f) - 0.229F^2(df) - 0.095F^4(df) + 1.957G^1(df) \\ - 0.019G^3(df) - 0.021G^5(df), \quad (7)$$

where $F^k(\ell_i\ell_j)$ and $G^k(\ell_i\ell_j)$ are the Slater parameters for the direct and exchange electrostatic interactions, respectively; and the spin-orbit interaction terms are not included. The other terms of these configurations have coefficients for the electrostatic interactions of comparable magnitude, except for the coefficients of the dipole exchange integrals, G^1, which are small. Therefore, if we apply the variational principle to these energy expressions to obtain the term-dependent HF equations, the resulting exchange functions are quite different for the ^{1}P terms as compared to all other terms of the configuration; however, the potential terms and exchange functions for the other terms are well approximated by the CAHF method.

The large positive dipole exchange interactions within the expressions for $V_{eff}(r)$ of the ^{1}P terms can produce pronounced double-well effective potentials for the excited d and f electrons in such configurations. The sensitivity of the radial wavefunctions to these barriers may, in turn, lead to large term-dependent variations which have significant effects on a number of important atomic properties. Extensive calculations for configurations of the type p^{5}d have been performed by Hansen.[20] One of the most extreme examples of these effects for d^{9}f configurations occurs in the excited configuration 4d^{9}5s^{2}5p^{6}6s^{2}4f of neutral Ba. In Fig. 7 we show the 4f electron HF effective potentials and radial wave functions for this case.[21] The configuration average (CA) effective potential, which provides an accurate representation of the potential for the nine terms other than ^{1}P, has a very small barrier and a deep inner well. The large difference between this potential and that shown in Fig. 2 for the 4d^{10}5s^{2}5p^{6}6s4f configuration of Ba is due to the reduction in shielding when an electron is promoted from the closed 4d subshell to the 6s subshell. The CA 4f wave function has collapsed into

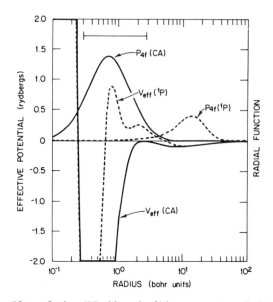

Fig. 7. Plot of the HF $4f_{CA}$ (solid curves) and $4f\ ^1P$ (dotted curves) effective potentials and radial wave functions for the the $4d^9 5s^2 5p^6 6s^2 4f$ configuration in neutral Ba (Ref. 21). A logarithmic scale is used for the radius. The bar indicates the range of radii over which the second antinode of the 4d radial wave function occurs.

the inner region, has a mean radius of 1.17 and a large overlap with the 4d electrons within the core. On the other hand, the ^1P effective potential has two distinct positive barriers; the large inner one is due to the positive exchange term, while the smaller outer one is due to the centrifugal term. However, contrary to the frozen-core term-dependent HF effective potential of Wendin and Starace,[22] the above potential, determined from a fully-relaxed, self-consistent HF calculation, does not go below zero in the region between the two barriers. These barriers cause the ^1P 4f wave function to remain in the outer-well region where it closely resembles a 4f hydrogenic wave function, has a mean radius of 17.5 and an extremely small overlap with the 4d radial wave function.

When one uses the $4f_{CA}$ wave function to calculate the position of the ^1P term it is found to lie above the $4d^9 5s^2 5p^6 6s^2$ ionization threshold because of the very large value of $G^1(4d,4f_{CA})$. However, as first pointed out by Hansen et. al.,[23] when one uses the proper $4f\ ^1P$ wave function, the $4d^9 4f\ ^1P$ term lies below the $4d^9$ ionization limit and has a binding energy close to that of a 4f hydrogenic electron.

Since all the $nf\ ^1P$ wave functions in the $4d^9 5s^2 5p^6 6s^2 nf$ configurations of neutral Ba lie in the outer hydrogenic region, and thus have very small overlaps with the 4d wave function, the photo-absorption transitions $4d^{10} \rightarrow 4d^9 nf\ ^1P$ will have very little oscillator strength. The oscillator strength is, instead, transferred to the continuum transition $4d^{10} \rightarrow 4d^9 \varepsilon f\ ^1P$, which is the cause of the well known giant resonance in the photoionization cross section of neutral Ba. This phenomenon will be discussed in detail by other authors in this volume, and therefore will not be pursued any further here.

In considering LS term dependence, it is important to note that one can approximate these effects by using CA radial wave functions and performing a configuration-interaction calculation which includes the other members of the Rydberg series. For example, Younger[24] has performed such a calculation for $Xe^{8+}(4d^94f)$ and was able to approximate the term-dependent 4f ^1P wave function from an expansion of CA wave functions of the form:

$$P_{4f}(^1P) \cong c_1 P_{4f}(CA) + c_2 P_{5f}(CA) + c_3 P_{6f}(CA), \qquad (8)$$

where c_1, c_2 and c_3 are the mixing coefficients in the configuration-interaction calculation. A similar calculation has been performed for $K^+(3p^53d\ ^1P)$,[14] but the Rydberg series members through n=11 had to be included before resonable agreement was obtained. Finally Wendin[25] and Wendin and Starace[22] have performed extensive CA, Rydberg series interaction calculations on the neutral Ba case just considered, in which very large basis sets had to be included.

By Brillouin's theorem,[26] configuration-interaction integrals which connect members of a Rydberg series, like the ones discussed above, will vanish when one uses the proper term-dependent HF wave functions. That is, correlations of the type represented by such interactions are already included in the zero-order, term-dependent wave functions. However, this is not true for CA wave functions. Nevertheless, if a complete set of CA wave functions are employed in a Rydberg series expansion, the resulting wave function must be identical to that obtained from the solution to the term-dependent HF equations. The primary advantage of the CA approach is that one can include LS term dependence and the effects of the spin-orbit interaction in a single multi-configuration calculation; the primary disadvantage is that many configurations may be required, including continuum states, before an accurate representation is obtained.

ELECTRON-IMPACT EXCITATION-AUTOIONIZATION

The study of electron-impact ionization increases our understanding of collisional dynamics and atomic structure and has important applications in laboratory and astrophysical plasmas. In the last decade, experimental and theoretical investigations have demonstrated the importance of the indirect mechanism involving the excitation of an inner-shell electron, followed by autoionization, to the total ionization process; in fact, for certain intermediate and heavy atomic sytems in lower stages of ionization, these indirect contributions can completely dominate the ionization cross section.[7,8,9]

The processes we wish to consider here are:

$$e + A^{q+} \rightarrow A^{(q+1)+} + e + e, \qquad (9)$$

$$e + A^{q+} \rightarrow (A^{q+})^* + e \qquad (10)$$
$$\hookrightarrow A^{(q+1)+} + e$$

where q is the charge of the atomic ion; Eq. (9) represents direct ionization, while Eq. (10) represents the excitation of an inner-shell electron to an intermediate state, followed by autoionization. The contribution of excitation-autoionization to the ionization cross section is given, to a very good approximation, by the the excitation cross section times a branching ratio for autoionization versus radiative stabilization. For relatively low stages of ionization, where the autoionizing rate is much higher than the radiative rate, this branching ratio is normally equal to unity, unless there are selection rules which inhibit autoionization from a particular level.

When the first step in Eq. (10) involves an excitation of the form $d^{10} \rightarrow d^9 f$, then the potential barrier effects and the resulting term dependence discussed in the last section become important in the determination of the size and shape of the ionization cross section. One of the best examples of this occurs in the Cd isoelectronic sequence.[27] Energy-level diagrams showing the $4d^{10}5s^2$ ground-state configurations and the excited $4d^9 5s^2 4f$ configurations for the Cd-like ions In^+, Sb^{3+}, and Xe^{6+} are given in Fig. 8. As can be seen, all of the 20 levels of $4d^9 5s^2 4f$ are autoionizing for the first two ions, while only the $4d^9 5s^2 4f$ $^1P_1$ level is above the ionization threshold in Xe^{6+}. The contribution of excitation-autoionization through the $4d^9 5s^2 4f$ configuration in In^+ is small. This can be understood by studying Fig. 9 where plots of the 4f effective potential and radial wave function for this ion are shown. The ^1P effective potential has both a large potential barrier and positive minimum in the inner well; as a result, the 4f ^1P wave function must remain in the outer well-region. However, the CA effective potential, which provides a fairly accurate representation for the other nine terms, has a much smaller barrier and the minimun in the inner well is now negative; nevertheless, the $4f_{CA}$ wave function still resides in the outer region, and is nearly identical to the 4f ^1P wave function. Thus the overlap of these wave functions with the 4d radial wave function is minimal, and the collision strength for the $4d^{10} \rightarrow 4d^9 4f$ electron impact excitation is extremely small.

As we move along the isoelectronic sequence, things begin to change, and by Sb^{3+}, term dependence sets in, as can be seen by examining Fig. 10. The increase in nuclear charge nearly wipes out the CA potential barrier, causing the $4f_{CA}$ wave function to drop into the inner region. However, the effective potential for the ^1P term still has

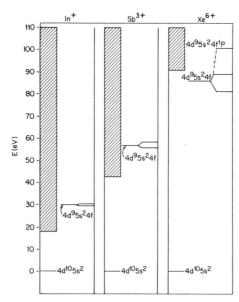

Fig. 8. Energy-level diagrams showing the $4d^{10}5s^2$ ground-state configurations, the $4d^9 5s^2 4f$ excited configurations and the ionization thresholds of In^+, Sb^{3+}, and Xe^{6+} (Ref. 27). Hatched regions indicate the ionization continuum for the next higher ionization stage.

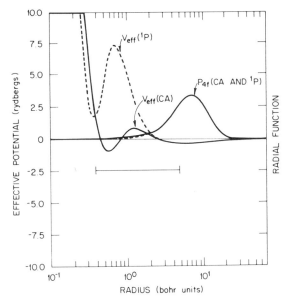

Fig. 9. Plots of the $4f_{CA}$ and $4f\ ^1P$ effective potentials and radial wave functions for the $4d^95s^24f$ configuration in In^+ (Ref. 27). The notation is the same as in Fig. 7.

a large positive barrier, which forces the $4f\ ^1P$ radial wave function to remain in the outer region. This has a pronounced effect on the size and shape of the $4d^{10}5s^2 \rightarrow 4d^95s^24f$ excitation cross section. The size of the dipole-allowed transition to the $^1P_1$ level depends on the overlap of the 4d and 4f ^1P wave functions which, in this ion, is quite small. However, the non-dipole excitations to the other 19 levels of $4d^95s^24f$ depend primarily on the size of the exchange contributions to the cross

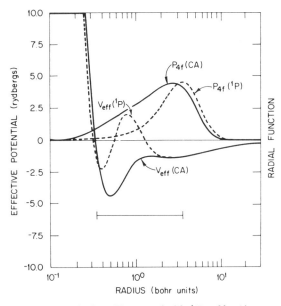

Fig 10. Plots of the $4f_{CA}$ and $4f\ ^1P$ effective potentials and radial wave function for the $4d^95s^24f$ configuration in Sb^{3+} (Ref. 27). The notation is the same as in Fig. 7.

section, and these contributions, are in turn, a function of the overlap of the 4d radial wave function with the outgoing scattered wave, as well as the overlap of the $4f_{CA}$ wave function with the incident wave. For electron energies near threshold, the incident wave has its first node in the barrier region, and thus there is a large overlap between its first antinode and the $4f_{CA}$ wave function which has a large amplitude in the inner-well region. However, as the energy increases, the first node of the incident wave moves further into the inner region and cancellation effects begin to decrease the magnitude of the collision strength. Therefore, in Sb^{3+} we would expect the excitation cross section for $4d^{10}5s^2 \rightarrow 4d^95s^24f$ to be dominated by the non-dipole transitions to the 19 levels other than $^1P_1$, and on the basis of the above discussion, its magnitude should fall off rapidly with electron energy.

The theoretical ionization cross section for Sb^{3+}, showing the excitation-autoionization contributions from both the $4d^95s^24f$ and $4d^95s^25f$ configurations is given in Fig. 11. The 5f configuration is also term dependent and the cross section is again dominated by the non-dipole transitions. The dipole collision strength has been transferred to higher members of the Rydberg series, and perhaps into the continuum. Thus, we would expect some of the dipole-allowed transitions to the $4d^95s^2nf$ $^1P_1$ (n>5) levels to make measurable contributions to the ionization cross section, and they should be included in a complete calculation. These general conclusions are supported by a recent crossed electron-ion beam experiment performed on Sb^{3+} at ORNL[28], which shows a large indirect contribution to the ionization cross section of the shape shown in Fig. 11, but with apparent contributions from higher members of the Rydberg series. In addition, the experimental results show evidence for the added indirect process of resonant recombination followed by double autoionization,[29] which will not be discussed here. Finally, it should be mentioned that large non-dipole dominated excitation-autoionization contributions are also observed in several members of the Xe isonuclear sequence in low stages of ionization.[29]

Plots of effective potentials and radial wave functions for Xe^{6+}, the last ion in the Cd sequence to be considered, are shown in Fig. 12. The

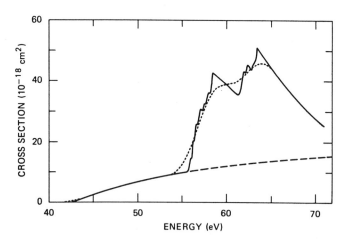

Fig. 11. Ionization cross section for Sb^{3+} (Ref. 27). Dashed curve, direct-ionization cross section calculated from the Lotz equation; solid curve, distorted-wave calculation of the indirect contribution due to $4d^{10}5s^2 \rightarrow 4d^95s^2nf$ (n=4,5) plus Lotz; dotted curve, solid curve convoluted with a 2-eV Gaussian to simulate the experimental energy spread.

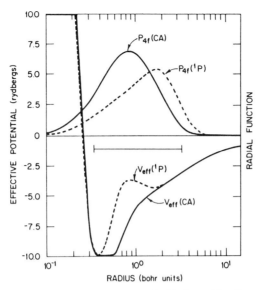

Fig. 12. Plots of the $4f_{CA}$ and $4f\ ^1P$ effective potentials and radial wave functions for the $4d^9 5s^2 4f$ configuration in Xe^{6+} (Ref. 27). The notation is the same as in Fig. 7.

additional increase in nuclear charge has caused the potential barrier for the ^1P effective potential to become negative and small, and the $4f\ ^1P$ wave function is on the verge of sliding completely into the inner-well region. For still higher stages of ionization the barrier will become completely insignificant, the $4f\ ^1P$ wave function will become indistinguishable from the $4f_{CA}$ wave function, and term-dependent effects will disappear. Nevertheless, for Xe^{6+}, these effects are still quite important and must be included in order to obtain accurate theoretical values of the ionization cross section. This is illustrated in Fig. 13, where we show the contribution of the $4d^{10} 5s^2 \rightarrow 4d^9 5s^2 4f\ ^1P_1$ excitation followed by autoionization to the ionization cross section. It is legitimate to treat the only autoionizing level of the $4d^9 5s^2 4f$ configuration, $^1P_1$, separately since intermediate-coupling calculations indicate that it is over 99% pure. As can be seen, the distorted-wave calculations show a factor-of-2 reduction in the cross section when the proper $4f\ ^1P$ wave function, rather than the $4f_{CA}$ wave function is used. The very small difference between the term-dependent, distorted-wave results and the results one obtains when the $4f\ ^1P$ wave function is employed in a two-state, close-coupling calculation indicates that it is term dependence, and not continuum coupling, which is important for this case.

The more complete results for Xe^{6+} calculated using the distorted-wave method are shown in Fig. 14, along with the crossed electron-ion beam measurements of Gregory and Crandall.[30] The agreement between experiment and theory is quite good. The $4d^9 5s^2 4f\ ^1P_1$ level at 100.4 eV makes the largest contribution to the ionization cross section. However, the indirect contributions from other levels are also important. The step at 116.4 eV is due to the $4d^9 5s^2 5f\ ^1P_1$ level. The first antinode of the $5f\ ^1P$ wave function overlaps more strongly with the second antinode of the $4d$ wave function than does the first antinode of the $5f_{CA}$ wave function, which has moved further inside. This increases the cross section to the $5f\ ^1P_1$ level when the proper term-dependent radial wave function is used. Thus some of the $4f\ ^1P$ collision strength has been transferred to $5f\ ^1P$. Also included in this calculation are the excitation-autoionization

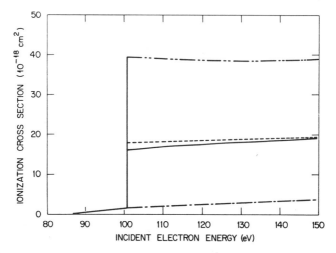

Fig. 13. Ionization cross section for Xe^{6+} showing the indirect contributions from the $4d^9 5s^2 4f$ $^1P_1$ level (Ref. 27). Dot-dash curve, direct-ionization cross section from the Lotz equation; solid curve, term-dependent, close-coupling calculation for the $4d \rightarrow 4f$ $^1P_1$ excitation plus Lotz; dotted curve, term-dependent, distorted-wave calculation of the $4d \rightarrow 4f$ $^1P_1$ excitation plus Lotz; chain curve, CA, distorted-wave calculation of the $4d \rightarrow 4f$ $^1P_1$ excitation plus Lotz.

contributions from the $4d^9 5s^2 5d$ and $4d^9 5s^2 6p$ configurations, which show no term dependence. Finally, we include in Fig. 14 the excitation cross section to the lower levels of $4d^9 5s^2 4f$ (marked O for optical transitions) which are below the ionization threshold and, of course, do not contribute to the ionization cross section.

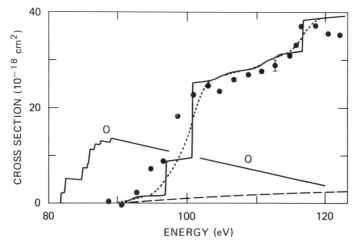

Fig. 14. Ionization cross section for Xe^{6+} (Ref. 27). Dashed curve, direct-ionization cross section calculated from the Lotz equation; solid curve, distorted-wave calculation of the excitations $4d \rightarrow 4f, 5f, 5d, 6p$ plus Lotz; dotted curve, solid curve convoluted with a 2-eV Gaussian to simulate the experimental energy spread; solid curve marked O is the total excitation cross section to non-autoionizing levels of $4d^9 5s^2 4f$; filled circles, experimental measurements, Ref. 30.

Potential-barrier and term-dependent effects can also be important for excitation-autoionization transitions involving excited d electrons near the beginning of the transition series of elements. This is especially true in transitions of the type:

$$e + np^6(n+1)s \rightarrow np^5nd(n+1)s + e$$
$$ \hookrightarrow np^6 + e \qquad (11)$$

in the singly ionized species Ca^+, Sr^+, and Ba^+. These are the first ions where excitation-autoionization was found experimentally to dominate the total ionization cross section.[31,32,33] Hansen[34,35] first showed that the $np^5nd(n+1)s$ configurations in these ions exhibit pronounced term dependence, and he predicted that this would have a significant effect on the cross section for the $np^6 \rightarrow np^5nd$ excitation. Since then, an R-matrix calculation[36] has demonstrated the importance of continuum coupling in this excitation process, and distorted-wave calculations[37] have confirmed Hansen's prediction regarding the significance of term dependence. However, to date, no calculation has been reported which includes both of these effects, and sizable disagreement between theory and experiment persists near the ionization threshold. Theoretical work on these ions, aimed at gaining a better understanding of the combined effects of term dependence and continuum coupling, is continuing.

DIRECT-IONIZATION CROSS SECTIONS

Potential-barrier and term-dependent effects in the continuum, which have been shown to be of great importance in certain photoionization transitions (see for example the review articles by Connerade[15] and Karaziya[38]), may also be significant in direct electron-impact ionization. Distorted-wave calculations have demonstrated that term dependence in the ejected-electron continuum can have a large effect on both the size and shape of single[39,40] and double electron-impact ionization cross sections.[41,42]

Let us first consider electron-impact ionization out of the 4d subshell within the Xe isonuclear sequence:[41,42]

$$e + Xe^{q+}(4d^{10}5s^25p^{6-q}) \rightarrow Xe^{(q+1)+}(4d^95s^25p^{6-q}) + e + e, \qquad (12)$$

where q is the ionization stage of the initial ion. For Xe^+, Xe^{2+}, and Xe^{3+}, all states of the configuration $4d^95s^25p^{6-q}$ are above the first ionization threshold, while for Xe^{4+}, only one third of the states of $4d^95s^25p^2$ are autoionizing. Thus for the first three ions, we would expect that ionization out of the 4d subshell should be followed by autoionization and contribute to the double-, rather than the single-, ionization cross section; in the case of Xe^{4+}, assuming a statistical distribution of the collision strength, about one third of the 4d ionization cross section will be observed as double ionization.

In all these ions, the 4d ionization is dominated by the $4d^9kf\ ^1P$ ejected-electron channel, and as one would expect from prior discussion, this channel is highly term dependent. In Fig. 15 we show the plots of the CA and ^1P effective potentials and continuum orbitals for Xe^+ and Xe^{2+} for an ejected-electron energy of 1 eV. For Xe^+ the CA effective potential has a small negative barrier, while the ^1P effective potential has large positive barrier. As a result, the first antinode of the kf_{CA} radial wave function is located in the inner-well region, where it has a large overlap with the second antinode of the 4d orbital; while the first antinode of kf ^{1}P is primarily in the outer-well region, where it overlaps only weakly with the second antinode of the 4d orbital. A strong overlap between the kf ^{1}P orbital and the 4d orbital will not occur until we reach higher ejected-electron energies. For Xe^{2+}, at the same ejected-electron energy, the situation is quite different. With the decreased shielding resulting from the removal of another 5p electron, the barrier in the CA effective potential has disappeared and the kf$_{CA}$ has moved far enough into the inner region that wave cancellation has begun to decrease its overlap with the 4d wave function. However, the decrease in the ^{1}P potential barrier has caused the kf ^{1}P continuum orbital to slide much further into the inner region, where it has a sizable overlap with the second antinode of the 4d wave function.

On the basis of these simple effective-potential arguments, we can make a number of predictions regarding the 4d ionization cross section for these two ions. In Xe$^+$, the use of the proper term-dependent orbital, rather than the configuration-average orbital, for the kf ^{1}P ejected-electron channel should lead to a reduction of the cross section near threshold (i.e., a delayed onset) and, most likely, an overall decrease in the magnitude of the cross section. However, in the case of Xe^{2+}, the 4d ionization cross section calculated with the term-dependent continuum function should be larger than that calculated with the CA orbital, even near threshold.

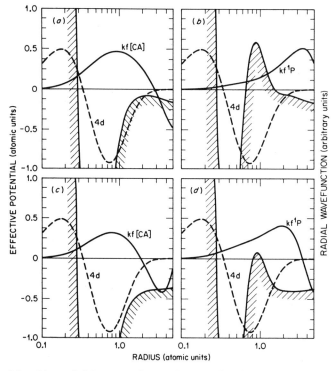

Fig. 15 CA and ^{1}P term-dependent effective potentials and radial wave functions for the f continuum electron ejected from the 4d subshell in Xe$^+$, (a) and (b), and Xe^{2+}, (c) and (d) (Ref. 42). Hatched solid curves, effective potentials; solid curves, kf continuum wave functions at an energy of 1 eV; dashed curves, 4d bound orbital. Note here that the effective potentials are in hartree atomic units, rather than rydbergs.

The experimental and theoretical results for double ionization in Xe⁺ through Xe⁴⁺ are shown in Fig. 16. The direct double-ionization cross section was estimated from the binary-encounter approximation of Gryzinski,[43] and as expected, it decreases rapidly with ionization stage. The 4d ionization-autoionization contributions, however, dominate the total double-ionization cross section for all 4 ions, and show significant term-dependent effects. The calculations which employ the kf ^1P term-dependent continuum functions for the $4d^9kf$ ^1P ejected-electron channel also include an estimate for the effects of ground-state correlation by incorporating the important $4d^{10} \longleftrightarrow 4d^84f^2$ configuration interaction. The difference between the CA and the term-dependent results for Xe⁺ and Xe²⁺ is about as expected from the potential-well arguments above and, in each case, the term-dependent results are closer to the experimental measurements.[44] The results for Xe³⁺ are similar to those for Xe²⁺; however, in all three ions the experimental threshold is below that predicted by theory. In the case of Xe⁴⁺, we also show curves for the 4d ionization cross sections multiplied by the fraction of states of the configuration $4d^95s^25p^2$ which autoionize (1/3). From a comparison of these scaled curves with the experimental measurements, it would appear that the collision strength is not distributed statistically.

Recently Younger[40] completed a theoretical study of photoionization and direct electron-impact ionization out of the 4d subshell for ten ions

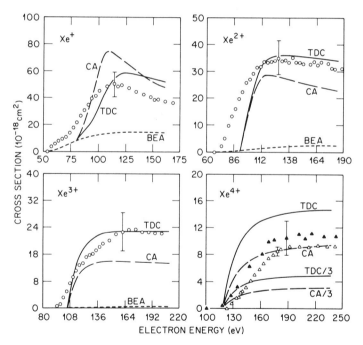

Fig. 16. Electron-impact double-ionization cross sections for (a) Xe⁺, (b) Xe²⁺, (c) Xe³⁺, and (d) Xe⁴⁺. Open circles and darkened triangles, experimental data (Ref 44); open triangles, experimental data (Ref 42); short-dashed curves labeled BEA, direct-double ionization using the binary-encounter approximation; long-dashed curve labeled CA, 4d ionization contribution using configuration-average kf radial wave functions plus direct-double ionization; solid curves labeled TDC, 4d ionization contribution using kf ^1P term-dependent radial wave functions with ground-state correlations plus direct-double ionization.

within the palladium isoelectronic sequence, with ground-state configurations $4d^{10}$. The cross sections are, as in the cases discussed above, dominated by the $4d^9 kf$ ^1P channel, and term dependence was found to be significant up to a charge state of +10. In an earlier calculation,[45] he demonstrated similar effects for electron-impact ionization out of the closed 5d subshell in Hg^+.

Term dependence in the ejected-electron continuum may also be important in electron-impact ionization out of closed p subshells. This was shown in a study of the argon isoelectronic sequence[39] where, for low stages of ionization, the 3p ionization cross section is dominated by the $3p^5 kd$ ^1P ejected-electron channel. In neutral argon, the term-dependent effective potential for the d continuum electron differs significantly from the CA effective potential and has a well defined barrier region. Thus at low ejected-electron energies, the kd ^1P partial wave, unlike the kd_{CA} wave, will not penetrate the barrier and will have a small overlap with the 3p orbital; however, at some higher ejected-electron energy the penetration will become appreciable. This behaviour, which is similar to that already discussed for the kf ^1P wave in Xe^+, is displayed in Fig. 17 where the phase shift, δ, is plotted as a function of ejected-electron energy for the CA and term-dependent kd partial waves. The phase shift for the kd_{CA} wave is large at very low energies, while the kd ^1P phase shift is quite small until the wave begins to penetrate the barrier at about 4 eV.

The ionization cross section for Ar I calculated using the CA and term-dependent kd partial wave for the $3p^5 kd$ ^1P ejected-electron channel are shown in Fig. 18. As can be seen, term-dependence is quite important for electron energies up to more than three times threshold. Also shown is a term-dependent calculation corrected for the effects of ground-state pair correlation $3p^6 \longleftrightarrow 3p^4 3d^2$. Term-dependence enhances the influence of this interaction by increasing the overlap of the kd continuum wave and the 3d correlation orbital.[39] For the higher members

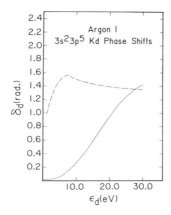

Fig. 17. Phase shifts, δ_d, for the kd partial waves in neutral argon as a function of ejected-electron energy, ε_d (Ref. 39). Dashed curve, from a distorted-wave calculation using the CA effective potential; solid curve, distorted-wave calculation using the ^1P effective potential.

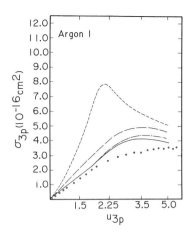

Fig. 18. 3p electron-impact ionization cross section, σ_{3p}, of neutral argon (Ref. 39). The incident electron energy in threshhold units is represented by u_{3p}. Short dash, calculated using the CA kd wave for the $3p^5kd$ ^1P ejected-electron channel; long dash, calculated using the term-dependent kd wave for the $3p^5kd$ ^1P ejected-electron channel; solid curve; term-dependent calculation with approximate ground-state correlation; dot-dash, solid curve plus 3s subshell ionization; open circles, experimental measurements (Ref. 46).

of this isoelectronic sequence the effects of term dependence decrease rapidly with ionization stage. Not only is the barrier height diminished, reducing the difference between the kd_{CA} and kd ^1P partial waves, but also, the relative importance of the kd ^1P exit channel to the total ionization cross section decreases.[39]

CONCLUSIONS

The sudden changes with atomic number in the binding energy of excited d and f electrons preceding the onset of transition and rare-earth series of elements can be explained in terms of one-electron effective potentials. The change in binding energy and the associated wave function collapse are most abrupt when the potential contains a centrifugal barrier which is positive and well defined. Thus the contraction of wave functions is more gradual for d electrons than for f electrons and occurs much less abruptly for ionized species than for neutrals.

Barriers in the effective potential for configurations of the type p^5d and d^9f are enhanced by an unusually large positive exchange interaction for the ^1P terms. This may lead to a large difference between the ^1P Hartree-Fock radial wave functions and the wave functions for all other terms of these configurations, especially near the beginning of the transition and rare-earth series of elements, where the collapse of the wave function is about to occur. This term dependence can have a significant influence on many atomic properties, including the excitation and ionization of both neutral and ionized species.

ACKNOWLEDGEMENTS

The authors would like to thank C. Bottcher and S. Younger for a number of very useful discussions. The collaborative efforts of the electron-ion beam experimental groups at ORNL and JILA are gratefully acknowledged. This work was supported by the Office of Fusion Energy, U. S. Department of Energy, under Contract No. DE-AC05-84OR21400 with Martin Marietta Energy Systems, Inc., and under Contract No. DE-FG05-86ER53127 with Auburn University.

REFERENCES

1. E. Fermi, in: "Quantentheorie und Chemie," H. Falkenhagen, ed., S. Hinzel-Verlag, Leipzig, 1928, p. 95.
2. M. Mayer, Phys. Rev. 60, 184 (1941).
3. R. Latter, Phys. Rev. 99, 510 (1955).
4. A. R. P. Rau and U. Fano, Phys. Rev. 167, 7 (1968).
5. D. C. Griffin, K. L. Andrew, and R. D. Cowan, Phys. Rev. 177, 62 (1969).
6. D. C. Griffin, R. D. Cowan, and K. L. Andrew, Phys. Rev. A 3, 1233 (1971).
7. R. A. Falk, G. H. Dunn, D. C. Griffin, C. Bottcher, D. C. Gregory, D. H. Crandall, and M. S. Pindzola, Phys. Rev. Lett. 47, 494 (1981).
8. D. C. Griffin, C. Bottcher, and M. S. Pindzola, Phys. Rev. A 25, 1374 (1982).
9. C. Bottcher, D. C. Griffin, and M. S. Pindzola, J. Phys. B 16, L65 (1983).
10. C. Froese Fischer, "The Hartree-Fock Method for Atoms," Wiley, New York (1977).
11. J. C. Slater, Phys. Rev. 81, 385 (1951).
12. F. Herman and S. Skillman, "Atomic Structure Calculations," Prentice-Hall, Inc., Englewood Cliffs, New Jersey (1963).
13. R. D. Cowan, Phys. Rev. 163, 54 (1967).
14. R. D. Cowan, "The Theory of Atomic Structure and Spectra," University of California Press, Berkeley (1981).
15. J. P. Connerade, Contemp. Phys. 19, 415 (1978).
16. V. Kaufman and J. Sugar, J. Res. Natl. Bur. Stand. 71A, 583 (1967).
17. R. A. Roig and G. Tondello, J. Opt. Soc. Am. 65, 829 (1975).
18. R. D. Cowan, J. Opt. Soc. Am. 58, 924 (1968).
19. D. H. Hartree and W. Hartree, Proc. Roy. Soc. A154, 588 (1936).
20. J. E. Hansen, J. Phys. B 5, 1083 (1972).
21. D. C. Griffin and M. S. Pindzola, Comments At. Mol. Phys. 13, 1 (1983).
22. G. Wendin and A. F. Starace, J. Phys. B 11, 4119 (1978).
23. J. E. Hansen, A. W. Fliflet, and H. P. Kelly, J. Phys. B 8, L127 (1975).
24. S. M. Younger, Phys. Rev. A 22, 2682 (1980).
25. G. Wendin, J. Phys. B 9, L297 (1976).
26. L. Brillouin, J. Phys. Radium 3, 373 (1932).
27. M. S. Pindzola, D. C. Griffin, and C. Bottcher, Phys. Rev. A 27, 2331 (1983).
28. D. C. Gregory and A. M. Howald, Phys. Rev. A 34 (1986).
29. D. C. Griffin, C. Bottcher, M. S. Pindzola, S. M. Younger, D. C. Gregory, and D. H. Crandall, Phys. Rev. A 29, 1729 (1984).
30. D. C. Gregory and D. H. Crandall, Phys. Rev. A 27, 2338 (1983).
31. B. Pert and K. T. Dolder, J. Phys. B 1, 872 (1968).
32. B. Pert, J. G. Stevenson, and K. T. Dolder, J. Phys. B 6, 146 (1973).
33. B. Pert and K. T. Dolder, J. Phys. B 8, 56 (1975).
34. J. E. Hansen, J. Phys. B 7, 1902 (1974).

35. J. E. Hansen, J. Phys. B **8**, 2759 (1975).
36. P. G. Burke, A. E. Kingston, and A. Thompson, J. Phys. B **16**, L385 (1983).
37. D. C. Griffin, M. S. Pindzola, and C. Bottcher, J. Phys. B **17**, 3183 (1984).
38. R. I. Karaziya, Sov. Phys. Usp. **24**, 775 (1981).
39. S. M. Younger, Phys. Rev. A **26**, 3177 (1982).
40. S. M. Younger, Submitted Phys. Rev. A (1986).
41. M. S. Pindzola, D. C. Griffin, and C. Bottcher, J. Phys. B **16**, L355 (1983).
42. M. S. Pindzola, D. C. Griffin, C. Bottcher, D. H. Crandall, R. A. Phaneuf, and D. C. Gregory, Phys. Rev. A **29**, 1749 (1984).
43. M. Gryzinski, Phy. Rev. **138**, A336 (1965).
44. C. Achenbach, A. Muller, E. Salzborn, and R. Becker, Phys. Rev. Lett. **50**, 2070 (1983).
45. S. M. Younger, private communication.
46. R. K. Asuni and M. V. Kurepa, J. Electron. Control **15**, 41 (1963).

RELATIVISTIC CORRELATIONS IN ATOMS

K. Dietz

Physikalisches Institut der Universität Bonn
Nussallee 12
D-5300 Bonn
West Germany

INTRODUCTION

Atoms are particularly well-suited objects when it comes to testing certain concepts of many-body theories. They play a unique role in this respect because of two constructively interfering reasons: first of all, the laws describing the interactions of their constituents are the ones best known in all of Physics; secondly, their structure is comparatively simple and amenable to concise theoretical treatment. Because of these two reasons, physically motivated many-body approximation schemes, ordered in a systematic hierarchy of precision, can be carefully tested; discrepancies between theory and experiment are due to many-body effects and are never masked by uncertainties in the constituent-interaction (needless to say, we leave out the very small hadronic contributions to atomic structure).

Many-body effects in atoms are solely produced by the electron-electron interaction which derives from the laws of Quantum Electrodynamics or, in a very good approximation from the repulsive Coulomb potential; in the general nomenclature they are named correlations. The intuitive meaning of the latter is clear on a naive basis, when it comes to precise notions confusingly many facets are found in the literature. We shall try to give precise definitions and to clarify underlying physical concepts. It has to be kept in mind, however, that an ultimate scope for categorising correlations in a physically relevant manner has not yet been set up. For instance, the quest for coherent effects (plasmons?) has not yet led to a definite answer. The corresponding category of correlations could be defined by considering the frequency spectrum of time-dependent correlations: pronounced peaks could sensibly be interpreted as coherent effects, a new type of correlations.

Our material is organised in two chapters: chapter 1 deals with a general introduction and discussion of g-Hartree mean-field theories, chapter 2 deals with applications. We emphasise the role of vacuum fluctuations and deformations of the Dirac sea in a consistent construction of mean-fields and give their explicit form in the g-Hartree theory.

1. GENERAL CONSIDERATIONS

Before we get to the subject of these lectures, correlations in atoms, we propose to discuss the notion of correlations in a slightly larger framework. To this end we start by considering an inhomogeneous gas of electrons in thermal equilibrium. For a proper definition of correlations, it suffices to take non-relativistic kinematics and a pure Coulomb potential for the electron-electron interaction. The case of Quantum Electrodynamics (QED) is then unambigously implemented in our way of reasoning, the theory of both relativistic and non-relativistic many-electron systems bear close conceptual and structural similarities.

The fundamental quantity which governs the dynamics and from which all physical properties can be derived is the action. For non-relativistic electrons in an external field which produces the inhomogeneity of the system we have

$$S = \int_0^T dt \int_\Omega d^3x \, \psi^*(\vec{x},t) \{i\partial_t - D\} \psi(\vec{x},t)$$

$$- \frac{1}{2} \int_0^T dt \int d^3x \int d^3x' \, |\psi(\vec{x},t)|^2 V(\vec{x},\vec{x}') |\psi(\vec{x}',t)|^2 \quad (1.1)$$

where

$$D = -\frac{\hbar^2}{2m} \Delta + \mu(\vec{x}) \quad (1.2)$$

is the free particle dispersion and

$$V(\vec{x},\vec{x}') = \frac{e^2}{4\pi|\vec{x}-\vec{x}'|} \quad (1.3)$$

the electron-electron interaction; $\psi(\vec{x},t)$ denotes the electron field, T is the length of a finite time-interval and Ω the volume in which our system is enclosed and which will eventually tend to infinity.

Thermal equilibrium is described by the partition function Z

$$Z = \text{tr } e^{-\beta H} \quad (1.4)$$

where H denotes the Hamiltonian to be derived from the action (1.1). Precisely speaking, Z is a function of the inverse temperature ß, the volume Ω and a functional of the external potential which plays the role of a space-dependent chemical potential.

$$Z = Z[\mu(\vec{x}); \beta, \Omega] \quad (1.5)$$

The procedure to actually calculate Z is well-known and follows ideas developed in exploring connections of euclidean field theory and statistical mechanics: one has to continue analytically to purely imaginary times $t \to -it$ to obtain the euclidean action S_E

$$S_E[\psi^*, \psi; \beta] := i S[\tilde{\psi}^*, \tilde{\psi}; T = -i\beta]$$

$$= \int_0^\beta d\tau \int_\Omega d^3x \; \psi^* \{-\partial_\tau - D\} \psi - \frac{1}{2} \int_0^\beta d\tau \iint |\psi|^2 V |\psi|^2 \tag{1.6}$$

where we indicated the functional dependence of the action on the electron field explicitly and

$$\psi(\vec{x}, \tau) := \tilde{\psi}(\vec{x}, -i\tau)$$

The partition function then has the following functional integral representation

$$Z = \int D\psi^* \int D\psi \; e^{S_E[\psi^*, \psi; \beta]} \tag{1.7}$$
$$\text{antiperiodic}$$

the functional integration has to be performed over antiperiodic, Grassmannian, i.e. anti-commuting, fields

$$\psi(\vec{x}, \tau) = -\psi(\vec{x}, \tau + \beta)$$

in order to comply with Fermi-Dirac statistics.

Unfortunately, we do not have space and time to explain the physical concepts leading to this representation and its connection with the usually more familiar notions of the method of Second Quantisation. Suffice it to say that the formalism presented here can be translated literally into the language of local field operators once the appropriate machinery is set up. However, we want to stress that the formulation of the theory in terms of functional integrals and the corresponding functional differential equation has considerable advantages. For the problems treated in these lectures we emphasise that a new criterion for the choice of mean-fields results; furthermore, the representation appears as a means of effective book-keeping for an expansion in terms of Feynman graphs describing higher-order correlations.

We now proceed to evaluate the representation (1.7). A Hubbard-Stratonovitch[1] transformation leads[2] to an equivalent form for Z

$$Z = \int D\varphi \; e^{\frac{1}{2} \int_0^\beta d\tau \int_\Omega d^3x \; \varphi(\vec{x},\tau) V^{-1} \varphi(\vec{x},\tau)} \; Z_0[\varphi + \mu; \beta, \Omega] \tag{1.8}$$
$$\text{periodic}$$

where

$$e^2 V^{-1}(\vec{x}, \vec{x}') = -\Delta_{\vec{x}} \delta(\vec{x}-\vec{x}') \tag{1.9}$$

and

$$Z_0[\mu+\varphi, \beta, \Omega] = \int_{\text{antiperiodic}} \mathcal{D}\psi^* \int \mathcal{D}\psi \; e^{S_E^{(0)}[\psi^*, \psi; \beta]} \tag{1.10}$$

with

$$S_E^{(0)} = \int_0^\beta d\tau \int_\Omega d^3x \; \psi^*(\vec{x},\tau)\{-\partial_\tau - D - \varphi\} \psi(\vec{x},\tau) \;. \tag{1.11}$$

Obviously, Z_0 is the partition function for a free electron gas in an external field $\mu + \varphi$ and the representation (1.8) tells us that correlations are brought into the game by weighting Z_0 with a Gaussian weight determined by the inverse electron-electron interaction[*]. The potential φ appears as a bosonic variable and we have to integrate over periodic configurations

$$\varphi(\vec{x}, \tau) = \varphi(\vec{x}, \tau+\beta) \;. \tag{1.12}$$

For time-independent $\varphi = \varphi_0(\vec{x})$, Z_0 can be computed[3,4]: introducing the spectrum

$$(D + \varphi_0) \psi_\alpha = \varepsilon_\alpha \psi_\alpha \;, \tag{1.13}$$

one finds

$$\ln Z_0 = \frac{\beta}{2} \sum_\alpha \varepsilon_\alpha + \sum_\alpha \ln(1 + e^{-\beta \varepsilon_\alpha}) \tag{1.14}$$

where the sums run over the spectrum. The second term is the familiar Fermi-distribution; the first term accounts for the energy of vacuum fluctuations, $\hbar\omega/2$ for each degree of freedom, and will be seen to yield only a trivial, infinite shift of the chemical potential in the non-relativistic case. In the relativistic case, however, it combines with the energy of the Dirac sea of occupied negative energy states and yields interesting effects[5,6].

The central idea[7] of our approach is now to determine the mean-field as a stationary point of the φ-integration in (1.8) which defines

---

[*] The Gaussian weight in (1.8) is defined by analytic continuation from attractive potentials.

the lowest order of an expansion in terms of fluctuations, i.e. deviations from the mean-field.

It can be shown[2)8)] that for thermal equilibrium (enforced by periodic and antiperiodic boundary conditions for bosons and fermions respectively) this stationary point is time-independent. It, therefore, can be deduced from the following equation which we gain by inserting (1.14) into (1.8) and collecting exponentials in the integrand

$$\frac{\delta}{\delta \varphi_0(\vec{x})} \left(\frac{\beta}{2} \int d^3x' \, \varphi_0(\vec{x}') V^{-1} \varphi_0(\vec{x}') + \frac{\beta}{2} \sum \varepsilon_\alpha + \sum \ln(1 + e^{-\beta \varepsilon_\alpha}) \right) = 0 \quad (1.15)$$

Since, for ψ_α normalised to unity,

$$\frac{\delta \varepsilon_\alpha}{\delta \varphi_0(\vec{x})} = |\psi_\alpha(\vec{x})|^2, \quad (1.16)$$

we have

$$\varphi_0(\vec{x}) = \int d^3x' \, V(\vec{x},\vec{x}') \left\{ \sum_\alpha n_\alpha |\psi_\alpha(\vec{x}')|^2 + \rho_0^{vac}(\vec{x}') \right\} \quad (1.17)$$

where

$$n_\alpha = \frac{1}{e^{\beta \varepsilon_\alpha} + 1} \quad (1.18)$$

is the Fermi distribution.

The stationary point is immediately recognised as the Hartree-field induced by Fermi-distributed occupied orbitals; inserting into (1.13) the Hartree equation obtains.

The vacuum contribution $\rho_0^{vac}(\vec{x})$ needs some further consideration. For the non-relativistic case we argue using admittedly formal manipulations which, however, reveal the essential point. From (1.16) we get

$$\rho_0^{vac}(\vec{x}) = -\frac{1}{2} \sum_\alpha |\psi_\alpha(\vec{x})|^2$$

Since the summation runs over the whole spectrum we employ the completeness relation to obtain

$$\rho_0^{vac}(\vec{x}) = -\frac{1}{2} \delta(0) \quad (1.19)$$

for, and this is the important point, <u>all</u> potentials $\varphi_0(\vec{x})$. Hence, vacuum fluctuations produce a trivial infinite shift which is consistently neglected. We remark already at this point that the relativistic case is qualitatively different: the energy of the filled Dirac sea has to be included and a space-dependent potential appears[6] which has to be determined selfconsistently.

To derive the Hartree equation from the general concepts which we explained above is certainly not a very astounding innovation although one point must be stressed here: the Hartree mean-field has been shown to fulfil the fundamental requirement of being a stationary point of the "classical" action (determining the statistical weight, the normalisation of which is the partition function). Our discussion on the following pages will hopefully help to elucidate the importance of requiring stationarity of mean fields.

The variable φ playing the role of a (time-dependent) potential in the representation (1.8) and which was introduced via the Hubbard-Stratonovitch[1] transformation can be generalised[2)9] to include contributions to the self-consistent mean-field which are generated by a bilocal charge correlation. Exchange interactions are introduced in this way. Essentially the same argumentation as presented above then leads to the stationary point

$$\varphi_0(\vec{x}) = \int d^3x' \, V(\vec{x}, \vec{x}') \sum_\alpha h_\alpha \left\{ g \, |\psi_\alpha(\vec{x}')|^2 - (1-g) \, \psi_\alpha^+(\vec{x}') \psi_\alpha(\vec{x}) * \right\} \qquad (1.20)$$

which exhibits the exchange convolution and contains a constant g which expresses the fact that the stationary point is not unique but rather given as a one-parameter set. Direct and exchange potentials are weighted by g and 1-g respectively.

Having determined the mean-field we expand the exponential in terms of deviations from the mean-field, i.e. in terms of fluctuations and calculate their contribution to the partition function perturbatively; the techniques for doing that are standard field theoretic repertoire. We set

$$Z = Z_0[\varphi_0 + \mu; \beta, \Omega] \, Z_{corr} \qquad (1.21)$$

where Z_{corr} is the contribution of fluctuations represented by Feynman graphs.

The equilibrium properties of our electron gas obtain from the grand canonical potential

$$K = -\beta^{-1} \ln Z = K_{g-H} + K_{corr} \qquad (1.22)$$

of which we depict the corresponding graphical expansion. We have

$$K_{g-H} := -\beta^{-1} \ln Z_0$$

$$= -\beta^{-1} \sum_\alpha \ln(1+e^{-\beta \varepsilon_\alpha}) - \frac{g}{2}\,\bigcirc\!\!\!\sim\!\!\!\bigcirc + \frac{1-g}{2}\,\bigcirc\!\!\!\!\sim\!\!\!\!\bigcirc$$

$$K_{corr} := -\beta^{-1} \ln Z_{corr}$$

$$= \frac{1-g}{2}\,\bigcirc\!\!\!\sim\!\!\!\bigcirc - \frac{g}{2}\,\bigcirc\!\!\!\!\sim\!\!\!\!\bigcirc$$

$$- \frac{g^2}{2}\,\bigotimes + g(1-g)\,\bigoplus\!\!\sim\!\!\bigcirc$$

$$- \frac{1}{2}\,\bigcirc\!\!\sim\!\!\bigcirc\!\!\sim\!\!\bigcirc\,(1-g)^2$$

$$- \frac{1}{4}\,\bigotimes + \frac{1}{4}\,\bigcirc\!\!\!\Longleftrightarrow\!\!\!\bigcirc + O(e^6). \qquad (1.23)$$

These graphs have to be computed with the following rules (integration $\int d\tau \int d^3x$ is understood):

$$\underset{\vec{x},\tau}{\bullet}\!\!\sim\!\!\sim\!\!\sim\!\!\underset{\vec{x}',\tau'}{\bullet} \qquad V(\vec{x},\vec{x}')\,\delta(\tau-\tau')$$

$$\underset{\vec{x},\tau}{\bullet}\!\!-\!\!-\!\!\underset{\vec{x}',\tau'}{\bullet} \qquad g_{\varphi_0}(\vec{x},\tau;\vec{x}',\tau') =$$

$$\sum_\alpha \psi_\alpha(\vec{x})\psi_\alpha^*(\vec{x}')\,e^{-\varepsilon_\alpha(\tau-\tau')}\,\{\Theta(\tau-\tau')-n_\alpha\} ;$$

$$(1.24)$$

the $\psi_\alpha(\vec{x})$ are the g-Hartree orbitals, details are found in ref. (2).

It is of central importance to realise that the energy (1.22) is g-independent, it is the relative magnitude of E_{g-H} and E_{corr} which is controlled by the choice of g. We surmise the possibility of suppressing the contribution of correlations to the energy by appropriately choosing g. Indeed it can be shown, either phenomenologically[2] or theoretically by using a Thomas-Fermi expansion[8] that there is a choice of g, $g=g_o$, such that the contribution of correlations vanishes

$$K_{corr}\big|_{g=g_o} = 0 \tag{1.25}$$

the g-Hartree part K_{g-H} is then the exact grand canonical potential of our electron gas.

It has to be clear that g_o is a function of ß and Ω and a functional of the space-dependent chemical potential

$$g_o = g_o[\mu(\vec{x}); \beta, \Omega] \tag{1.26}$$

The set of solutions of the g-Hartree equation, i.e. the equation obtained by inserting (1.20) into (1.13), for $g=g_o$ will be dubbed "optimal basis" and is denoted as follows

$$g=g_o : \quad \{\gamma_\alpha^{(o)}\} \quad \text{optimal basis}$$

$$\{\varepsilon_\alpha^{(o)}\} \quad \text{set of eigenvalues.}$$

The optimal basis can be thought of as a realisation of an independent quasi-particle basis in the sense of the concept originally conceived by Landau[10]. Quasiparticle spectra can be established under various premises: the classical theory of Fermi liquids was introduced[10] in such a way that the entropy and the particle density are approximated by an ideal gas of fermionic excitations; in other thermodynamic quantities, interactions between these quasiparticles play in general a non-negligible role. The g-Hartree theory yields an optimal independent particle basis which determines the grand canonical potential <u>exactly</u>. We have

$$K = -\beta^{-1} \ln Z_o \big|_{g=g_o}$$

$$= -\beta^{-1} \sum_\alpha \ln(1+e^{-\beta \varepsilon_\alpha^{(o)}}) - \frac{1}{2}\int d^3x \sum_\gamma n_\gamma \gamma_\gamma^*(\vec{x}) V_{g_o}(\vec{x}) \gamma_\gamma(\vec{x}) \tag{1.27}$$

where

$$V_{g_o}(\vec{x}) = \varphi_o(\vec{x})\big|_{g=g_o} \tag{1.28}$$

However, we should not hesitate to emphasise that thermodynamic quantities derived from this representation of K do show in general contributions from fluctuations. As an example let us consider the electron density

$$n(\vec{x}) = \frac{\delta K}{\delta \mu(\vec{x})} . \tag{1.29}$$

Taking into account all functional dependences on the chemical potential $\mu(\vec{x})$ (of course, the basis $\{\psi_\alpha\}$ and the spectrum $\{\varepsilon_\alpha\}$ do depend on $\mu(\vec{x})$) we find

$$n(\vec{x}) = \sum_\alpha n_\alpha |\psi_\alpha^{(0)}(\vec{x})|^2 + \frac{\beta B}{2} \frac{\delta g_0}{\delta \mu(\vec{x})}$$

$$B = \int d^3x \int d^3x' \, V(\vec{x},\vec{x}') \sum_{\alpha,\gamma} n_\alpha n_\gamma \cdot$$

$$\left\{ |\psi_\alpha^{(0)}(\vec{x})|^2 |\psi_\gamma^{(0)}(\vec{x}')|^2 + \psi_\alpha^{(0)}(\vec{x}) \psi_\alpha^{(0)}(\vec{x}')^* \psi_\gamma^{(0)}(\vec{x}') \psi_\gamma^{(0)}(\vec{x})^* \right\}. \tag{1.30}$$

We see that the observed electron density consists of the quasiparticle density

$$n_Q(\vec{x}) := \sum_\alpha n_\alpha |\psi_\alpha^{(0)}(\vec{x})|^2 \tag{1.31}$$

and a correlation density

$$n_{corr}(\vec{x}) = \frac{\beta B}{2} \frac{\delta g_0}{\delta \mu(\vec{x})} \tag{1.32}$$

determined by the functional dependence of g_0 on $\mu(\vec{x})$.

Further thermodynamic relations can now be established. The equation of state for the inhomogeneous electron gas now derives from the g-optimised K as follows.

The two equations

$$N := \int d^3x \, n(\vec{x}) = N[\mu(\vec{x}); \beta, \Omega] \tag{1.33}$$

and (P is the pressure of the system)

$$P\Omega = K[\mu(\vec{x}); \beta, \Omega] \tag{1.34}$$

define functionals $\mu(\vec{x}) \to N$ and $\mu(\vec{x}) \to P$ from which the Fermi-energy ε_F of the quasiparticle distribution defined by

$$\mu(\vec{x}) = \varepsilon_F + \mu_0(\vec{x})$$

$$\mu_0(\vec{x}) \underset{|\vec{x}| \to \infty}{\longrightarrow} 0 \qquad (1.35)$$

has to be eliminated to obtain a relation

$$F[\mu_0(\vec{x}); P, \Omega, \beta, N] = 0 \qquad (1.36)$$

the equation of state.

On the other hand, we can make use of the local structure of (1.30) which defined for each $\vec{x}$ a functional $\mu(\vec{y}) \to n(\vec{x})$

$$n(\vec{x}) = n[\mu(\vec{y}); \vec{x}]$$

and invert this relation to obtain

$$\mu(\vec{x}) = \mu[n(\vec{y}); \vec{x}]$$

Inserting in (1.34) we arrive at a new functional

$$P\Omega = K = \tilde{K}[n(\vec{x}); \beta, \Omega] \qquad (1.37)$$

which for the ground state, i.e. $\beta \to \infty$, is the energy-density functional $E[n(\vec{x})]$ which received broad attention in the literature[11].

These elimination procedures are extraordinarily complicated in general and one can only hope to construct $E[n(\vec{x})]$ from the microscopic interactions by contriving a specific scheme of successive approximations. We have shown that a systematic Thomas-Fermi expansion is particularly useful and have established[8] an asymptotic expansion of the energy-density functional.

In the preceding paragraphs we tried to explain general notions on correlations in an inhomogeneous electron gas which are important in our approach. We pointed our in particular that optimal single-particle basis' exist in which one thermodynamic potential is exactly given: it is represented by an ideal Fermi gas of quasiparticle excitations (up to the obvious energy content of the mean field). We chose the grand canonical potential K for the construction of an optimal basis. It was furthermore important to note that equilibrium observables, e.g. the electron density, do receive contributions from correlations which are parametrised by the functional $g_0[\mu(\vec{x}); \beta, \Omega]$. The hope is, however, that having determined an optimal quasiparticle basis for a carefully chosen quantity, the energy, say, the correlations remaining in other physical observables of interest are significantly suppressed in this optimal basis.

To test this assertion we focus our attention on atoms and discuss the g-Hartree theory of atoms in the next chapter.

2. THE g-HARTREE THEORY OF ATOMS

Approaching the proper subject of this chapter, we must generalise the theory considered up to now to the fully relativistic case[2] the dynamics of which is determined by the QED-action

$$S_{QED}[\psi^*,\psi;A_\mu] = \int_0^T dt \int_\Omega d^3x \left\{ -\frac{1}{4} F_{\mu\nu} F^{\mu\nu} \right.$$

$$\left. + \psi^*(x)(i\partial_t - \mathcal{D} + e\gamma_0\gamma_\mu A^\mu)\psi(x) \right\} \quad (2.1)$$

where

$$\mathcal{D} = -i\gamma_0\vec{\gamma}\cdot\vec{\partial} + \gamma_0 m + \mu(\vec{x}) \quad (2.2)$$

Fixing a covariant, ghost-free gauge, the partition function obtains

$$Z_{QED} = \int_{antiperiodic} D\psi^* \int D\psi \int_{periodic} DA_\mu \; e^{iS'_{QED}[\psi^*,\psi;A_\mu]} \Bigg|_{T=-i\beta} \quad (2.3)$$

$$S'_{QED}[\psi^*,\psi;A_\mu] = \int d^4x \left\{ \psi^*(x)(i\partial_t - \mathcal{D} + e\gamma_0\gamma_\mu A^\mu)\psi(x) \right.$$

$$\left. + \frac{1}{2} \int d^4x' \; A^\mu(x) D_{\mu\nu}^{-1} A^\nu(x') \right\} \quad (2.4)$$

includes the gauge-fixing term and $D_{\mu\nu}(x,x')$ is the photon propagator in the corresponding gauge. In Feynman gauge, for instance, we have

$$D_{\mu\nu}^{(F)} = g_{\mu\nu} \frac{1}{\Box + i\varepsilon} \delta(x-x'). \quad (2.5)$$

It is instructive to perform the A_μ-integration and to find

$$Z_{QED} = \int D\psi^* \int D\psi \; e^{iS_{eff}[\psi^*,\psi]} \Bigg|_{T=-i\beta} \quad (2.6)$$

with

$$S_{eff} = \int d^4x\, \psi^*(x)(i\partial_t - \mathcal{D})\psi(x)$$
$$- \frac{e^2}{2}\int d^4x \int d^4x'\, j^\mu(x) D_{\mu\nu}(x,x') j^\nu(x') \qquad (2.7)$$

and

$$j_\mu = \psi^* \gamma_0 \gamma_\mu \psi(x).$$

We immediately recognise that S_{eff} is a covariant generalisation of the non-relativistic action (1.1): instead of the pure Coulomb propagator

$$\Delta(\vec{x},\vec{x}') = \frac{1}{4\pi}\frac{1}{|\vec{x}-\vec{x}'|}$$

we have its covariant extension $D_{\mu\nu}(x,x')$ which introduces the retardation required by relativity; at the same time the charge density is supplemented by space components such that the 4-vector current appears.

The vector potential A_μ plays the role of a covariant extension of the potential variable $\varphi(x)$ in the representation (1.8). The very same reasoning as above leads to the relativistic g-Hartree equation

$$(\mathcal{D} - e\gamma_0\gamma^\mu \varphi^{(g)}_\mu(\vec{x}))\psi_\alpha(\vec{x}) = \varepsilon_\alpha \psi_\alpha(\vec{x}) \qquad (2.8)$$

with

$$\varphi^{(g)}_\mu(\vec{x}) = -e\int d^3x'\, D_{\mu\nu}(\vec{x},\vec{x}') \sum n_\alpha \cdot$$
$$\{g\,\psi_\alpha^*(\vec{x}')\gamma_0\gamma^\nu \psi_\alpha(\vec{x}') - (1-g)\,\psi_\alpha^*(\vec{x}')\gamma_0\gamma^\nu \psi_\alpha(\vec{x}')*\}, \qquad (2.9)$$

the $\psi_\alpha(\vec{x})$ are Dirac spinors and $D^{(0)}_{\mu\nu}$ is the equal-time restriction of $D^{(F)}_{\mu\nu}$

$$D^{(0)}_{\mu\nu}(\vec{x},\vec{x}') = g_{\mu\nu}\frac{1}{4\pi|\vec{x}-\vec{x}'|}. \qquad (2.10)$$

As before, the g-Hartree potential turns out to be time-independent for equilibrium; we have to keep in mind, however, that this time-independence pertains to the specific gauge chosen since, of course, time-dependence is a gauge-dependent notion. Retardation does not contribute to the stationary point $\varphi_\mu^{(g)}(\vec{x})$: since non-trivial solutions of (2.8) are generated by the inhomogeneity $\mu(\vec{x})$, the external potential, questions of covariance and of compatibility of non-retardation and the covariant gauge chosen are out of place.

The computation of the partition function proceeds as before: in the Feynman rules (1.24) the Dirac spinor solutions of (2.8) have to be used, the graphical expansion of $\ln Z_{g-H}$ and $\ln Z_{corr}$ remain the same, of course.

In our discussion of atoms, we particularly emphasise the ground state, i.e. the limit $\beta \to \infty$. The notion of a ground state is, however, to be understood on a more general level to include (besides the thermodynamic ground state - a filled Fermi sphere of optimal quasiparticle levels) states corresponding to excited states characterised by appropriate sets of occupation numbers $\{n_\alpha\}$. The energy of these states can be shown, using our formalism, to develop an imaginary part which signals instability and gives the decay rate. The latter is, however, small enough to be neglected and to warrant that the approximation of quasi-stability is a viable one. Each of the states $\{n_\alpha\}$ generates by the g-Hartree equation (2.8), in which the occupation numbers are inserted, an optimal quasiparticle basis such that the total energy - the grand canonical potential at vanishing temperature - is exactly given by the g-Hartree contribution. Each set of occupied levels, $\{n_\alpha\}$ yields an optimal basis and a total ground state energy

$$E = K\big|_{\beta^{-1}=0} = E[n_\alpha]$$

$$= E_{g-H} + E_{corr} \qquad (2.11)$$

where, obviously,

$$E_{g-H}[n_\alpha;g] = K_{g-H}\big|_{\beta^{-1}=0}, \quad E_{corr}[n_\alpha;g] = K_{corr}\big|_{\beta^{-1}=0}$$

The optimal g-value, g_0, is obtained from

$$E_{corr}[n_\alpha;g]\big|_{g=g_0} = 0, \qquad (2.12)$$

the chemical potential $\mu(\vec{x})$ is a fixed quantity and equal to the nuclear Coulomb-potential, the volume Ω is supposed to approach infinity, $\Omega \to \infty$. Hence, g_0 remains as a functional of the distribution of occupied levels

$$g_0 = g_0[n_\alpha] \qquad (2.13)$$

We are now ready to apply our formalism to atomic configurations. There are two ways to exploit equation (2.12), semi-phenomenologically and in an ab initio manner to predict atomic levels from the two constants in the QED-Lagrangian, the electron mass and the fine-structure constant. The following two subsections will deal with these questions.

a) Semi-phenomenological g-Hartree Theory

A semi-phenomenological search for correlations in the g-Hartree theory proceeds on rather simple lines. Our central question is here to get insight in correlations remaining in an optimal g-Hartree basis. More precisely speaking, we propose to study correlations in physical observables - oscillator strengths, spin-orbit splittings - in an optimal basis $\{\psi_\alpha^{(o)}, \varepsilon_\alpha^{(o)}\}$ in which the transition energy is free of correlations since g_o is determined by*)

$$\varepsilon_{exp} = \left\{ E_{g\text{-}H}[n_\alpha^{(i)}; g] - E_{g\text{-}H}[n_\alpha^{(f)}; g] \right\}\bigg|_{g=g_o} \quad (2.14)$$

for the transition

$$\{n_\alpha^{(i)}\} \longrightarrow \{n_\alpha^{(f)}\},$$

where ε_{exp} is the experimental value for the transition energy. Another way of expressing the same thing is to say that g_o is determined such that the contribution of correlations (expressed in terms of the graphical expansion (1.23)) to the transition energy, vanishes

$$\left\{ E_{corr}[n_\alpha^{(i)}; g] - E_{corr}[n_\alpha^{(f)}; g] \right\}\bigg|_{g=g_o} = 0 \quad (2.14')$$

i) As a first application, we calculated[12] the oscillator strengths for the Principal Series of the alkali atoms Li, Na, K using the corresponding optimal basis. The results are shown in Figs. 1 - 3, experiments are taken from ref. (13). All three figures demonstrate that the g-Hartree oscillator strengths closely follow the experimental curve and constitute a substantial improvement of Dirac-Hartree-Fock (DHF) calculations. Even if the latter are shifted to agree with the large N asymptotics - a procedure which has no theoretical justification - deviations still persist: the slope of the shifted DHF-strengths does not fit the experimental data whereas the g-Hartree predictions agree in absolute value and slope.

A first conclusion can already be formulated: an energy-optimised g-Hartree basis leaves little room, in the examples considered, for correlations in the oscillator strength where by "correlations" we again

---

*) Solving the g-Hartree equation (2.8) one has to bear in mind that the spectral sum runs over both negative and positive energy states. In the present context we neglect the former, for a discussion we refer to the end of this chapter.

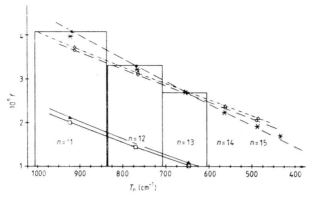

Figure 1. The oscillator strengths for the principal series of lithium, $2s\,^2S_{1/2} \to np\,^2P_{3/2}$. Experimental data are compared with g-Hartree ($g = 0.851\,719$) and DF (both $V(N)$ and $V(N-1)$) values. ●, experimental; *, g_0-Hartree; △, $V(N)$-DF (corrected for the limit); ○, $V(N-1)$-DF (corrected for the limit); ▲, $V(N-1)$-DF; □, $V(N)$-DF. Note that the correction of the DF values for the limit is an arbitrary procedure, so that the slopes of the DF and g-Hartree calculations may be compared.

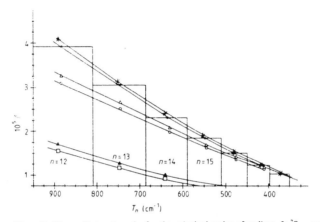

Figure 2. The oscillator strengths for the principal series of sodium, $3s\,^2S_{1/2} \to np\,^2P_{3/2}$. g-Hartree results are obtained for $g = 0.956\,929$. ●, experiment; *, g_0-Hartree; ○, $V(N)$-DF (corrected for the limit); △, $V(N-1)$-DF (corrected for the limit); ▲, $V(N-1)$-DF; □, $V(N)$-DF.

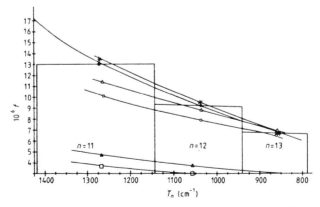

Figure 3. The oscillator strengths for the principal series of potassium, $4s\,^2S_{1/2} \to np\,^2P_{3/2}$. g-Hartree results are obtained for $g = 0.974\,213$. ●, experiment; *, g_0-Hartree; ○, $V(N)$-DF (corrected for the limit); △, $V(N-1)$-DF (corrected for the limit); ▲, $V(N-1)$-DF; □, $V(N)$-DF.

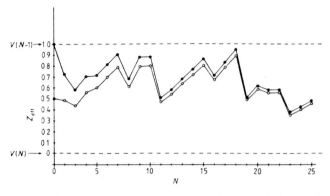

Figure 4. The effective charge, calculated from equation 2.15, for both virtual (○) and occupied (●) orbitals, using g-values obtained by fitting to experimental ionisation energies (cf equation 2.14)

understand the contribution of fluctuations, i.e. field configurations deviating from the g-Hartree optimised mean-field, to the oscillator strength which can be expressed in terms of a graphical expansion like the one in (1.23).

For the purpose of further illustration we compute the effective charge Z^* seen by the excited (bound) states

$$Z^* = \begin{cases} Z+1-g_o(N+1) & \text{for occupied states} \\ Z-g_o N & \text{for virtual orbitals} \end{cases} \quad (2.15)$$

and compare it with $V^N(Z^*=0)$ and $V^{N-1}(Z^*=1)$ charges for a selection of elements. From Fig. 4 we see that all Z^*_N lie between 0 and 1: the g-Hartree potential interpolates between the V^N and V^{N-1} DHF potentials[14].

ii) We have studied the 5d → nf transitions in Ba^+ which generate and are generated by a strikingly structured double-well[15]. It can be surmised that in the DHF-picture Brückner DHF double-well potential is built up by complicated correlations which appear as the determining factor in tuning its critical properties: we expect a sensitive test of the optimal g-Hartree potential.

In the actual calculation we determined g_o by fitting the quantum defects of the 5d → nf transitions for each n and computed their intensities J_{5n} and spin-orbit splittings.

Unfortunately, there are no absolute measurements of the intensities J_{5n}; the dip in the observed intensity distribution[17] is, however, reproduced in the g-Hartree results showing that the 5f orbital is the most strongly localised as is expected from the occurence of the dip at n=5.

A spectacular agreement is found if we compare the g-Hartree predictions for the spin-orbit splitting with experiment. This comparison is displayed in Fig. 5 where we also plotted the DHF and the multi-configuration DHF results which are both seen to fail rather badly, in particular for the critical 4f and 5f transitions.

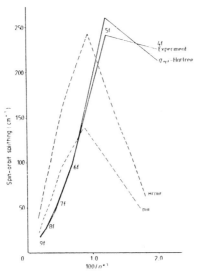

Figure 5 Comparison between g_{opt}-Hartree calculations of the spin-orbit splitting for the nf series of Ba II and the experimental values of Roig and Tondello (1975). Hartree-Fock results and MCDHF results are also displayed.

Our conclusion is again that correlations are suppressed in an optimised g-Hartree basis.

b) Ab initio calculations of atomic levels

Equation (2.12) can be used as the starting point of an ab initio calculation of energies: the expansion of E_{corr} in terms of fluctuations leads to a power series which is ordered in powers of the electron-electron interaction V_{int} and which is depicted in (1.23). Now, let $E_{corr}^{(k)}$ denote this series up to the k-th order, then

$$E_{corr}^{(k)}[h_\alpha; g] = 0 \qquad (2.16)$$

is an equation for g, its solution is an approximate value for g_o. From the latter E_{g-H} is calculated by solving the g-Hartree equation and we have

$$E = E_{g-H} + O(V_{int}^{k+1})$$

where the expansion coefficients in front of the (k+1)st and higher orders are believed to be small since they are calculated in an optimised basis.

In Table 1 we present our results[18] (k=2) for first row atoms: relativistic total energies are compared with experiment, perfect agreement is found. Since, however, the experimental situation does not allow for a precise comparison of theory and experiment, we collate our results of a non-relativistic g-Hartree calculation with CI-calculations[19] and find an impressive agreement.

Table 1. Total relativistic g-Hartree energies $E^{\text{g-Hartree}}(g_o^{(2)})$ are compared with experimental data taken from Scherr et al (1962 SSM) and Bashkin and Stoner (1975 BS) (au) (ref. (18))

Atom	$E^{\text{g-Hartree}}(g_o^{(2)})$	$E_{\text{tot}}^{\text{exp}}$	
		BS	SSM
Be	−14.6697	−14.6761	−14.6693
B	−24.6615	−24.6711	−24.6591
C	−37.8501	−37.8754	−37.8574
O	−75.1248	−75.1491	−75.1125
F	−99.8222	−99.8582	−99.8088
Ne	−129.0720	−129.1119	−129.0602

Table 2. Non-relativistic correlation energies $E_{\text{nonrel}}^{\text{corr}}$ are compared with the results of CI calculations by Bunge and Peixoto (1970) and Sasaki and Yoshimine (1974) (au)(ref. (18))

Atom	$E_{\text{nonrel}}^{\text{corr}}$	$E_{\text{CI}} - E_{\text{HF}}$
Be	−0.0938	−0.0943
B	−0.1250	−0.1247
C	−0.1452	−0.1450
O	−0.2600	−0.2580
F	−0.3209	−0.3220
Ne	−0.3838	−0.3850

Perhaps of greater relevance for a critical confrontation of g-Hartree predictions to experimental data are ab initio determinations of transition energies

$$\varepsilon_{if} = E[n_\alpha^{(f)}] - E[n_\alpha^{(i)}] \qquad (2.17)$$

where $\{n_\alpha^{(f)}\}$ and $\{n_\alpha^{(i)}\}$ are the final and initial atomic configurations respectively. One way of calculation would be to compute ab initio total g-Hartree energies for the initial and final state and subtract, hardly a feasible method. Our treatment of the problem is tailored such that this subtraction is already performed on the level of theory.

We write

$$\varepsilon_{if} = E_{g-H}[n_\alpha^{(f)}; g] + E_{corr}[n_\alpha^{(f)}; g]$$

$$- E_{g-H}[n_\alpha^{(i)}; g] - E_{corr}[n_\alpha^{(i)}; g]$$

$$=: \sum n_\alpha^{(f)} \varepsilon_\alpha^{(f)} - \sum n_\alpha^{(i)} \varepsilon_\alpha^{(i)} + \hat{E}[n_\alpha^{(f)}, n_\alpha^{(i)}; g] \qquad (2.18)$$

and then proceed as follows[20]. Suppose we have a g-Hartree basis for the initial configuration $\{n_\alpha^{(i)}\}$; the corresponding g-Hartree potential then is not a stationary point for the final configuration $\{n_\alpha^{(f)}\}$. Nonetheless, $E[n_\alpha^{(f)}]$ can be expanded in terms of deviations from the g-Hartree basis corresponding to the initial state $\{n_\alpha^{(i)}\}$; terms linear in these fluctuations appear and have to be taken into account in the graphical expansion.

Following this procedure, we have only one set of g-Hartree eigenvalues $\{\varepsilon_\alpha^{(i)}\} = \{\varepsilon_\alpha^{(f)}\} =: \{\varepsilon_\alpha\}$ and (2.18) becomes

$$\varepsilon_{if} = \sum (n_\alpha^{(f)} - n_\alpha^{(i)}) \varepsilon_\alpha + \hat{E}[n_\alpha^{(f)}, n_\alpha^{(i)}; g] \qquad (2.18')$$

Let $\hat{E}^{(k)}$ be the expansion of $\hat{E}$ in terms of the electron-electron interaction up to k-th order, then g_0 is determined by the requirement

$$\hat{E}^{(k)}[n_\alpha^{(f)}, n_\alpha^{(i)}; g] = 0. \qquad (2.19)$$

The transition energy then becomes

$$\varepsilon_{if} = \sum (n_\alpha^{(f)} - n_\alpha^{(i)}) \varepsilon_\alpha ; \qquad (2.20)$$

this equation is to be seen as an exact version of Koopmans' theorem[21].

Our results (k=2) are presented in Table 3. The values obtained from this g-Hartree approach compare very favourably with the values predicted by a bona fide second order DHF-Goldstone(DHFG) calculation; in comparing the g-Hartree predictions with experiment an estimated computational error of 0.3 to 0.5 eV (which becomes larger for outer shells) has to be accounted for together with experimental uncertainties. Inspecting Table 3 we observe an erratic pattern when comparing second order DHFG calculations with experiment: rather good values emerge sometimes, while, frequently, large deviations show up. DHFG-correlations appear to be strongly configuration-

Table 3. Differences between experiment and theory (g-Hartree and DHFG, 2. order) in eV (full details and references are to be found in ref. (20) for the ionisation energies of elements between He and Ar.

Element	Level	g-Hartree	DHFG, 2. order
He	1s	0.147	0.167
Li	1s	−0.1	−1.44
	2s	0.032	0.052
Be	1s	−0.3	−1.1
	2s	−0.197	−0.547
Ne	1s	−0.39 ∼ −0.49	
	2s	0.24 ∼ 0.296	−0.424 ∼ −0.48
	$2P_{1/2}$	0.502	0.962
	$2P_{3/2}$	0.505	1.065
Mg	1s	−0.1 ∼ −0.7	
	2s	0.3	2.0
	$2P_{1/2}$	0.02 ∼ 0.05	−0.78 ∼ −0.75
	$2P_{3/2}$	0.04 ∼ 0.07	−0.63 ∼ −0.66
Ar	1s	−0.2	
	2s	0.3 ∼ 0.37	−2.43 ∼ −2.5
	$2P_{1/2}$	−0.33 ∼ −0.55	−1.0 ∼ −1.24
	$2P_{3/2}$	−0.38 ∼ −0.55	−0.88 ∼ −1.05
	3s	0.14 ∼ 0.2	−0.06 ∼ 0.0
	$3P_{1/2}$	0.507	1.337
	$3P_{3/2}$	0.520	1.46

dependent; the behaviour of the ab initio g-Hartree values is much more consistent, computational and experimental errors almost allow for smooth behaviour of eventually remaining g-Hartree correlations.

c) The effect of negative energy states

Up to now we neglected negative energy states in the g-Hartree potential (we remind the reader that the summations in the potential (2.9) run over the spectrum of the Dirac equation). To understand their role in the simplest case, the Hartree case, let us set g=1 and go to (1.17) where we imply relativistic summation.

To get a viable relativistic theory, we have to apply Dirac's prescription of an occupied negative energy sea. After proper charge conjugation

$$n_\alpha^{(+)} = n_\alpha \qquad \varepsilon_\alpha > 0$$

$$n_\alpha^{(-)} = 1 - n_\alpha \qquad \varepsilon_\alpha < 0$$

we obtain

$$\varphi_0(\vec{x}) = \int d^3x' \, V(\vec{x},\vec{x}') \left\{ \sum_{\varepsilon_\alpha > 0} n_\alpha^{(+)} |\psi_\alpha(\vec{x}')|^2 - \sum_{\varepsilon_\alpha < 0} n_\alpha^{(-)} |\psi_\alpha(\vec{x}')|^2 \right.$$

$$\left. + \rho_0^{vac} + \sum_{\varepsilon_\alpha < 0} |\psi_\alpha(\vec{x}')|^2 \right\}$$

$n_\alpha^{(+)}$ and $n_\alpha^{(-)}$ are, of course, particle and antiparticle Fermi distributions. The total vacuum contribution

$$\rho^{vac}(\vec{x}) = \rho_0^{vac}(\vec{x}) + \sum_{\varepsilon_\alpha < 0} |\psi_\alpha(\vec{x})|^2$$

$$= -\frac{1}{2} \left(\sum_{\varepsilon_\alpha > 0} |\psi_\alpha(\vec{x})|^2 - \sum_{\varepsilon_\alpha < 0} |\psi_\alpha(\vec{x})|^2 \right) \qquad (2.21)$$

is seen to be composed of the density induced by vacuum fluctuations and by the contribution of the Dirac sea. It is a non-trivial relativistic part of the Hartree potential and has, in principle, to be determined selfconsistently, at least for large nuclear charges Z.

Equation (2.21) is straightforwardly extended to include exchange contributions and the fully relativistic g-Hartree potential appears. For a detailed discussion see refs. (5) and (6).

3. CONCLUSIONS

We have studied the subject of correlations in both relativistic and non-relativistic theories. A new mean-field equation, the g-Hartree equation, was derived within the framework of a functional formulation of Quantum Field Theory. All relations established in this report are exact relations, approximations made in actual computations have been indicated.

We have discussed correlations in atomic states and did not find any particular necessity to distinguish between typically relativistc effects, so-called QED-effects and many-body correlations: Our starting point was the full QED action and correlations are derived from it without any approximations. The freedom in the choice of g-Hartree potentials and even of optimised potentials does not allow for a clear-cut distinction among these categories of correlations since the latter do depend on this choice.

We have, furthermore, demonstrated the efficiency of g-Hartree optimised potentials for the suppression of large correlation contributions: in the examples shown, optimised g-values, $g=g_0$, are seen to subsume correlations which then enter non-perturbatively into the optimised quasi-particle basis determined by solving the g-Hartree equation for $g=g_0$.

REFERENCES

1. R.L. Stratonovich, Sov. Phys. Dokl. 2 (1958) 416.
2. K. Dietz, O. Lechtenfeld and G. Weymans, J. Phys. B 15 (1982) 4301, 4315.
3. C. Bernard, Phys. Rev. D 9 (1974) 3312.
4. K. Dietz and Th. Millack, Physica, to be published.
5. K. Dietz and G. Weymans, J. Phys. B 17 (1984) 4801.
6. K. Dietz, Zeitschr. f. Phys. D, to be published.
7. R. Dashen, B. Hasslacher and A. Neveu, Phys. Rev. D10 (1974) 4114, 4130.
8. K. Dietz and G. Weymans, Physica 131A (1985) 363.
9. K. Dietz, "On the Relativistic Theory of Inhomogeneous Many-Electron Systems", AIP Conference Proceedings No. 136.
10. L.D. Landau, Sov. Phys. JETP 3 (1956) 920. A detailed account is found in: E.M. Lifshitz and L.P. Pitaevskii, Statistical Physics (Pergamon, New York, 1980).
11. For a review see S. Lundquist and N.H. March, eds. "The Theory of the Inhomogeneous Electron Gas", Plenum, New York, 1983.
12. J.P. Connerade, K. Dietz and G. Weymans, J. Phys. B 18 (1985) L309.
13. W.L. Wiese, M.W. Smith and B.M. Glennon, 1966 Atomic Transition Probabilities, vol. 1; W.L. Wiese, M.W. Smith and B.M. Miles, 1969 Atomic Transition Probabilities, vol 2 (Washington D.C.: US Govt. Printing Office).
14. H. P. Kelly, S.T. Carter and B.E. Norum, Phys. Rev. A25(1982) 2052.
15. For a review see: J.P. Connerade and K. Dietz, Imperial College Report.
16. J.P. Connerade, K. Dietz, M.W.D. Mansfield and G. Weymans, J. Phys. B 17 (1984) 1211.
17. R.A. Roig and G. Tondello, J. Opt. Soc. Am 65 (1975) 829.
18. K. Dietz and G. Weymans, J. Phys. B 17 (1984) 2987.
19. C.F. Bunge and E.M. A. Peixoto, Phys. Rev. A 1 (1970) 1277; F. Sasaki and M. Yoshimine, Phys. Rev. A 9 (1974) 17.
20. K. Dietz, M. Ohno and G. Weymans, J. Phys. B, to be published.
21. T. Koopmans, Physica 1 (1934) 104.

ATOMIC MANY BODY THEORY OF GIANT RESONANCES

Hugh P. Kelly and Zikri Altun

Department of Physics
University of Virginia
Charlottesville, VA 22901

INTRODUCTION

During the past decade there has been great interest in the subject of photoionization, both by experimenters[1] and by theorists.[2,3] Much of the interest has centered on the large resonance structures which are observed in many cross sections. In particular, the rare gases show evidence of broad resonance-like structures, both for the outer $(np)^6$ subshells and also for the inner $(4d)^{10}$ subshell in the case of xenon. Similar structure has been also observed in other atoms with large filled subshells such as the $4d^{10}$ subshell in atomic barium which has been studied extensively.[4-13] It is now generally believed[2] that calculations of these cross sections by different theoretical methods which take electron correlations into account give good agreement with experiment. However, a very recent photoemission experiment by Becker et al.[14] for the $4d^{10}$ subshell of xenon indicates that the many-body calculations are not in close agreement with the photoemission results but are in reasonable agreement with the experimental results for total absorption which includes multiple excitations, i.e., satellite structure. Preliminary photoemission results[15] for $4d^{10}$ subshell of barium also indicate a similar situation. It is important for our understanding of the photoionization process that we be able to calculate accurately such cross sections as those for the $4d^{10}$ subshells of xenon and barium.

In this paper the use of many-body perturbation theory (MBPT) to include effects of electron correlations is discussed.[16-18] The various physical processes contributing to the broad photoionization cross sections of the rare gases are studied in terms of the relevant many-body diagrams. Use of the random phase approximation with exchange (RPAE) is discussed by Amusia and Cherepkov.[3] Calculations using the relativistic RPAE are reviewed by Johnson.[19]

In addition, many-body perturbation theory (MBPT) is used to study resonances which are due to excitation of bound states degenerate with the continuum. Very interesting "giant resonance" structure can occur when an inner shell electron is excited into a vacant open-shell orbital of the same principal quantum number. A particular example which is studied is the neutral manganese atom $3p^6 3d^5 4s^2 (^6S)$, in which the spins of the five 3d electrons are aligned. A very large resonance occurs in the 3d and 4s cross sections due to $3p \to 3d$ excitation near 51 eV, and calculations of this resonance by MBPT and RPAE are discussed. A second example of this type of resonance occurs in open-shell rare-earth atoms with configurations $4d^{10}4f^n5s^25p^66s^2$. Calculations and experimental results will be discussed for the case of europium with a half-filled subshell $4f^7$.

THEORY

It is readily shown[20] that the photoionization cross section

$$\sigma(\omega) = 4\pi \frac{\omega}{c} \operatorname{Im} \alpha(\omega), \tag{1}$$

where $\alpha(\omega)$ is the frequency-dependent polarizability, given by

$$\alpha(\omega) = -\sum_k |\langle\psi_k| \sum_{i=1}^N z_i |\psi_0\rangle|^2 \times \left\{ \frac{1}{E_o - E_k - \omega} + \frac{1}{E_o - E_k + \omega} \right\}. \tag{2}$$

Atomic units ($e=\hbar=m_e=1$) are used in Eq.(2) and throughout the rest of this article. In Eq.(2) $|\psi_o\rangle$ is the exact initial many-particle state and $|\psi_k\rangle$ is an exact excited many-particle state. Since $E_o - E_k + \omega$ can vanish, we add to it a small imaginary part $i\eta$ and note that

$$\lim_{\eta \to 0} (D+i\eta)^{-1} = PD^{-1} - i\pi\delta(D), \tag{3}$$

where P represents principal value integration. The summation $\sum_k$ represents a sum over bound states and integration over the continuum with $\sum_k \to (2/\pi) \int dk$, where the factor $2/\pi$ results from the normalization of continuum states according to

$$R_k(r) \to \cos[kr + \delta_\ell + (q/k)\ln 2kr - \tfrac{1}{2}(\ell+1)\pi] \tag{4}$$

as $r \to \infty$, where $V(r) \to q/r$.

Using Eq. (3) we find that

$$\sigma(\omega) = (8\pi\omega/ck) |\langle\psi_k| \sum z_i |\psi_o\rangle|^2 . \tag{5}$$

Perturbation theory may be used to calculate the many-particle matrix element $\langle\psi_k| \sum z_i |\psi_o\rangle$. We start with an arbitrary single-particle potential $V_i = V(r_i)$ which can be non-local but should be Hermitian. It is desirable that V_i should take as good account as possible of the effects of the N-1 other electrons on the ith electron. We write

$$H = H_o + H', \tag{6}$$

where

$$H_o = \sum_{i=1}^N (T_i + V_i), \tag{7}$$

and T_i is the sum of kinetic energy and nuclear interaction for the ith electron. The perturbation

$$H' = \sum_{i<j} v_{ij} - \sum_i V_i , \tag{8}$$

where $v_{ij} = e^2/r_{ij}$.

The correlated ground state wavefunction

$$\psi_o = \sum_L \left(\frac{1}{E_o^{(o)} - H_o} H' \right)^n \phi_o , \tag{9}$$

where

$$H_o \Phi_o = E_o^{(o)} \Phi_o, \quad (10)$$

and $\sum_L$ indicates a sum over "linked" terms only.[16,17] The expression for ψ_k is similar with E_o and Φ_o being replaced by $E_k^{(0)}$ and Φ_k where

$$H_o \Phi_k = E_k^{(\ )} \Phi_k. \quad (11)$$

The "length" form of the dipole matrix element $\langle \psi_k | \Sigma z_i | \psi_o \rangle$ is related to the "velocity" form $\langle \psi_k | \sum_i d/dz_i | \psi_o \rangle$ by

$$\langle \psi_k | \Sigma z_i | \psi_o \rangle = (E_o - E_k)^{-1} \langle \psi_k | \sum_i \frac{d}{dz_i} | \psi_o \rangle, \quad (12)$$

where ψ_k and ψ_o are eigenstates of H with eigenvalues E_k and E_o. The many-particle matrix elements of Eq. (12) may be calculated by perturbation theory using a complete set of single-particle states ϕ_n which satisfy

$$(T+V)\phi_n = \varepsilon_n \phi_n. \quad (13)$$

The N orbitals lowest in energy are occupied in Φ_o, and the others are called excited states. The perturbation expansion for $\langle \psi_k | \Sigma z_i | \psi_o \rangle$ may be derived from Eq. (9) and the corresponding equation for ψ_k. It is, however, simpler to consider the many-body perturbation expansion for Im $\alpha(\omega)$ in terms of diagrams. The diagrams for Im $\alpha(\omega)$ may then be factored into a sum of diagrams giving the exact $\langle \psi_k | \Sigma z_i | \psi_o \rangle$ times its complex conjugate plus corrections due to normalization diagrams.[21] In calculations so far, the effects of the normalization diagrams have been less than 5%.

It is desirable to choose V as the usual Hartree-Fock potential R_{HF} for orbitals ϕ_n in Φ_o. For excited states the potential is not unique, since we may write[22-26]

$$V = R_{HF} + (1-P)\Omega(1-P), \quad (14)$$

where Ω is an arbitrary Hermitian operator and

$$P = \sum_{n=1}^{N} |n\rangle\langle n|. \quad (15)$$

For excited states, Ω is often chosen so that $R_{HF}+\Omega$ is the appropriate LS-coupled Hartree-Fock potential for the excited states.

In Fig. 1 are shown diagrams for the perturbation expansion of $\langle \psi_k | \Sigma zi | \psi_o \rangle$ for a transition in which an electron in an orbital labelled p is excited to k. Dashed lines ending in a solid dot represent the dipole interaction z (d/dz for the "velocity" case), and the other dashed lines represent interaction with v_{ij}. We use Goldstone diagrams which are read from bottom to top corresponding to right to left in the perturbation terms. The diagram of Fig. 1 (a) represents $\langle k|z|p \rangle$. Diagrams 1 (a), (b), and (c) are first-order in the correlation interaction H' of Eq. (8). Diagram 1 (b) represents correlations in the initial state and is given by

$$\sum_{k'} \frac{\langle q|z|k'\rangle\langle kk'|v|pq\rangle}{\varepsilon_p + \varepsilon_q - \varepsilon_k - \varepsilon_{k'}}, \quad (16)$$

where the summation is over excited states. There is also an exchange term in which $\langle kk'|v|pq \rangle$.

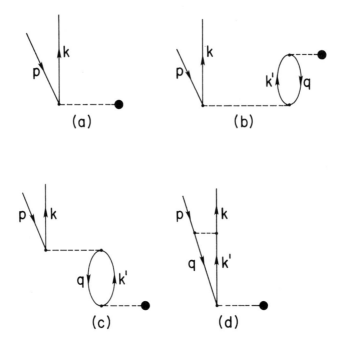

Fig. 1. Diagrams contributing to $\langle \psi_k | \Sigma z_i | \psi_0 \rangle$, which describes the transition $p \to k$. Dashed line ending in large dot represents interaction with z (length) or d/dz (velocity). Other dashed lines are Coulomb interactions. (a) $\langle k|z|p \rangle$ (b) First-order correction due to ground state correlations. There is also a corresponding exchange diagram. (c) First order correlation in excited state (d) Exchange diagram corresponding to (c). Diagrams are read from bottom to top.

Diagrams 1 (c) and (d) represent correlation to first order in the final state. Diagram 1 (c) is given by

$$\sum_{k'} \frac{\langle kq|v|pk' \rangle \langle k'|z|q \rangle}{\varepsilon_q - \varepsilon_{k'} + \omega + i\eta}, \qquad (17)$$

and diagram 1 (d) is its exchange, in which $\langle kq|v|pk' \rangle$ is replaced by $-\langle qk|v|pk' \rangle$. Diagram 1 (c) produces a resonance when q is in an inner shell, k' is a bound excited state, and ω is such that the denominator vanishes (when $\eta \to 0$). Higher-order diagrams may be summed[27] geometrically to all orders to give a real shift $\Delta(\omega)$ and an imaginary term $i\Gamma/2$ to the denominator of Eq. (17). In addition to the first-order final-state correlations shown in diagrams 1 (c) and (d) there are final-state interactions with passive unexcited states as shown in diagrams 2 (a) and (b) of Fig. 2 and also in diagrams 2 (c) and (d) when $p = q$. Diagram 2 (e) represents final state interaction with $-V$ of Eq. (8). For closed shell atoms, the LS-coupled Hartree-Fock potential V cancels the diagrams 2 (a), (b), and (c) and also diagram 2 (d) and its exchange when q is in the same subshell as p and k and k' have the same orbital angular momentum. If we limit interactions to a given subshell, then in first order in H' the only contribution comes from the ground state correlation diagram of Fig. 1 (b).

Higher-order diagrams are shown in Fig. 3. Diagrams 3 (a) and 3 (b) are examples of diagrams with excitation and deexcitation of particle-

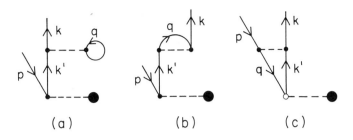

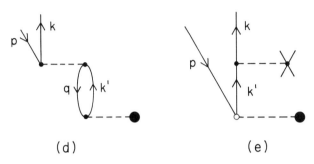

Fig. 2. First-order correlations in excited state. (a) and (b) are direct and exchange interactions with passive unexcited states q. (c) and (d) couple q → k' excitation with p → k excitation. (e) interaction with -V represented by cross. Potential V can be chosen to cancel (a), (b) and (c) and (d) when q is in same subshell as p.

hole pairs (and their exchange) and are included in the random phase approximation with exchange (RPAE) method.[3] We note also that diagram 3 (a) with p, q, and r all in the same subshell is cancelled by interactions with -V when the LS-coupled Hartree-Fock potential is used to calculate excited states. Diagram 3 (c) represents corrections to the hole state p. When p = r, the diagram is part of a geometric series which may be summed to shift the energy ε_p and therefore shift the threshold of the photoionization cross section. Diagram 3 (d) represents a modification of the interaction in Fig. 1 (d). In particular, when p = q, we interpret diagram 3 (d) as contributing to the relaxation which occurs when an electron is removed. Diagram 3 (e) accounts for the "polarization" effects on the orbital p. There is a similar diagram with the interaction on the hole line labelled p as well as exchange diagrams. Diagram 3 (e) causes single excitations, and it is included in the "Brueckner orbitals".[16] Diagram 3 (f) accounts for "polarization" effects on the final state. The diagrams of Fig. 3 do not exhaust the totality of diagrams with interactions with H' to second order. There are higher-order diagrams in the ground state correlations and diagrams with different time orderings of the dipole interaction. Nevertheless, it is expected that the major physical effects are represented by diagrams such as those of Fig. 3. Examples of these effects are given in the following sections.

CLOSED SHELL ATOMS

An excellent example of the use of MBPT in photoionization calculations is provided by calculations of the photoionization of neutral argon

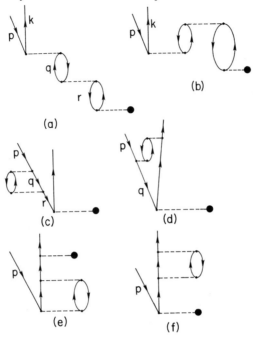

Fig. 3. Higher-order diagrams contributing to p → k transition. (a) Coupling of single excitations. When p, q, r are in same subshell, this diagram is cancelled by LS-coupled Hartree-Fock potential. (b) Second order RPAE diagram. There is also a diagram with the dipole interaction between the two Coulomb interactions. (c) interaction on hole line. (d) relaxation diagram. There is also a diagram with the order of the two interactions is reversed. (e) polarization of orbital p. interaction can also occur on hole line p. (f) polarization diagram for final state orbital. All diagrams also have exchange counterparts.

as shown in Fig. 4. The data are the experimental total absorption results by Madden, Ederer and Codling,[30] and by Samson[31] (above 37 eV). Calculations using the Herman-Skillman potential[2] peak much closer to threshold and reach a maximum greater than 60 Mb, in poor agreement with expeirment. Calculations by Ammsia, Cherepkov, and Chernysheva[28] using the RPAE gave very good agreement with experiment for argon and the other rare gases. Since this is an infinite-order theory in certain classes of terms (excitation and deexcitation of particle-hole pairs) it was interesting to examine the contributions of perturbation theory, order by order.

It was found to be very important to couple the final state to be $3p^5 kd(^1P)$ or $3p^5 ks(^1P)$ since the initial $3p^6(^1S)$ after absorbing a photon must produce a final ^1P state in the dipole approximation. The potential is determined by

$$\langle \psi_k^{(0)} | \sum_{i<j} v_{ij} - \sum_i V_i | \psi_k^{(0)} \rangle = 0 , \qquad (18)$$

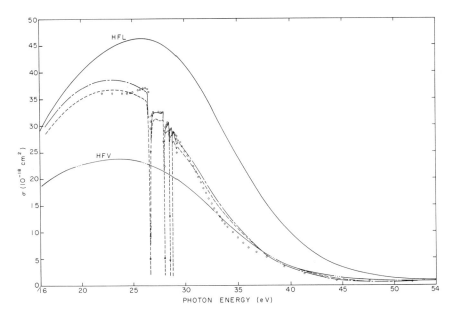

Fig. 4. Photoionization cross section for the neutral argon atom. HFL and HFV are the Hartree-Fock length and velocity cross sections. Dot-dashed (dashed) line is the calculated length (velocity) cross section including higher-order terms. Circles are experimental data from ref. 30 (below 37 eV) and from ref. 31 (above 37 eV). Resonances are due to 3s → np excitations.

where $\psi_k^{(0)}$ is $3p^5$ (kd or ks) (^1P). As discussed, this includes diagrams (a), (b), and (c) of Fig. 2, as well as diagram (d), of Fig. 2 for all orbitals q in the 3p subshell and k and k' with the same ℓ. Of course, p is in the 3p subshell as well. The lowest-order results for "length" and "velocity" are the Hartree-Fock curves labelled HFL and HFV, respectively.[32] In the next order, the only contribution within the 3p subshell is from the ground state correlation diagram of Fig. 1 (b), and including this diagram brings the length and velocity curves into close agreement.[32] Including the second-order RPAE diagram of Fig. 3 (and also the similar diagram with the dipole interaction above the lowest correlation interaction), lowered the calculated curves in region 30-50 eV. The results of this low-order calculation[32] in Fig. 4 are in excellent agreement with experiment. From a physical point of view, the coupling $3p^5kd$(^1P) results in a large repulsive exchange term which causes the cross section to peak at higher energy and be broader than a potential with attractive exchange.

The resonance structure is due to 3s → np excitations and was calculated by including diagrams (c) and (d) of Fig. 1 with q = 3s, and also by including diagram (a) of Fig. 3 with p = 3p, q = 3r, and r = 3p. Only the imaginary contribution ($-i\pi\delta$) from the r excitation was included. The main effect of diagram 3 (a) is to cancel diagram 1 (a) at the resonance.[32]

The photoionization of cadmium $4d^{10}5s^2$ provides a good illustration of many-body effects in a large closed subshell. The final states resulting from photoionization of a 4d electron are $4d^9kf(^1P)$ and $4d^9kp(^1P)$. Results of Hartree-Fock length and velocity calculations[33]

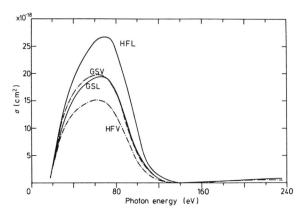

Fig. 5. Cadmium 4d → kf cross section calculated with dipole-length (full curve) and dipole-velocity (chain curve) matrix elements, ref. 33. Lowest order (Hartree-Fock) results labelled HFL, HFV. First-order results labelled GSL, GSV.

are shown in Fig. 5. Ground state correlations among the 4d electrons were included[33] by diagram (b) of Fig. 1, and the resulting curves are shown in Fig. 1 labelled GSL (length) and GSV (velocity). These curves are very close to the geometric mean of the HFL and HFV curves as suggested by Hansen.[34] Inclusion of the second-order RPAE diagram of Fig. 3 (b) (and the other time-ordering of the dipole interaction) results in the curve labelled GMS in Fig. 6. The geometric mean of length and velocity results has been taken. For comparison, the curve labelled GMU in Fig. 6 is the geometric mean of the Hartree-Fock length and velocity results. We note that these Hartree-Fock calculations were carried out with "unrelaxed orbitals." That is the continuum electron is

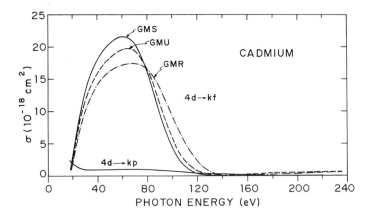

Fig. 6. Cadmium 4d → kf and 4d → kp cross sections. Curves shown are the geometric mean of length and velocity results. GMS includes second-order RPAE diagrams. GMU is lowest order unrelaxed results. GMR is lowest order results using relaxed orbitals. Calculations from ref. 33.

calculated in the field of $4d^9$ (^2D), but the 4d orbitals are taken from the Hartree-Fock calculation of the neutral ground state $4d^{10}$. This is appropriate at sufficiently high energies where the photoejected electron leaves the atom so quickly that the resulting ion has not had time to relax. A rough estimate is $t < t_B$, where $t = d/v$ and $t_B = \hbar/\varepsilon_{4d}$. The "size" of the atom is taken to be d and v is the velocity of the outgoing electron. At low energies, we expect that a reasonable account of relaxation is provided by calculating continuum states in the relaxed field of the $4d^9$ ion. This type of relaxed result is shown in Fig. 6 and labelled GMR. Again, the geometric mean has been taken for length and velocity results. A more accurate method to include relaxation is to calculate the appropriate relaxation diagrams such as Fig. 3 (d) with p = q = 4d. From Fig. 6 it can be seen that for cadmium the effects of the second-order RPAE diagrams such as Fig. 3 (b) and relaxation are opposite and tend to cancel. A similar situation exists for barium, although the effects are more dramatic. In Fig. 6 is also shown the 4d → kp cross section including coupling with 4d → kf described by Fig. 1 (c) with k = kp and k' = kf. This coupling causes the 4d → kp cross section to be larger near 60 eV than the lowest-order result.

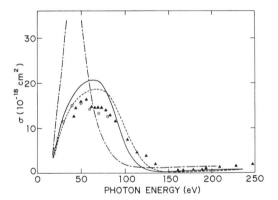

Fig. 7. Cadmium photoionization cross section. Experimental points from Marr and Austin, ref. 36, X; Cairns et al., ref. 37, 0; Codling et al., ref. 38, ▲; Chain curve is Herman-Skillman calculation by McGuire, ref. 35. Full curve is geometric mean of length and velocity results from the first-order MBPT calculation for the 4d → kf cross section and for the 4d → kp cross section, ref. 33. Broken curve includes relaxation effects.

In Fig. 7 the calculated results[33] are compared with earlier calculations by McGuire[35] using the Herman-Skillman potential. Comparison is also made with experiment.[36-38] Both unrelaxed and relaxed calculations are shown. Beyond 100 eV, the calculated curves are below experiment which contains contributions from the 4p-subshell cross section which was not calculated. It is noted that below 100 eV the calculations are higher than experiment, with the relaxed results giving better agreement than the unrelaxed results.

The photoabsorption from $4d^{10}$ subshells has been discussed in terms of double well potentials by a number of authors.[39-42] Barium is an example of an atom with a $4d^{10}$ subshell which is particularly sensitive to the details of the potential used to calculate the continuum states. A number of calculations[8-13] and experiments[4-7] have been carried out on photoionization of the $4d^{10}$ subshell of barium. In Fig. 8 are shown some

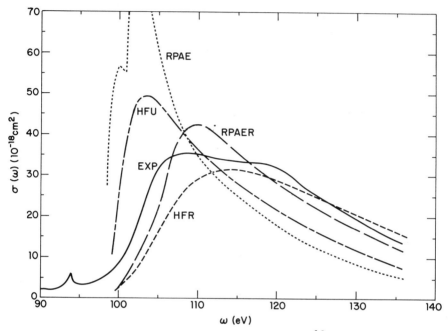

Fig. 8. Photoionization cross section for the $4d^{10}$ subshell of barium. Expt, experiment by Hecht and Lindau, ref. 6. ....RPAE calculation by Fliflet et al., ref. 10. — — — —RPAER, random-phase approximation using relaxed orbitals by Amusia et al., ref. 11. — - — HFU, Hartree-Fock geometric mean calculation with unrelaxed states, ref. 13. ----- HFR, Hartree-Fock geometric mean calculation with relaxed states, ref. 13.

calculations which illustrate some of the different effects which contribute to this cross section along with a comparison with the experiment by Hecht and Lindau.[6] The curve labelled HFU is the geometric mean of the lowest order Hartree-Fock[13] $4d^9(kf$ and $kp)(^1P)$ length and velocity results. The calculations are unrelaxed, i.e., calculated with the potential using orbitals of the ground state.[13] The curve HFU is very close to that in which ground state correlations are explicitly included. Similarly, the curve labelled HFR is the geometric mean of the lowest-order results obtained when the continuum state potential uses relaxed orbitals of the $4d^9$ ion. The curve labelled RPAE is the usual random phase approximation with exchange result.[10] There has been a statistical inclusion of spin-orbit effects. The poor agreement with experiment by the RPAE as compared with HFU is evidence that inclusion of the higher-order RPAE terms in this case unbalances the perturbations expansion, particularly because relaxation effects are not included. The result labelled RPAER is an RPAE calculation by Amusia et al.[11] in which the usual excited orbitals are replaced by relaxed orbitals. Similar results have also been obtained by Wendin.[9] The experimental curve is normalized to the peak of the time-dependent ideal density (TDLDA) calculation by Zangwill and Soven[12] which is shown in Fig. 9. The TDLDA

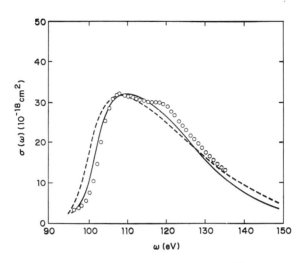

Fig. 9. Photoionization cross section of the $4d^{10}$ subshell of Ba. ———, TDLDA calculation by Zangwill and Soven, ref. 12. - - - -, Hartree-Fock calculation with relaxed states and threshold at 93.8 eV as in TDLDA calculation, from ref. 13. O experimental results by Hecht and Lindau, ref. 6. The experimental curve is normalized to the peak of the calculated TDLDA curve.

calculation is essentially an RPAE calculation using orbitals calculated in the local density approximation (LDA) in which the asymptotic potential lacks the Coulomb tail seen by Hartree-Fock orbitals. The TDLDA calculates the threshold to be at 93.8 eV as compared with the experimental threshold at 99 eV. In Fig. 9 the HFR calculation[13] has been shifted to start at 93.8 eV and is seen then to give close agreement with experiment and with the TDLDA results. It will be interesting, however, to compare all these calculations with photoemission experiments[15] which do not include processes of multiple excitation which are, of course, included in a measurement of total photoionization.

Very recently, Becker et al.[14] have measured the photoemission cross section for the $4d^{10}$ subshell of xenon. They have found that the photoemission cross section is noticeably smaller than the cross section for total absorption. For example, the peak of the photoemission cross section is approximately 22 Mb, whereas the peak of the total absorption cross section[43] is 28.5 Mb. In Fig. 10 are shown the recent photoemission results by Becker et al.[14] along with a number of theoretical calculations.[11,12,28,44] The various calculations including relaxation are in reasonable agreement with experiment on the low energy

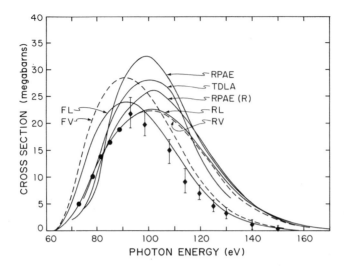

Fig. 10. Photoionization cross section of the $4d^{10}$ subshell of xenon. O and ◆ are different sets of data from the photoemission experiment by Becker et al., ref. 14. Curves labelled FL and FV are Hartree-Fock length (L) and velocity (V) plus ground state correlations with excited states calculated with frozen core orbitals, ref. 44. Curves labelled RL and RV are HF plus ground state correlations using relaxed orbitals. Curves labelled RPAE and RPAER use frozen core orbitals and relaxed ionic core orbitals, respectively, to calculate excited states used in RPAE calculations, ref. 11.

side of the peak cross section but are considerably higher than experiment at higher energies. All calculations except for the low-order MBPT with relaxation peak higher than experiment. Very recent calculations[44] have explicitly calculated the relaxation diagrams as shown in Fig. 3 (d) and the diagram like 3 (d) which the interaction with the hole and particle lines occur in the opposite order. The results are almost identical with those already obtained by use of MBPT with relaxed orbitals shown in Fig. 10. It is desirable to investigate the effects of other higher-order diagrams as shown in Fig. 3 such as the polarization diagrams of Fig. 3 (e) and (f). In the case of Fig. 3 (f) we should consider the coupling between single photoionization and the other channels in which a second electron is also excited either to a bound state or to the continuum.

OPEN SHELL ATOMS

In the previous section we have discussed broad, resonance-like absorption in closed shell atoms, particularly in the $4d^{10}$ subshell. These cross sections may be described by a single channel, i.e., $4d \to \varepsilon f$ transitions. A different type of "giant resonance" absorption is found to occur in open shell atoms. Striking examples of such resonances have been observed[45-51] in transition metal atoms with configurations $3p^6 3d^n 4s^2$ or $3p^6 4d^n 4s$. There is very large resonance structure due to $3p^6 3d^n 4s^2 \to 3p^5 3d^{n+1} 4s^2$ resonances in the 4s and 3d subshell cross sections. For atoms such as chromium, there are also $3p \to 4s$ resonances.[51]

The lowest-order contribution to the $3p \to 3d$ resonances is given by the diagrams of Fig. 1 (c) and (d) with p = 3d or 4s, q = 3p, and k' = 3d. Note that k' represents an unfilled 3d orbital in the ground state. Manganese is a particularly simple system for calculations since the ground state $3p^6 3d^5 4s^2$ (^6S) consists of a half-filled shell of 3d electrons with all spins aligned. The resonance due to $3p \to 3d$ excitation in Mn was first calculated by Davis and Feldkamp[52] using Hartree-Slater continuum orbitals and also successfully calculating the spin-orbit structure. The $3p \to 3d$ resonance was also calculated by Amersia et al.[53] using the closed-shell RPAE formalism and taking approximate account of multiplet structure by separating electrons into spin "up" and "down" groups as in earlier MBPT calculations[54] for Fe. Garvin et al.[55] also studied the $3p \to 3d$ resonance using MBPT and including spin-orbit splitting. Results of calculations and experiment are shown in Fig. 11.

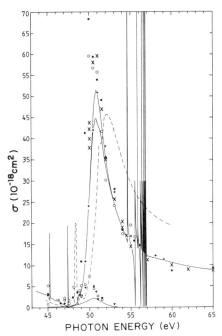

Fig. 11. Resonance structure in 3d and 4s subshell cross sections of atomic manganese. The highest full curve is the 3d cross section calculated by Garvin et al., ref. 55, by the coupled equations method of ref. 56 from 50 eV to 65 eV. The solid curve just below is from ref. 55 including spin-orbit effects and extends from 43 eV to 53 eV. The broken curve is a calculation by Davis and Feldkamp, ref. 52. Plus signs are from the RPAE calculation, ref. 53. Experimental data for the 3d cross section are from ref. 47 (open circles), ref. 49 (full circles), and ref. 48 (crosses). Lowest full curve is the 4s subshell MBPT calculation from ref. 55. Measurements for 4s are from ref. 47 (squares) and ref. 49 (triangles). Experimental data were normalized to fit the calculation of ref. 55 at energies above 53 eV.

The experimental results are all from photoemission experiments and have been normalized to fit the MBPT calculation at energies above 53 eV. In the MBPT calculations, Garvin et al. used two different techniques. The upper solid curve was calculated[55] using the coupled equations method[56] in which 3d → εf, εp excitations were coupled with 3p excitations to all orders. The solid curve just below it was the result of the interacting resonance method[57] and included spin-orbit effects.[55] The solid curve in the bottom of the figure is the 4s cross section, and it clearly shows the effect of the 3p → 3d resonance although the peak of the calculated cross section is too low compared with experiment by more than a factor of two.

Garvin et al.[55] calculated the position of the 3p → 3d resonance by first calculating the difference between self-consistent Hartree-Fock energies for the $3p^53d^64s^2$ excited state and the ground state (53.4 eV). They calculated second-order excited state correlation energies and estimated other pair correlation energies among 3p and 3d electrons from calculations[58] for Fe, the neighboring atom, and from previous calculations of three-body effects.[59] The corrected energy value of 50.4 eV is in good agreement with experiment and illustrates how calculations on one atom may be used to estimate correlation energies in neighboring atoms.

The MBPT calculations[55] predicted resonance structure due to $3p^53d^54s^2(^7P)$ nd and ns Rydberg transitions leading to the $3p^53d^54s^2(^7P)$ edge at 57 eV. However, these resonances which would normally be expected have not been observed in the photoemission experiments. Because of spin considerations, these states cannot decay (in the ^7P case) by one 3d electron being ejected. However, it has been speculated[60] that the decay from a 4s electron filling the 3p hole and a 3d or 4s electron being ejected may cause a sufficient width to eliminate the resonance. In the case of the resonance series leading to the $3p^53d^54s^2(^5P)$ edge at 74.3 eV, calculations[55] showed that the broadening was so great as to eliminate the resonance series. In this case, a 3d electron can fill the 3p hole while a second 3d electron is ejected. This result is shown in Fig. 12. It should be noted that resonances are expected[61] in the photoionization-with-excitation cross sections before both the ^7P and ^5P edges, and this should be reflected also in the total absorption cross section.

Studies by Schmidt et al[62] have shown that there is a significant discrepancy between the total absorption measurements and the photo-emission measurements above 52 eV, and that this discrepancy is due to large satellite (photoionization-with-excitation) effects.

Calculations of the angular distribution of 3d photoejected electrons[55] and experiments[48,49] also show a resonance due to 3p → 3d excitation, although the calculated resonance appears sharper and more prominent than the measurements. Calculations of the 3p → 3d resonance structure in the 3d cross section of Ni have been ca-ried out by Combet-Farnoux and Ben-Amar[63] using R- matrix theory.

It is expected that there should also be giant resonance structure in other open-shell atoms such as those with $4p^64d^n5s^2$ configurations. Calculations[64] were recently carried out for Yttrium (Y) which has the configuration $4p^64d5s^2$. As expected, there is very large resonance structure in the 4d and 5s cross sections due to 4p → 4d excitations. The 4d cross section is the dominant one, and 4d → εf transitions are much larger than 4d → εp. There is multiplet structure due to $^1D, ^1G$, and ^3F multiplets of $4d^2$. The two largest resonances are due to $4d^2$ (^1G) and (^3F), the total coupling being $4p^54d^2$ (^2F). There are no published

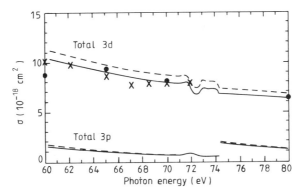

Fig. 12. Photoionization cross sections of atomic manganese. Calculated dipole length (full curves) and velocity (broken curves) for the total 3d and 3p cross sections, ref. 55. X experimental results from ref. 48. ● experimental results from ref. 49. The second 3p threshold (^{5}P) lies just above 74 eV.

experimental results on the free Y atom, but results on the solid by Weaver and Olson[65] indicate considerable absorption in the region of the calculated resonances. However, the region of large absorption extends beyond the region of calculated resonances. Very recent data on the photoionization of atomic yttrium indicate qualitative confirmation of the calculated resonance structure.[66]

A particularly interesting example of giant resonance structure due to core excitation into a partially filled subshell occurs in the case of atomic europium. The ground state is $4d^{10}4f^75s^25p^66s^2$ ($^8S_{7/2}$). That is, there is a half-filled shell of 4f electrons with all spins aligned. A number of experiments[67] have shown greatly increased absorption near 140 eV. A very recent photoemission experiment by Becker et al.[68] gives a clear picture of the photoionization, and illustrates the necessity of measuring the individual partial cross sections. Calculations have been carried out by Amusia, Sheftel, and Chernysheva[69] using the RPAE method and dividing the electrons into two classes, spin up and spin down as in their previous calculation on atomic manganese.[53] Their calculations indicate a very large resonance in the $4f \rightarrow \varepsilon g$ cross section due to $4d \rightarrow 4f$ excitation. The lowest-order contribution to this resonance is given by diagrams (c) and (d) of Fig. 1 with p = 4f, k = εg and εd, q = 4d, and k' = 4f. The experimental results by Becker et al.[68] also showed significant resonance structure due to $4d^94f^8$ in the 4d cross section. Recently, MBPT calculations have also been carried out by Pan, Carter, and Kelly[70] using MBPT and using LS-coupling. These calculations show that splitting due to LS-coupling is much greater than spin-orbit effects. Many higher-order diagrams as shown in Fig. 3 were considered. The effects of higher-order RPAE diagrams as in Fig. 3 (b) and relaxation diagrams in Fig. 3 (d) tended to cancel. The calculations showed the level $4d^9$ (^{2}D) $4f^8$ (^{7}F) (^{8}P) to lie at approximately 141 eV and above the $4d^94f^7$ (^{9}D) calculated threshold at 139.6 eV. This then places the $4d^94f^8$ (^{8}P) level above the $4d^94f^7$ (^{9}D) threshold and causes a resonance in the $4d^94f^7$ (^{9}D) εf, εp cross sections. The other 4d threshold ($4d^94f^7(^7$D)) is calculated by ΔSCF to be at 160 eV using LS-coupling. Pan et al.[70] calculated the range of the spin-orbit splitting of the $4d^9 4f^7(^9D)$ ionic level in LSJ-coupling as 4.4 eV.

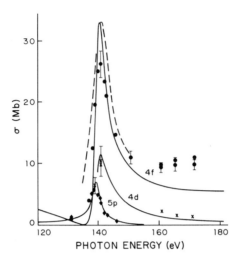

Fig. 13. Partial cross sections for atomic europium. Solid lines are MBPT calculations by Pan et al, ref. 70. — — — — RPAE calculation by Amusia et al, ref. 69. o 4f, X 4d, ♦ 5p measurements by Becker et al., ref. 68. 4f experiment normalized to MBPT calculation.

Results of the calculations by Amusia et al.[69] and by Pan et al[70] are shown in Fig. 13 along with the photoemission results by Becker et al.[68] The total photoemission results are not normalized, although the relative normalizations of the partial cross sections were measured. The deviation of the calculated MBPT 4f cross section from experiment above 155 eV may be due to inclusion of satellites in the 4f photoemission measurement.[71] In Fig. 14 results are presented for the 5s and 6s subshell cross sections.

For each subshell except for 4f, there will be ^9L and ^7L cross sections in LS-coupling where ^9L and ^7L are the ionic multiplets due to coupling ^2L of the subshell with ^8S of the open $4f^7$ subshell. The ratios of the ^9L to ^7L cross sections are given in Fig. 15 for the 5s, 5p, and 6s subshells in the region of the 4d → 4f resonance.

Similar calculations could be carried out for other rare earth atoms. However, the calculations will be more complicated due to the complexity of the multiplet structure when the 4f subshells is not half-filled.

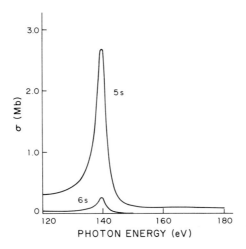

Fig. 14. Partial cross sections (5s and 6s) of atomic europium calculated by Pan et al., ref. 70.

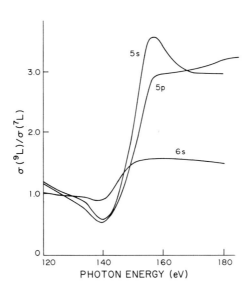

Fig. 15. Ratio of ^9L to ^7L cross sections for atomic europium for 5s, 5p, and 6s subshell cross sections, calculated by Pan et al., ref. 70. ^9L and ^7L refer to the LS-state of the ionic core.

ACKNOWLEDGEMENTS

We wish to acknowledge support from the U.S. National Science Foundation and helpful discussions with W. Wijesundera and Cheng Pan. One of us (Z.A.) also wishes to thank Marmara University, Istanbul, Turkey, for partial support.

REFERENCES

1. J. A. R. Samson, 1982, in "Handbuch der Physik XXXI," S. Flügge and W. Mehlhorn, eds., Springer-Verlag, p. 123.
2. A. F. Starace, ibid., p. 1.
3. M. Ya. Amusia and N. A. Cherepkov, Case Studies in Atomic Physics 5:47 (1975).
4. R. Rabe, K. Radler, and H. W. Wolff, in: "VUV Radiation Physics," E. E. Koch, R. Haensel, and C. Kunz, eds., Pergamon Vieweg, Braunschweig (1974).
5. J. P. Connerade and M. W. D. Mansfield, Proc. R. Soc. London A341:267 (1974).
6. M. H. Hecht and I. Lindau, Phys. Rev. Lett. 47:821 (1981).
7. T. B. Lucatorto, T. J. McIlrath, J. Sugar, and S. M. Younger, Phys. Rev. Lett. 47:1124 (1981).
8. G. Wendin, Phys. Lett. 46A:119 (1973).
9. G. Wendin, Phys. Lett. 51A:291 (1975).
10. A. W. Fliflet, R. L. Chase, and H. P. Kelly, J. Phys. B7:L443 (1974).
11. M. Ya. Amusia, V. K. Ivanov, and L. V. Chernysheva, Phys. Lett. 59A:191 (1976).
12. A. Zangwill, in Atomic Physics 8, I. Lindgren and A. Rosen, eds., Plenum (1983), p. 339.
13. H. P. Kelly, S. L. Carter, and B. E. Norum, Phys. Rev. A25:2052 (1982).
14. U. Becker, T. Prescher, E. Schmidt, B. Sonntag, and H.-E. Wetzel, Phys. Rev. A33:3891 (1986).
15. U. Becker et al, private communication.
16. K. A. Brueckner, Phys. Rev. 97:1353 (1955); 100:36 (1955).
17. J. Goldstone, Proc. Roy. Soc. London A239:267 (1957).
18. H. P. Kelly, Adv. Chem. Phys. 14:129 (1969).
19. W. R. Johnson, in Atomic Physics 8, I. Lindgren and A. Rosen eds., Plenum (1983), p. 149.
20. U. Fano and J. W. Cooper, Rev. Mod. Phys. 40:441 (1968).
21. H. P. Kelly and A. Ron, Phys. Rev. A5:168 (1972).
22. L. M. Frantz, R. L. Mills, R. G. Newton, and A. M. Sessler, Phys. Rev. Lett. 1:340 (1958).
23. B. A. Lippman, M. H. Mittleman, and K. M. Watson, Phys. Rev. 116:920 (1959).
24. R. T. Pu and E. S. Chang, Phys. Rev. 151:31 (1966).
25. H. J. Silverstone and M. L. Yin, J. Chem. Phys. 49:2026 (1968).
26. S. Huzimaga and C. Arnau, Phys. Rev. A1:1285 (1970).
27. H. P. Kelly, 1976, in "Photoionization and Other Probes of Many-Electron Interactions," ed. F. J. Wuilleumier, Plenum, New York.
28. M. Ya. Amusia, N. A. Cherepkov, and L. V. Chernysheva, Sov. Phys. JETP 33:90 (1971).
29. T. Ishihara and R. T. Poe, Phys. Rev. A6:111 (1972).
30. R. P. Madden, D. L. Ederer, and K. Codling, Phys. Rev. 177:136 (1969).
31. J. A. R. Samson, Advan. At. Mol. Phys. 2:177 (1966).
32. H. P. Kelly and R. L. Simons, Phys. Rev. Lett. 30:529 (1973).
33. S. L. Carter and H. P. Kelly, J. Phys. B11:2467 (1978).
34. A. E. Hansen, Molec. Phys. 13:425 (1967).

35. E. J. McGuire, Phys. Rev. 175:20 (1968).
36. G. V. Marr and J. M. Austin, Proc. R. Soc. A310:137 (1969).
37. R. B. Cairns, H. Harrison, and R. I. Schoen, J. Chem. Phys. 51:5440 (1969).
38. K. Codling, J. R. Hamley, and J. B. West, J. Phys. B11:1713 (1978).
39. J. P. Connerade, Contemp. Phys. 19:415 (1978).
40. D. Griffin, this volume and references therein.
41. K. T. Cheng and W. R. Johnson, Phys. Rev. A28:2820 (1983).
42. K. T. Cheng and C. Froese-Fischer, Phys. Rev. A28:2811 (1983).
43. R. Haensel, G. Keitel, P. Schreiber, and C. Kunz, Phys. Rev. Phys. Rev. 188:1375 (1969).
44. Z. Altun and H. P. Kelly, to be published.
45. J. P. Connerade, M. W. D. Mansfield, and M. A. P. Martin, Proc. Roy. Soc. A350:405 (1976).
46. R. Bruhn, B. Sonntag, and H. W. Wolff, Phys. Lett. 69A:9 (1978).
47. R. Bruhn, E. Schmidt, H. Schröder, and B. Sonntag, Phys. Lett. 90A:41 (1982).
48. P. H. Kobrin, U. Becker, C. M. Truesdale, D. W. Lindle, H. G. Kerkhoff, and D. A. Shirley, J. Elect. Spectrosc. Relat. Phenom. 34:129 (1984).
49. M. O. Krause, T. A. Carlson, and A. Fahlmann, Phys. Rev. A30:1316 (1984).
50. E. Schmidt, H. Schröder, B. Sonntag, H. Voss, and H. E. Wetzel, J. Phys. B16:2961 (1983).
51. R. Bruhn, E. Schmidt, H. Schröder, and B. Sonntag, J. Phys. B15:2807 (1982).
52. L. C. Davis and L. A. Feldkamp, Phys. Rev. A17:2012 (1978).
53. M. Ya. Amusia, V. K. Ivanov, and L. V. Chernysheva, J. Phys. B14:L19 (1981).
54. H. P. Kelly and A. Ron, Phys. Rev. A5:168 (1972).
55. L. J. Garvin, E. R. Brown, S. L. Carter, and H. P. Kelly, J. Phys. B16:L269 (1983).
56. E. R. Brown, S. L. Carter, and H. P. Kelly, Phys. Rev. A21:1237 (1980).
57. A. W. Fliflet and H. P. Kelly, Phys. Rev. A10:508 (1974).
58. H. P. Kelly and A. Ron, Phys. Rev. A4:11 (1971).
59. E. R. Cooper and H. P. Kelly, Phys. Rev. A7:38 (1973).
60. B. Sonntag, private communication.
61. L. J. Garvin, E. R. Brown, S. L. Carter, and H. P. Kelly, J. Phys. B16:L643 (1983).
62. E. Schmidt, H. Schröder, B. Sonntag, H. Voss, and H. E. Wetzel, J. Phys. B18:79 (1985).
63. F. Combet-Farmoux and M. Ben Amar, Phys. Rev. A21:1975 (1980).
64. W. Wijesundera and H. P. Kelly, to be published.
65. J. H. Weaver and C. G. Olson, Phys. Rev. B14:3251 (1976).
66. M. W. D. Mansfield, to be published.
67. M. W. D. Mansfield and J. P. Connerade, Proc. Roy. Soc. A352:125 (1976), and refs. therein.
68. U. Becker et al, to be published.
69. M. Ya. Amusia, S. I. Sheftel, and L. V. Chernysheva, Sov. Phys. Tech. Phys. 26:1444 (1982).
70. Cheng Pan, S. L. Carter, and H. P. Kelly, to be published.
71. U. Becker, private communication.

GIANT RESONANCES IN ATOMS AND IN FLUORINE CAGE MOLECULES

M.W.D. Mansfield

University College
Cork
Ireland

1. INTRODUCTION

Giant resonances in the photoabsorption spectra of atoms occur in the extreme ultraviolet region of the electromagnetic spectrum. In order to observe absorption spectra in this region it is necessary to generate columns of atomic vapour which will often be very hot and chemically aggressive, and to contain them without solid windows between two regions of high vacuum, the spectrometer and the light source, usually an electron synchrotron. The technical problems are often formidable so that although it had long been recognised that giant resonances in solid lanthanides were essentially atomic phenomena (Fomichev et al. 1967, Dehmer et al. 1971) earlier investigations of giant resonances in atoms were limited to the more managable elements which precede the transition rows, the inert gases, alkali and alkaline earth elements.

Over the last ten years the technical problems of containing transition metal vapours have been steadily overcome so that it is now possible to follow the evolution of giant resonances in atoms and in some associated free molecules through almost the entire lanthanide row and through most of the first transition row. It is therefore an appropriate time to collect together and compare results.

In the following section I will discuss the spectra of transition row atoms in order of decreasing localisation (Smith and Kmetko 1983) viz. $4d \rightarrow f$, $5d \rightarrow f$, $3p \rightarrow d$, $4p \rightarrow d$ and $5p \rightarrow d$. I will tend to avoid discussion of the giant resonances themselves because their profiles and interpretation will be discussed comprehensively by other contributors. Instead I will concentrate on the detailed analyses which have been attempted of the discrete structure which usually accompanies giant resonances in atoms. Interpretation of this structure can provide accurate determinations of thresholds for inner shell excitation in atoms and can also be used to anticipate structure which may overlie the giant resonances and distort their profiles.

While the emphasis in this review will be on photoabsorption by atoms and fluorine cage molecules it should be noted that complementary information on decay channels in particular is becoming available from photoelectric and electron impact experiments. Unfortunately I will only have time to make passing reference to this work but I draw your attention to the seminar of M. Richter who will present some of the latest results from DESY in this field.

2. 4d → f RESONANCES IN THE LANTHANIDE TRANSITION ROW

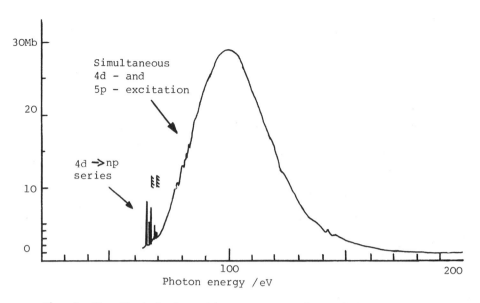

Fig. 1 The 4d photoabsorption spectrum of atomic xenon (after Haensel et al. 1969a)

I will begin by considering 4d → f excitation in Xe I (Z = 54), well before the 4f wave function has collapsed in an atomic ground state (Z = 58). The 4d-subshell absorption spectrum of Xe I (Figure 1) has been studied extensively by many workers (Lukirskii and Zimkina 1963, Codling and Madden 1964, Ederer 1964, Madden and Codling 1965, Samson 1966, Haensel et al. 1969a). The most striking feature of the spectrum is the 'delayed onset' of continuous absorption whereby the photoionization cross-section rises from a low value directly above the 4d-ionization thresholds to a maximum about 40 eV above threshold. This phenomenon was explained within the independent particle model by Cooper (1964) in terms of the exclusion of bound and low energy continuum f-electrons from the atomic core by the centrifugal barrier for f-electrons. In consequence only 4d → np transitions are observed below the 4d ionization thresholds (Codling and Madden 1964) and it is not until the continuum f-electrons acquire sufficient kinetic energy to penetrate the barrier that the 4d → f oscillator strength begins to rise. The broad maximum may be regarded as a virtual state of the inner potential well for f-electrons (Connerade 1984), a precursor of the giant 4d → f resonances which we will encounter in the lanthanides.

Shortly after the Xe I experiments very strong 4d-subshell resonances were reported in the spectra of lanthanide metals (Zimkina et al. 1967, Haensel et al. 1970). These were soon recognised as essentially atomic $4d^{10} 4f^N - 4d^9 4f^{N+1}$ transitions (Formichev et al 1967) and extensive calculations of $4d^{10} 4f^N - 4d^9 4f^{N+1}$ transitions in free lanthanide ions were performed to support this interpretation (Dehmer et al 1971, Starace 1972, Sugar 1972, Dehmer and Starace 1972, Starace 1974). It was not at that time technically feasible to demonstrate the atomic character of the 4d-subshell resonances through direct comparisons between the solid and atomic spectra so that experimental interest centred on the 4d → f spectrum of atomic Ba (Z = 56) where Wendin (1973a, b) had predicted a giant collective resonance.

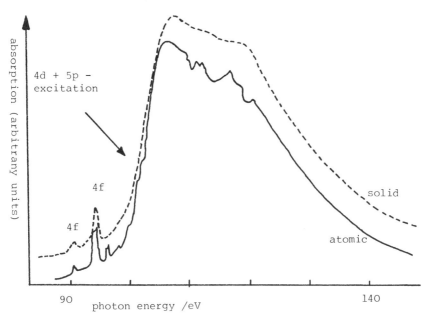

Fig. 2. The 4d photoabsorption spectra of atomic and solid barium (after Rabe et al. 1974)

The Ba I 4d-subshell photoabsorption spectrum (figure 2) was recorded independently by 3 groups at about the same time (Connerade and Mansfield 1974, Rabe et al. 1974, Ederer et al. 1975). I will discuss the barium spectrum at some length because it illustrates well many of the points I will be stressing in discussing discrete structure in the lanthanide spectra.

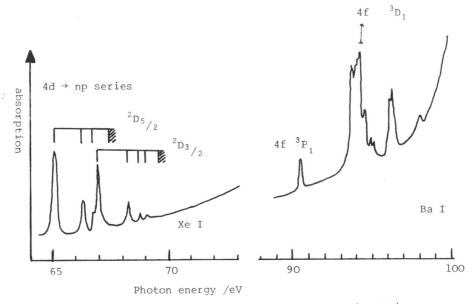

Fig. 3. Comparison of the Xe I and Ba I 4d photo-absorption spectra below ionization threshold (after Haensel et. al. 1969a and Rabe et al. 1974)

In figure 3 the discrete structure beneath the Ba I 4d-subshell thresholds is compared with the equivalent structure in Xe I. It is immediately apparent that the two regular Rydberg series observed in Xe I do not persist into the Ba I spectrum, at first sight a surprising result because the two simple excitation schemes, viz. Xe I $4d^{10} \to 4d^9$ np

and Ba I $4d^{10} 6s^2 \to 4d^9 6s^2$ np

differ only in the prescence of a closed $6s^2$ subshell in Ba. Two distinct reasons may be given for the greater complexity of the Ba I spectrum:-

(i) <u>4f collapse</u>: While in Ba I the 4f wave function does not collapse when a valence electron is excited (Griffin et al. 1969) the removal of an inner shell electron increases the central field sufficiently for 4f collapse to occur in the $4d^9$ 4f $6s^2$ configuration. Consequently two $4d^9$ 4f $6s^2$ levels accessible from the ground state, $^3D_1$ and $^3P_1$, become bound states of the inner potential well for f electrons and give rise to two quite strong absorption lines. The two lines may be easily identified in figure 2 where the atomic and metallic spectra are compared because, unlike the $4d \to np$ lines, the final states are localized in the barium core and are little affected by the metallic environment. The collapse of the 4f wave function means in effect that the cross over point between localized and itinerant behaviour in the quasi-periodic table of Smith and Kmetko (1983) is shifted to lower Z elements for core excited states. Thus 4d excited barium is expected to show many of the features of 4d excited lanthanides.

(ii) <u>Double excitation through 5d - 6s mixing</u>: The position of barium in the periodic table is doubly critical because two wave functions, 4f and 5d, are about to collapse into the core. The complete collapse of the 5d wave function is delayed until after the 4f subshell is filled at Z = 70 but in barium it is nonetheless very sensitive to small changes in the central field.

Analysis of the 5p-subshell absorption spectrum of Ba I (Connerade et al. 1979) has clearly established that, when the 5p-subshell is excited, one or both valence electrons are often excited at the same time. In effect inner shell excitation shatters the valence shell. A similar effect is to be expected therefore when 4d electrons are excited to np states, the end result being the appearance of $4d^{10} 6s^2 \to 4d^9 (6s^2 + 5d6s + 5d^2)$ np lines. A general interpretation of the discrete Ba structure below threshold in these terms was suggested by the original investigators and detailed (unpublished) multiconfiguration calculation support this interpretation, excitation to $4d^9$ 5d 6s 6p levels in particular producing supernumerary lines.

In figure 4 the calculated average energies of Ba I configurations in which one or both 6s electrons have been excited to 5d states are compared for the Ba I ground state and for 4d excited configurations in which the 4d electron has been excited to 4f, 6p or continuum states. The HXR program (Cowan 1981) was used for these calculations. It is interesting to note that for the 4f states, which are localized in the atomic core, the separations between the $4d^9$ ($6s^2$ + 5d 6s + $5d^2$) 4f configurations are similar to those between the $4d^{10}$ ($6s^2$ + 5d 6s + $5d^2$) configurations and that in consequence 5d - 6s mixing is small. For the non-localized 6p and continuum state however the 5d wave function collapses quite sharply leading to extensive 5d + 6s configuration mixing.

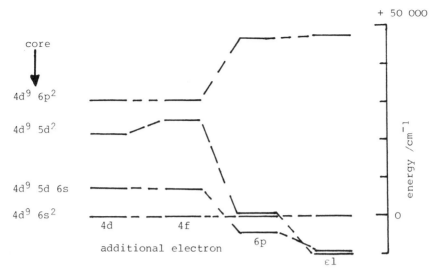

Fig. 4. Comparison of calculated average energies of the Ba I configurations $4d^9$ ($6s^2$, 5d 6s, $5d^2$, $6p^2$)nl where nl = 4d, 4f, 6p and εl (Ba II thresholds). Energies are referred to $4d^9$ $6s^2$ nl in each case.

5d - 6s mixing complicates the Ba I 4d-subshell spectrum below threshold considerably and introduces many supernumerary thresholds in addition to the two $4d^9$ $6s^2$ limits. According to calculation each Ba II $4d^9$ $6s^2$ level is expected to share half its character with $4d^9$(5d 6s + $5d^2$ + $6p^2$) levels, principally $4d^9$ 5d 6s, producing altogether 13 levels with more than 1% $4d^9$ $6s^2$ purity spread over an energy range of 86,000 cm^{-1}. It is therefore unrealistic to expect to observe Rydberg series to well defined $4d^9$ $6s^2$ limits and in these circumstances comparisons with ab initio calculations are required to deduce thresholds from the observed Ba I spectrum. Strong mixing in the limits is a familiar situation for inner shell excitation in the heavier alkaline earth elements, Ca, Sr and Ba because in each case a d wave function is on the point of collapsing below the valence s-subshell. In analyzing the very similar case of 3d-subshell excitation in Sr I Mansfield and Connerade (1982) show that the acute sensitivity of d - s mixing to the amount of nuclear charge seen by the d electron can largely invalidate analysis of an inner shell spectrum by comparison with valence excitation in the next element (the Z + 1 method). Comparisons with ab initio calculations are required in these cases.

We now turn to the region above the 4d thresholds - to the giant resonance itself. The configuration average calculation of $4d^9$ 4f $6s^2$, which as we have already noted describes the transitions to $^3P_1$ and $^3D_1$ levels well, predicts a very strong $4d^{10}$ $6s^2$ $^1S_0 \rightarrow 4d^9$ $6s^2$ 4f $^1P_1$ transition which is thrown above threshold by the very large 4d - 4f exchange interaction. However, as Hansen et al. (1975) have pointed out, the contribution of the exchange energy is so great that the $^1P_1$ state is no longer well described by the configuration average 4f wave function. When the energy of the $^1P_1$ state is minimized separately in the Hartree-Fock procedure (Fliflet et al. 1975) the transition to $^1P_1$ is calculated to occur below threshold and to possess negligible oscillator strength.

This situation provoked much discussion as to whether the giant resonance should be labelled 4f ^{1}P or ϵf ^{1}P, a discussion which has been clarified by Wendin and Starace (1978). Here I will use Connerade's (1982) description of the state as a virtual state of the inner well which cannot be decoupled from the continuum of which it is a constituent part.

The giant resonance in Ba I (figure 2) shows clear discrete structure superposed on its peak (Connerade and Mansfield 1974, Rabe et al. 1974). Connerade and Mansfield suggest that the structure may be due to 'displaced giant resonances' built on the highly mixed $4d^9$ ($6s^2$ + 5d 6s + $5d^2$) thresholds. The validity of this explanation will depend on the extent to which the virtual state is localized in the inner well. If it is as highly localized as the $4d^9$ 4f $6s^2$ configuration average state it is unlikely to induce 5d - 6s mixing according to the earlier discussion of d - s mixing. If however it is not well localized it might indeed be expected to induce s - d mixing, as was the case for 4d → np, ϵl excitation. The spread of the discrete structure matches the calculated spread of thresholds with $4d^9 5s^2$ character quite well so that to this extent Connerade and Mansfield's interpretation is supported by observation. It should be noted that there is additionally a small probability of excitation to the $4d^9$ 4f ($5p^2$ ^{1}S)^{1}P$_1$ state through $6s^2$ + $6p^2$ mixing in the Ba I ground state. Calculations indicate that this transition will be weak however.

Before moving on from Ba I we should also take note of the discrete structure which occurs on the long wavelength side of the giant resonance and which is attributed to simultaneous excitation of the 4d - and 5p-subshells in figure 2. Equivalent structure is also evident in the Xe I (figure 1) and Cs I (figure 5) spectra. Calculations of simultaneous 4d and 5p excitation show that structure due to this type of transition could extend to the peak of the giant resonance and could provide an alternative explanation for the structure which is attributed to 'displaced giant resonances' above. However it is clear that the prominent structure due to simultaneous 4d - and 5p - excitation in Xe I and Cs I does not extend to the peaks of the giant resonances themselves so that the 5d - 6s mixing interpretation of the structure at the peak of the Ba I giant resonance seems more probable.

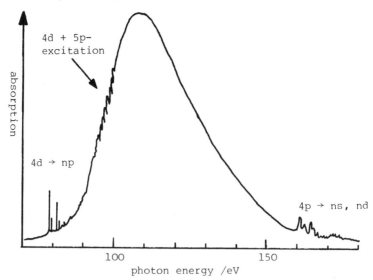

Fig. 5. The 4d photo-absorption spectrum of atomic Cs (after Petersen et al. 1976)

The Cs I (Z = 55) 4d-subshell spectrum has been reported by Petersen et al. (1976). The Cs I spectrum generally resembles the Xe I spectrum more than that of Ba I. Calculations show that minor 5d - 6s mixing complicates the 4d → np discrete spectrum a little but that the $4d^9$ 6s limits are quite pure. This result is borne out by the observation of Rydberg series. The 4f wave function is beginning to collapse in the $4d^9$ 4f $6s^2$ case and some weak 4d → 4f lines may be present below threshold. As noted above discrete structure due to simultaneous 4d - and 5p excitation is present on the long wavelength side of the giant resonance but otherwise its profile is free of structure. Recent determinations of Cs I photoelectron spectra and partial photoionization cross-sections (Prescher et al. 1986) show that direct 4d ionization is the dominant decay process, confirming the non-localized nature of the virtual f-state in Cs I.

The 4d → f photo absorption spectra of the lanthanide atoms, with the sole exception of the unstable element Pm (Z = 61), have now been reported. References for these experiments are as follows, Ce (Wolff et al. 1976), Eu (Mansfield and Connerade 1976), La, Pr, Nd, Sm, Tb, Dy, Ho, Er, Tm (Radtke 1979a, b) and Gd (Connerade and Pantelouris 1984). Radtke (1979b) has made systematic comparisons between the atomic and metallic (Zimkina et al. 1967) lanthanide spectra. He divided the lanthanides into 2 groups:

Group 1: Elements for which there is a close correspondence between the spectra in the solid and vapour phases. These are lanthanides which are either trivalent in the solid and have an atomic ground state which includes a 5d electron (La, Ce, Gd, Lu) or which are divalent in the solid without a 5d electron in the atomic ground state (Eu, Yb. The atomic and solid spectra of La are compared in figure 6 as an example of an element of this group.

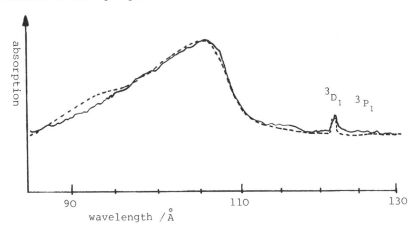

Fig. 6. Comparison of the 4d photoabsorption spectra of La I (solid line) and solid La (dashed line) (after Radtke 1979 a)

Group 2: Elements for which there is a close resemblance between the spectrum of the atom with nuclear charge Z and the metallic spectrum of the element with nuclear charge Z + 1. These are lanthanides which are trivalent in the solid and which do not possess a 5d electron in the ground state. The spectrum of atomic Er (Z = 68) is compared with that of solid Tm (Z = 69) in figure 7 as an example of an element of this group.

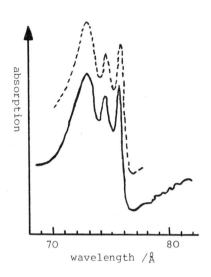

Fig. 7. Comparison of the 4d photoabsorption spectra of Er I (solid line) and solid Tm (dashed line) (after Radtke 1979 b)

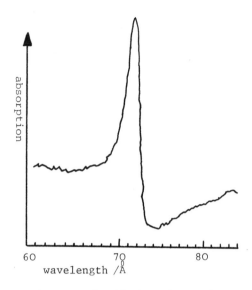

Fig. 8. The 4d photoabsorption spectrum of Tm I (after Radtke 1979 a)

Radtke's categorisation is consistent with the interpretation of the photo-absorption spectra of lanthanide metals in terms of $4d^{10}\ 4f^N \rightarrow 4d^9\ 4f^{N+1}$ transitions in the trivalent or divalent ions (Dehmer et al. 1971, Sugar 1972, Starace 1974). The close correspondence between the metallic and atomic spectra also means that Sugar's (1972) detailed interpretation of structure in the solid spectra in terms of $4d^9\ 4f^{N+1}$ multiplet splittings will also be valid in the atomic spectra. In the case of the $4d^9\ 4f^8$ configuration which occurs in the Eu and Gd metals Sugar found it necessary to simplify his calculations by considering only the $4d^{10}\ 4f^7\ ^8S \rightarrow 4d^9\ (4f^8\ ^7F)$ transitions. He justified this restriction because in LS coupling only octet states can be reached from the ^8S ground state and also because only the ^7F state of $4f^8$ has sufficiently high multiplicity to produce octet states on addition of the nine 4d electrons. Mansfield and Connerade (1976) have pointed out that this approximation is equivalent to assuming that the seven ground state 4f electrons remain spectators in any 4d → 4f transitions, retaining their ground state total angular momenta. This is an example of the spectator model which will be used again later.

Radtke (1979a) was also able to demonstrate that the profiles of 4d → 4f resonances in atomic lanthanides become progressively narrower with increasing Z. Comparison of the giant resonance in Tm I (Z = 69) (figure 8) with that in La I (Z = 57) (figure 6) demonstrates this effect. The effect is partly obscured by the varying multiplet splittings of the $4d^9\ 4f^{N+1}$ configuration but the trend is nonetheless clear. It is readily understandable in terms of increasing localisation of the 4f state with increasing Z (Connerade 1984). The giant resonance evolves from a virtual state of the inner well for low Z to a bound state below threshold for higher Z.

In view of the earlier discussion of atomic barium the close correspondence which has been noted between the atomic and metallic lanthanide spectra is perhaps surprising because it suggests that the two processes which served to complicate the Ba I spectrum are not important in the atomic lanthanides. These were:-

(i) $4d^{10} \to 4d^9$ np excitation and (ii) 5d - 6s mixing.

I will discuss these points in turn.

(i) A possible 4d → np series in Eu I was suggested by Mansfield and Connerade (1976). The series was assigned

$$4d^{10}\ 4f^7\ 6s^2\ {}^8S \to 4d^9\ (4f^7\ {}^8S)\ 6s^2\ ({}^{7\text{or}\,9}D)\ np\ {}^8P$$

in accordance with the prediction of the spectator model but the appearance of just one simple np series in such a potentially complex situation must be considered fortuitous. In view of the absence of obvious $4d^9\ 4f^N$ np series in any of the other lanthanide spectra Mansfield and Connerade's proposed np assignments must still be regarded as very tentative. The absence of 4d → np transitions for Z = 70 is not unexpected because, according to the assignments of Fano and Cooper (1968), a transfer of oscillator strength from the 4d - to the 4f-subshell is likely when the latter subshell is completed. It is nonetheless surprising that 4d → np transitions seem to disappear so quickly after barium because calculations indicate that there is little interaction between the 4d → np and 4d → nf channels. Possibly 4d → np oscillator strength is distributed between so many $4d^{10}\ 4f^N \to 4d^9\ 4f^N$ np lines that they tend to be lost in the very strong $4d^{10}\ 4f^N \to 4d^9\ 4f^{N+1}$ spectrum.

(ii) 5d - 6s mixing is well established in the 5p-subshell spectrum of La I (Connerade et al. 1979) so that its apparent absence in the 4d-subshell spectrum of La I (figure 6) is at first sight surprising. An explanation may be found in the earlier discussion of the structure overlying the peak of the Ba I giant resonance (figure 2). In that case it was suggested that this structure could be attributed to 5d - 6s mixing if the virtual f state was not very localized in the inner well. The lack of any equivalent structure in La I (figure 6) therefore implies that the virtual state has become so well localized that 4d → f excitation has little effect on the valence shell. In the case of 5p → 5d excitation (Connerade et al. 1979) however the final state wave function will overlap the valence subshell and 5d - 6s mixing is expected. 5p-subshell excitation is discussed further in section 5.

One further point should be made in discussing possible structure in the lanthanide 4d subshell spectra namely that, at the temperatures used to produce atomic vapours, levels other than the ground state will be thermally excited in some lanthanides. Absorption from excited states will therefore be superposed on the ground state photo-absorption spectrum. However because the energy splittings of the initial states is small in comparison with the energies of 4d transitions the overall appearance of the spectrum should not be changed much by thermal excitation.

In summary the following picture emerges as we follow the giant 4d → f resonance from Xe through the lanthanide atoms. In Xe and Cs the virtual f state is not well localized in the inner well so that simultaneous 4d - and 5p-subshell excitation accompanies 4d - excitation. Simultaneous excitation of the 4d - and 6s-subshells, i.e. 5d - 6s mixing cannot of course occur in Xe I and is of minor importance in Cs I because the 5d wave function has not collapsed. In Ba I however 5d collapse has started and the f-state is still not well localized so that both simultaneous 4d - and 5p - excitation and 5d - 6s mixing accompany the giant resonance. In La I the f state is still a virtual state of the inner well but is sufficiently well localised that 4d → f excitation has little effect on the outer regions of the atom and structure due to 5d - 6s mixing is

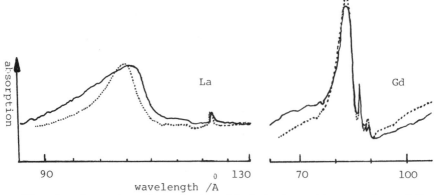

Fig. 9. Comparison between the 4d photoabsorption spectra of La and Gd vapours (solid lines) and of La F_3 and Gd F_3 vapours (dotted lines) (after Connerade et al. 1980b and Connerade and Pantelouris 1984)

not observed. With increasing Z the f-state becomes more localized ending up as a bound state of the inner well with very little interaction with either the outer regions of the atom or the atomic environment.

Before leaving the lanthanide row I will also consider 4d → f giant resonances in lanthanide fluorine cage molecules. In figure 9 comparisons are made between the atomic La (Z = 57) and Gd (Z = 64) spectra and those of La F_3 and Gd F_3 vapours (Connerade et al. 1980b, Connerade and Pantelouris 1984). The presence of a fluorine cage is expected to reinforce the centrifugal barrier for f-electrons, the electronegative F atoms drawing electrons away from the nucleus (Dehmer 1972). This leads to increased localisation of the f-state in the molecular case. The narrowing of the giant resonance in La F_3 (figure 9) is evidence of such a process. The giant resonance in Gd F_3 on the other hand is very similar to the atomic giant resonance (figure 9) indicating that in atomic Gd the f-state is so well localized in the inner well that the addition of a fluorine cage has little effect. On this evidence it is likely that the atomic profiles of giant resonances in other lanthanides will be little affected by a molecular environment.

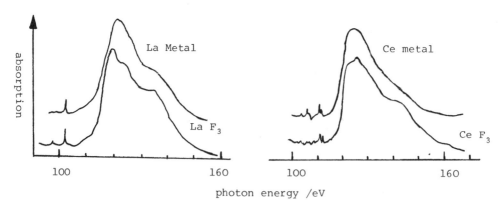

Fig. 10. Comparisons between the 4d photoabsorption spectra of solid La and Ce and of solid La F_3 and Ce F_3 (after Suzuki et al. 1975)

Comparisons between giant resonances in solid La and Ce and in solid La F_3 and Ce F_3 (figure 10) (Suzuki et al. 1975) also indicate that f-state localization in the lighter lanthanides can be affected by a change in the atomic environment. In La the giant resonance is broader in the fluoride and acquires a prominent shoulder on its short wavelength wing. The resulting profile is reminiscent of the giant resonance in Ba I (figure 2). A similar effect is also evident in Ce I. Suzuki et al. (1975) also show that the $4d^9\,4f^{N+1}$ line structure below threshold in La and Ce is essentially the same for all the solid halides indicating that these states are well localized for all lanthanides. Finally it should be noted that the broadening of the giant resonance in solid La F_3 and Ce F_3 is evidence that the fluorine cage model does not apply in the crystalline environment.

3. 5d → f RESONANCES IN THE ACTINIDE TRANSITION ROW

Only two actinide elements Th (Z = 90) and U (Z = 92) are readily accessible experimentally so that it is not yet possible to establish trends in this row. 5d excitation has been studied in the free fluorine cage molecules, U F_4 (Connerade et al. 1980a) and Th F_4 (Connerade et al. 1980b) and also in atomic U (Pantelouris and Connerade 1983). This data has been collected together in figure 11 where comparisons are made between Th F_4 and metallic Th and between UF_4 and metallic and atomic U. Two features of these spectra are surprising in the light of the preceding discussion of f localization in the lanthanides. These are:-

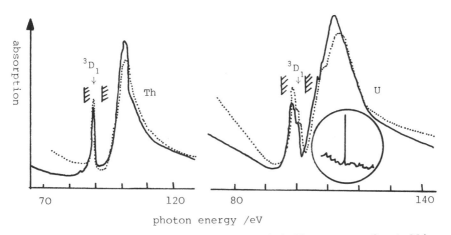

Fig. 11. Comparison between the 5d-subshell spectra of metallic Th and U (solid lines) and of Th F_4 and UF_4 (dotted lines). Inset - the spectrum of U I (after Connerade et al. 1980a, b and Pantelouris and Connerade 1983).

(1) The increase in breadth of the giant resonance between Th F_4 and UF_4. This indicates that the f state is more localized in Th in apparent contradiction of the expected trend towards greater localization with increasing Z (Smith and Kmetko 1983).

Connerade et al. (1980b) note that the most likely configurations of the U^{4+} and Th^{4+} ions are $6p^6\,6d^2$ and $6p^6$ respectively. They therefore suggest that the UF_4 giant resonance is broadened by splitting in the $6d^2$ configuration, an effect which does not occur in Th F_4. There is clear structure on the peak of the UF_4 giant resonance and, as in the case of Ba I 4d → f excitation which was discussed in the previous section, this would seem to indicate that the f-states are not completely localized in the atomic core. The 5d thresholds in the solid (Fuggle et al. 1974) are indicated in figure 11 for both Th and U. It is notable that in both the metals and the molecules the giant resonances lie above these thresholds. The $^3D_1$ states also lie above the lower 5d thresholds and the $^3D_1$ resonances are much broader than the triplet resonances in La (figure 6) which lie below both 4d thresholds. Connerade et al. (1980b) suggest that the widths of the ^3D resonances in Th and U may be controlled by tunnelling through the potential barrier.

(2) The second and probably greater surprise is the observation of a single sharp line in the atomic spectrum of U (figure 11) which contrasts strongly with the broad 5d → f resonances observed in the metal and molecule. As noted by Pantelouris and Connerade (1983) the sharpness of the atomic resonance indicates a strongly localized f state below threshold. This conflicts with the expected decrease of localization when a fluorine cage is removed. If the interpretation of the U I line as a discrete resonance below threshold is correct it also follows that the atomic thresholds are about 25 eV lower in energy than the solid U thresholds (Fuggle et al. 1974). It is not yet possible to test this conclusion by comparison with reliable calculated values of the atomic thresholds because, as Wendin and Del Grande (1985) point out, one electron calculations are inadequate in this region and have to be replaced by methods which include polarization of the 5d-subshell. The atomic result in U is hard to understand and an independent check of the atomic threshold energies would be valuable.

In summary on the evidence available it appears that the localization of 5f states in the actinides is not as marked as that of 4f states in the lanthanides and that 5f localization can be modified substantially by changes in the atomic environment.

3. 3p → d RESONANCES IN THE FIRST TRANSITION ROW

The potential barrier for d electrons is smaller than that experienced by f electrons and consequently d electrons are never completely excluded from the atomic core, even for elements such as Ar (Z = 18) in which the 3d wave function has not started to collapse. As a result 3p → nd Rydberg series are observed in Ar I (Beutler 1935, Huffman et al. 1965) although the lines are quite weak with most 3p→d oscillator strength going into the continua above the 3p ionization thresholds.

For the next two elements in the periodic table K (Z = 19) and Ca (Z = 20) there is evidence that the 3d wave function has started to collapse when the 3p-subshell is excited (Mansfield 1975a, Mansfield and Newsom 1977). In both elements 3p → 3d giant resonances are beginning to form but 3d collapse is not easy to follow for reasons which are outlined below.

In configuration average calculations for both K I (Mansfield 1975a) and Ca I (Mansfield and Newson 1977) the 3d wave function collapses when the 3p-subshell is excited and most 3p → d oscillator strength is calculated to reside in transitions to states based on $3p^5 3d\ ^1P$. These are the K I $3p^5 3d\ (^1P)\ 4s\ ^2P$ levels and the Ca I $3p^5 3d\ 4s^2\ ^1P$ level. However, as in the Ba I $4d^9 4f 6s^2\ ^1P$ case which was discussed in section 2 the exchange interaction between the open core and the collapsed wave function is so great that the configuration average description is no longer appropriate. The energies of $3p^5 3d\ ^1P$ based states have therefore been minimized separately in the Hartree-Fock procedure and these calculations place transitions to $3p^5 3d\ ^1P$ based levels below threshold with reduced oscillator strength. The reduction in oscillator strength is much less severe than in the equivalent Ba I case (Fliflet et al. 1975) however and the lines are expected to remain the strongest in the spectrum. Very strong lines are observed at the calculated wavelengths but it is difficult to use the usual methods of multiconfigurational calculation to estimate the redistribution of oscillator strength following separate minimization because an important assumption of the theory has been seriously violated, namely that the radial integrals are independent of the angular momentum coupling. The K I and Ca I 3p-subshell spectra are also complicated by extensive 3d - 4s mixing which occurs as the 3d wave function collapses into the region of the valence shell. The very strong $3p^5$ 3d ^1P based lines - the incipient giant resonances - show structure due to mixing and supernumerary series to 3d - 4s mixed limit systems are clearly in evidence.

Analysis of the 3p-subshell spectra of K I and Ca I is quite complicated therefore and I will turn instead to the 2p-subshell spectra to illustrate the main features of 3d collapse. The exchange interaction between the open core and the collapsed wave function is substantially smaller for 2p excitation than for 3p excitation. Level splittings are determined mainly by the 2p spin-orbit splitting, a very stable quantity, so that detailed analysis is much more straigthtforward than for the 3p-subshell.

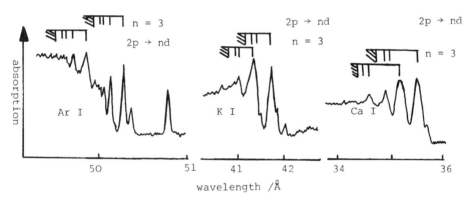

Fig. 12. The 2p-subshell spectra of Ar I, K I and Ca I
(after Nakamura et al. 1968, Mansfield 1975b, 1976)

The 2p-subshell spectra of Ar I (Nakamura et al. 1968), K I (Mansfield 1975b) and Ca I (Mansfield 1976) are compared in figure 12. The K I and Ca I identifications are based on recent multiconfigurational calculations and differ a little from those given originally. Figure 12 illustrates well the principal feature of 3d collapse in the first transition row. These are:

(1) As the 3d wave function collapses 2p → d oscillator strength is taken from the continuum and appears in very strong 2p → 3d transitions - giant resonances.

(2) Prominent 2p → nd series are observed even after the 3d wave function has collapsed into the inner well. Thus although nd (n > 3) states are eigenstates of the outer well they are still able to penetrate the atomic core sufficiently to produce 2p → nd transitions.

(3) Because the 3d and nd (n > 3) states are eigenstates of different potential wells the normal progression of intensity and effective quantum number in a Rydberg series is not observed by the 2p → 3d lines which are anomalously intense and possess anomalously high quantum defects.

Multiconfigurational calculations show that some 3d - 4s mixing occurs for 2p-subshell excitation in K I and Ca I. Weak supernumerary lines are predicted and observed. Structure is also predicted in the broad 2p → 3d lines but this is not resolved. Both 2p → 3d lines lie below the main 2p thresholds in both K I and Ca I although the shorter wavelength resonance in Ca I lies above some of the additional thresholds produced by 3d - 4s mixing.

The essentially atomic origin of 3p → 3d giant resonances in the photo-absorption spectra of first row transition metals (Sonntag et al. 1969) was proposed by Fano and Cooper (1969). Sonntag et al. 1969's data is reproduced in figure 13. The resonances tend to become narrower with increasing Z, indicating increasing localization, but the trend is not as clear as it was in the lanthanides. The resonances' profiles also show a pronounced asymmetry which has been interpreted (Fano and Cooper 1969, Davis and Feldkamp 1976) in terms of interference between 3p → 3d resonances and background absorption by outer shells. This effect would also seem to indicate that 3d states are not completely localized.

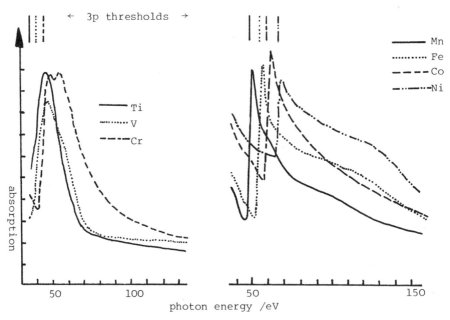

Fig. 13. The 3p-subshell spectra of first transition row metals (after Sonntag et al. 1969)

The first experiments on the 3p-subshell photoabsorption spectra of first transition row elements in atomic form were performed on Mn I (Connerade et al. 1976) and Cr I (Mansfield 1977). The measurements were hindered by the presence of a third order background continuum so that considerably more structure was revealed when these experiments were repeated by Bruhn et al. (1978, 1982a). Bruhn et al. also suggested that the wavelength measurements reported in the earlier work were in error and a recent Mn I experiment in Bonn has supported Bruhn et al.'s (1978) measurements. On this evidence the Cr I measurements of Bruhn et al. (1982a) should also be preferred.

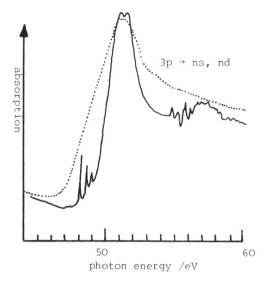

Fig. 14. The 3p-subshell spectra of Mn I (solid line) and metallic Mn (dashed line) (after Bruhn et al. 1978)

The 3p-subshell photoabsorption spectrum of Mn I is shown in figure 14 (Bruhn et al. 1978). Connerade et al. (1976) analyzed the spectrum through the spectator model assuming that, in a first approximation, the five 3d electrons retain their $^6S_{\frac{5}{2}}$ ground state coupling when the 3p-subshell is excited. This model leads to the expectation of very strong transitions to $3p^5\ 3d^6\ 4s\ ^6P$ levels followed by $3p^5\ (3d^5\ ^6S)\ 4s^2\ (^7P)$ ns, nd series at shorter wavelengths. The spectrum was assigned accordingly and some additional weaker lines at long wavelength were interpreted in terms of a breakdown of the $\Delta J_c = 1$ selection rule for <u>core</u> excitation, which ignores the five 3d electrons. $\Delta J_c = 2, 3$ leads to transitions to $3p^5\ 3d^6\ 4s^2\ ^6D$, ^6F levels and the weak lines were assigned accordingly. Bruhn et al. (1982) broadly confirmed this picture but more sensitive measurements of the longer wavelength lines led them to revise Connerade et al.'s (1976) ^6D and ^6F assignments on the basis of Davis and Feldkamp's (1978) theoretical predictions. New ^6F lines were identified and ^4P assignments were introduced. Comparison with the spectrum of metallic Mn (Sonntag et al. 1969) shows that only the giant resonance persists in the solid and that its shape is modified substantially.

The 3p photo absorption spectrum of Mn I has attracted a good deal of theoretical interest (Davis and Feldkamp 1978, Amusia et al. 1981, Garvin et al. 1983) because Mn possesses a half-filled 3d-subshell in its ground with the spins of the five electrons parallel. This quasi-closed shell is particularly amenable to treatment by the RPAE method so that Mn is a useful test case of the theory for open shell atoms.

Theoretical investigations of the Mn I photo absorption spectrum have been aided by extensive photo-emission measurements in which the different channels contributing to absorption may be distinguished. (Bruhn et al. 1982b, Schmidt et al. 1985). The strong coupling of the $3p \to 3d$ and $3d \to \varepsilon f$ excitation channels produces resonant enhancement of the 3d photo-emission lines. Schmidt et al. (1985) note that it is difficult to obtain reliable cross-sections from photographically registered spectra. The Mn I photoabsorption spectrum has recently been recorded photoelectrially (Koberle and Kung 1985). Their results indicate that the photographic profile shown in figure 14 may be distorted and this should be born in mind in assessing the profiles of giant resonances presented in this paper.

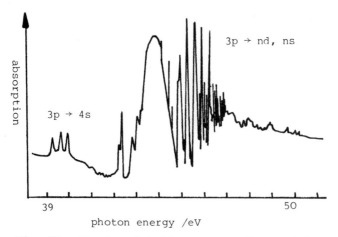

Fig. 15. The 3p-subshell spectrum of Cr I (after Bruhn et al. 1982a)

The 3p-subshell spectrum of Cr I (Bruhn et al. 1982a) is shown in figure 15. Mansfield (1977) analyzed the Cr I spectrum through the spectator model. Cr is unusual among first row transition elements in that it possesses only one 4s electron in its ground state so that the spectator model was extended by assuming that the five 3d electrons and the 4s electron retain their $^7S_3$ coupling when the 3p-subshell is excited. Following this model the three long wavelength lines were assigned as 3p → 4s transitions to $3p^5$ $3d^5$ $4s^2$ ^7P levels and the discrete structure at short wavelengths was analyzed as 3p → ns, nd series converging on the $3p^5$ $(3d^5$ $^6S)(^7P)$ $4s$ ^8P limits. Mansfield had difficulty identifying the $3p^6$ $3d^5$ $4s$ ^7S → $3p^5$ $3d^6$ $4s$ ^7P giant resonance however which, from comparison with Hartree - Fock calculations, he expected to lie above the ^8P thresholds. When Bruhn et al. (1982a) re-photographed the Cr I spectrum they were able to show that the giant resonance lies below threshold. At shorter wavelengths they were able to extend observations of the Rydberg series considerably. Figure 16 shows how accurately the ionization thresholds can be determined from Bruhn et al.'s spectrum. Some weak lines are also observed which are not predicted by the spectator model. In some cases these can be tentatively identified as resulting from breakdown of the $\Delta J_c = 1$ core selection rule. Bruhn et al. (1982a) supplemented their Cr I absorption measurements with photo-emission measurements.

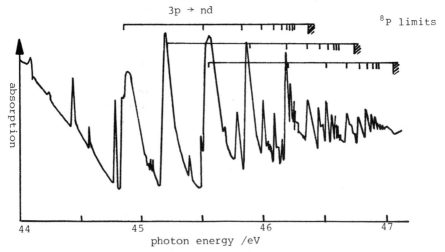

Fig. 16. The 3p-subshell spectrum of Cr I between 44 and 47 eV (after Bruhn et al. 1982a)

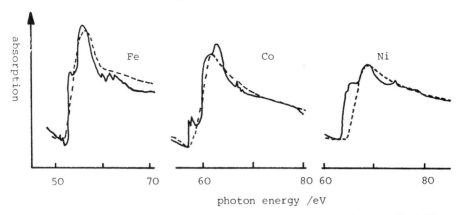

Fig. 17 The 3p-subshell spectra of Fe I, Co I and Ni I (solid lines) and of metallic Fe, Co and Ni (dashed lines) (after Bruhn et al. 1979)

The 3p-subshell photo absorption spectra of the last three atoms in the first transition row, Fe, Co and Ni (Bruhn et al. 1979) are compared with the corresponding metallic spectra in figure 17. In all three atoms short wavelength structure occurs in the atom which can be identified broadly as 3p → ns, nd lines converging on 3p thresholds although detailed analyses of these lines have not yet been attempted. Bruhn et al. compared the widths of the Fe and Co giant resonances with calculated $3p^5 3d^{N+1}$ multiplet splittings and found reasonable agreement. It should be noted that in all three elements non-ground state levels acquire substantial populations through thermal excitation at the temperatures used and that the atomic spectra are modified significantly by absorption from excited states.

We are now in a position to compare the 3p-subshell atomic absorption spectra of five consecutive first transition row elements, viz. from Z = 24 to 28. Discrete structure arising from transitions to nd (n > 3) levels is observed in all these spectra but tends to diminish in intensity with increasing Z. Discrete structure is particularly strong in the Cr I spectrum where it appears to take oscillator strength from the giant resonance. The giant resonance lies below the main 3p thresholds in all five atoms and its distance below threshold tends to increase with increasing Z. These trends are all consistent with increasing localization of the 3d state with increasing Z.

Comparisons between the atom and metallic spectra do not show the close correspondence which was noted in the lanthanides. This is consistent with the partly itinerant character of the d electron in the solid. It is to be expected that 3p → d resonances will be modified in the presence of a solid environment. The spectra of solid first transition row halides (Nakai et al. 1974) also show that the 3p → d resonances are modified by changes in the solid environment.

The 3p photo absorption spectra of the three lighter first transition row atoms, Sc (Z = 21), Ti (Z = 22) and V (Z = 23) have not yet been reported. On the evidence presented above 3p → 3d giant resonances in these atoms would be expected to be weaker and broader than in the heavier atoms and may also lie above the main 3p thresholds. The spectra should be very rich in discrete structure which may be complicated by 3d - 4s mixing.

5. 4p → d AND 5p → d RESONANCES IN THE SECOND AND THIRD TRANSITION ROWS

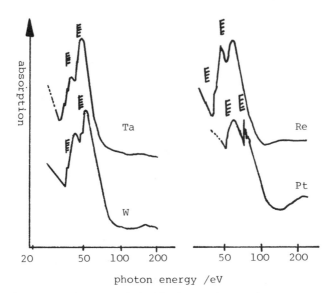

Fig. 18. The 5p-subshell photo absorption spectra of metallic Ta, W Re and Pt (after Haensel et al. 1969b)

Although 5p → d giant resonances have been observed in the photo absorption spectra of four metals of the third transition row (Haensel et al. 1969b) (figure 18) very little work has been done on 4p → d and 5p → d excitation in atoms of the second and third transition rows. The discussion will therefore be limited to the elements preceding these transition rows and to some very recent atomic absorption experiments on Y I (Z = 39) and Lu I (Z = 71) the first elements in the second and third rows.

4d collapse has been studied in the 4p-subshell atomic photo-absorption spectra of the elements which precede the second transition row. Rb (Beutler 1934, Connerade 1970, Mansfield 1973) and Sr (Mansfield and Newson 1981). Analyses of both spectra are complicated by effects which have already been mentioned in the discussion of the 3p-subshell spectra of K I and Ca I. These are the very large exchange interaction between the open core and the collapsed wavefunction which necessitates separate minimization of $4p^5$ 4d ^1P based levels and s - d configuration mixing, 4d - 5s mixing in this case. Furthermore the ζ_{4p} spin-orbit interaction is much larger than ζ_{3p} leading to a wider distribution of 4p → 4d oscillator strength. Analysis of the Rb I $4p^5$ 4d 5s configuration (Mansfield 1978) shows that there are many strong 4p → 4d lines but that none merits description as a giant resonance. In Sr I (Mansfield and Newson 1981) a group of very strong lines is observed in the expected position of the $4p^5$ 4d $5s^2$ ^1P line but in general the main Rb I and Sr I 4p → 4d lines are not as strong as their K I and Ca I 3p → 3d counterparts. d collapse does not seem to advance as quickly as it did in the first transition row.

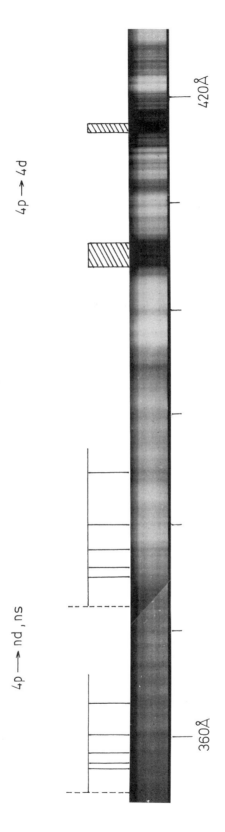

Figure 19. The 4p-subshell photoabsorption spectrum of YI.

109

The 4p-subshell photoabsorption spectrum of Y I is reproduced in figure 19. As far as I am aware this is the only data available at present which shows 4p → 4d excitation in the second transition row. Only a very tentative analysis has been performed so far. 4p → 4d oscillator strength seems to go into two main groups of lines. At shorter wavelengths two rather diffuse series emerge to limits separated by the spin-orbit splitting of the 4p core. Thermal excitation is expected to populate the Y I 4d $^2D_{\frac{3}{2},\frac{5}{2}}$ states to a similar extent and some lines in the observed spectrum appear to be doubled as a result.

On the evidence accumulated so far it seems reasonable to expect that the 4p photo absorption spectra of second row transition atoms will be richer in structure than the 3p spectra of first transition row elements and that the $4p^5$ spin-orbit splitting will be observable in the profiles of the giant resonances.

5d - 6s configuration mixing occurs quite strongly in the 5p photo-absorption spectrum of Ba I (Connerade et al. 1979) with the result that 5p → d oscillator strength is spread quite thinly amongst a very large number of lines. As the 4f-subshell fills 5d collapse is reversed to some extent and 5d - 6s mixing decreases (Tracy 1977, Connerade and Tracy 1977, Connerade et al. 1979). This finding is consistent with the results of Brewer (1971) who has shown that as the 4f-subshell fills in the lanthanide row the $5p^6\ 4f^N\ 6s\ 5d$ and $5p^6\ 4f^N\ 5d^2$ levels rise relative to the $5p^6\ 4f^N\ 6s^2$ levels. As 5d - 6s mixing decreases the 5p-subshell spectra simplify, the open 4f subshell electrons acting as spectators during the 5p-excitation process.

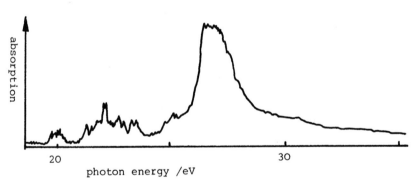

Fig. 20   The 5p-subshell photoabsorption spectrum of Sm I (after Tracy 1977)

From Z = 59 (Pr) upwards lanthanide 5p-subshell spectra are dominated by very strong asymmetric resonances which lie above the $5p^5\ ^2P_{\frac{3}{2}}$ based thresholds. The Sm I (Z = 62) spectrum is reproduced in figure 20 as an example of this type of spectrum. The position of the peak of the asymmetric resonance relates more to the $5p^5\ ^2P_{\frac{3}{2}}$ threshold than to the $5p^5\ ^2P_{\frac{1}{2}}$ level on which it is built. For the heaviest lanthanides (Z = 66 to 70) 5p → 5d excitation built on the $^2P_{\frac{1}{2}}$ level also produces strong lines as shown in the Yb I spectrum (figure 21). The analysis shown in figure 21 has been changed slightly from that of Tracy (1977). The 5p-subshell spectra of the heaviest lanthanides such as Yb I therefore show two giant resonances separated by the $5p^5$ spin-orbit spliting.

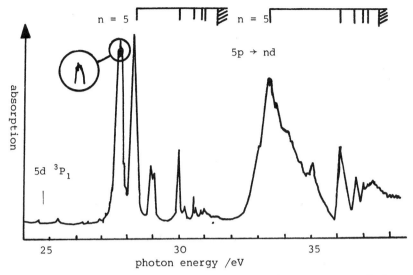

Fig. 21. The 5p-subshell photo-absorption spectrum of Yb I (after Tracy 1977)

Lu (Z = 71) is the only third transition row element for which an atomic photoabsorption spectrum has been recorded. The experiment was performed at low densities of Lu vapour and showed a number of broad diffuse absorption bands. Analysis is at a very early stage. The effects of 5d - 6s may re-appear at the beginning of the third transition row as the 5d wave function finally collapses.

The 5p-subshell spectra of solid Ta, W, Re and Pt (Haensel et al. 1969b) are presented in figure 18. They show giant resonances with double peaks which are separated by the spin-orbit splitting of the $5p^5$ core.

There is so little data available on giant resonances in the second and third transition rows that it is only possible to speculate on their appearance. The d states in the second and third transition rows are expected to be less well localized than in the first transition row. Moreover the spin-orbit interaction of the open p-subshell will be substantially greater. As a result it is likely that second and third row giant resonances will be broader than their first row counterparts and that they will be accompanied by more extensive and complex discrete structure.

6. CONCLUSION

A theme which has emerged consistently in the preceding sections is that the prominence of the discrete structure which accompanies giant resonances is related to the degree to which the final d or f state is localized. According to this picture excitation to a well-localized final state hardly disturbs the outer shells of the atom and produces little discrete structure apart from multiplet splittings in the well-localized state. Furthermore when the final state is well localized discrete structure is only slightly affected by changes in the atomic environment.

It has been shown that discrete structure does tend to diminish with increasing localization as described in the quasic-periodic table (Smith and Kmetko 1983) and that when the atomic core is excited the boundary between localized and itinerant behaviour shifts to lower Z elements.

REFERENCES

Amusia M. Ya., Dolmatov V.K. and Ivanov V.K. (1981), J. Phys. B: At Mol. Phys. 16, L753
Beutler H. (1934) Z. Phys. 91, 131
Beutler H. (1935) Z. Phys. 93, 177
Brewer L. (1971) J. Opt. Soc. Am. 61, 1666
Bruhn R., Sonntag B. and Woff H.W. (1978) Phys. Lett. 69A, 9
Bruhn R., Sonntag B. and Woff H.W. (1979) J. Phys. B.: At. Mol. Phys. 12, 203
Bruhn R., Schmidt E., Schröder H. and Sonntag B. (1982a) J. Phys. B: At. Mol. Phys. 15, 2807
Bruhn R., Schmidt E., Schröder H. and Sonntag B. (1982b) Phys. Lett. 90A, 41
Codling K. and Madden R.P. (1965) Appl. Opt. 4, 1431
Connerade J.P. (1970) Ap. J. 159, 685
Connerade J.P. (1982) New trends in atomic physics, Les Houches Session XXXVIII
Connerade J.P. (1984) J. Phys. B: At. Mol. Phys., 17, L165
Connerade J.P. and Mansfield M.W.D. (1974) Proc. R. Soc. Lond. A. 341, 267
Connerade J.P., Mansfield M.W.D. and Martin M.A.P. (1976) Proc. R. Soc. Lond. A350, 404
Connerade J.P., Mansfield M.W.D., Newsom G.H., Tracy D.H., Baig M.A. and Thimm K. (1979) Phil. Trans. R. Soc. Lond. A290, 327
Connerade J.P., Mansfield M.W.D., Cukier M. and Pantelouris M. (1980a) J. Phys. B: At. Mol. Phys. 13, L235
Connerade J.P., Pantelouris M., Baig M.A., Martin M.A.P. and Cukier M., (1980b) J. Phys. B: At. Mol. Phys. 13, L357
Connerade J.P. and Pantelouris M. (1984) J. Phys. B: At. Mol. Phys. 17, L173
Connerade J.P. and Tracy D.H. (1977) J. Phys. B: At. Mol. Phys., 10, L235
Cooper J.W. (1964) Phys. Rev. Lett. 13, 762
Cowan R.D. (1981) The Theory of Atomic Structure and Spectra, (California: University of California Press)
Davis L.C. and Feldkamp L.A. (1976) Solid St. Commun. 19, 413
David L.C. and Feldkamp L.A. (1978) Phys. Rev. A. 17, 2012
Dehmer J.L. (1972) J. Chem. Phys. 56, 4496
Dehmer J.L., Starace A.F., Fano U., Sugar J. and Cooper J.W. (1971) Phys. Rev. Lett. 26, 1521
Dehmer J.L. and Starace A.F. (1972) Phys. Rev. B. 5, 1792

Ederer D.L. (1964) Phys. Rev. Lett. 13, 760
Ederer D.L., Lucatorto T.B., Saloman E.B., Madden R.P. and Sugar J. (1975) J. Phys. B: At. Mol. Phys. 8, L21
Fano U. and Cooper J.W. (1968) Rev. Mod. Phys. 40, 441
Fano U. and Cooper J.W. (1969) Rev. Mod. Phys. 41, 724
Fliflet A.W., Kelly H.P. and Hansen J.E. (1975) J. Phys. B: At. Mol. Phys. 12, L268
Fomichev V.A. Zimkina T.M., Gribovskii C.A. and Zhukova I.I. (1967) Sov. Phys. - Solid State 9, 1163
Fuggle J.C., Burr A.F., Watson L.M., Fabian D.T. and Lang W. (1974) J. Phys. F: Met. Phys. 4, 335
Garvin L.J., Brown E.R., Carter S.L. and Kelly H.P. (1983) J. Phys. B: At. Mol. Phys. 16, L269
Griffin D.C., Andrew K.L. and Cowan R.D. (1969) Phys. Rev. 177, 62
Haensel R. Keitel G., Schreiber P. and Kunz C. (1969a) Phys. Rev. 188, 135
Haensel R., Radler K., Sonntag B. and Kunz C. (1969b) Sol. St. Commun. 7, 1495
Haensel R. Rabe P. and Sonntag B. (1970) Sol. St. Commun. 8, 1845
Hansen J.E., Fliflet A.W. and Kelly H.P. (1975) J. Phys. B: At. Mol. Phys. 8, L127
Huffman R.E., Tanaka Y. and Larabee J.C. (1963) J. Chem. Phys. 39, 902
Koberle H. and Kung Th. (1985) Diplomarbeiten, Universität Hamburg
Lukirskii A.P. and Zimkina T.M. (1963) Bull. Acad. Sci. USSR Phys. Ser. 27, 808
Madden R.P. and Codling K. (1964) Phys. Rev. Lett. 12, 106
Mansfield M.W.D. (1973) Ap. J. 183, 691
Mansfield M.W.D. (1975a) Proc. R. Soc. Lond. A346, 539
Mansfield M.W.D. (1975b) Proc. R. Soc. Lond. A346, 555
Mansfield M.W.D. (1976) Proc. R. Soc. Lond. A348, 143
Mansfield M.W.D. (1977) Proc. R. Soc. Lond. A358, 253
Mansfield M.W.D. (1978) Proc. R. Soc. Lond. A 364, 135
Mansfield M.W.D. and Connerade J.P. (1976) Proc. R. Soc. Lond. A 352, 125
Mansfield M.W.D. and Connerade J.P. (1982) J. Phys. B: At. Mol. Phys. 15, 503
Mansfield M.W.D. and Newsom G.H. (1977) Proc. R. Soc. Lond. A 357, 77
Mansfield M.W.D. and Newsom G.H. (1981) Proc. R. Soc. Lond. A 377, 431
Nakai S., Nakamori H., Tomita A., Tsutumi K., Nakamura H. and Sugiura C. (1974) Phys. Rev. B. 9, 1870
Nakamura M., Sasamuma M., Sato S., Watanabe M., Yamashita H., Iguchi Y., Ejiri A., Nakai S., Yamaguchi S., Sagawa T., Nakai Y. and Oshio T. (1968) Phys. Rev. Lett. 21, 1303
Pantelouris M. and Connerade J.P. (1983) J. Phys. B: At. Mol. Phys. 16, L23
Petersen H., Radler K., Sonntag B. and Haensel R. (1975) J. Phys. B: At. Mol. Phys. 8, 31
Prescher Th., Richter M., Schmidt E., Sonntag B. and Wetzel H.E. (1986) to be published
Rabe P., Radler K. and Wolff H.W. (1974) IV International Conference on Vacuum Ultraviolet Radiation Physics Hamburg. (Pergamon, Braunschweig)
Radtke E.-R. (1979a) J. Phys. B: At. Mol. Phys. 12, L71
Radtke E.-R. (1979b) J. Phys. B: At. Mol. Phys. 12, L77
Samson J.A.R. (1966) J. Opt. Soc. Am. 55, 935
Smith J.L. and Kmetko E.A. (1983) J. Less-Common Met. 90, 83
Sonntag B., Haensel R. and Kunz C. (1969) Sol. State Comm. 7, 597
Starace A.F. (1972) Phys. Rev. B. 5, 1773
Starace A.F. (1974) J. Phys. B: At. Mol. Phys. 7, 14

Sugar J. (1972) Phys. Rev. B $\underline{5}$, 1785
Suzuki S., Ishii, T. and Sagawa T. (1975) J. Phys. Soc. Japan $\underline{38}$, 156
Tracy D.H. (1977) Proc. R. Soc. Lond. $\underline{A357}$, 485
Wendin G. (1973a) Phys. Lett. $\underline{46}$ A, 101
Wendin G. (1973b) Phys. Lett. $\underline{46}$ A, 119
Wendin G and Del Grande N.K. (1985) Phys. Script. $\underline{32}$, 286
Wendin G. and Starace A.F. (1978) J. Phys. B: At. Mol. Phys. $\underline{11}$, 4119
Woff H.W., Bruhn R., Radler K. and Sonntag B. (1976) Phys. Lett. $\underline{59A}$, 67
Zimkina T.M. Fomichev V.A. Gribovskii S.A. and Zhukova I.I. (1967)
    Sov. Phys. - Solid State $\underline{9}$, 1128

GIANT RESONANCES IN THE 4d SUBSHELL PHOTOABSORPTION SPECTRA OF Ba, Ba^+ and Ba^{++}

K. Nuroh*, E. Zaremba and M.J. Stott
Department of Physics
Queen's University
Kingston, Ontario
K7L 3N6 Canada

ABSTRACT

Evidence of giant resonances in the ionized series Ba, Ba^+, Ba^{++} has been shown in the calculations of the photoabsorption spectra near the 4d ionization threshold by employing the time-dependent local density approximation (TDLDA). It is shown that for positive excitation frequencies, TDLDA is essentially equivalent to the conventional random phase approximation (RPA) which has been used so successfully in calculating photoabsorption of closed shell atoms.

1. Introduction

As early as around the mid-1930's the discussion of collective excitations in finite systems had cropped up[1,2] to explain the inelastic scattering events of electrons by atoms. The atom was then modelled as a spherical fluid capable of sustaining plasmon (charge density) oscillations, just as the well established situation in some extended systems. This conjecture seemed to have been abandoned until it was revisited by Brandt and Lundquist[3]. The idea was supported with realistic calculations in terms of the charge density of the atom and the appropriate response functions. This was to be continued by Amusia and co-workers[4] on the one hand and Wendin[5] on the other with model calculations. It had also been observed that models of independent one electron excitations used so successfully in explaining experimental data[6,7] involving deeper core states had to give way to many-body

---

*Present address :
Department of Mathematics, University of Benin,
Benin City, Nigeria

techniques when the excitations involved electronic states of the intermediate subshells of atoms.

A lot of such many-body techniques had been 'borrowed' from nuclear physics and concepts like renormalization were being applied to atomic system[8]. Naturally the concept of 'giant resonances' used in the description of collective excitations in cross sections of nuclei[9] found its way in atomic physics[5], because of the similarities between nuclear and atomic excitations in terms of the shell model of the atom.

There are abundance of theoretical models in the literature[10] which describe the general spectral features quite well. There are, however, conflicting opinions when it comes to interpreting the spectra. We describe one such model based on the theory or linear response and the density functional formalism.

2. Linear Response and TDLDA

The key quantity which enters our subsequent discussion is the density of an arbitrary system of electrons. Given such a system one would like to determine, for instance, its ground state density $n^o(r)$. Also if the system is externally perturbed, it will be desired to have a knowledge of the deviation of the density $\delta n(r)$ from its ground state value. The ground state properties are effectively studied using the Kohn-Sham (KS) density functional formalism[11] in which the ground state energy of the system is shown to be a functional of the density, $E^o[n]$, and attains its minimum value for the correct ground state density $n^o(\vec{r})$. Upon application of an external perturbation $V^{ext}(\vec{r})$ which couples to the electronic density, the ground state energy functional assumes a new one

$$E[n] = E^o[n] + \int d\vec{r}\, n(\vec{r})\, V^{ext}(\vec{r}) \tag{2.1}$$

This functional is then minimized by the density $n(\vec{r})$ which is related to the ground state according to

$$n(\vec{r}) = n^o(\vec{r}) + \delta n(\vec{r}) \tag{2.2}$$

In the KS approach, ground-state properties are obtained by solving a Hartree-like equation (atomic units throughout, $e = m = \hbar = 1$)

$$[-\tfrac{1}{2}\nabla^2 + V^{eff}(\vec{r})]\, \varphi_i(\vec{r}) = \varepsilon_i\, \varphi_i(\vec{r}) \tag{2.3}$$

The density is then constructed from the set of single-particle orbitals $\varphi_i(\vec{r})$ according to

$$n(\vec{r}) = \sum_{i\,(occ)} |\varphi_i(\vec{r})|^2 \tag{2.4}$$

The effective potential in (2.3) is a functional of the density and is given by :

$$V^{eff}(\vec{r}) = V^{ext}(\vec{r}) + V(\vec{r}) + \int d\vec{r}\,' \frac{n(\vec{r}\,')}{|\vec{r}-\vec{r}\,'|} + \frac{\delta E_{xc}^0[n]}{\delta n(\vec{r})} \tag{2.5}$$

$$= V^{ext}(\vec{r}) + V(\vec{r}) + \phi(\vec{r}) + V_{xc}(\vec{r})$$

Here $\phi(\vec{r})$ is the electronic Hartree potential and $V_{xc}(\vec{r})$ is the exchange-correlation potential obtained from the assumed exchange-correlation energy functional $E_{xc}^0[n]$.

In the absence of $V^{ext}(r)$ denote the effective potential by $V_0^{eff}(\vec{r})$ and given by

$$V_0^{eff}(\vec{r}) = V(\vec{r}) + \phi^0(\vec{r}) + V_{xc}^0(\vec{r}) \tag{2.6}$$

Then if the applied field $V^{ext}(\vec{r})$ is sufficiently small $\delta n(\vec{r})$ can be assumed small and a perturbative expansion of (2.5) can be made in terms of $\delta n(\vec{r})$. Writing

$$V^{eff}(\vec{r}) = V_0^{eff}(\vec{r}) + V^{SCF}(\vec{r}), \tag{2.7}$$

the change $V^{SCF}(\vec{r})$ in the effective potential to the lowest order in $V^{ext}(\vec{r})$ is then given by

$$V^{SCF}(\vec{r}) = V^{ext}(\vec{r}) + \int d\vec{r}\,' \frac{\delta n(\vec{r}\,')}{|\vec{r}-\vec{r}\,'|} + \int d\vec{r}\,' V'_{xc}(\vec{r},\vec{r}\,')\delta n(\vec{r}\,') \tag{2.8}$$

where

$$V'_{xc}(\vec{r},\vec{r}\,') = \left.\frac{\delta^2 E_{xc}[n]}{\delta n(\vec{r})\delta n(\vec{r}\,')}\right|_{n=n^0(\vec{r})} \tag{2.9}$$

A popular approximation to the exchange-correlation energy is the local density approximation (LDA), in which V'_{xc} is simply a function of the density and (2.9) becomes

$$V'_{xc}(\vec{r},\vec{r}\,')_{LDA} = \left.\frac{dV_{xc}(n)}{dn}\right|_{n=n^0(\vec{r})} \delta(\vec{r}-\vec{r}\,') \tag{2.10}$$

Thus for the applied external field has been considered to be time-independent and hence the induced quantities $\delta n(\vec{r})$ and $v^{CSF}(\vec{r})$ are static as well. If we now ask for time-dependence of the applied external field, the discussion above is still valid except that (2.8) be modified to exhibit time-dependence, namely

$$V^{SCF}(\vec{r},t) = V^{ext}(\vec{r},t) + \int d\vec{r}' \frac{\delta n(\vec{r}',t)}{|\vec{r}-\vec{r}'|} + \int d\vec{r}' V'_{xc}(\vec{r},\vec{r}')\delta n(\vec{r}',t) \qquad (2.11)$$

The ensuing discussion will be based on (2.11) where the LDA version of the exchange-correlation energy (i.e. (2.10) is used. As such the formulation is termed time-dependent local density approximation (TDLDA).

The induced density can be viewed as the response of the KS system to the self-consistent potential (2.11). Thus it can be expressed within the framework of linear response theory[12] as

$$\delta n(\vec{r},t) = \int d\vec{r}' \chi^0(\vec{r},\vec{r}',t) V^{SCF}(\vec{r}',t) \qquad (2.12)$$

where $\chi^0(\vec{r},\vec{r}',t)$ is the non-interacting time-dependent density response function to be constructed from the single particle orbitals of (2.3) (with $V_0^{eff}(\vec{r})$ replacing $V^{eff}(\vec{r})$ to constitute the inhomogeneous KS system. In this picture all the many-body effects are contained in $V^{SCF}(\vec{r},t)$. Indeed (2.11) and (2.12) constitute an integral equation for $\delta n(\vec{r},t)$ which when solved determines $V^{SCF}(\vec{r},t)$ via (2.11). However, this can be effectively done if an explicit expression for the response function is known. It then becomes convenient to work with frequency dependent quantities. Thus (2.11) and (2.12) are Fourier transformed in time to obtain respectively

$$V^{SCF}(\vec{r},\omega) = V^{ext}(\vec{r},\omega) + \int d\vec{r}' \frac{\delta n(\vec{r}',\omega)}{|\vec{r}-\vec{r}'|} + \int d\vec{r}' V'_{xc}(\vec{r},\vec{r}')\delta n(\vec{r}',\omega) \qquad (2.13)$$

and

$$\delta n(\vec{r},\omega) = \int d\vec{r}' \chi^0(\vec{r},\vec{r}',\omega) V^{SCF}(\vec{r}',\omega) \qquad (2.14)$$

Equations (2.13) and (2.14) provide the ω-dependent integral equation for the induced density.

3. Response functions for spherically symetric system

The response function in (2.12) or (2.14) is the retarded density-density response function $D^R(\vec{r},\vec{r}',\omega)$ which is complex. Thus we may write

$$\chi^0(\vec{r},\vec{r}',\omega) = \text{Re}D^R(\vec{r},\vec{r}',\omega) + i\,\text{Im}D^R(\vec{r},\vec{r}',\omega) \qquad (3.1)$$

It is, however, convenient for analytical purposes to deal with the time-ordered response function $D^T(\vec{r},\vec{r}',\omega)$ which is related to the retarded counterpart according to [13]

$$\text{Re}D^R(\vec{r},\vec{r}',\omega) = \text{Re}D^T(\vec{r},\vec{r}',\omega) \qquad (3.2)$$

and

$$\text{Im}D^R(\vec{r},\vec{r}',\omega) = \text{sgn}(\omega)\,\text{Im}D^T(\vec{r},\vec{r}',\omega) \qquad (3.3)$$

For a system of N non-interacting particles the time ordered response function can be expressed in terms of time-ordered Green's functions as [13]

$$D^T(\vec{r},\vec{r}',t-t') = -2i\,G(\vec{r},t;\vec{r}',t')G(\vec{r}',t';\vec{r},t)$$
$$= -2i\,G(\vec{r},\vec{r}';t-t')G(\vec{r}',\vec{r},t'-t) \qquad (3.4)$$

The factor 2 appears since G is taken to be diagonal in the particle spin. The time Fourier transform of (3.4) gives

$$D^T(\vec{r},\vec{r}',\omega) = -i\int_{-\infty}^{\infty}\frac{d\omega'}{\pi}\,G(\vec{r},\vec{r}',\omega')G(\vec{r}',\vec{r},\omega'-\omega) \qquad (3.5)$$

For a fermion systems of particles the time-ordered Green's function is expressible in the form:

$$G(\vec{r},\vec{r}',\omega) = \sum_j \varphi_j^*(\vec{r}')\varphi_j(\vec{r})\left[\frac{\theta(\varepsilon_j-\mu)}{\omega-\varepsilon_j+i\delta} + \frac{\theta(\mu-\varepsilon_j)}{\omega-\varepsilon_j-i\delta}\right] \qquad (3.6)$$

where μ is the Fermi level. It has also been shown[14] that the fermion Green's function can be expressed in terms of retarded (G^+) and advanced (G^-) one particle Green's functions

$$G(\vec{r},\vec{r}',\omega) = G^+(\vec{r},\vec{r}',\omega)\theta(\omega-\mu) + G^-(\vec{r},\vec{r}',\omega)\theta(\mu-\omega) \qquad (3.7)$$

where G^+ and G^- have their usual spectral representations

$$G^\pm(\vec{r},\vec{r}',\omega) = \sum_j \frac{\varphi_j(\vec{r})\varphi_j^*(\vec{r}')}{\omega-\varepsilon_j\pm i\delta} \qquad (3.8)$$

The final expression for $D^T(\vec{r},\vec{r}',\omega)$ we intend to derive may be obtained by substituting for $G(\vec{r},\vec{r}',\omega)$ from either (3.6) or (3.7) into (3.5). However, the derivation via (3.7) is rather cumbersome and will not be discussed here but will be reported elsewhere.

Substitution of (3.6) into (3.5) will result essentially in contour integration over four separate terms. In two of these terms, the simple poles will all lie in the upper half or lower half of the complex ω'-plane. As such the contour is closed in the opposite half plane to give no contribution to D^T. If the residue theorem is then applied to the two remaining terms we get

$$D^T(\vec{r},\vec{r}',\omega) = 2 \sum_{ij} \frac{\varphi_i(\vec{r})\varphi_i^*(\vec{r}')\varphi_j(\vec{r}')\varphi_j^*(\vec{r})}{\omega + \epsilon_j - \epsilon_i + i\delta} \theta(\epsilon_i-\mu)\theta(\mu-\epsilon_j)$$

$$+ \frac{\varphi_i(\vec{r})\varphi_i^*(\vec{r}')\varphi_j(\vec{r}')\varphi_j^*(\vec{r})}{\epsilon_i - \omega - \epsilon_j + i\delta} \theta(\mu-\epsilon_i)\theta(\epsilon_j-\mu)] \quad (3.9)$$

By interchanging the dummy indices in the first term of (3.9) and using the relation

$$f_i = \theta(\mu - \epsilon_i), \quad 1-f_i = \theta(\epsilon_i - \mu) \quad (3.10)$$

between the Fermi occupation numbers f_i and the unit step function θ, (3.9) takes the form

$$D^T(\vec{r},\vec{r}',\omega) = 2 \, [\sum_i f_i \varphi_i^*(\vec{r})\varphi_i(\vec{r}') \sum_j \frac{\varphi_j^*(\vec{r}')\varphi_j(\vec{r})}{\epsilon_i+\omega-\epsilon_j+i\delta}$$

$$+ \sum_i f_i \varphi_i(\vec{r})\varphi_i^*(\vec{r}') \sum_j \frac{\varphi_j^*(\vec{r})\varphi_j(\vec{r}')}{\epsilon_i-\omega-\epsilon_j+i\delta}] \quad (3.11)$$

Finally, according to (3.1)-(3.3) and (3.8), (3.1) takes the form

$$\chi^0(\vec{r},\vec{r}',\omega) = 2 \sum_{\substack{i \\ (occ)}} [\varphi_i^*(\vec{r})\varphi_i(\vec{r}')G^+(\vec{r}',\vec{r},\epsilon_i+\omega)$$

$$+ \varphi_i(\vec{r})\varphi_i^*(\vec{r}')G^+(\vec{r},\vec{r}',\epsilon_i-\omega)] \quad (3.12)$$

The usefulness of (3.12) for χ^0 lies in getting an accurate expression for the retarded Green's function G^+. Instead of using expression (3.8) for G^+, it can directly be obtained as the solution of the differential equation

$$[-\tfrac{1}{2}\nabla^2 + V_0^{eff}(\vec{r}) - E]G^+(\vec{r},\vec{r}',E) = -\delta(\vec{r} - \vec{r}') \quad (3.13)$$

with the appropriate out-going wave boundary conditions.

3.1. Response function for Spherical Symmetry

In application to systems which exhibit spherical symmetry, it becomes convenient to work with sperical harmonics. We concentrate for the moment on the first term of the response function (3.12) and write

$$\chi^+(\vec{r},\vec{r}',\omega) = \sum_{\substack{i \\ (occ)}} \varphi_i^*(\vec{r})\varphi_i(\vec{r}')G^+(\vec{r},\vec{r}',\epsilon_i+\omega) \quad (3.14)$$

Since the spherical harmonics form a complete set, the single particle orbitals may be represented as

$$\varphi_i(\vec{r}) = \frac{1}{r} U_{n_i \ell_i}(r) Y_{L_i}(\hat{r}) \qquad (3.15)$$

Here L is equivalent ot the set (ℓm) of the angular and magnetic quantum numbers. Accordingly both χ^+ and G^+ will be functions of the radial variables r, r' and the angle between the vectors $\vec{r}$ and $\vec{r}'$, and may be expressed respectively as

$$\chi^+(\vec{r},\vec{r}',\omega) = \sum_L \chi_L^+(r,r',\omega) Y_L(\hat{r}) Y_L^*(\hat{r}') \qquad (3.16)$$

and

$$G^+(\vec{r},\vec{r}',\omega) = \sum_L G_\ell^+(r,r',\omega) Y_L(\hat{r}) Y_L^*(\hat{r}') \qquad (3.17)$$

If the expansions (3.16) and (3.17) and the representation (3.15) are put in (3.14) we obtain

$$\chi_\ell^+(r,r',\omega) = 2\pi \sum_i^{(occ)} \frac{U_{n_i \ell_i}(r)}{r} \frac{U_{n_i \ell_i}(r')}{r'} \sum_{\ell'} C_{\ell_i \ell' \ell} G_{\ell'}^+(r,r',\varepsilon_i + \omega) \qquad (3.18)$$

with

$$C_{\ell_i \ell' \ell} = \frac{(2\ell_i+1)}{4\pi} \frac{(2\ell'+1)}{4\pi} \int_{-1}^{1} dx\, P_{\ell_i}(x) P_{\ell'}(x) P_\ell(x) \qquad (3.19)$$

The integral of the product of three Legendre polynomials is related to the Clebsch-Gordon coefficients which in turn is related to 3-j symbol as

$$\tfrac{1}{2} \int_{-1}^{1} dx\, P_\ell(x) P_{\ell'}(x) P_{\ell''}(x) = \begin{pmatrix} \ell & \ell' & \ell'' \\ 0 & 0 & 0 \end{pmatrix}^2 \qquad (3.20)$$

In particular with $\ell''=1$, using a formula by Gaunt[15] or the expansion of the 3-j symbols[16] yields the result

$$\tfrac{1}{2} \int_{-1}^{1} dx\, P_\ell(x) P_{\ell'}(x) P_1(x) = \begin{pmatrix} \ell & \ell' & 1 \\ 0 & 0 & 0 \end{pmatrix}^2$$

$$= \frac{\ell+1}{(2\ell+1)(2\ell+3)} \delta_{\ell', \ell+1} + \frac{\ell}{(2\ell-1)(2\ell+1)} \delta_{\ell', \ell-1} \qquad (3.21)$$

On using (3.21), expression (3.19) becomes

$$C_{\ell_i \ell' 1} = \frac{1}{8\pi^2} [(\ell_i+1)\delta_{\ell', \ell_i+1} + \ell_i \delta_{\ell', \ell_i-1}] \qquad (3.22)$$

We note that if the second term of (3.12) is included, we come out with the same coefficient (3.19) and similar forms for the radially dependent quantities as in (3.18). Consequently, taking the dipole ($\ell=1$) component of (3.18) and combining with (3.12) using (3.22), we finally obtain

$$\chi_1^0(r,r',\omega) = \frac{1}{2\pi} \sum_{i \atop (occ)} \frac{U_{n_i \ell_i}(r)}{r} \frac{U_{n_i \ell_i}(r')}{r'} \qquad (3.23)$$

$$\times [\{ \ell_i G^+_{\ell_i-1}(r',r,\varepsilon_i+\omega) + (\ell_i+1)G^+_{\ell_i+1}(r',r,\varepsilon_i+\omega) \}$$
$$+ \{ \ell_i G^+_{\ell_i-1}(r',r,\varepsilon_i-\omega) + (\ell_i+1)G^+_{\ell_i+1}(r',r,\varepsilon_i-\omega) \}]$$

Of course, in arriving at (3.23) we have assumed the multipole expansion of the response function $\chi^0(r,r',\omega)$. From (3.8) and (3.15) we obtain the spectral representation

$$G^+_\ell(r,r',E) = \sum_j \delta_{\ell,\ell_j} \frac{1}{rr'} \frac{U_{n_j \ell_j}(r) U_{n_j \ell_j}(r')}{E - \varepsilon^0_{n_j \ell_j} + i\delta} \qquad (3.24)$$

As has been observed earlier, the arduous task of performing the infinite summation over the single-particle radial orbitals can be circumvented since $G^+_\ell(r,r',E)$ is the solution of the inhomogeneous radial equation

$$[-\frac{1}{2r^2}\frac{d}{dr}(r^2 \frac{d}{dr}) + \frac{\ell(\ell+1)}{2r^2} + V_0^{eff}(r) - E] G^+_\ell(r,r',E) = -\frac{1}{r^2}\delta(r-r') \qquad (3.25)$$

which satisfies the appropriate boundary conditions at infinity and the origin. Following the earlier observations[14], if E corresponds to a bound state energy then $G^+_\ell(r,r',E)$ can be expressed in terms of solutions to the radial homogeneous equation at energy $E = k^2/2$:

$$[-\frac{d^2}{dr^2} + \frac{\ell(\ell+1)}{r^2} + 2 V_0^{eff}(r) - k^2] U_{\ell k}(r) = 0 \qquad (3.26)$$

Then the Green's function is given by

$$G^+_\ell(r,r',\omega) = \frac{2}{W[\phi_{\ell k}, \chi^{(1)}_{\ell k}]} \frac{\phi_{\ell k}(r) \chi^{(1)}_{\ell k}(r')}{r r'} \qquad (3.27)$$

Here $\chi^{(1)}_{\ell k}(r)$ is the solution to (3.26) which behaves asymptotically for $r \to \infty$ as $r h^{(1)}_\ell(kr)$ and $\phi_{\ell k}(r)$ the solution which is regular at the origin, and W refers to the Wronskian of the two solutions.

If E does not correspond to a bound state energy, (3.27) is further simplified by normalizing $\phi_{\ell k}(r)$ such that it behaves asymptotically as $r[\gamma h^{(1)}_\ell(kr) + h^{(2)}_\ell(kr)]$. In this case we obtain

$$G^+_\ell(r,r',E) = -ik\, R_\ell(r;k) R^{(1)}_\ell(r';k) \qquad (3.28)$$

where

$$R_\ell(r;k) = \phi_{\ell k}(r)/r; \quad R^{(1)}_\ell(r;k) = \chi^{(1)}_{\ell k}(r)/r \qquad (3.29)$$

For **negative energies**, k is purely imaginary and $G_\ell^+(r,r',\omega)$ is **real**. We thus have a prescription for calculating the dipolar response function (3.23) exactly by essentially getting a numerical solution of (3.26).

3.2. Atomic dipole polarizability and absorption cross-section

The application of a uniform frequency dependent electric field $\vec{E}(\omega)$ corresponds to the external potential

$$V^{ext}(\vec{r},\omega) = \vec{E}(\omega)\cdot\vec{r} = E(\omega)\cdot\frac{4\pi}{3} r \sum_{m=-1}^{+1} Y_{1m}(\hat{r}) Y_{1m}^*(\hat{E}) \quad (3.30)$$

For a spherical atom, the induced density can be expressed as

$$\delta n(\vec{r},\omega) = -\alpha(\vec{r},\omega)\vec{E}(\omega)\cdot\vec{r} = -E(\omega)\cdot\frac{4\pi}{3}\alpha(r,\omega)\sum_{m=-1}^{+1} Y_{1m}(\hat{r})Y_{1m}^*(\hat{E}) \quad (3.31)$$

Using these results in (2.14) we obtain the linear integral equation for the atomic dipole polarizability $\alpha(r,\omega)$ as

$$\alpha(r,\omega) = -\int_0^\infty dr' r'^3 \chi_1^0(r,r',\omega)$$
$$+ \int_0^\infty dr' r'^2 \int_0^\infty dr'' r''^2 \chi_1^0(r,r',\omega)\gamma_1(r',r'')\alpha(r'',\omega) \quad (3.32)$$
$$+ \int_0^\infty dr' r'^2 \int_0^\infty dr'' r''^2 \chi_1^0(r,r',\omega)V_{xc,1}^-(r',r'')\alpha(r'',\omega)$$

where

$$\gamma_\ell(r,r') = \frac{4\pi}{2\ell+1}\frac{r_<^\ell}{r_>^{\ell+1}} \quad (3.33)$$

and in general we have that

$$V_{xc,\ell}^-(r,r') = \int d\hat{r} \int d\hat{r}' Y_L^*(\hat{r}) V_{xc}^-(\vec{r},\vec{r}') Y_L(\hat{r}') \quad (3.34)$$

However, if the LDA is invoked via (2.10), the dipole component is given by

$$V_{xc,1}^-(r,r') = \frac{1}{r^2}\delta(r-r')\frac{dV_{xc}(n)}{dn}\bigg|_{n=n^0(r)} \quad (3.35)$$

On the other hand the application of the perturbation (3.30) gives rise to the induced dipole moment of the atom

$$\vec{p}(\omega) = -\int d\vec{r}\, \vec{r}\, \delta n(\vec{r},\omega) \quad (3.36)$$

Using (3.31) we infer from (3.36) that

$$\vec{p}(\omega)\cdot\vec{E}(\omega) = E^2(\omega)\int dr\, r\, \alpha(r,\omega)(\hat{r}\cdot\hat{E})^2 \quad (3.37)$$

But the dynamic polarizability $\alpha(\omega)$ is related to the induced dipole moment and the applied electric field by

$$\vec{p}(\omega) = \alpha(\omega) \vec{E}(\omega) \tag{3.38}$$

Substituting this value of $\vec{p}(\omega)$ in (3.37) gives

$$\alpha(\omega) = \frac{4\pi}{3} \int_0^\infty dr\, r^3 \alpha(r,\omega) \tag{3.39}$$

Finally, the photoabsorption cross section is calculated as

$$\sigma(\omega) = 4\pi \frac{\omega}{c} \operatorname{Im}\alpha(\omega) = 4\pi \alpha_f a_0^2 \left(\frac{\omega}{2}\right) \operatorname{Im}\alpha(\omega) \tag{3.40}$$

Here α_f is the fine structure constant and a_0 is the Bohr radius. If the input frequency ω is in Rydberg unit of energy then the final form of (3.40) gives the cross section in megabarns (Mb).

3.3. Partial cross section

By virtue of (3.23), (3.32) and (3.39) the photoabsorption cross section (3.40) contains contributions from all electronic bound states of the atom consistent with the dipole excitations. For some atomic systems and in energy regions around certain excitation thresholds it is found that particular isolated excitations dominate the spectrum. Then only small differences in the cross section are observed if it is calculated with contributions from all bound states or when all other bound states are frozen with the exception of one at the level of (3.23) to give the partial cross section. The 4d→nf and 5p→nd transitions[17] in the rare earth atoms are good examples of such a situation. It is thus wortwhile to investigate the underlying physics with such an assumption in mind.

Suppose we make the following ansatz about the effective driving field, namely

$$V^{SCF}(\vec{r},\omega) = P_1(\cos\theta)\, V^{SCF}(r,\omega) \tag{3.41}$$

then the induced density (2.14) becomes

$$\delta n(\vec{r},\omega) = P_1(\cos\theta) \int dr'\, r'^2 \chi_1^0(r,r',\omega) V^{SCF}(r',\omega) \tag{3.42}$$

and (3.24) for the response function can be taken as

$$\chi_1^0(r,r',\omega) \sim \sum_{\substack{i \\ (\text{occ})}} \frac{U_{n_i \ell_i}(r)}{r} \frac{U_{n_i \ell_i}(r')}{r'} G^+_{\ell_{i+1}}(r,r',\varepsilon_i - \omega) \tag{3.43}$$

Substitution of (3.43) in (3.42) and using the spectral representation of G^+ then gives for the induced density

$$\delta n(\vec{r},\omega) = \frac{P_1(\cos\theta)}{r^2} \sum_{\substack{j \\ (occ)}} \sum_n U_{n_j\ell_j}(r) U_{n\ell_{j+1}}(r) \frac{\int dr' U_{n_j\ell_j}(r') V^{SCF}(r',\omega) U_{n\ell_{j+1}}(r')}{\varepsilon_j - \varepsilon_n - \omega + i\delta} \quad (3.44)$$

In particular if 4d→nf transitions are dominating (3.44) becomes

$$\delta n_{4d \to f}(\vec{r},\omega) = \frac{P_1(\cos\theta)}{r^2} \sum_n U_{4d}(r) U_{nf}(r) \frac{\int dr' U_{4d}(r') V^{SCF}(r',\omega) U_{nf}(r')}{\varepsilon_{4d} - \varepsilon_{nf} - \omega + i\delta} \quad (3.45)$$

Alternatively, we may rewrite (3.45) as

$$\delta n_{4d \to f}(\vec{r},\omega) = \frac{P_1(\cos\theta)}{r^2} \sum_n C_{nf}(\omega) U_{4d}(r) U_{nf}(r)$$

$$= \frac{P_1(\cos\theta)}{r^2} \alpha_{4d \to f}(r,\omega) \quad (3.46)$$

Here the frequency dependent complex coefficients $C_{nf}(\omega)$ are defined by ($\omega_{nf} = \varepsilon_{4d} - \varepsilon_{nf}$)

$$C_{nf}(\omega) = \frac{<nf|V^{SCF}(r',\omega)|4d>}{\omega_{nf} - \omega + i\delta} = \int_0^\infty dr \frac{\alpha_{4d \to f}(r,\omega)}{U_{4d}(r)} U_{nf}(r) \quad (3.47)$$

On the other hand, for a single 4d→f* transition, $\alpha_{4d \to f}$ would have the form

$$\alpha_{4d \to f^*}(r,\omega) = U_{4d}(r) U_{f^*}(r) \frac{<f^*|r'|4d>}{\omega^*_f - \omega + i\delta} \quad (3.48)$$

Comparison of (3.46) and (3.47) identifies $U_{f^*}(r)$ as

$$U_{f^*}(r) = \frac{1}{N(\omega)} \sum_n U_{nf}(r) \, \text{Im} C_{nf}(\omega) \quad (3.49)$$

where $N(\omega)$ is a normalization factor. The normalized hybridized wave function (3.49) corresponds to a one-dimensional Schrödinger equation from which can be constructed the effective radial potential

$$V_{f^*}(r,\omega_{f^*}) - \omega_{f^*} = (U_{f^*})^{-1} \tfrac{1}{2} d^2 U_{f^*}/dr^2 \quad (3.50)$$

The above analysis will be used in discussing some features in the 4d-edge excitation spectrum of Ba^{++} in the next section.

4. Application to photoionization in Ba, Ba$^+$ and Ba^{++}

A deeper insight into the dynamics of photoionization is perceived if (2.11) is rewritten as

$$V^{SCF}(\vec{r},t) = V^{ext}(\vec{r},t) + g \int d\vec{r}' V(\vec{r},\vec{r}') \delta n(\vec{r}',t) \quad (4.1)$$

where

$$V(\vec{r},\vec{r}') = \frac{1}{|\vec{r}-\vec{r}'|} + V'_{xc}(\vec{r},\vec{r}') \quad (4.2)$$

Here g is a coupling constant which affords the opportunity to follow the evolution of the system from the independent particle or non-interacting limit (g=0) to the fully interacting limit (g=1). It is in this way the photoabsorption cross sections have been calculated by first evaluating the dipole response function (3.23). This is then substituted into (3.32) to solve the integral equation $\alpha(r,\omega)$. Finally, via (3.39) and (3.40) we obtain the photoabsorption cross section. As described earlier[14] an expansion mess of points were used so as to describe as accurately as possible the variations of wave functions and hence the induced density near the core of the atom. This usually meant obtaining the response function in matrix form whose order was $\sim$ 100 x 100.

The results of such a calculation for atomic Ba near the 4d-edge excitation are shown in Fig. 1. As a function of the coupling constant (g), it is seen how the cross section is modified from its nearly sharp resonance character (g = 0.2) to broad resonance (g = 0.5) and finally to the giant dipole resonance when there is full interaction between the

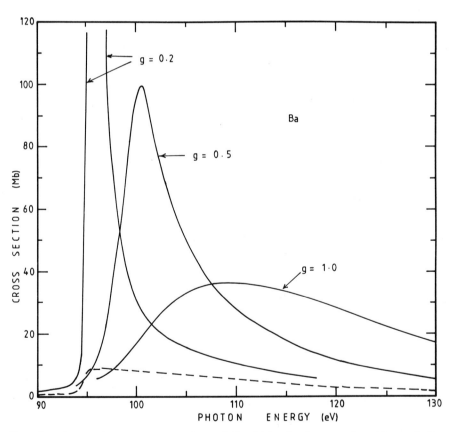

Fig. 1. Absorption cross section of Ba (4d-edge) as a function of the coupling constant, g.
--- non interacting cross section.

electrons (g = 1.0). As has been pointed out earlier[18], the photoionization phenomenon in this energy region is mainly due to the primary 4d→4f excitation which is pushed far into the continuum, autoionizes with it and manifests itself into the giant resonance. This picture is even more illuminating if the movement of the 4f-state resonance is followed with the variation of the coupling constant g. This is depicted in Fig. 5 where the 4f-resonance is located at a photon energy 92.3eV below the 4d→4f excitation threshold (93.7eV) when the interaction between the electrons is turned off (g = 0). As the interaction is gradually increased (g >0), the 4f-resonance moves across the threshold and finally emerges at photon energy 107.9eV (g = 1), around the maximum calculated value of the cross section. The overall features in the cross section (g = 1) is in agreement with the one calculated by Zangwill and Soven[19] using TDLDA.

The calculation of the cross section near the 5p-excitation edge has also been presented in Fig. 2. Again it compares well in its overall characteristics to the 5p-level partial absorption cross section calculated earlier[19] except the appearance of the sharp resonance at around 18.6eV below the 5p-threshold (19.3eV) and the dip in the higher

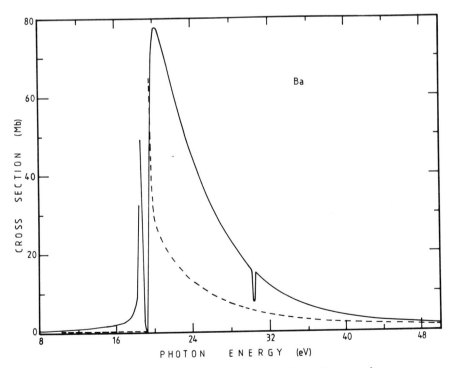

Fig. 2. Absorption cross section of Ba (5p-edge)
--- non interacting cross section

energy part of the spectrum. However, the present calculation is about twenty-fold as intense. This is not surprising though as all bound state electrons contribute to the construction of the response function used in (3.32) for the solution of $\alpha(r,\omega)$. Since the 5s→6p excitation edge is at energy 30.3eVn, the dip in the higher energy end of the spectrum may be attributed to this edge. Why a dip but not a spike is rather puzzling. To be able to determine the root cause of this will need an elaborate investigation in that energy region. Following similar arguments as in the 4d-edge case, the broad resonance may be considered as the broadening of the primary 5p→6d excitation at energy 16.6eV, after it has gone past the threshold to interact with the continuum. Equally well, the sharp resonance at energy 186eV may be due to the primary 5p→5d excitation. But to offer an interpretation as to how it comes about is difficult. Since the calculation is spin-independent, the possibility of the two peaks arising from the $5p_{1/2}$ and $5p_{3/2}$ spin-orbit levels is ruled out.

Figures 3 and 4 show the respective photoabsorption spectra of Ba^+ and Ba^{++}. In Fig. 5 we represent the variation of the f-states with the coupling constant g, for Ba^+ and Ba^{++}. Unlike the Ba situation they still lie below threshold even when the interaction is fully turned on, manifesting themselves in the structures below threshold. In Ba^+, the TDLDA calculation reveals two discrete structures below the 4d-excitation

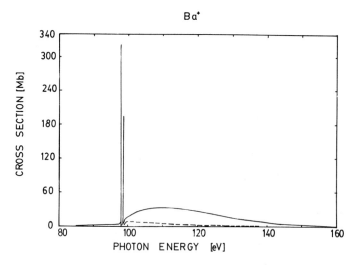

Fig. 3. Absorption cross section of Ba^+ (4d-edge)
--- non interacting cross section.

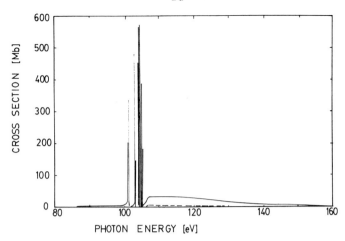

Fig. 4. Absorption cross section of Ba^{++} (4d-edge)
--- non interaction cross section.

threshold at energy 99.4eV. These are positively identified with the 4d→4f,5f transitions. The energies at which these resonances occur are shown in Table 1 together with their calculated absolute intensity strengths. In Ba^{++} the TDLDA calculation reveals six discrete structures below the 4d-excitation threshold (106.1eV). Five of these structures have been positively identified with the 4d→nf (n = 4,5,6,7,8) transitions. These resonant f-state excitation energies and their corresponding calculated absolute intensities are shown in Table 1 as well. The sixth resonance, which lies very close in energy to the 6f-resonance derives from a 4d→np (n ≥ 6) transition. Now, the KS values for the 4d→np (n = 6,7,8,9,10) excitation energies are respectively 97.1, 101.8, 103.5, 104.29, and 104.8eV. Without the knowledge of these energies, the most obvious guess would be the lowest lying 6p-state, in line with the assignments of the f-state resonances. But it is not expected that this moves in energy from its non-interacting value of 97.1eV to the fully interacting value of 104.4eV. On account of the resonance coinciding with the 4d→9p excitation energy, we therfore, tentatively label this resonance as '9p' as shown in Table 1 and Fig. 6. For comparison with experiment, we refer to the beautiful experimental results of Lucatorto et al.[20].

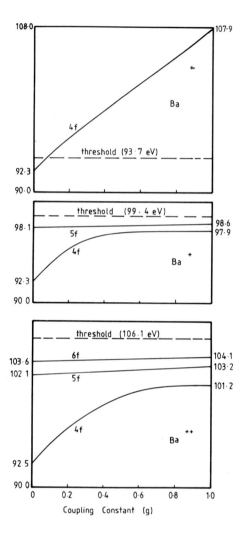

Fig. 5. Positions of f-state resonances as a function of coupling constant, g in Ba, Ba$^+$ and Ba^{++}.

TABLE 1 — Ba$^+$ and Ba^{++} discrete structures

	Designation	Energy (eV) 'non-interacting' spectrum	Energy (eV) 'interacting' spectrum	Intensity (Mb)
Ba$^+$	4f	92.3	97.9	322.4
	5f	98.5	98.62	196.4
Ba^{++}	4f	92.5	101.2	335.1
	5f	102.1	103.2	478.6
	6f	103.6	104.1	454.4
	'9p'	104.28	104.3	563.2
	7f	104.33	104.7	570.7
	8f	104.8	105.0	386.6

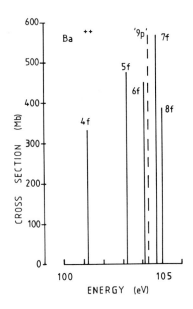

Fig. 6. Discrete structures of Ba^{++}.

Many interpretations have been given for the appearance of the discrete structure below threshold in the Ba^{++} spectrum. There is the discrete state interacting with the continuum background by Wendin[21] using the Fano formation[22]. There is the wave function collapse by Connerade and Mansfield[23], which had been used successfully earlier in analysing some properties of the rare-earth elements[24]. There is the collective mode one based on the harmonic-oscillator model of Nuroh et al.[25], and also the quantum defect theory based explanation of Connerade[26]. The wave-function collapse explanation hinges on the existence of a two-well potential for the f-state level. In this regard there is some parallelism in our work and those of Connerade and Mansfield[23] and Wendin and Starace[18]. For if (3.49) is used in calculating the wave function U_{f*} and (3.50) the corresponding potential at ω = 103.2eV, the wave-function is nodeless as expected of a 4f-like wave function (Only the values n = 3,4 and 5 were used in the summation). The potential is a two-well one with a deeper inner valley and a shallow outer one (see Fig. 7). This is more or less the picture found earlier[18,23].

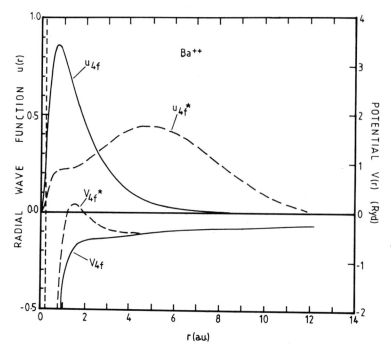

Fig. 7. Final 4f radial wave functions, U(r), in the independent-particle (dashed line) and fully interacting (solid line) calculations. Corresponding effective potentials, V(r), are also shown.

5. TDLDA versus RPA

Calculated photoionization cross sections based on the random phase approximation (RPA)[27] have yielded results which compare well with experiment just like the TDLDA calculations reported here. It then becomes interesting to find what relationship there is between RPA and TDLDA based calculations. We have seen that the quantity which determines the cross section is essentially $\text{Im}\alpha(\omega)$. In the one-electron approximation, once the self-consistent field $V^{SCF}(\vec{r},\omega)$ is known this quantity is then given by

$$\text{Im}\alpha(\omega) \sim \sum_{mi} |<m|V^{SCF}(\vec{r},\omega)|i>|^2 \, \delta(\varepsilon_m - \varepsilon_i - \omega) \qquad (5.1)$$

To evaluate the matrix element in (5.1) in TDLDA, we use (2.13) and (2.14) neglecting the exchange-correlation part of the static interactions between the electrons and write

$$V^{SCF}(\vec{r},\omega) = V^{ext}(\vec{r},\omega) + \int d\vec{r}\,'\,\frac{\delta n(\vec{r}\,',\omega)}{|\vec{r}-\vec{r}\,'|}$$
$$= V^{ext}(\vec{r},\omega) + \int d\vec{r}\,'\,\frac{1}{|\vec{r}-\vec{r}\,'|}\int dr'' \chi^0(\vec{r}\,',\vec{r}\,'',\omega)V^{SCF}(\vec{r}\,'',\omega) \quad (5.2)$$

Consequently, the matrix element in (5.1) becomes

$$<m|V^{SCF}(\vec{r},\omega)|i> = <m|V^{ext}(\vec{r},\omega)|i>$$
$$+ <m|\int d\vec{r}\,'\,\frac{1}{|\vec{r}-\vec{r}\,'|}\int d\vec{r}\,''\chi^0(\vec{r},\vec{r}\,'',\omega)V^{SCF}(\vec{r}\,'',\omega)|i> \quad (5.3)$$

To proceed further, we need to insert the expression for the independent particle response function. In (3.12), since the summation is over bound states, the energy argument of $G^+(r,r',\varepsilon_j - \omega)$ is wholly negative for $\omega > 0$ and hence renders this Green's function real. Thus it does not matter if this retarded Green's function is replaced by its advanced counterpart. Hence (3.12) may be written as

$$\chi^0(\vec{r},\vec{r}\,',\omega) \sim \sum_{j \leq F} \varphi_j^*(\vec{r})\varphi_j(\vec{r}\,') \sum_n \frac{\varphi_n(\vec{r})\varphi_n^*(\vec{r}\,')}{\varepsilon_j + \omega - \varepsilon_n + i\delta}$$
$$+ \sum_{j \leq F} \varphi_j(\vec{r})\varphi_j^*(\vec{r}\,') \sum_n \frac{\varphi_n^*(\vec{r})\varphi_n(\vec{r}\,')}{\varepsilon_j - \omega - \varepsilon_n - i\delta} \quad (5.4)$$

Here we adopt the notation of Amusia and Cherekpov[28] where F denotes the Fermi level separating occupied and unoccupied states. Substitution of (5.4) into (5.3) gives

$$<m|V^{SCF}(\vec{r},\omega)|i> = <m|V^{ext}(\vec{r},\omega)|i>$$
$$+ <m|\int d\vec{r}\,'\,\frac{1}{|\vec{r}-\vec{r}\,'|}\int d\vec{r}\,''\sum_{\substack{j \leq F \\ n}}\frac{\varphi_j^*(\vec{r}\,')\varphi_j(\vec{r}\,'')\varphi_n(\vec{r}\,')\varphi_n^*(\vec{r}\,'')}{\varepsilon_j + \omega - \varepsilon_n + i\delta}V^{SCF}(\vec{r}\,'',\omega)|i>$$
$$+ <m|\int d\vec{r}\,'\,\frac{1}{|\vec{r}-\vec{r}\,'|}\int d\vec{r}\,''\sum_{\substack{j \leq F \\ n}}\frac{\varphi_j(\vec{r}\,')\varphi_j^*(\vec{r}\,'')\varphi_n^*(\vec{r}\,')\varphi_n(\vec{r}\,'')}{\varepsilon_j - \omega - \varepsilon_n - i\delta}V^{SCF}(\vec{r}\,'',\omega)|i> \quad (5.5)$$

Because of the structure of the energy denominators in (5.5), the summation index $\underline{n}$ over all the single particle states reduces to summation over only unoccupied states. Thus in terms of the bra and ket notation, (5.5) is equivalent to

$$<m|V^{SCF}(\vec{r},\omega)|i> = <m|V^{ext}(\vec{r},\omega)|i>$$
$$+ \sum_{\substack{j \leq F \\ n > F}}[\frac{<n|V^{SCF}(\omega)|j><mj|V|in>}{\omega - \varepsilon_n + \varepsilon_j + i\delta} - \frac{<j|V^{SCF}(\omega)|n><mn|V|ij>}{\omega + \varepsilon_n - \varepsilon_j + i\delta}] \quad (5.6)$$

The interaction matrix element is the usual two-body direct Coulomb matrix element defined by

$$<mj|V|in> = \int d\vec{r}\int d\vec{r}\,'\varphi_m^*(\vec{r})\varphi_j^*(\vec{r}\,')\frac{1}{|\vec{r}-\vec{r}\,'|}\varphi_i(\vec{r})\varphi_n(\vec{r}\,') \quad (5.7)$$

Except for the sign (+) of the infinitesimal factor in the energy denominator of the second term of (5.6), it is precisely the effective matrix element used in the conventional RPA calculations of photoionization cross sections (see eq. (19) of Ref. 28). This is then the contact between RPA and TDLDA.

6. Concluding remarks

The foregoing calculations and analysis have demonstrated that TDLDA, an explicitly many-body based calculation is capable of producing photoionization spectra quite accurately. It has also been shown that neglecting exchange and correlation effects, TDLDA is equivalent to RPA. Indeed, it the exchange-correlation interaction is neglected in the calculation of the cross sections reported, we find very little differences with the full calculation. This observation runs parallel with RPA and random phase approximation with exchange (RPAE) calculations. In most situations[17,28], what the retention of the exchange interactions do is to give refinement in the calculation of photoionization cross sections without changing much of the spectral shapes.

Acknowledgements

This work has been supported by the Natural Sciences and Engineering Research Council Canada.

References

1. F. Bloch, Z. Phys. **18** (1933) 363.
2. M. Jensen, Z. Phys. **106** (1937) 620.
3. W. Brandt and S. Lundquist, Phys. Rev. **132** (1963) 2135 ; Ark. Fys. **38** (1965) 399 ; W. Brandt, L. Eder and S. Lundquist, J. Quant. Spectr. Rad. Transfer **7** (1967) 411.
4. M. YA. Amusia, Phys. Lett. **14**(1965) 36 ; Phys. Lett.**24A**(1967) 394 M. Ya. Amusia, N.A. Cherekpov and L.V. Chernysheva, Zh. Eskp. Tesr. Fiz. **60** (1971) 160 ; (Sov. Physics - JETP **33** (1971) 90).
5. G. Wendin, Phys. Lett. **37 A** (1971) 445 ; Phys. Lett. **46 A**(1973)119.
6. J.W. Cooper, Phys. Rev. Lett. **13** (1964) 762.
7. U. Fano and J.W. Cooper, Rev. Mod. Phys. **40** (1968) 441.
8. see e.g. K. Nuroh, Ph.D. thesis, Chalmers University, Gothenburg, Sweden (1978).
9. G. Brown, Unified Theory of Nuclear Models and Forces (North Holland Publ. Co., 1971).
10. Many-Body Theory of Atomic Systems, Proceedings of Nobel Symposium 46, eds. I. Lindgren and S. Lundquist, Physica Scripta **21**, No. 3/4 (1980).
11. W. Kohn and L.J. Sham, Phys. Rev. **140** (1965) A1133.
12. S. Lundquist and G. Mukhopadhay, Physica Scripta **21** (1980) 503 ; K. Nuroh and E. Zaremba (to be published).

13. A.L. Fetter and J.D. Walecka, Quantum Theory of Many-Particle Systems (McGraw-Hill, New York, 1971).
14. M.J. Stott and E. Zaremba, Phys. Rev. **A 21** (1980) 12.
15. E.U. Condon and G.H. Shortley, The Theory of Atomic Spectra, Chpt. VI, § 8 (Cambridge University Press, 1963).
16. I. Lindgren and J. Morrison, Atomic Many-Body Theory (Springer-Verlag, Berlin).
17. G. Wendin, Application of Many-Body Problems to Atomic Systems, Les Houches (1982).
18. G. Wendin and A.F. Starace, J. Phys. **B 11** (1978) 411.
19. A. Zangwill and P. Soven, Phys. Rev. **A 21** (1980) 1561.
20. T.B. Lucatorto, T.J. McIrath, J. Sugar and S.M. Younger, Phys. Rev. Lett. **47** (1981) 1124.
21. G. Wendin, J. Phys. **B 5** (1972) 110.
22. U. Fano, Phys. Rev. **124** (1961) 1866.
23. J.P. Connerade and M.W.D. Mansfield, Phys. Rev. Lett. **48** (1982) 131.
24. M.G. Mayer, Phys. Rev. **60** (1941) 184.
25. K. Nuroh, M.J. Stott and E. Zaremba, Phys. Rev. Lett. **49** (1982) 862.
26. J.P. Connerade, J. Phys. **B 16** (1983) L 257.
27. G. Wendin, J. Phys. **B 6** (1973) 42 ; M. Ya. Amusia, N.A. Cherekpov, R.K. Janev and Dj. Zivanovic, J. Phys. **B 7** (1974) 1435.
28. M. Ya. Amusia and N.A. Cherekpov, Case Studies in Atomic Physics **5** (1975) 47.

GIANT RESONANCES IN THE TRANSITION REGIONS OF THE PERIODIC TABLE

Charles W. Clark and Thomas B. Lucatorto

Radiation Physics Division
National Bureau of Standards
Gaithersburg, MD 20899

INTRODUCTION

In the transition regions of the periodic table of the elements, atomic d or f orbitals undergo a fairly sudden change from hydrogenic to fully collapsed form. This transition involves a large reduction in the mean orbital radius - by about 95% for the 4f orbital - and results in corresponding qualitatitive changes in physical processes sensitive to orbital size (e.g. excitation cross sections, bonding character). It is caused by a shift, as the nuclear charge Z increases, in the close balance between repulsive centrifugal and attractive atomic forces on the electron. The balance can also be tilted *within* a given element in the transition region, for instance by a change in the occupancy of its core or valence orbitals, or by the formation of a molecular bond. Transition region elements are thus characterized by an unusual sensitivity of gross orbital properties to external perturbations; and, from the standpoint of theoretical representation, to the effects of electron correlation, LS term dependence, and special relativity.

This paper reports some experimental and theoretical work directed towards exploring this sensitivity. We take the approach of tracing physical processes along isoelectronic, isonuclear, and isoionic sequences which span particular transition regions. The experimental work described here consists of soft x-ray photoabsorption studies of alkaline earth atoms and ions in the gas phase. It is based upon techniques of time-resolved sequential laser and soft x-ray excitation, which enable us to obtain the subvalence photoabsorption spectra of ground and excited states of an atom and its ions. Our theoretical work is based primarily upon single- and multiconfiguration Hartree-Fock calculations, with particular attention to effects of orbital term dependence.

The main points emphasized here are as follows. Giant resonances, associated with orbital collapse, are manifested in some of the principal mechanisms of excitation of transition region elements. Their properties can be altered by external means, specifically by photoexcitation of atomic valence electrons. The basic physics governing their behavior is contained within the independent electron model, but *not* in the average-of-configuration approximation which usually accompanies it. We believe it is unnecessary to invoke a picture of "collective excitation" to understand the giant resonances in these regions. Electron correlations do play an important role, but not because they cause extensive mixing of different configurations; rather, their effects are amplified by the delicate balance of independent-electron forces.

THE TRANSITION REGIONS

The character of d and f orbitals of ground state atoms can be inferred from figures 4 and 1 of the paper of Griffin et al. elsewhere in this volume.[1] These show the effective principal quantum numbers n* of the orbitals as a function of nuclear charge Z. Note particularly the curves for 3d and 4f, the most tightly bound orbitals for $l = 2$ and 3. For $Z \leq 19$ and 57 respectively, n* remains close to its hydrogenic value of 3 or 4. This means that the corresponding orbitals are essentially the same as those of the hydrogen atom, which are very diffuse wavefunctions with mean radii of 18 a_0 and 10.5 a_0 respectively. As Z increases past the critical value, n* makes a sudden drop and the orbital contracts to a size of about 1 a_0. This is the radius characteristic of that of the filled subshells of the same principal quantum number n (3s, 3p and 4s, 4p, 4d respectively).

This transition has been explained within the framework of familiar atomic models by Fermi,[2] Goeppert Mayer,[3] and Rau and Fano,[4] and is dealt with in other papers in this volume.[1,5] These descriptions all portray the transition as occuring when the attractive potential for the electron (arising from the sum of Coulomb interactions with the nucleus and the other electrons) overcomes the repulsive centrifugal potential $V(r) = l(l+1)/2r^2$, forming a potential well in the interior of the atom which can support a bound state. However, the particular value of Z at which the transition occurs depends upon the atomic model used. This is not surprising in view of the very rapid variation of n* seen in the figures cited above. It also suggests that the point of transition might vary with *real* changes in the effective atomic potential (in contrast to changes in approximate atomic models), which can be made, for example, by altering the electronic configuration.

It seems appropriate to identify transition regions for orbital contraction, consisting of those elements in which the contracting orbital appears in forms between hydrogenic and fully collapsed, depending upon the electronic configuration, ionization stage, or local chemical environment. For 3d contraction the transition region appears to consist of K, Ca, and Sc (Z = 19 - 21), and for 4f contraction, Cs, Ba, La, and Ce (Z = 55 - 58). The transition region concept is also useful in understanding the contraction of 4d, 5d, and 5f orbitals; we will not be discussing those cases in detail, except to note that the transition region for 5d may be said in a certain sense to extend from Cs past Yb (Z = 55 ~ 70). The transition regions thus contain some of the favorite elements of experimental atomic physics, the alkalis and alkaline earths.

A manifestation of orbital contraction phenomena in the 4f transition region is seen in Fig. 1, which shows the photoabsorption spectra of Ba, Ba$^+$, and Ba^{2+}, at photon energies in the vicinity of the 4d ionization threshold. The familiar "giant resonance" of barium appears to undergo a radical change of character as the valence (6s) electrons are removed. The roles of different basic physical mechanisms in the evolution of this resonance are now fairly well understood, and we shall use this case as the main focus of discussion in this paper.

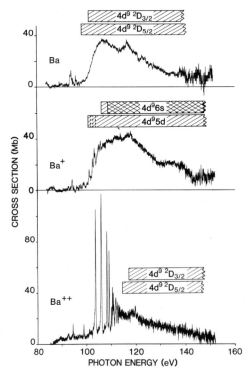

Fig. 1. Photoabsorption by Ba, Ba$^+$, and Ba^{++} near the 4d giant resonance, from ref.6.

EXPERIMENTAL STUDY OF PHOTOABSORPTION BY EXCITED ATOMS AND IONS

Most giant resonance phenomena occur at energies well above typical atomic ionization potentials, so their study by atomic photoabsorption techniques has been done at vacuum ultraviolet (VUV) or soft x-ray wavelengths. A typical application of such techniques in the continuous region of the spectrum involves the measurement of the net absorption of radiation by a gas column of density $\sim 10^{15}$ atoms cm^{-3} and length of 10 - 100 cm. Such measurements on ground state neutral atoms can be performed with a variety of different VUV sources. However, since it is difficult to maintain a continuous population of excited states or ions in such a setting, the method used in our laboratory for studying these species involves pulsed VUV measurements on a pulsed laser-excited vapor. An alternative approach for ions is steady-state photoionization measurements using intersecting ion and photon beams.[7]

The resonant laser-driven ionization (RLDI) scheme produces a large population of ions by irradiating a dense atomic vapor (10^{15} atoms cm^{-3}) with a long-pulse (1μs) laser tuned to an atomic resonance transition. Early in the laser pulse, resonant photoabsorption causes a buildup of population in the upper state of the transition. Some of these excited atoms become ionized, either by multiphoton absorption, energy-pooling collisions with other excited atoms, or associative ionization.[8] This produces a small population of free "seed" electrons. These electrons are rapidly heated by superelastic collisions with the excited atoms, which are in turn re-excited by the laser. When the electron population becomes sufficiently hot, electron impact ionization of the ground and excited atoms thus occurs in the form of a chain reaction. As a function of time after the start of the pulse, the population of ground state atoms decreases more or less monotonically, the population of the excited state increases to a maximum and then falls to zero, and the population of ions increases from zero to as much as 100%. Examples of the time evolution are shown in Fig. 2 for the case of calcium vapor in a heat pipe oven, excited by a laser tuned to the $4s^2\ ^1S \rightarrow 4s4p\ ^3P^o$ intercombination transition.[9]

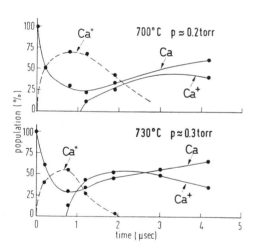

Fig. 2. Populations of Ca, Ca*, and Ca$^+$ vs. time after the laser pulse. Temperatures and pressures in the heat pipe are indicated. From ref. 9.

By suitable timing of a VUV pulse with respect to the laser pulse, the absorption spectra of various mixtures of the ground, excited, and ionized atoms can be obtained. The pulsed VUV continuum source used to obtain the spectra of Fig. 1 was a uranium anode spark; the current version of our apparatus uses instead a laser-produced plasma source, which consists of a Nd:YAG laser focused tightly on a heavy metal target.[10] The spectrum of Ba in Fig. 1 was obtained by photographic measurement of absorption of the VUV radiation by a column of Ba vapor in a heat pipe oven. The spectrum of Ba$^+$ was recorded by delaying the VUV pulse with respect to a pulse from a dye laser tuned to the 553.7 nm resonance line of Ba ($6s^2\ ^1S \rightarrow 6s6p\ ^1P^o$), which completely ionizes the Ba vapor. In the course of ionization, a free electron population with a temperature of approximately 4000 °K is created, and as a result some 35% of the Ba$^+$ in the plasma resides in the low-lying metastable 5d ^{2}D states.

The depicted absorption spectrum of Ba$^+$ is thus a composite of those of the 6s and the 5d states. A second laser tuned to the 493.5 nm resonance line of Ba$^+$ ($6s\ ^2S \rightarrow 6p\ ^2P^o_{1/2}$) produces virtually complete ionization of the Ba$^+$. The absorption spectrum of Ba^{2+} was thus obtained by firing the VUV pulse after the two laser pulses.

The RLDI technique has the advantage of permitting measurements on ground states, excited states, and ions in the same apparatus. On the other hand, it does sometimes produce a distribution of initial states (as in Ba$^+$) which cannot easily be changed. The basic underlying mechanisms of RLDI are not entirely understood, and the specific roles of associative ionization and energy pooling collisions are subjects of active research. Nevertheless, the technique has been found to be applicable to a wide range of elements. It is currently being applied in our laboratory[9] to studies of 3p $\rightarrow$ 3d giant resonance photoabsorption in Ca, Mn, and their excited states and ions, but we restrict our discussion here to the case of Ba.

THE 4d $\rightarrow$ f GIANT RESONANCE IN THE BARIUM ISONUCLEAR SEQUENCE

Photoabsorption by barium in the 100 eV energy range is dominated by excitations of the 4d shell.[11] The broad peak seen in Ba at 105 eV is a typical member of the family of giant resonances in 4d photoexcitation of neutral atoms, which occur in elements with a filled 4d shell and vacancies in the 4f shell. They occur as broad features (widths > 10 eV) that lie *above* the 4d ionization limit, and it is widely agreed that they are associated in some way with the continuum f wave. The details of this association remain somewhat less a matter of consensus, particularly as concerns the role of independent particle vs. collective behavior.[12,13]

Photoexcitation of the 4d Shell in Lighter Atoms

We believe that the evolution, as atomic number Z increases, of the 4d giant resonance in neutral atoms can be understood roughly as follows. In the electric dipole approximation the final-state orbital has p or f symmetry, and the bulk of the oscillator strength is in the f channel. At low Z, the effective potential for an f electron in the vicinity of the residual ion core (r < 5 a_0) is completely dominated by the centrifugal term, and it corresponds to a purely repulsive force (i.e. the net potential has no local minimum in this region). This situation is almost entirely independent of the electronic configuration of the core, and of the way in which the core and the excited electron are coupled. The f wave is thus kept out of the core at low energies, giving a "delayed onset" of the photoabsorption cross section for 3d or 4d electrons: the cross section is small at threshold, and increases with energy as the f wave penetrates the core, where it overlaps the initial state. (A corresponding phenomenon in the discrete spectrum is the *increase* in quantum defects of nf Rydberg states as the principal quantum number n increases, due to greater penetration of the core.) The cross section must fall off at high energy in the usual way, so it goes through a maximum. Thus the photoabsorption cross section exhibits a broad (~ 50 eV) peak structure which is not associated with any resonance in the final state channel.

As Z increases, the strength of the attractive atomic field increases, whereas that of the centrifugal potential remains constant. At some value of Z, in the vicinity of silver[14] (Z =47), the atomic field becomes sufficiently strong to create a local minimum in the effective potential in the core region, i.e. an "inner well." The minimum value of the potential is quite sensitive to the ion core configuration and the mutual coupling of the core and the f electron. On the other hand, the effective potential at large distances from the core is determined primarily by the net charge of the residual ion and by the centrifugal potential. Thus, along an atomic sequence of constant ionization stage (e.g. singly charged ions with a 4d core hole), the depth of the inner well increases with Z, while the potential at long range remains more or less constant. Eventually the inner well becomes capable of supporting a *shape resonance*, a state in which the f electron is temporarily confined (~ 10^{-14} sec) in the inner well. This takes place somewhere before Xe (Z=54). Again, the energy and the width of the shape resonance depend strongly upon the electronic configuration and symmetry. A good illustration of this is contained in the paper of Cheng and Johnson,[15] which shows scattering eigenphaseshifts for the three J = 1 channels of Xe $4d^9\varepsilon f$, as computed in the relativistic random phase approximation. There are three possible LS compositions of these channels: $^3P^o$, $^3D^o$, and $^1P^o$. Two of the channels are composed of the triplets, and they each display a narrow shape resonance (as manifested by an increase of the eigenphaseshift by π radians) at energies within a few eV of the Xe $4d^9$ threshold. The third channel is almost entirely $^1P^o$ in character, and exhibits a broad shape resonance about 30 eV above the threshold. The underlying mechanism of this LS term dependence will be discussed below. Effective potentials which give rise to shape resonances in the analogous $^1P^o$ channels of Ba, Ba$^+$, and Ba^{2+} are shown in Fig.3.

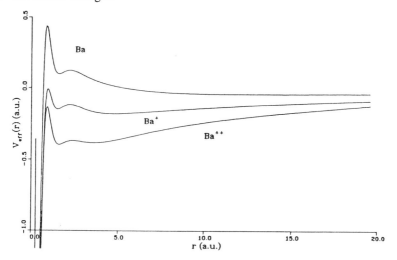

Fig. 3. Effective potentials for the f electron in $4d^9\varepsilon f(^1P^o)5s^25p^66s^n$ states of Ba, Ba$^+$, and Ba^{2+} (n = 2, 1, 0 respectively). These are obtained by inverting the Hartree-Fock 4f orbital calculated for the respective states, to find an equivalent local potential (see eq. 2 below). The potential in the inner region drops to -4 a.u. at r ~ 0.5 a_0 in each case, and increases as $6r^{-2}$ as r → 0. The sharp barrier structures at r ~ 1 a_0 derive from exchange repulsion of the f electron by the 4d hole; the softer bumps around 2 a_0 reflect the influence of the 5s and 5p shells.

The inner well of the potential is located in the same spatial region as the 4s, 4p, and 4d electrons. Thus the f wave shape resonance has a large overlap with these electrons, and so will exert a strong influence on atomic N shell excitation processes. In particular, the dipole matrix element $<4d^{10}|\underline{D}|4d^9\varepsilon f>$, which governs photoabsorption by the 4d shell, is largest at energies near the $4d^9\varepsilon f$ $^1P^o$ shape resonance. It has been shown[11] that the broad absorption peak of Ba in Fig. 1 can be accounted for within the Hartree-Fock approximation with inclusion of initial-state correlation; the continuum final states in this approach are essentially the same as those that would be obtained from the potential of Fig.3. More recently, the role of similar shape resonances in the electron impact ionization of Cs$^+$ has been elucidated, again within the Hartree-Fock approximation.[16] It should be noted that such descriptions do not invoke mechanisms of "collective" excitation of electrons, but are rather based on a picture of a single electron moving in a (nonlocal) atomic potential.

Distinction between Shape Resonances and "Collapsed" States

The inner well of the potential deepens as Z increases, and at some point it will become sufficiently deep that the inner well shape resonance becomes a bound state (it may still decay via Auger processes, if it is associated with a core hole). Another way of stating this is that for Z greater than some critical value, the most tightly bound f orbital (the 4f) will lie predominantly in the inner well. This pheneomenon is usually referred to as the "collapse" of the 4f orbital. Again, the value of Z at which this occurs is variable. The above-mentioned $4d^9\varepsilon f5s^25p^6$ triplet shape resonances of Xe become bound in Cs$^+$(Z=55), the next member of the isoelectronic sequence. For neutral atoms without core holes, the transition occurs at La (Z=57). The $4d^9\varepsilon f$ $^1P^o$ shape resonance in neutral atoms does not become bound at least until Ce (Z=58).

The observable consequences of orbital collapse arise from the insensitivity of the collapsed orbital to variations in the atomic environment. The chemical similarity of most of the rare earths, whose ground-state electronic configurations differ only in the occupation number of the 4f shell, reflects the inability of the 4f orbital to participate in molecular bonding. In inner-shell excitation processes, orbital collapse is revealed by the weak dependence of particular excitations upon the state of the atom. One such example can be seen clearly in Fig. 1: a sharp absorption feature that occurs at ~ 95 eV in Ba, Ba^+, and Ba^{2+}. This is associated with the excitation $4d^{10}5s^25p^66s^n \rightarrow 4d^94f(^3D^o_{J=1})5s^25p^66s^n$. In contrast to all other features visible in this figure, the 95 eV line does not shift as the ionization stage changes. This can be understood in terms of the size of the atomic orbitals, which is indicated by Table 1. The 4d and 4f orbitals lie well inside the valence shell. If the outer electrons are considered as a spherical distribution of electric charge, which vanishes in the vicinity of the 4d and 4f orbitals, then by classical electrostatics the difference in potential energy between 4d and 4f electrons is independent of the total charge outside. Many other states of the $4d^94f$ configuration of Ba are also collapsed, notably the $^3P^o_{J=1}$ state, which is excited in photoabsorption, though too weakly to be discerned in Fig. 1. However, the bulk of absorption oscillator strength is associated with states whose properties evidently are *not* determined solely by the inner well; for the "giant resonance" is seen to change dramatically with the ionization stage.

Table 1. Mean radii $<r>$, in atomic units, of nl orbitals of Ba $4d^94f(^3D^o_{J=1})5s^25p^66s^2$ as computed in the Hartree-Fock approximation. The hydrogenic value of $<r>$ is $[3n^2 - l(l+1)]/2Z_{eff}$, so $<r> = 18\, a_0$ for an uncollapsed 4f orbital in a neutral atom. Note that the hydrogenic value of $<r>$ *decreases* with increasing l for fixed n; this behavior is recovered in inner shells as Z increases, so that eventually (Z ~ 86 in the Hartree-Fock approximation) the 4f orbital becomes the most compact orbital of the N shell.

nl	$<r>$	nl	$<r>$	nl	$<r>$
4s	0.70	5s	1.71	6s	5.23
4p	0.72	5p	1.93		
4d	0.79				
4f	1.15				

LS Term Dependence of Orbital Collapse

The 4f orbital of Ba $4d^94f(^1P^o)5s^25p^66s^2$ is not collapsed. It lies in the broad outer well of Fig. 3, and is essentially hydrogenic in character. Thus there is more than an order of magnitude difference in the mean size of the 4f orbital, between different LS terms of the same configuration. We now review the physical mechanisms which underlie this differential orbital collapse; these are also discussed elsewhere.[1,12,17]

The non-relativistic Hartree-Fock approximation represents an atomic wavefunction as a product of one-electron wavefunctions, appropriately antisymmetrized to conform to the Pauli principle and to be an eigenfunction of total orbital and spin angular momenta L and S. The one-electron radial orbitals $P_{nl}(r)$ each satisfy a Schrödinger-type equation, in which the interaction with electrons in other orbitals $n'l'$ is represented by local and nonlocal potentials $Y(nl, n'l'; r)/r$ and $X(nl, n'l'; r')/r$:[18]

$$Y(nl, n'l'; r)\, P_{nl}(r) = \sum_k A_k Y^k(n'l', n'l'; r)\, P_{nl}(r), \qquad (1a)$$

$$X(nl, n'l'; r)\, P_{nl}(r) \equiv \sum_k B_k Y^k(nl, n'l'; r)\, P_{n'l'}(r), \qquad (1b)$$

with

$$Y^k(nl, n'l'; r) = r \int dr' (r_<^k / r_>^{k+1}) P_{nl}(r') P_{n'l'}(r'), \quad (1c)$$

where $r_<, r_>$ are the lesser and greater of r, r'. The local potential Y comes from the multipole expansion of the classical electrostatic potential of the charge distribution of the $n'l'$ electron subshell, and X is its exchange counterpart. The coefficients A_k and B_k depend upon the way in which the nl and $n'l'$ electrons are coupled. These coefficients are displayed for all the LS terms of the $4d^9 4f$ configuration in Table 2.

Table 2. Coefficients of the local and nonlocal potentials of the interaction between 4d and 4f electrons, for all LS terms of the $4d^9 4f$ configuration. The monopole coefficient $A_0 = 9$ common to all terms is omitted. Common denominators have been extracted for clarity; thus $B_1(^3P^o) = -3/7$. Note that the L-pole exchange potential (represented by B_L) is identical and attractive for all terms except $^1L^o$, for which it is *repulsive*.

	$7 B_1$	$105 A_2$	$21 B_3$	$693 A_4$	$2541 B_5$
$^3P^o$	-3	-24	-4	-66	-550
$^1P^o$	11	-24	-4	-66	-550
$^{3,1}D^o$	-3	-6	-4	99	-550
$^3F^o$	-3	11	-4	-66	-550
$^1F^o$	-3	11	4	-66	-550
$^{3,1}G^o$	-3	15	-4	22	-550
$^3H^o$	-3	-10	-4	-3	-550
$^1H^o$	-3	-10	-4	-3	150

The net Hartree-Fock potential of course contains contributions from all other electrons in the atom, but if these are all in closed shells their net contribution is independent of the LS term. Because of the nonlocal character of the exchange potential it is impossible to predict the consequences of the term dependence indicated by Table 2. Its net effect can only be assessed by carrying out self-consistent field calculations for all orbitals $P_{nl}(r)$ for each LS term. However, from such orbitals one can easily construct an effective potential $V_{eff}(r)$ by inverting the Schrödinger equation (in a.u.):

$$V_{eff}(r) = \varepsilon_{nl} - [d^2 P_{nl}(r) / dr^2] [2 P_{nl}(r)]^{-1}, \quad (2)$$

where ε_{nl} is the diagonal orbital energy. The potential $V_{eff}(r)$ has meaning only insofar as it is a local potential which reproduces the orbital $P_{nl}(r)$. However, it does often happen that the V_{eff} generated by the orbital of the lowest member of a Rydberg series (e.g. the 4f), also reproduces the Hartree-Fock orbitals for higher members of the series with reasonable accuracy. This has been found to be the case for the $4d^9 4f$ configuration for elements in the 4f transition region.

Figure 4 shows effective potentials obtained for the $^3P^o$, $^3D^o$, and $^1P^o$ terms of the $4d^94f\ 5s^25p^6$ configuration of Ba^{2+}. The $^1P^o$ potential is dramatically different from those of the triplets. From Table 2 it is seen that the only difference in the Hartree-Fock calculations for the $^3P^o$ and $^1P^o$ states is the weight of the dipole exchange potential. Thus it is reasonable to attribute the difference in effective potentials to the d - f dipole exchange interaction. The repulsive effect of this interaction in the $^1P^o$ channel is represented as a barrier in V_{eff} that peaks at ~ 1 a_0, which is at the edge of the 4d shell (see Table 1). The $^3P^o$ and $^3D^o$ potentials for Ba^{2+} hold collapsed states, as they do in Ba. The 4f orbital of the $^1P^o$ term is found to lie predominantly in the outer well, as again is the case in Ba. However, because of the greater long-range Coulomb potential of Ba^{2+}, the 4f orbital overlaps the 4d shell much more than in the neutral.

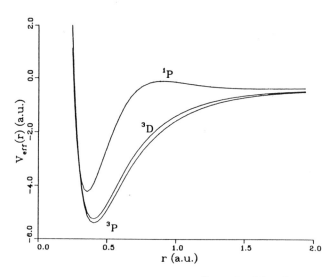

Fig. 4. Effective potentials for the f electron in $4d^94f\ 5s^25p^6\ ^3P^o$, $^3D^o$, and $^1P^o$ states of Ba^{2+} as indicated. From ref. 19.

Coulombic Contraction vs. Orbital Collapse

The principal mechanism for the qualitative difference in the absorption spectra of Fig. 1 is the increase in the long-range Coulomb potential. The effect of this increase, as depicted in Fig. 3, is to lower the outer well of the potential while still retaining a substantial barrier between inner and outer wells. Thus the shape resonance in Ba is pulled down into the discrete spectrum but is not converted into a collapsed state. It is manifested by an increased *tunneling* of the nf Rydberg orbitals into the inner potential well.

This effect is further illustrated in Fig. 5, which shows the $^1P^o$ 4f orbitals of Ba, Ba^+, and Ba^{2+} plotted on a scale which eliminates variations associated only with contraction in the Coulomb field. The 4f orbital in a purely Coulombic potential - Z_{eff}/r is described by $P_{4f}(r) = (\ Z_{eff}\ 2^{-9}\ /\ 8!)^{1/2}\ \rho^4\ e^{-\rho/4}$, with $\rho = Z_{eff} r$. This orbital is very close to that of neutral Ba in Fig. 5, for which $Z_{eff} = 1$ is the net charge seen by the 4f electron at large distances. The $^1P^o$ 4f orbitals of Ba^+ and Ba^{2+} ($Z_{eff} = 2$ and 3) are seen to depart somewhat from hydrogenic behavior, but they cannot be said to be qualitatively different on this scale. The collapsed orbitals of the $^3P^o$ and $^3D^o$ terms are hardly affected by the change in the Coulomb field; they attain their maximum amplitude at r ~ 1 a_0, as indicated in Table 1.

The barium isonuclear sequence thus exhibits an effect which is not seen in the *isoionic* sequence of neutral atoms (ground state Xe, Cs, Ba,...). It has a counterpart in valence-excited states in the Cs isoelectronic sequence, as will be discussed below. The "charge-dependent wavefunction collapse" in the Xe isonuclear sequence discussed by O'Sullivan[20] appears to consist primarily of Coulombic contraction, since the size of the orbitals scales in the manner described above.

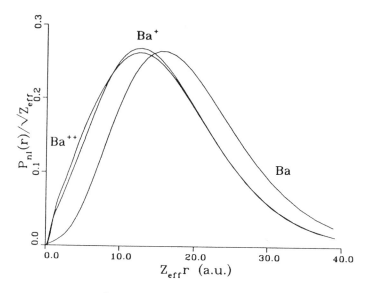

Fig. 5. 4f orbitals for the $4d^94f(^1P^o)5s^25p^66s^n$ states of Ba, Ba$^+$, and Ba^{2+}, plotted as a function of the scaled hydrogenic radius $\rho = Z_{eff}r$. Interaction of the 4f electron with the unpaired 6s electron in Ba$^+$ has been approximated by a spherical average.

Quantitative Evaluation of the Discrete 4d Absorption Spectrum of Ba^{2+}

Calculations of continuum photoabsorption by neutral Ba near the 4d threshold have been carried out in a variety of approximations.[11,21-23] Several of these have achieved quite good agreement with experimental values of the total cross section, though there are still some discrepancies with respect to the individual electron subshell contributions.[24] In particular, it has been found that good results for the total cross section can be obtained within the Hartree-Fock approximation,[11] providing contributions of initial-state correlations to the transition matrix element are taken into account. As we have emphasized above, in this framework the system is described in terms of a single electron escaping from an ion, and the picture of a collective oscillation of the 4d shell does not emerge from the formal development of the calculation.

The corresponding interpretation of the sharp lines of the Ba^{2+} absorption spectrum would be that they reflect the excitation of Rydberg series, in which a single electron is excited to a large orbit about the residual ion. As such states are the substance of standard atomic structure theory, their analysis would appear to be a good test of the utility of such theory in the description of giant resonance phenomena. We now discuss briefly some calculations,[19] done in a multiconfiguration Hartree-Fock approximation, of the oscillator strengths and energies of these sharp absorption lines.

As discussed above, there is a great difference between the Hartree-Fock f orbitals for $^3P^o$ and $^3D^o$ states, and those for $^1P^o$ states. These states are, however, mixed by the spin-orbit interaction, and L and S will generally not be good quantum numbers. The bulk of the spin-orbit effect is due to the $4d^9$ ionic core, the $J = 5/2$ and $J = 3/2$ states of which are separated by about 2.5 eV. This core splitting has a pronounced effect on the discrete spectrum, though it does not greatly affect the total photoabsorption cross section in the continuous region. The treatment of the spin-orbit interaction simultaneously with strong LS term dependence is a problem which has not yet been solved in a general and practical manner. It was dealt with in this example by the following means. An orbital basis was obtained by solving[25] the multiconfiguration Hartree-Fock (MCHF) equations for coupled $4d^9 nf(LS)5s^2 5p^6$ and $4d^9 5s^2 5p^5 5d^2$ states for each LS over a range of n values (the reasons for including interaction with the $5p^5 5d^2$ configuration are discussed below). The resulting radial orbitals of different LS and n are in general *not* orthogonal, because of the difference in the effective potentials: for instance the $4f\ ^1P^o$ orbital has a large radial overlap with the $5f\ ^3P^o$ and $^3D^o$ orbitals (orthogonality of the associated *states* is of course enforced through the LS coupling scheme). The spin-orbit interaction in this basis was approximated by the contribution from the 4d hole alone, the part due to the f electron being small in comparison. Different terms of different *configurations* are thus allowed to interact, in contrast to the usual approach of diagonalizing the spin-orbit interaction within each configuration. The energy spectrum was computed by diagonalization of the matrix of MCHF energies and spin-orbit interaction; the results are shown in comparison to the experimental data in Fig.6.

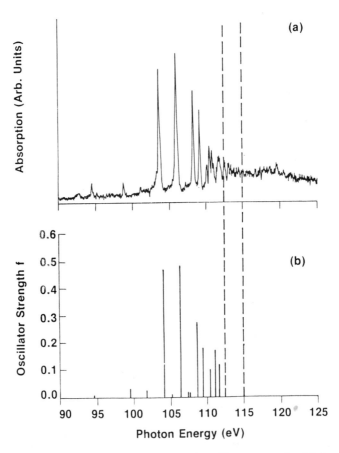

Fig. 6. Experimental absorption (a) and calculated oscillator strengths (b) for 4d excitation of Ba^{2+}. Dashed lines indicate the 4d ionization thresholds as inferred from the quantum defects of the $4d^9 np$ series, some members of which are visible in this spectrum. Limited energy range of calculated results reflects truncation of orbital basis set. From ref. 19.

The Role of Electron Correlation

Two types of electron correlation phenomena are found to be of importance in the absorption spectrum of Ba^{2+}. As in the continuous absorption spectrum of Ba (and other elements in this region of the periodic table), correlations in the initial state affect the distribution of oscillator strength. The most significant of these can be described in terms of the admixture of a $4d^84f^2$ configuration with the $4d^{10}$ ground state. Its corresponding excitations of the type 4f → 4d tend to reduce the intensity of 4d → nf transitions from the dominant configuration. However, MCHF calculations give the $4d^84f^2$ configuration a weight of only ~ 2% in the ground state. Thus the initial state is reasonably well approximated by a single configuration.

Correlations in the final state affect both the relative energies and oscillator strengths of the lines. Here again MCHF calculations give a relatively small weight to correlating configurations in the total wavefunction, but the effects of correlation are amplified by the potential barrier that governs the wavefunction of the f electron in the dominant configuration. The $5p^55d^2$ configuration mentioned above is coupled to the dominant $5p^6nf$ configuration by the dipole interaction. Although this coupling is not strong in the sense of inducing large configuartion mixing, it provides an effective polarization potential which reduces the height of the barrier. This permits greater penetration of the core by the f wave.

In the language of conventional atomic structure, the Ba^{2+} spectrum presents a pronounced case of intermediate coupling. Of the lines apparent in the spectrum of Fig. 6, only the collapsed $4d^94f$ $^3P^o$ and $^3D^o$ states and the $4d^9np$ states are characterized by more or less pure coupling schemes. The first few strong lines have been designated in ref. 19 as $4d^9nf$ $^1P^o$ states, but this is based on an observed regularity in the $^1P^o$ component of their total wavefunctions rather than on any predominance of $^1P^o$ character. This regularity does show that the strong lines are indeed part of a one-electron Rydberg series, though one which perhaps cannot be given a simple and accurate name.

Other accounts of this spectrum have been given. Connerade and Mansfield[26] employed a Hartree-Slater basis as well as some term-dependent wavefunctions in a single configuration approach. Cheng and Froese Fischer[27] performed term dependent calculations which resulted in a gross oscillator strength distribution similar to that reported here, but they did not attempt to treat the spin-orbit interaction quantitatively. Nuroh et al.[28] utilized the time-dependent local density approximation, and have asserted the importance of collective effects in interpreting their results; however, their calculation does not incorporate the proper form of the long-range potential nor the spin-orbit interaction.

PHOTOABSORPTION BY NEUTRAL ATOMS AFTER BARIUM

Photoabsorption experiments on the next two members of the Ba^{2+} isoelectronic sequence, La^{3+} (Z=57) and Ce^{4+} (Z=58), have been performed in Dublin,[29] though the results have not yet been published. In the model described above, the long-range Coulomb attraction and the depth of the atomic inner well both increase monotonically along the isoelectronic sequence, and so the $4d^94f$ $^1P^o$ state contracts into the inner well. The 4d absorption oscillator strength is thus progressively concentrated in the $4d^{10}$ ^1S → $4d^94f$ $^1P^o$ line. This view is borne out by term dependent Hartree-Fock calculations along the isoelectronic sequence,[17,27] though the effects of spin-orbit coupling remain to be investigated in detail.

The situation is more complicated in the *isoionic* sequence of ground-state neutral atoms. Quantum defect analysis shows that in neutral Ba, the lowest-lying f orbital - the 4f orbital of the configuration $5p^66s4f$ - is hydrogenic, and in neutral La, the lowest-lying f orbital is collapsed, although it is not occupied in the ground state configuration $5p^65d6s^2$. One might therefore suppose that since the 4f orbital of La $4d^{10}5s^25p^64f6s^2$ is collapsed, the 4f orbital in the presence of a 4d core hole - for example La $4d^95s^25p^64f6s^25d$ - should also be collapsed. A 4d electron screens the Coulomb field of the nucleus, and its removal ought to deepen the inner well. This does indeed happen for most of the $4d^94f$ states, which appear as relatively sharp features below the 4d ionization threshold in photoabsorption and electron energy loss experiments.[30,31] However, term dependendent calculations of the $4d^94f\ ^1P^o$ state of neutral La give a hydrogenic result, and an effective potential which holds a shape resonance above the 4d threshold, in accord with experimental results. Again, this is a consequence of the repulsive dipole exchange interaction; it is remarkable that this repulsion is sufficiently strong to overcome the Coulomb attraction of the core hole. Thus the "giant resonance" in 4d excitation of La is a shape resonance similar to that of Ba.

As one moves further into the lanthanide sequence, the 4f orbital is occupied in the ground state. The 4d photoabsorption process is then described as $4d^{10}4f^n \rightarrow 4d^94f^{n+1}$, and the multiplicity of final states increases greatly. The occurence of giant resonances in this process was analyzed in terms of an average-of-configuration model by Dehmer et al.,[32] who identified the following mechanism. The oscillator strength for these transitions is concentrated in final states which have primarily $4d^94f(^1P^o)4f^n$ parentage, by electric dipole selection rules. The repulsive 4d - 4f exchange interaction puts such states at much higher energies than the majority of $4d^94f^{n+1}$ states, and in fact they are capable of autoionizing to continuua of the type $4d^94f^n\varepsilon f$, where the residual $4d^94f^n$ state has no appreciable $4d^94f(^1P^o)4f^{n-1}$ parentage. The electric monopole interaction which drives this autoionization can be very strong, and this gives the giant resonances their large width. This model has been quite succesful in interpreting experimental data, especially in reproducing the multiplet structure of low-lying $4d^94f^{n+1}$ states.

Orbital term dependence could alter the predictions of this model in a qualitative way. However, we know of no positive evidence of substantial term dependence in $4d^94f^n$ states for Ce and heavier atoms. We have carried out MCHF calculations for $4d^94f^2$ states of neutral La and Ce, and have found no great qualitative differences between the 4f orbitals associated with states of predominantly $4d^94f\ ^1P^o$ parentage and other states. However, such calculations are probably too restrictive to preclude entirely the possibility of shape-resonant behavior in the rare earths.

GIANT RESONANCES IN OTHER MODES OF EXCITATION

We have discussed 4d photoabsorption in terms of initial and final states of atomic systems, which lend themselves to description in more or less familiar terms. In this approach one views "giant resonance" states as static entities which exist independently of specific excitation mechanisms. It is thus appropriate to examine the properties of the states associated with giant resonance phenomena, outside the familiar context of electric dipole excitation of inner electron shells.

The nf Rydberg series of Ba⁺

There are many similarities in the properties of states of atoms with the same values of $Z + c^*$, where c^* is the number of inner shell holes. If Z increases in step with the filling of holes, the monopole potential on an outer electron stays constant. Furthermore, for small values of c^*, the change in local charge density associated with a change in c^* is relatively small: for example, removal of a 4d electron from a closed shell decreases the number of N shell electrons by 1 / 18. Thus along a sequence of this type one often finds slow variation of properties of outer electrons, such as the quantum defects of the Rydberg series Xe $5p^5nl$ and Cs $5p^6nl$ (ref. 33).

Considerable attention has been given to the Rydberg series of Ba⁺ $4d^{10}5s^25p^6nf$ as an example of the atomic double-well potential in the 4f transition region.[34-37] An interesting correspondence can be made between this Rydberg series and its *core-excited* counterpart in Cs⁺ $4d^95s^25p^6nf$, which also illustrates some of the effects of final-state correlation in 4d excitation of Ba²⁺. Figure 7 shows the 4f orbital of Ba⁺, as computed in several approximations. The bimodal structure of this orbital is quite similar to that of the 4f orbital of Cs⁺ $4d^95s^25p^64f$, as computed in the Hartree-Fock average-of-configuration approximation.[17,38] Although that orbital of Cs⁺ does not represent any given LS term of the $4d^94f$ configuration very well, it provides the appropriate point of comparison with the state of Ba⁺, in which the 4d - 4f interactions are indeed spherically averaged. The quantum defect of the Cs⁺ orbital is 0.129, compared to 0.097 for Ba⁺.

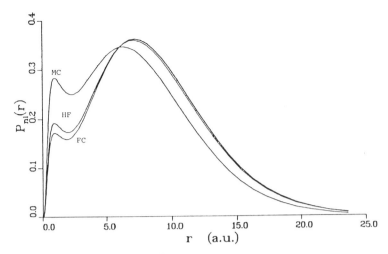

Fig. 7. 4f orbitals for Ba^+ $4d^{10}5s^25p^64f$. FC: computed in the frozen core of Ba^{2+}; HF: Hartree-Fock solution with full core relaxation; MC: Multiconfiguration Hartree-Fock, including interaction with $5p^55d^2$ configuration.

The "amplification" of small correlation effects is shown quite strikingly by the orbital labelled MC. This was obtained by solving the MCHF equations for the coupled configurations $5p^64f$ and $5p^55d^2$, the admixture which was found to be of importance in the 4d photoabsorption of Ba^{2+}. The resulting state has only a 1.2% content of $5p^55d^2$, yet the f orbital of the dominant configuration is able to penetrate much further into the core. Similar results are obtained for the 5f orbital. The effect this correlation has upon observable quantities is summarized in Table 3.

It is evident that further treatment of correlation, and probably also relativistic effects, is required to bring the quantum defects of the f series into agreement with experiment. Calculations in the g-Hartree approximation, in which the parameter g is adjusted for each state in order to obtain the correct quantum defect, reproduce quite well the experimental values of the fine structure intervals.[36,37] It is interesting to note that to obtain the increase in quantum defect from n = 4 to n = 5, the value of g was changed only from 0.997 to 0.995; this presumably also reflects the great sensitivity of the f wave to small changes in an effective potential.

Table 1. Quantum defects μ and fine structure intervals Δ (in cm^{-1}) as computed for the 4f and 5f states of Ba^+ in frozen core (FC), Hartree-Fock (HF), and multiconfiguration Hartree-Fock (MC) approximations, in comparison to experimental values.

		FC	HF	MC	Experiment
4f	μ	0.093	0.097	0.169	0.314
	Δ	50.1	62.7	189.6	224.7
5f	μ	0.228	0.276	0.456	0.646
	Δ	107.5	231.7	313.8	240.7

Another obvious challenge to our "structural" approach to giant resonance phenomena is the elucidation of multiplet features. The selection rules of atomic photoabsorption have caused attention to be focused upon processes of dipole excitation, but distinctive effects are also present in higher multipole processes.

Electron energy loss spectroscopy at primary energies of 0.1 - 2 keV has provided a wealth of information about multiplet structure of giant resonance states.[31,39] For example, many of the LS terms and fine structure of the $4d^9 4f$ configuration of La have been identified by this means, whereas photoabsorption reveals only the three $J = 1$ states. In particular, the $^1F^o$ and $^1H^o$ states of this configuration can be excited by electric multipole interaction at large impact parameters, and they appear with significant intensity even at relatively high incident energies. These states are collapsed in La, in accordance with Hartree-Fock calculations. However, similar states in Ba or Cs may appear as shape resonances. As seen in Table 2, the higher multipole potentials can also be repulsive, and the direct quadrupole and exchange octupole potentials are particularly effective in repelling the f electron from the core.[38]

The shape resonance in 4d photoabsorption in the f transition region has been attributed to the repulsive dipole exchange potential in the $4d^9 \varepsilon f\ ^1P^o$ channel, which is also the channel that carries all the $4d^{10} \rightarrow 4d^9 \varepsilon f$ dipole transition strength in the absence of spin-orbit coupling. An analogous statement can be made about $4p^6 \rightarrow 4p^5 \varepsilon f$ and $4s^2 \rightarrow 4s \varepsilon f$ transitions, which can be driven respectively by electric quadrupole (or hexadecapole) and octupole excitation: the final state channel, respectively ^1D (or ^1G) and $^1F^o$, is subject to a repulsive exchange interaction of the same multipole order. Hartree-Fock calculations indicate that these potentials can give rise to shape resonances and to substantial orbital term dependence in transition region elements.[38] Some $4p \rightarrow 4f$ excitations have been identified in electron energy loss spectroscopy of the lanthanides.[40]

CONCLUSION

We have presented a description of a limited variety of giant resonance phenomena, with emphasis on their structural features, in terms of familiar (to us) concepts of atomic spectroscopy and collision physics. An alternative viewpoint, articulated in this volume by Wendin and Zangwill, prefers to consider giant resonance phenomena in terms of the dynamical response of the atom to an external field. These two approaches produce more or less identical results when they are actually applied in practice, at a similar level of approximation. Whether the differences in their outlooks amount to more than matters of taste, may be decided on issues of multiple electron excitation and ionization processes, an emerging field of "giant," if not always resonant, phenomena which poses formidable challenges to both.

ACKNOWLEDGEMENTS

Tom McIlrath, Chris Cromer, John Cooper, Wendell Hill, Bernd Sonntag, Jack Sugar, Andy Weiss, and Steve Younger have contributed to various aspects of the work reported here. Support from the U.S. Air Force Office of Scientific Research, under contract ISSA 86-0023, is gratefully acknowledged.

REFERENCES

1. D. C. Griffin, R. D. Cowan, and M. S. Pindzola, this volume.
2. E. Fermi, Leipziger Vorträge :95 (1928)
3. M. Goeppert Mayer, Phys.Rev. 60:184 (1941)
4. A. R. P. Rau and U. Fano, Phys.Rev. 167:7 (1968)
5. J.-P. Connerade, this volume.
6. T. B. Lucatorto, T. J. McIlrath, J. Sugar, and S. M. Younger, Phys.Rev.Lett. 47:1124 (1981)
7. I. C. Lyon, B. Peart, J. B. West, A. E. Kingston, and K. Dolder, J.Phys.B 17:L345 (1984); J. B. West, private communication (1986); F. Wuilleumeier, private communication (1986)
8. T. B. Lucatorto and T. J. McIlrath, Appl.Opt. 19:3948 (1980)
9. B. F. Sonntag, C. L. Cromer, J. M. Bridges, T. J. McIlrath and T. B. Lucatorto, in : "Short Wavelength Coherent Radiation: Generation and Applications," American Institute of Physics, New York (1986)
10. C. L. Cromer, J. M. Bridges, J. R. Roberts, and T. B. Lucatorto, Appl.Opt. 24:2996 (1985)
11. H. P. Kelly, S. L. Carter, and B. E. Norum, Phys.Rev.A, 25:2052 (1982)
12. J. E. Hansen, Phys.Scr. 21:510 (1980)
13. G. Wendin, this volume.
14. M. O. Krause, W. A. Svensson, T. A. Carlson, G. Leroi, D. L. Ederer, D. M. P. Holland, and A. C. Parr, J.Phys.B 18:4069 (1985)
15. K. T. Cheng and W. R. Johnson, Phys.Rev.A 28:2820 (1983)
16. S. M. Younger, Phys.Rev.Lett. 56:2618 (1986)
17. T. B. Lucatorto, T. J. McIlrath, W. T. Hill III, and C. W. Clark, in: "X-Ray and Atomic Inner-Shell Physics - 1982," B. Crasemann, ed., American Institute of Physics, New York (1982)
18. C. Froese Fischer, "The Hartree-Fock Method for Atoms," John Wiley and Sons, New York (1977)
19. C. W. Clark, J.Opt.Soc.Am.B 1:626 (1984)
20. G. O'Sullivan, J.Phys.B 15:L765 (1982)
21. G. Wendin, Phys.Lett.46A:119 (1973); and this volume.
22. M. Ya. Amusia, V. K. Ivanov, and L. V. Chernysheva, Phys.Lett.59A:191 (1976)
23. A. Zangwill and P. Soven, Phys.Rev.Lett. 45:204 (1980); A. Zangwill, this volume.
24. H. P. Kelly, this volume
25. using a modified version of the code of C. Froese Fischer, Comp.Phys.Commun. 14:145 (1978)
26. J.-P. Connerade and M. W. Mansfield, Phys.Rev.Lett. 48:131 (1982)
27. K. T. Cheng and C. Froese Fischer, Phys.Rev.A 28:2811 (1983)
28. K. Nuroh, M. J. Stott, and E. Zaremba, Phys.Rev.Lett. 49:862 (1982)
29. G. O'Sullivan, private communication (1986)
30. E.-R. Radtke, Ph.D. Thesis, Universität Bonn (1980)
31. F. P. Netzer, G. Strasser, and J. A. D. Matthew, Phys.Rev.Lett. 51:211 (1983)
32. J. L. Dehmer, A. F. Starace, U. Fano, J. Sugar, and J. W. Cooper, Phys.Rev.Lett. 26:1521 (1971)
33. V. Kaufman, J. Sugar, C. W. Clark, and W. T. Hill III, Phys.Rev.A 28:2876 (1983)
34. D. C. Griffin, K. L. Andrew, and R. D. Cowan, Phys.Rev. 177:62 (1969); D. C. Griffin, R. D. Cowan, and K. L. Andrew, Phys.Rev.A 3:1233 (1971)
35. J.-P. Connerade and M. W. Mansfield, Proc.Roy.Soc.Lond.A 346:565 (1975)
36. J.-P. Connerade, K. Dietz, M. W. Mansfield, and G. Weymans, J.Phys.B 17:1211 (1984)
37. K. Dietz, this volume
38. C. W. Clark, to be published
39. J. A. D. Matthew, this volume
40. J. A. D. Matthew, F. P. Netzer, C. W. Clark, and J. F. Morar, to be published

GIANT RESONANCES IN HEAVY ELEMENTS : OPEN - SHELL EFFECTS,
MULTIPLET STRUCTURE AND SPIN - ORBIT INTERACTION

Françoise Combet Farnoux

Laboratoire de Photophysique Moléculaire du CNRS
Université Paris-Sud, Bâtiment 213
91405 Orsay (France)

INTRODUCTION

As soon as 1964, Ederer (1) pointed out experimentally an absorption maximum beyond the 4d thresholds of xenon. Such a behaviour was confirmed some years later by Haensel et al. (2). Meanwhile, this effect characterised as the "delayed onset" of continuous absorption was explained in terms of a non Coulombic central potential model by several theoreticians, e.g. Cooper (3) , Combet Farnoux and Heno (4) , Fano and Cooper (5) , Manson and Cooper (6) , who pointed out the importance of the angular momentum potential barrier for the outgoing (l=3) electron , in photoionisation of nd subshells of many heavy elements.

In fact, the effect of the $l(l+1)/r^2$ repulsive term in the Schrödinger equation tends to keep electrons away from the core so that at threshold, the continuum function hardly penetrates the core and the spatial overlap between the initial and final state functions is very small. As the kinetic energy of the outgoing electron increases, its ability to penetrate the barrier increases also and the overlap becomes larger. This is going on until the negative part of the continuum wave function overlaps the core function, when phase cancellation sets in and the overlap integral, proportional to the photoionisation cross sections falls off.

A review over heavy elements which display such delayed onsets of absorption peaking at different energies above the np and nd thresholds has been published by Combet Farnoux (7). These calculations, as those of Manson and Cooper (6) which concerned lighter elements, were achieved with the non Coulombic central potential of Herman and Skillman (8), the same to describe both intial and final states. Numerous examples of the delayed onset have been measured by several groups using synchrotron radiation, and it was remarkable to notice the close similarity between spectra obtained from vapors and solid state samples. Unfortunately, in most cases, we had to notice also a certain quantitative discrepancy with the results of the calculations mentioned above. A central potential model able to point out the existence of an absorption maximum more or less delayed from threshold according to the choice of this potential does not provide with a sufficient agreement with the experimental results when positions and amplitudes are compared. As a consequence, as soon as 1970, in atomic theory, it was clear that in VUV photoionisation, the escaping electron cannot be well described by an eigenfunction of a zero-order Hamiltonian including the same central potential as the Hamiltonian relative to the initial state. Improved calculations were proposed, starting from either a time-dependent picture of the photoionisation process or a stationary-state formulation within the framework of spectroscopic methods. Both aspects are equivalent and allow to understand why resonances similar to those observed beyond thresholds can be observed below them in some cases and the importance of many-electron effects in the theoretical treatment of resonances in the photoabsorption spectra of the XUV energy range.

The giant resonances whose study is the aim of this paper are a probe of these many-electron effects in so far as they are relevant to transitions (either in the discrete or in the continuous spectrum) involving filled and empty orbitals with similar spatial dimensions and resulting in large radial overlaps and large polarizabilities. As examples, the $4d \rightarrow 4f$ and $5p \rightarrow 5d$ transitions in the rare-earths series are strongly influenced by many-electron effects as emphasized in several papers (9 ,10 ,11 ,12 ,13 ,14) which clearly show why central field models break down in the description of the corresponding photoionisation cross sections, as well as in that of other quantities involving the same core level.

In the present paper devoted to giant resonances in heavy atoms, I will explain first their various origins for atomic numbers corresponding to the progressive filling of the 5p (6p) and 4f (5f) subshells, i.e. for the atoms included between Ag (Au) and the end of the lanthanide (actinide) series, as well as the characteristics of other giant resonances, not so well studied up to now, and corresponding to the filling of the nd subshells in various transition metal series. The other Section will emphasize the treatment of such resonances, especially when the system considered (atom or ion) is an open-shell system. Moreover, the influence of spin-orbit effects will be discussed, in so far as they are able to precipitate the collapse of the excited electron and to change the coupling as a consequence.

THE VARIOUS TYPES OF GIANT RESONANCES

Although the central potential models introduced up to now in the calculations are not adequate to describe quantitatively the resonances, a realistic non Coulombic potential as that of Herman and Skillman (8) can be of great help to explain their origin since it takes into account the shell structure of the atoms and introduces an averaged exchange term. Such a potential is able to account for both the filling and the contraction of d or f orbitals in the long sequences of the periodic table. This orbital contraction leads to consider the wavefunction collapse when investigating some atomic effects connected with the overlap of two wavefunctions, e.g. transition probabilities. As recently emphasized by Connerade (15) , the wavefunction collapse hinges on a basic difference between the properties of long-range (Coulomb) and short-range potentials. The long-range potential possesses an infinite number of bound states (Rydberg states) converging on a series limit which separates the discrete spectrum from the continuum, whereas by contrast, the short-range potential is able to trap a finite and variable number of bound states fairly independent on environment.

Let us remark that when solving the Schrödinger equation, both the long and short range potentials are effective potentials because they include the centrifugal term and are described by the following expression:

$$V_{eff}(r) = - V_{HS}(r) + \frac{l(l+1)}{r^2} \quad (1)$$

$V_{HS}(r)$ representing the Herman and Skillman (8) potential. Curves $V_{eff}(r)$ plotted for l=2, 3 and some elements by Rau and Fano (16) are shown in Fig.1.

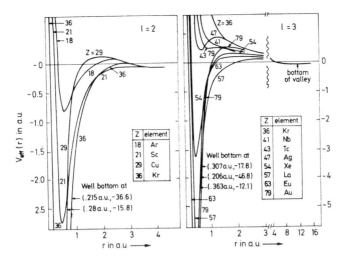

Fig. 1: The effective potential $V_{eff}(r)$ defined in formula (1) and plotted versus coordinate r for l=2 and l=3 (from reference (16)).

The curves in Figure 1 allow to understand these cases when the effective potential displays a double-valley, i.e. an inner well and an outer one separated by a positive barrier, so that both situations mentioned above may coexist. When the inner well is too shallow to support a new bound state, a pseudo-state of short lifetime develops in the continuum, becoming sharper and moving downwards in energy, when the inner well is becoming deeper. In fact, the pseudo-state is simply a loop in the oscillatory wavefunction of the continuum which becomes resonant in the well and whose amplitude grows. As the well becomes deeper, i.e. more binding, the lifetime of the pseudo-state increases, its resonant behavior occurs at lower energy and eventually crosses over into the discrete spectrum, where it turns into a true bound state. This general evolution is in agreement with the observation of giant resonances located either beyond threshold, as in Xe (4d absorption maximum delayed from threshold), or below threshold, as for example, the transition $4d \rightarrow 4f$ in lanthanum. In other words, we should distinguish 2 types of giant resonances :

The shape resonances
===================

They correspond to a large overlap between the core level and the pseudo-state orbitals. Let us recall that a shape resonance is a phenomenon associated with the motion of a single particle with energy close to the edge of a deep potential well. It occurs for an electron-ion or electron-atom mainly when it exhibits a centrifugal barrier because otherwise the edge of an atomic field would be too diffuse. As discussed extensively in the paper of Fano and Cooper (5) and in references therein, many situations exist wherein the potential well inside the barrier can support a narrow band of quasibound states at some energy near or below the top of the barrier. The conventional way of observing this resonance phenomenon is to tune the total energy so that the net well inside the barrier is attractive enough to support a quasibound pseudo-state.

For instance, in Xe (Z=54) the 4f orbital in the excited configurations $4d^9 4f 5s^2 5p^6$ is hydrogenic (i.e. uncollapsed), as are all successive f orbitals in the Rydberg series $4d^9 nf$, $n > 4$. As the overlaps of bound f orbitals with the 4d orbital are thus very small, most of the $4d \rightarrow f$ oscillator strength is associated with transitions to continuum states. The centrifugal

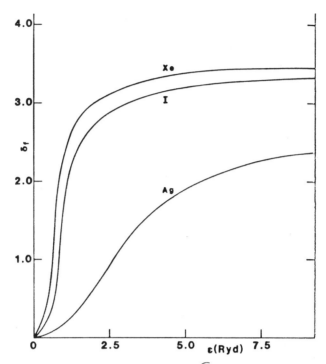

Fig.2 : Variation of the phase-shift $\delta_f(\varepsilon)$ versus the photoelectron energy ε, calculated in a central potential model (Herman and Skillman). δ_f is in radian. (From reference (18)).

barrier prevents very low energy continuum f waves from penetrating the inner well so that maximum $4d - \varepsilon f$ overlap is reached when the continuum energy exceeds the barrier height. For ε much greater than the barrier intensity, however, the overlap decreases because of the oscillatory behavior of the continuum wave in the inner well. Thus, for atoms without collapsed f orbitals, the double-well structure is responsible for a delayed onset of the photoionization continuum and for a peak in the photoabsorption cross section at energies of the order of the barrier ($\simeq$ 1 Ry.) above threshold. This resonant character of the free wave solution of the Schrödinger equation can be pointed out via the determination of the phase-shift $\delta_1(\varepsilon)$ which reflects the non Coulombic character of the central potential $V(r)$ into which the photoelectron escapes, according to the following normalisation condition:

$$P_{\varepsilon 1} \lim_{r \to \infty} (r) = \varepsilon^{-\frac{1}{4}} \sin(\varepsilon^{\frac{1}{2}} r - \frac{\pi}{2} 1 + \frac{z}{\sqrt{\varepsilon}} \ln 2\sqrt{\varepsilon} r + \arg \Gamma(1+1-\frac{iz}{\sqrt{\varepsilon}}) + \delta_1(\varepsilon)) \quad (2)$$

the continuum wave functions being usually normalized per unit energy range.

Indeed, the increase of $\delta_f(\varepsilon)$ by π between $\varepsilon = 0$ and $\varepsilon = \varepsilon_o$, ($\varepsilon_o$ value close to the barrier height) indicates that $P_{\varepsilon 1}(r)$ has gone through a resonance or that, alternatively, the first node of this wave function has moved from outside to inside the atom. This change of π for the phase-shift, which characterizes a shape resonance, is accompanied by a rapid variation of the "enhancement factor" $C_1(\varepsilon)$, i.e. the normalisation constant allowing to factorize the dipole-matrix element $R_{\varepsilon 1}$ into 2 factors:

$$R_{\varepsilon 1} = C_1(\varepsilon) \overline{R_{\varepsilon 1}} \quad (3)$$

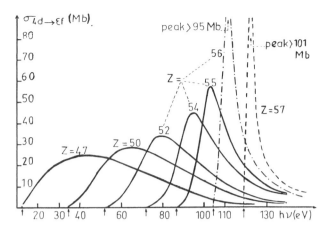

Fig. 3: Partial cross sections $\overline{\sigma_{4d \to \varepsilon f}}$ calculated from Hartree-Fock continuum wave functions by Combet Farnoux (19).

For inner subshells, $\overline{R_{\varepsilon l}}$ is almost constant over a wide range of energies since the potential $V(r)$ is large in that region and the dependence of both matrix elements and cross sections is almost completely determined by the variation of the normalisation factor $C_l(\varepsilon)$ beyond threshold. Let us note that $C_l(\varepsilon)$ depends on the behavior of the potential $V(r)$ over its entire range and that consequently the shape of the cross section for an inner subshell may be determined essentially by the behavior of the potential outside the subshell. The normalisation constant varies rapidly when $\delta_l(\varepsilon)$ does, the relation between both quantities involving the Jost function (17).

In order to illustrate this dependence so typical of shape resonances, Figure 2 displays the variation of $\delta_l(\varepsilon)$ calculated (18) for 3 atoms : Ag, I and Xe, using the Herman and Skillman (8) potential. It is of interest to notice that for I and Xe, there is a change of π over a few Rydbergs beyond threshold, whereas for Ag, the smoothly increasing curve will never reach π. In relation with Figure 1 which shows a difference between $V_{eff}(l=3)$ for Ag and Xe since there is no negative inner well for Ag, a sharpening up of $4d \to \varepsilon f$ resonances when going from Ag to Xe has to be expected: it corresponds to the increase of the nuclear charge which allows to an εf continuum wave to mimic the behaviour of a collapsed 4f orbital. As long as there is no inner well attractive enough to support a pseudo 4f state, there may exist a broad bump due to the variation of the overlap of the 4d core level and f continuum orbitals but no giant shape resonance. This distinction is illustrated by Figure 3 which displays the $4d \to \varepsilon f$ cross sections (the $4d \to \varepsilon p$ cross sections are negligible relatively to the others) of Ag (Z=47), Sn (Z=50), Te (Z=52), Xe (Z=54), Cs (Z=55), Ba and La (Z=56 and 57) calculated by Combet Farnoux (19) in an independent-particle model since the εf continuum wave function is determined as the solution of an Hartree-Fock equation, both ground and ionic states being described by the ground state Herman and Skillman wave functions. The sharpening up of the peak when Z increases from 47 up to 57 is in favor of the existence of a giant resonance only for elements around xenon. Moreover, let us remark the spectacular giant peaks of Ba (Z=56) and La (Z=57): their existence beyond threshold in the continuous spectrum is in agreement with experiment

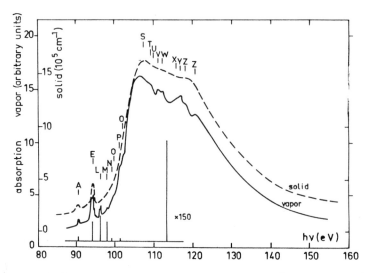

Fig.4: Photoabsorption cross-section for solid (dashed curve) and gaseous Ba (solid curve) in 4d excitations region (from Rabe et al. (20)).

as Figure 4 allows to point it out for Ba for which photoabsorption measurements (20) (21) display a striking similarity between solid and gaseous Ba spectra (20) on the one hand, and between 4d photoabsorption spectra of Ba, Ba^+ and Ba^{2+} (21) on the other hand.

This case of Ba deserves a special paragraph, in so far as despite the first evidence by Combet Farnoux (19) of a giant resonance in the continuous $4d^9 \varepsilon f\ ^1P$ channel as soon as 1972, the labelling by Connerade and Mansfield (22) of the "quasi resonance" above the 4d photoionization threshold as "$4d \rightarrow 4,\varepsilon f$" transitions to the continuum and not as a broadened discrete structure was challenged by Ederer et al.(11) who argued on the basis of an analogy with the lanthanides that this resonance would be more aptly labelled $4d^{10}\ ^1S^e \rightarrow 4d^9 4f\ ^1P^o$, the $4d^9 4f\ ^1P^o$ level having been pushed up into the continuum where it could autoionize strongly, as described by Dehmer et al. (9) and Starace (23). According to these authors, photoionisation of the 4d subshell in Ba should be understood as a $4d \rightarrow 4f$ transition inside the atom, a scattering process being the origin of the "leaking" into the associated continuum of this 4f level. All these explanations did not take into account the previous calculations of Combet Farnoux (19) which had introduced the diagonalization of the intrachannel interactions within the particular channel $4d^9 \varepsilon f\ ^1P^o$. The consideration of the ^1P term is essential in both discrete and continuous spectra as shown in 1975 by Hansen et al. (12) who performed a separate variational calculation for the $4d^9 4f\ ^1P^o$ level in Ba (L-S Hartree-Fock method) leading to conclusions different from those of earlier treatments in which all the $4d^9 4f$ levels were described from the integrals and parameters obtained for the average configuration. In other words, via different routes (calculations in the continuous spectrum on the one hand, separation of the various terms of a discrete configuration on the other hand) both Combet Farnoux (19) and Hansen et al.(12) were able

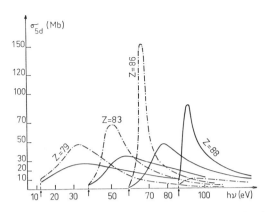

Fig.5: Total photoionisation cross section relative to the 5d subshell for Au, Bi, Rn, and Ra calculated with a central potential model (dot-dashed line) and with a continuum Hartree-Fock wave function (solid line) (From reference 19).

to establish that the $4d^9 4f\ ^1P^o$ wave function is not localized in the inner valley of a double-valley potential which is just slightly too shallow to support a collapsed 4f wave function but that the $4d^9 \varepsilon_o f\ ^1P^o$ continuum wave function for ε_o, a positive energy of the order of magnitude of the barrier is resonant in the inner valley and closely resembles the lowest energy eigenfunction that the well could have supported had it been slightly deeper. Moreover, the consequence of the transfer of the oscillator strength to the continuum $4d^9 \varepsilon f\ ^1P^o$ lies in the quenching of the transition $4d^{10}\ ^1S^e \rightarrow 4d^9 4f\ ^1P^o$, as Fliflet et al.(24) has proved it.

Returning to the historical point of view of this problem of giant resonances in atomic physics, the Ba case has certainly played a pioneering role, since the term giant dipole resonance was proposed by Wendin (25) for Ba to emphasize the very close relationship between collective resonances in atoms and in nuclei. However, excepting a great sensitivity to the term considered for the excited configuration $4d^9 4f$ (the radial functions associated to the triplet terms are collapsed whereas the function associated with the optically allowed ^1P term is uncollapsed), the situation in a Hartree-Fock term-dependent basis is similar to that in Xe and Cs, when dealing with the 4d subshell oscillator strength of Ba : the optically allowed bound 4f state has little oscillator strength because of the small 4d-4f overlap, and the bulk of the absorption occurs in the $4d^9\ \varepsilon f\ ^1P$ continuum. However, in order to unify the various interpretations proposed and mentioned above (9)(11)(12)(19)(22)(23) Wendin and Starace (14) have demonstrated an equivalence between the first order approximation wave functions derived by applying perturbation theory to the average Hartree-Fock basis and the term-dependent Hartree-Fock approximation. As a conclusion, the term-dependent Hartree-Fock basis accounts for the intrachannel coupling to all orders within the single particle picture and seems to be a very suitable model for this problem.

A similar evolution for the 5d subshell photoionisation cross sections may be predicted and can be seen in Figure 5, where results obtained for 4 heavy elements (Au, Pb, Rn and Ra) by Combet Farnoux (19) using both models mentioned above for the 4d subshell cross section (the Herman and Skillman central potential on the one hand and the Hartree-Fock continuum wave function in the $5d^9 \varepsilon f\ ^1P$ channel on the other hand) can be compared. It is remarkable to point out that for the first 3 atoms, Au (Z=79), Bi (Z=83) and Rn (Z=86) the second model pushes the shape resonance up to higher energies and broadens it whereas an opposite situation occurs for Ra (Z=88). In other words, the diagonalization of the intrachannel matrix within the $5d^9 \varepsilon f\ ^1P$ channel leads generally to a broadening of the resonance which is shifted to higher energies except in cases when the Hartree-Fock continuum basis is term-dependent, as it is the case for Ra located beyond the rare gas Rn in the periodic table, on the analogy of the 4d subshell in Ba.

Despite a certain similarity with the series of 4d elements discussed above, more important spin-orbit effects are likely to precipitate the localisation (collapse) of the empty 5f levels, and consequently to lead to a sligthly different 5d subshell behavior when the atomic number increases inside the 5d series elements. For example, the empty 5f levels start to become important as very broad resonances in the continuum as soon as gold and mercury (Z= 79 and 80) influencing the dynamics of the $5d \rightarrow 5f, \varepsilon f$ transitions , as shown by Keller and Combet Farnoux (26) and the structure of 5s and 5p holes , as explained by Wendin (27) . Such a difference between the 4d behavior in Ag and the 5d in Au could be predicted in Figure 1 which presents clearly different shapes of the inner well in both elements: the negative inner well of gold is evidently more attractive than the positive short-range effective potential in silver. Such effects increase dramatically with increasing atomic number Z due to the sharpening up of the εf shape resonance in the continuum, until the 5f levels become occupied in the potential of two 5d holes, as emphasized by Wendin (27) and finally also in the ground state of the 5f elements (actinides). Taking into account the fact that Figure 1 did not introduce any spin-orbit effect, an extra difference with the 4d series elements lies in the larger value of the spin-orbit splitting for the 5d hole. Although in the case of giant shape resonances the spin-orbit splitting is small or comparable to the exchange interaction, (L-S limit) this parameter has a short-range influence and strengthens the attractive effect of the inner well, more or less according to the resonant component considered. There results a large transfer of the oscillator strength and the low-energy resonant component becomes weak in comparison with the resonant component at higher energy. We will come back later to this problem of relativistic effects upon the giant resonances in atoms, but as soon as now, we can assert that the 5d absorption in thorium and uranium (28)(29)(30) are examples of this type of behavior in the same manner as the 4d absorption in Ba and La (20) .

In fact, as shown in Figure 5, Combet Farnoux's calculations (19) do not go further than Ra (Z=88) since experiments were scarce 15 years ago. However, for Ra, the great sensitivity of theoretical results to the choice of the model lets us expect a certain similarity with behavior of elements beyond xenon. The sharpening up of the 5d resonance when replacing the central potential model by a term-dependent Hartree-Fock approximation supports the previous statement of an average 5f orbital collapsing into the inner well of a double-valley potential, whereas the $5d^9 5f\ ^1P$ discrete wave function is eigenfunction of the outer valley (i.e.uncollapsed) which implies a transfer of the major part of the 5d oscillator strength into the optically

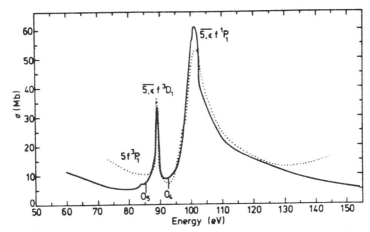

Fig.6 : Comparison between experimental data for the absorption spectrum of metallic thorium (solid line: Cukier et al.(28) and dotted curve: Connerade et al.(29)). The dotted curve corresponds to relative cross sections. Experimental $O_4 O_5$ thresholds have been measured by Fuggle et al.(31). (From reference 29).

allowed channel $5d^9 \varepsilon f\ ^1P$, the rest of this oscillator strength being shared between the continuum channel $5d^9 \varepsilon f\ ^3D$ and the discrete state $5d^9 5f\ ^3P$. Such a distribution has been observed in both metallic thorium by Cukier et al.(28) and ThF_4 vapour by Connerade et al.(29) as can be seen in Figure 6.

We have discussed extensively the origin of the giant shape resonances when excitation or ionisation of a d subshell is involved, i.e. when the effective potential relative to the final excited state displays a double-valley profile, a more or less high barrier separating both inner and outer wells. Giant shape resonances have also been observed when the ionisation takes place in a p subshell, although because of the more diffuse edge of the atomic field, the shape of the potential barrier and the particular features of the collapse are highly sensitive to the shell structure of the atom, as it has been discussed by Griffin et al.(32). For these reasons, each spectrum would deserve a particular discussion and we will mention here only some studies which emphasize that the collapse of d-wave functions is more gradual than the collapse of f-wave functions and the competition between the filling of the s and d subshells favours s-d mixing effects which complicate the observed spectra. It would be particularly interesting to study both experimentally and theoretically the influence of core excitation on d wave function collapse for a wide range of elements near the beginning of the transition metal series. Among some studies involving rather light elements, let us mention the observation of the 3d wave function collapse under 2p and 3p subshell excitation in K by Mansfield (33) and in Ca by Mansfield (33), Connerade (34) and Mansfield and Newsom (35). As an example of a heavy atom, the 5p spectrum of Ba obtained by Connerade et al.(36) and interpreted by Rose et al. (37) shows the existence of 14 series which converge to 12 limits (instead of 4 series converging to the 2 limits $5p^5\ ^2P_{1/2}$ and $5p^5\ ^2P_{3/2}$) within the framework of a strong s-d mixing. This observation is not in agreement with the previous prediction by Wendin (38) of one giant absorption resonance in this spectrum. Another interesting study has been achieved by Cukier et al.(39) using fast electron energy loss spectroscopy and provides with a comprehensive study of the 3p-6p spectra in the 30-60 eV range, Z varying between 21 and 92.

The main characteristic of these giant shape resonances lies in their occurring within a single excitation channel, without any dependence for their existence on any interchannel coupling; however, the multiplet effects due to the collapse of the excited wavefunction need to be taken into account.

The decaying discrete resonances

We are interested now in a second type of giant resonances which appear below thresholds, by contrast with the giant shape resonances which we have discussed above and which belong to the continuous spectrum, beyond thresholds. They correspond to these discrete transitions $nl \rightarrow n\,l+1$ related directly to the collapse of the $n\,l+1$ electron, which may hold the greater part of the oscillator strength. Giant resonances corresponding to these large discrete transitions are observed in the absorption spectra, but more details concerning the dynamics of these transitions and the origin of the many-electron effects which affect both their position and intensity have been obtained these last ten years by photoelectron spectroscopy and photoemission techniques which allow to separate the various decay processes in the various corresponding channels. We will not present here all the theoretical problems arising from resonant photoemission and will only discuss the relative importance of the various decay channels to account for the broad width of some resonances. However, 2 types of such collapsed resonances have to be considered:

a) the collapse of the $n\,l+1$ orbital exists in the initial ground state configuration : $nl^{4l+2} n(l+1)^N$

The corresponding giant resonances involve the intense transitions:

$$nl^{4l+2} n(l+1)^N \xrightarrow{h\nu} nl^{4l+1} n(l+1)^{N+1} \quad (4)$$

This important case concerns the 3 transition metals series via the transitions :
$$np^6 nd^N \xrightarrow{h\nu} np^5 nd^{N+1} \quad (5)$$
and also both lanthanides and actinides series via the transitions:
$$nd^{10} nf^N \xrightarrow{h\nu} nd^9 nf^{N+1} \quad (6)$$
n being 4 or 5 according to the series considered.

Although the general theory of these giant resonances concerning both their origin and their broadening is well known now and often used within the framework of the multichannel scattering theory (23)(40)(41), there does not exist a single universal model able to provide us with absolute cross sections in perfect agreement with experiment for all the elements of these various series. As soon as 1970 (42), experiments on the 4d-shell photoabsorption spectra of the trivalent lanthanides showed a prominent peak about 6-18 eV above threshold. In many lanthanides, this main peak appears to be composed of several peaks that are not completely resolved. The magnitude of this main structure varies greatly: its experimental width at half-maximum varies between 6 and 26 eV, and its strength decreases to zero with increasing atomic number ($57 \lesssim Z \lesssim 71$), i.e. as the 4f subshell fills. Also present in the absorption spectra are numerous weak lines in the vicinity of the 4d-shell threshold. At lutetium (Z=71) all these absorption features disappear. Among many more recent experimental results for 4d photoabsorption spectra, let us mention those obtained by Radtke (43) for the elements from Z=59 up to 68.

Since all photoabsorption strength from the 4d (5d) shell in the lanthanides (actinides) should go the 4f (5f) subshell, the excellent overlap of the 4f (5f) orbital with the 4d (5d) orbital would also imply a strong exchange interaction between the $4f^{N+1}$ ($5f^{N+1}$) subshell configuration and the 4d (5d) vacancy. This interaction is so strong that it splits the term levels of the $4d^9 4f^{N+1}$ ($5d^9 5f^{N+1}$) configuration by about 20 eV, pushing up some levels far above threshold. These levels then autoionize to the various channels having the core configuration $4d^9 4f^N$ ($5d^9 5f^N$). The levels of $4d^9 4f^{N+1}$ ($5d^9 5f^{N+1}$) are regarded as entirely separate from the Rydberg series $4d^9 4f^N nf$. Besides having different level structures, these two configurations differ in that the 4f orbital is bound inside the potential barrier whereas the nf orbital ($n \geq 5$) is not. The final channels are thus considered to start with n=5 configuration.

The situation is sligthly different for the various transition metals series, although the same comprehensive theory (23) which combines elements from theoretical spectroscopy, collision theory and the theory of autoionization could be applied to the transition metal series. In fact, after collapse of the 4f orbital, the $\overline{4d}$ n,εf channel is dominated by transitions either to 4f or to resonantly localized continuum states of 4f character within the inner valley, whereas the $\overline{np}$ n,εd channels are not quite so dramatically affected and are dominated by transitions to the collapsed nd levels, but transitions to higher nd levels being also observable. As a consequence, the resonances np $\rightarrow$ nd occur below the corresponding threshold so that the decay broadening mechanism is rather different in both cases. In effect, the exchange interaction is not large enough to raise nd levels above threshold in the configuration average approximation and the resonances can thus be labelled np $\rightarrow$ nd without ambiguity. The mechanism of direct autoionization into underlying continua, which is successfully applied to the $4d^9 4f^{N+1}$ states in the lanthanides, as emphasized by Dehmer et al.(9) and Starace (23) is not valid anymore to explain the width of the np $\rightarrow$ nd resonances in the transition elements; the existence of decaying channels by autoionisation on the one hand, and by Auger effect (super-Coster-Krönig transitions) on the other hand has been pointed out by several authors. As soon as ten years ago, both experimentalists (44)(45) and theoreticians (46)(47) started the investigation of 3p-subshell excitation in the first-row transition elements. For a review of various references to such studies, I will send back the reader to 2 recent papers (48) (49), the first one (48) presenting only a brief review of both experimental and theoretical results on the resonant photoemission below 3p thresholds in first series transition elements and the second one (49) devoted to photoabsorption, photoelectron and ejected electron spectra of atomic 3d metals and rare-earths in the energy range of the strong $3p \rightarrow 3d$ and $4d \rightarrow 4f, \varepsilon f$ excitations. With these various experimental techniques, the Z-dependence of the strength, shape and decay channels of the giant resonances has been pointed out and the various decay mechanisms have been well understood. It is clear that for 3d elements autoionization leading to the emission of a 3d electron dominates the decay, whereas for low Z rare-earths, the $4d \rightarrow 4f, \varepsilon f$ resonances mainly decay via the emission of a 4d electron, by contrast with higher Z elements for which the emission of a 4f electron takes over.

b) there is no collapse of the n l+1 orbital in the initial ground state configuration, but this excited orbital is contracted in the excited final state : $nl^{4l+1} n\ l+1$

This category contains cases where the initial ground state is described by a closed-shell configuration: nl^{4l+2}. These cases arise because the collapse, instead of appearing in a series of neutral atoms, may occur in an isoelectronic series, upon a change of Z or for a given atom, upon an increase in the ionization degree or a change in the configuration. According to the variation of the effective potential which the excited electron n l+1 moves in, i.e. according to the variation of Z in an isoelectronic series or the variation of the ionization degree for the same Z, the excited electron wave function is an eigenfunction of the outer well (uncollapsed) or of the inner well (collapsed) or of both (in these cases, an hybrid wave function may arise which is distributed in both wells and whose principal maximum has two humps) wells. In this latter case, for example for Ba^+, because of differences in the effective potential due to its dependence on the radial wave function of the n l+1 excited electron, the

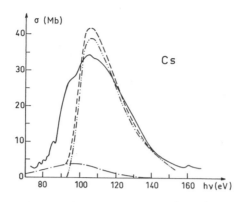

Fig.7 : Cross sections calculated for Cs (Z=55) by Amusia (52):
—··—··— 4d subshell —·—·— outer subshells — — — — sum,
compared with the yield spectrum of Cs (solid line) (53).

wave function of the 5f orbital is pulled into the inner well more than the 4f orbital. A two-humped wave function has also been found for the 4f electron of the excited configuration $4d^9 4f^2$ of CsI when the excited subshell involves several electrons. There may be a comparatively rapid change in the localisation (collapse) of the electron, even in cases where the potential barrier is absent (for example, the case of an electron with $l \geq 2$ in the field of a strongly ionized atom) if Z is close to the atomic number at which the considered electron appears in the initial configuration.

Up to now, there do not exist many experimental studies about the photoabsorption of ions, but the recent absorption spectra measured (21) for Ba, Ba^+ and Ba^{2+} have shown that in Ba and Ba^+ most of the oscillator strength is in the continuum, but that in Ba^{2+} several very strong discrete transitions can be observed, an interpretation having been done in terms of the partial collapse of the nf bound states in Ba^{2+}. It is remarkable that as soon as 1973, F.Combet Farnoux and M.Lamoureux (50) using independent-particle models had already drawn attention to an important modification of the spectrum when the ionization degree reaches the 2 value, in the vicinity of the 2p and 3p thresholds for rather small atomic numbers. Indeed, this alteration was consisting in a shift of the maximum of the cross section towards the discrete spectrum, reflecting a certain contraction of the continuum wave function even for small values of the kinetic energy of the ejected electron. More recently, sophisticated calculations have been proposed for the photoionization of the inner 4d shells for Xe-like ions (51) pointing out that along the isoelectronic sequence, the inner well becomes deeper and deeper as nuclear charge increases, and that at some point, the inner well being deep enough to support a bound state of its own, the 4f orbital is suddenly collapsed into the inner well.

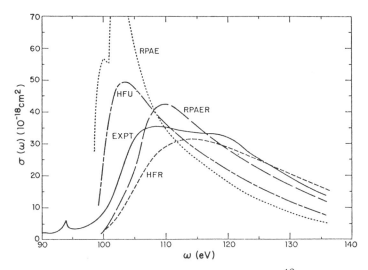

Fig.8 : Photoionization cross section for the $4d^{10}$ subshell of Ba.
Expt: experiment (54) ..... RPAE by Fliflet et al.(55) — — —
RPAER : RPAE with relaxed states by Amusia et al.(56) — - —
and - - - - Hartree-Fock with unrelaxed and relaxed states (57)

THEORETICAL TREATMENT OF GIANT RESONANCES

The Section above devoted to the origin of giant resonances has emphasized the importance of an average Hartree-Fock basis in the cases when there is little difference between the various radial functions of a given configuration. However, a term-dependent Hartree-Fock procedure in which a separate radial function is calculated for each term has to be introduced when dealing with cases in regions of the periodic table which are near points of shell collapse. Most of the theoretical treatments proposed to interpret giant resonances are based on a closed-shell initial atomic system and are devoted to the improvement of the description of the final excited configuration $nl^{4l+1}n'l+1$ and more especially of the ^1P term of this configuration because it corresponds to a large exchange term leading to an important increase or to the appearance of a positive potential barrier. Combet Farnoux (19) has obtained the curves representing σ_{4d} and σ_{5d} and displayed in Figures 3 and 5 within the framework of this approximation, but even for these cases where a sufficient qualitative agreement with experiment has been obtained, the introduction of many-body effects (in particular the ground-state correlations) in the calculations has improved this agreement .

Examples of this improvement obtained via the RPAE method are shown in Figures 7 and 8 for Cs and Ba respectively, both elements belonging to the critical cases mentioned above. In Figure 7, we can compare the theoretical results of Amusia (52) obtained from a RPAE model with experimental results provided by a yield technique (53). Although a certain discrepancy is visible it is not as important as if we compare with Hartree-Fock results (19). In Figure 8 which displays several calculations in both RPAE methods (55) (56) and Hartree-Fock approximations (57) and the experimental results of Hecht and Lindau (54) a pretty good agreement between several theoretical curves and the experiment is in favour of using relaxed orbitals in the basis, as well in the RPAE calculations as in the term-dependent Hartree-Fock procedure used by Kelly et al.(57) .

We will not discuss further the models used to describe the giant resonances in closed-shell systems in order to emphasize an open-shell case.

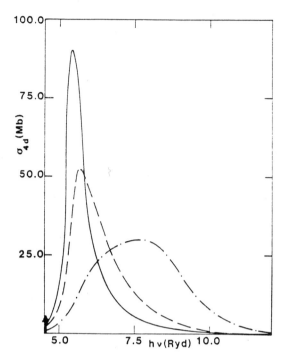

Fig.9: Total photoionisation cross section σ_{4d} of iodine determined with 3 models: ——— model 0 ; – – – – model I –·–·–·– model II   (From reference (58).)

A) Open-shell structure effects

Combet Farnoux and Ben Amar (58) have recently published a theoretical study of both inner and outer shell ionisation for atomic iodine (Z=53), the element just before xenon in the periodic table. Despite the absence of experimental results up to now, it was of great interest to perform atomic calculations relative to photoionisation of the inner 4d, 5s and outer 5p subshells of iodine. In particular, because of the $5p^5$ open subshell in the initial ground state configuration, it was impossible to extend to the ionisation of the $4d^{10}$ subshell in iodine the conclusions obtained from the different models used in case of the closed-shell xenon. Thus, it was of interest to know whether the important intrachannel interactions within the single $4d^9 \varepsilon f\ ^1P$ channel in xenon were transferred to large interchannel interactions between the various $4d^9 5p^5 \varepsilon f$ channels in iodine.

Considering the initial ground state $^2P^o$ for iodine, we have to take into account 3 continuum final states $^2P^e$, $^2D^e$ and $^2S^e$ in the photoionisation process of the 4d subshell which can be written:

$$4d^{10}5s^2 5p^5 + h\nu \rightarrow 4d^9 5s^2 5p^5\ ^{3,1}P^o, ^{3,1}D^o, ^{3,1}F^o + \varepsilon(f,p) \qquad (7)$$

According to (7) which mentions 6 ionic final states and 2 values for the angular momentum of the photoelectron, up to 12 channels have to be considered in the treatment of a continuum final state. Indeed, because of the L-S selection rules, only 4 channel-contributions have to be added for the $^2S^e$ final state, whereas 8 and 12 have to be taken into account respectively for the $^2P^e$ and $^2D^e$ final states. As it can be seen in Figure 9, we have performed σ_{4d} calculations using 3 different models which we will refer to as models 0, I and II, in the order of an increasing sophistication.

Model 0 means a zero-order approximation, i.e. a description of both initial and final states by eigenfunctions of an Hamiltonian H^o involving a non-Coulombic central field in Herman and Skillman approximation. Such a model 0 does not take into account any term dependence, only the average field being considered and the open-shell effect being neglected.

Model I is the equivalent of the term-dependent Hartree-Fock approximation used to account for the shape resonances in the $nl^{4l+1} \varepsilon l+1$ ^1P channel as discussed above. In other words, it is an intrachannel model which considers each channel independently from the others. The total cross section is the sum of all partial cross sections calculated within each channel.

Model II introduces the interchannel interactions, i.e. it involves the coupling of all channels opened by the ionisation of the 4d subshell. Up to 12 channels have to be coupled in the description of the $^2D^e$ final state. Both intrachannel and interchannel couplings have been introduced in our calculations relative to iodine via the resolution of the Lippmann-Schwinger equations, our numerical techniques having been previously developed (59). Let us remark that our model II couples both $4d^9 5p^5 \varepsilon f$ and $4d^9 5p^5 \varepsilon p$ channels and that Figure 9 presents the total 4d photoionisation cross section only beyond the $^3P^o$ threshold. The solid line, the dashed line and the dot-dashed one represent respectively our results (58) in the three models. It is clear that the successive introduction in the calculations of intrachannel and interchannel interactions alters dramatically the results of the zero-order central field model. The maximum of the shape resonance is lowered and its position is shifted towards higher energies. A broadening of the shape resonance is pointed out, similar to the broadening discussed in our previous work (59) on xenon.

This study for iodine ionized in the inner 4d subshell is the first one. The similarity of results with xenon is striking but rather expected if we refer to Fig.2 where we can see that the phase-shift $\delta_f(\varepsilon)$ changes of π over the same energy range for both elements. Up to now, no calculation with ground state correlations taken into account is available, but considering that for elements before xenon in the periodic table, the 4d shape resonance is altered slowly up to the giant resonance of xenon, our results in model II seem quite realistic, in contrast with the results of model I which emphasize that diagonalizing interactions within each channel followed by a summation of the various results is not at all equivalent to a single diagonalization of these same interactions within the $4d^9 \varepsilon f$ ^1P channel with the 5p open subshell as a spectator. Anyway, this approximation of an open-shell spectator should be valid only when its interaction with the excited electron may be neglected, which is certainly not the case in this region of near-collapse.

This is why these 3 calculations of σ_{4d} for iodine are especially instructive: a model coupling all the channels relative to the ionic configuration $4d^9 5p^5$ is the only one valid to interpret with a reasonable quantitative agreement the atomic iodine 4d spectrum and to find a giant resonance very similar to that one exhibited by the 4d spectrum of xenon. In other words, for an open-shell atom, a model coupling the various channels has to be used in order to reach results of the same degree of reliability as those obtained from a model introducing only the intrachannel interactions for a neighbour closed-shell atom.

B) <u>Spin-orbit interactions</u>

If the RPAE methods have been used with success when treating closed-shell atoms, the relativistic RPAE (RRPAE) has allowed to introduce the j-j coupling in photoionisation calculations for heavy atoms, for instance in case of mercury (60). However, such methods are so sophisticated that it is pretty hard at the end to separate the respective influence of both many-electron effects and spin-orbit interactions. It is normal that both play a

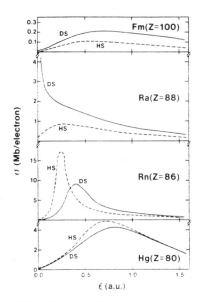 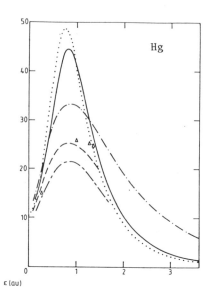

Fig.10 : on the left, photoionization cross section per electron for the $5d_{3/2} \rightarrow \epsilon f_{5/2}$ transition in Hg, Rn, Ra and Fm. The solid curves are DS results (61) and the dashed curves are the non relativistic Herman and Skillman $5d \rightarrow \epsilon f$ results (61) on the right, relativistic total cross section of 5d shell(26) compared with non relativistic calculations(26) .... Herman and Skillman model; -.-.-.- Hartree-Fock - - - - Dirac-Fock

mutual role in the collapse of the excited orbital since the spin-orbit has a short-range effect which can modify the attractive well. Figure 10 shows relativistic theoretical calculations of the 5d subshell cross section for several heavy atoms and their comparison with non-relativistic results. It is of great interest to remark on the left part of Fig.10 that for Hg (Z=80) Rn (Z=86) and Ra (Z=88) the differences between Herman-Skillman and Dirac-Slater results are exactly of the same type as the differences which I have mentioned above in Fig.5 when I compared the results of my calculations with both models: Herman and Skillman on the one hand, and term-dependent Hartree-Fock on the other hand. This remark leads to an important conclusion for these heavy elements since it means that the spin-orbit effects may have as much influence on the collapse of the excited orbital as an electrostatic interaction (61).

On the right part of Fig.10, it is clear that, in contrast with higher Z elements, the spin-orbit effects are not so important as the introduction of the intrachannel interactions in the allowed transition channel. The explanation lies in the remark already mentioned above in this paper, about the fact that for mercury, the influence of the 5f empty levels is just starting to appear.

Taking into account the difficulty to introduce into the theoretical models both correlation and relativistic effects in order to determine the positions and intensities of giant resonances of heavy elements, and more especially of those mentioned above, very close to the edge of collapse of the excited orbital, Wendin (62)(63) has used a nonrelativistic random-phase approximation to introduce exchange interaction between excitations from spin-orbit-split-core levels. Because he uses a local-density basis to evaluate the diagram expansion for the photoionization amplitude, within the usual RPA, corresponding calculations relative to La, Th and U (62) are referred in the literature as LDRPA calculations. They show that for giant dipole excitations, a narrow peak (triplet excitation) is present on the low energy side of the giant resonance. This LRDA method works well for these cases where autoionization (direct recombination) is the dominant mode of decay. For example,further studies of uranium (63) have shown that in case of the 5d subshell of uranium, the spin-orbit splitting leads to a one-electron picture of essentially two statistically weighted 5d-5f transitions (j-j coupling). In the RPA these transitions give rise to 2 new normal modes a low-energy one wherethe 5d shells oscillate out of phase and a high-energy one where they oscillate in phase. In the first mode the absorption strength is relatively weak, whereas in the second one, which becomes pushed to high energy, the absorption strength becomes 10 times larger, giving rise to the giant resonance.

In addition to their influence upon heavy atoms, spin-orbit effects play a role in the change of coupling within an isoelectronic series, as a consequence of electron collapse. In an isoelectronic series, the spin-orbit interaction increases in proportion to Z^4, while the electrostatic interaction increases only in proportion to Z, but because of the electron collapse, the importance of the electrostatic interaction increases, as a consequence of the increasing overlap between the core level and the collapsed electron orbitals. This is why a non monotonic change in the type of coupling has to be often expected when going along an isoelectronic series: it is the case for the isoelectronic series KrI $4p^54d$ and XeI $5p^55d$ (L-S tends toward jj).

REFERENCES

1. D. L. Ederer, Phys. Rev. Lett., 13, 760 (1964)
2. R. Haensel, G. Keitel, P. Schreiber, and C. Kunz, Phys. Rev., 188, 1375 (1969)
3. J.W. Cooper, Phys. Rev. Lett., 13, 762 (1964)
4. F. Combet Farnoux and Y. Heno, C. R. Acad. Sc. Paris, 264B, 138 (1967)
5. U. Fano and J.W. Cooper, Rev. Mod. Phys., 40, 441 (1968)
6. S.T. Manson and J.W. Cooper, Phys. Rev., 165, 126 (1968)
7. F. Combet Farnoux, J. de Physique, 30, 521 (1969)
8. F. Herman and S. Skillman, Atomic Structure Calculations (Prentice Hall Inc., New Jersey) 1963
9. J.L. Dehmer, A.F. Starace, U. Fano, J. Sugar, and J.W. Cooper, Phys. Rev. Lett., 26, 1521 (1971)
10. A.F. Starace, J. Phys., B7, 14 (1974)
11. D.L. Ederer, T.B. Lucatorto, E.B. Salomon, R.P. Madden, and J. Sugar, J. Phys., B8, L127 (1975)
12. J.E. Hansen, A.W. Fliflet, and H.P. Kelly, J. Phys., B8, L127 (1975)
13. J.P. Connerade and D.H. Tracy, J. Phys., B10, L235 (1977)
14. G. Wendin and A.F. Starace, J. Phys., B11, 4119 (1978)
15. J.P. Connerade, Contemporary Phys., 19, 415 (1978) J.P. Connerade, J., of the Less Common Metals, 93, 171 (1983)
16. A.R.P. Rau and U. Fano, Phys. Rev., 167, 7 (1968)
17. J.R. Taylor, "Scattering Theory: the quantum theory on non relativistic collisions," Wiley, New York, 1972
18. M. Ben Amar, Theses Doctorat d'Etat, Orsay, 1984 (unpublished)
19. F. Combet Farnoux, Proc. Int. Conf. on Inner Shell Ionization Phenomena, Atlanta, GA, April 17-22, 1972, Vol. 2, U.S. Atomic Energy Commission, Oak Ridge, TN, p. 1130
20. P. Rabe, K. Radler, and H.W. Wolff, "Vacuum Ultraviolet Rad. Phys." ed. by E.E. Koch, R. Haensel, and C. Kunz (Pergamon Vieweg) 1974, p.247
21. T.B. Lucatorto, T.J. McIlrath, J. Sugar, and S.M. Younger, Phys. Rev. Lett., 47, 1124 (1981)
22. J.P. Connerade and M.W.D. Mansfield, Proc. Roy. Soc, A341, 267 (1974)

23. A.F. Starace, Phys. Rev., B5, 1773 (1972)
24. A.W. Fliflet, H.P. Kelly and J.E. Hansen, J. Phys., B8, L256 (1975)
25. G. Wendin, Phys. Lett., 46A, 119 (1973)
26. F. Keller and F. Combet Farnoux, J. Phys., B12, 2821 (1979)
    F. Keller and F. Combet Farnoux, J. Phys., B15, 2657 (1982)
27. G. Wendin, Photoelectron Spectra in "Structure and Bonding", Springer-Verlag, 45, 1 (1981)
28. M. Cukier, P. Dhez, B. Gauthe, P. Jaegle, C. Wehenkel and F. Combet Farnoux, J. de Phys. (Paris), 39, L315 (1978)
29. J.P. Connerade, M. Pantelouris, M.A. Baig, M.A.P. Martin and M. Cukier, J. Phys., B13, L357 (1980)
30. J.P. Connerade, M.W.D. Mansfield, M. Cukier, and M. Pantelouris, J. Phys., B13, L235 (1980)
31. J.C. Fuggle, A.F. Burr, L.M. Watson, D.I. Fabian, and W. Lang, J. Phys. F. (Metal Phys.) 4, 335 (1974)
32. D.C. Griffin, K.L. Andrew, and R.D. Cowan, Phys, Rev., 177, 62 (1969)
33. M.W.D. Mansfield, Proc. Roy. Soc., A346, 539 (1975)
    M.W.D. Mansfield, Proc. Roy. Soc., A346, 555 (1975)
34. J.P. Connerade, Proc. Roy. Soc., A347, 575 (1976)
35. M.W.D. Mansfield and G.H. Newsom, Proc. Roy. Soc. A357, 77 (1977)
36. J.P. Connerade, M.W.D. Mansfield, G.H. Newsom, D.H. Tracy, M.A. Baig, and K. Thimm, Phil. Trans. Roy. Soc., A290, 327 (1979)
37. S.J. Rose, J.P. Grant, and J.P. Connerade, Phil. Trans. Roy. Soc. A296, 527 (1979)
38. G. Wendin, "Vacuum Ultraviolet Rad. Phys." ed. by E.E. Koch, R. Haensel, and C. Kunz (Pergamon Vieweg) 1974, P. 225
39. M. Cukier, B. Gauthe and C. Wehenkel, J. Phys. (Paris), 41, 603 (1980)
40. A.F. Starace, Phys. Rev., A16, 231 (1977)
41. F. Combet Farnoux, Phys. Rev., A25, 287 (1982)
42. R. Haensel, P. Rabe, B. Sonntag, Solid State Comm. 8, 1845 (1970)
43. E.R. Radtke, J. Phys., B12, L77 (1979)
44. J.P. Connerade, M.W.D. Mansfield, and M.A.P. Martin, Proc. Roy. Soc., A350, 405 (1976)
45. R. Bruhn, B. Sonntag, and H.W. Wolff, J. Phys. B12, 203 (1979)
46. L.C. Davis and L.A. Feldkamp, Phys. Rev., A17, 2012 (1978)
47. F. Combet Farnoux and M. Ben Amar, Phys. Rev., A21, 1975 (1980)
48. F. Combet Farnoux, Zeitschrift für Physik D 2, 337(1986)
49. M. Meyer, Th. Presher, E. von Raven, M. Richter, E. Schmidt, B. Sonntag, and H.E. Wetzel, Zeitschrift für Physik D 2,347 (1986)
50. F. Combet Farnoux and M. Lamoureux, Phys. Lett., 43A, 183 (1973)
    F. Combet Farnoux and M. Lamoureux, J. Phys. B9, 897 (1976)
51. K.T. Chang and W.R. Johnson, Phys. Rev., A28, 2820 (1983)
52. M. Ya. Amusia, "Vacuum Ultraviolet Radiation Physics" ed. by E.E. Koch, R. Haensel, and C. Kunz (Pergamon Vieweg) 1974, p. 205
53. H. Petersen, Phys. Status Solidi (b) 72, 591 (1975)
54. M.H. Hecht and I. Lindau, Phys. Rev. Lett., 47, 821 (1981)
55. A.W. Fliflet, R.L. Chase, and H.P. Kelly, J. Phys., B7, L443 (1974)
56. M. Ya. Amusia, V.K. Ivanov, and L.V. Chernysheva, Phys. Lett., 59A, 191 (1976)
57. H.P. Kelly, S.L. Carter, and B.E. Norum, Phys. Rev. A25, 2052 (1982)
58. F. Combet Farnoux and M. Ben Amar, J. of Electron Spectr. and Related Phenomena 41, 67 (1986)
59. M. Ben Amar and F. Combet Farnoux, J. Phys. B16, 2339 (1983)
60. W.R. Johnson, V. Radojevic, P. Deshmukh, and K.J. Cheng, Phys. Rev., A25, 337 (1982)
61. B.R. Tambe and S.T. Manson, Phys. Rev. A30, 256 (1984)
62. G. Wendin, Phys. Rev. Lett., 53, 724 (1984)
63. G. Wendin, Phys. Scripta, 32, 286 (1985)

GIANT RESONANCES AS A PROBE OF

COLLECTIVE EXCITATIONS IN ATOMS AND SOLIDS

Göran Wendin

Institute of Theoretical Physics
Chalmers University of Technology
412 96 Göteborg, Sweden

1. INTRODUCTION TO COLLECTIVE EFFECTS

Collective effects involve coherent motion of a large number of particles under the influence of macroscopic mean fields. When it comes to electrons in a metal, or other types of plasmas, it is well known and accepted that there are collective elementary excitations, the so called plasmons, which govern the response of the metal to external perturbations. It is then natural to ask the question whether a finite electronic system also may be characterized in terms of collective behaviour. This is an old question, and the aswer is by now quite clear: It can indeed show collective behaviour. The "problem" is, however, that a finite system is ameanable to different types of treatments which in principle give the correct answers. This will naturally lead to a number of different views and interpretations of the same phenomenon, and even to different "schools" of thought.

It is my personal opinion that in atomic physics the issue has been settled long ago by the observation that suitable applications of many-electron response theory is equivalent to term-dependent Hartree-Fock theory[1-5]. This proves that the physical content is the same and that the rest is a question of taste, physical insight, computational efficiency, etc. Nevertheless, the collective view of the atomic dynamics is still rather foreign to the thinking of many atomic physicists, who tend to believe that it represents an artificial concept. For solid state physicists, however, the collective view is quite natural, making possible a unified approach to many-electron excitations in the core- and valence regions of molecules, solids and adsorbates.

I do not intend to continue this somewhat philosophical (or rather, semantic) debate here. A number of the general aspects of the problem that I would like to put forward, including the work of other people, have already been discussed in refs. 6-8. In this paper I shall instead try to add some new points of view and also to present a number of recent results from different applications.

In the remaining part of this section I shall present some simple pictures of <u>linear response</u> which hopefully will clarify the similarities and differences between infinite and finite systems and the relative importance of collective behaviour.

The many-electron response can be accounted for in typically two ways, through an effective wave function or through an effective screened driving field. We shall first (Sec. 2) develop the effective one-electron wave function approach and calculate the associated one-electron effective potential for atomic giant dipole transitions connected with collective behaviour.

We then apply the effective driving field formulation (Sec. 3) to descriptions of resonant photoemission (RPES) in giant dipole resonance regions for systems which show important spin-orbit splitting. We shall first discuss the case of 5d-f transitions in uranium and then present some new, preliminary results for 5p-d transitions in ytterbium within the LDRPA. The discussion will be directed towards metallic systems, and will consider effects of additional screening by the electron gas and of delocalization of atomic levels. We shall extract a model which displays the most important aspects of branching and decay in resonant photoelectron emission.

In Sec. 4 we shall discuss resonant electron impact excitation of X-ray emission (bremsstrahlung isochromat spectra, BIS) in the 3d- and 4d-regions of La metal, and demonstrate the close analogy between resonant BIS (RBIS) and RPES.

Finally, in Sec. 5 we shall use the effective driving field formulation to describe multiphoton ionization (MPI), including resonant MPI (REMPI). This involves the <u>non-linear</u> response of the system. In particular, we shall present new results concening 2-photon 1-electron ionization of the 5p-shell of Xe, which is dominated by a 5p-d giant dipole resonance.

Linear Response and Collective Effects

An external potential v_{ext} applied to a thick metallic film will induce a charge displacement δn (polarization) and an induced potential v_{ind}

$$v_{ind} = V \delta n \qquad (a)$$
$$= V \chi v_{ext} \qquad (b) \qquad (1)$$
$$= V \chi^0 v_{eff} \qquad (c)$$

χ is the full dielectric linear susceptibility and V is the Coulomb interaction. χ^0 is the susceptibility of the system of independent electrons. The step from Eq. (1b) to Eq. (1c) is a critical one: In Eq. (1b) many-electron polarization is represented by the full response function, which can be calculated with the help of e.g. term-dependent "one-electron" wave functions; in Eq. (1c) many-electron polarization is represented by the effective (total) perturbing potential.

The effective perturbing potential v_{eff} is given by

$$v_{eff} = v_{ext} + v_{ind} \qquad (a)$$
$$\equiv \varepsilon^{-1} v_{ext} \qquad (b) \qquad (2)$$

$\varepsilon^{-1} \equiv \varepsilon^{-1}(\omega)$ is the inverse dielectric function and is given by

$$\varepsilon^{-1} = 1 + V\chi \qquad (3)$$

Finally, Eq. (1c) and (2a) may be combined to give an integral equation for the effective potential

$$v_{eff} = v_{ext} + V \chi^0 v_{eff} \qquad (4)$$

This describes the random phase approximation (RPA) and its different variations (RPAE, RRPA, LDRPA, TDLDA, etc.) depending on the precise definition of χ^0. Eq. (4) may also be written as

$$v_{eff}\, \varepsilon = v_{ext} \tag{5}$$

with the dielectric function $\varepsilon(\omega)$ given by

$$\varepsilon = (1 - v\chi^0) \tag{6}$$

For a homogeneous system, like the infinite electron gas, $\varepsilon^{-1} \equiv 1/\varepsilon$. This also applies to the bulk of the metal film. The relation $\varepsilon(\omega_p) = 0$ determines the plasma frequency ω_p in the bulk of the film. For $\omega < \omega_p$, $v_{eff} = 0$, i.e. the field does not penetrate and there is total reflection.

For an inhomogeneous system, Eq. (2b) must be generalized to

$$v_{eff}(\mathbf{r}; \omega) = \int d\mathbf{r}'\, \varepsilon^{-1}(\mathbf{r}, \mathbf{r}'; \omega)\, v_{ext}(\mathbf{r}'; \omega) \tag{7}$$

and may then be applied to metal surfaces[9] or to small metal particles[10,11], or to real atoms, molecules and solids. For a large metal sphere the one-electron valence levels will effectively form a continuum and the response may be calculated using classical physics (Maxwell's equations). Making the sphere smaller, the one-electron spectrum will develop discrete levels, and the plasmon will become an envelope of one-electron levels. As long as the distances between these levels are much smaller than the plasma frequency, the modifications can only be seen on a fine scale. However, making the sphere of atomic dimensions one can no longer separate the collective behaviour from the one-electron behaviour.

It should be noted that an electron gas confined to a small sphere (small "metal particle") does not provide any model of atomic behaviour since it does not take into account the atomic central potential and therefore neglects the atomic shell structure. The atomic potential will act as an "impurity" in the electron gas. It will modify the one-electron wave functions (spherical Bessel functions → atomic radial functions) and energies, introducing the atomic shell structure. The induced charge distribution will then change character, from induced charges on the surface of the particle to induced charges on the surface of the atomic shell.

Moreover, if the particle is considerably larger than the atom, there will be two coexisting screening mechanisms, involving the electron gas and the split-off atomic shell: The associated excitations (plasmon and (collective) shell excitation) will be coupled via the effective mean field, described through Eq. (4) with

$$\chi^0 = \chi^0_{elgas} + \chi^0_{atom} \tag{8}$$

In particular, due to the coupling to the collective plasmon mode, the atomic excitations will be shifted towards higher energies. In a very crude model (two-level atom[6]; homogeneous electron gas), the dielectric function in Eq. 6 takes the form

$$\varepsilon(\omega) = 1 + \frac{\omega_o^2}{\omega_d^2 - \omega^2} - \frac{\omega_{pl}^2}{\omega^2} \tag{9}$$

where $\omega_o = 2\omega_d V_{dd}$ describes the collective atomic shift (ω_d is the transition energy; V_{dd} is the electron-hole "exchange" interaction[6]). In the region of the atomic excitation, an approximate solution to the equation $\varepsilon(\omega) = 0$ is given by

$$\omega^2 = \omega_d^2 + \omega_o^2 \frac{\omega_d^2 + \omega_o^2}{\omega_d^2 + \omega_o^2 - \omega_{pl}^2} \qquad (10)$$

In metals like La-Lu, the plasma frequency ω_{pl} lies in the range 10-15 eV and the 5p-core excitations in the 25-35 eV range.[12] The atomic 5p-d giant dipole resonance can then be shifted by at least 1-2 eV towards higher energies, which may be important since the resonance structure is lying in the region of the atomic ionization thresholds.

Conditions for Giant Dipole Resonances (GDR) and Collective Effects in Atomic Shells

This problem has been discussed in some detail in refs. 6-8. To have strong collective effects we need large induced fields, i.e. large charge displacements (polarization) δn (Eq. 1). The largest polarization is obtained for transitions between orbitals which are localized in the same region of space, i.e. for <u>transitions within a main shell</u>. The possibility of one-electron transitions across the Fermi level in the partially filled 6s/5d conduction band is the basis for collective excitations of the electron gas. In an analogous manner, the condition for collective effects in atomic subshells is that they belong to <u>open main shells</u> so that there exist transitions to essentially empty subshells within the same main shell, leading to very large polarizability.

An open main shell guarantees the presence of collapsed (or collapsible) empty orbitals for atomic numbers $Z \geq Z_o$, giving rise to maximum overlap between occupied [n(l-k)] and empty (nl) orbitals and to large exchange integrals $G^k(n(l-k);nl)$. In the case dipole (k=1) transitions, the magnitude is in fact comparable to the direct electron-hole interaction $F^o(n(l-k);nl)$. Physically this means that the induced potential (many-electron dynamics) is about as strong as the direct electron-hole interaction energy (independent electron property). Incidentally, this also reveals the inherent limitations on the degree of collectivity of an atomic system: The localization of the core-hole establishes a large value of the direct electron-hole interaction $F^o(nl;n(l+k))$ and prevents the exchange interaction from dominating.

It should be noted that the collective giant dipole resonance behaviour is important for a wide range of atomic numbers around the value Z_o for which an empty orbital collapses and begins to fill in the ground state. In the case of 4d-f transitions, in the presence of a spherically averaged 4d-core hole the 4f-orbital collapses in Ba (Z_o=56). However, collective behaviour (polarization, electron-hole "exchange" effects, term-level dependence, etc) plays an essential or important role for the elements in a wide range of atomic numbers, say, $42 < Z < 65$.

According to the above definition, the simplest collective atomic system is the 2s-shell of atomic beryllium! This highlights the nature of collective effects in atoms: It is not primarily a question of the number of electrons in the shell but rather the strength of the polarization (induced dipole moment) and the associated shifts of the independent-electron excitation energies. The $G^1(2s;2p)$ exchange integral in Be is large, and one therefore has to work within the RPAE or, equivalently, use LS-dependent Hartree-Fock theory with inclusion of initial state correlations, in order to describe the 2s-p excitation-ionization channel in a reasonable way.

In the case of a heavy many-electron atom like Ba, this means that the 6s, 5p, and 4d-shells all show pronounced collective behaviour associated with giant dipole resonances. The absolute magnitude of the shifts will increase with the number of electrons in the shell (but not the shift per electron). Another effect is of course that the increase of angular momentum of the final states (6s→p; 5p→d; 4d→f) shifts the weight from the discrete spectrum (6s→p) to the continuum (4d→f; delayed onset).

Dipole resonances fulfilling the above conditions may be referred to as <u>giant dipole resonances (GDR)</u> in analogy with the language of nuclear physics.[13]

In the same way, one may also characterize 4p→4f,ϵf transitions as <u>giant quadrupole resonances</u>, and one can expect to find large corrections to independent-electron models of average Hartree-Fock type.

It follows that 3d-4f transitions in La and Ce and similar cases <u>do not really represent GDR's</u>. The transitions take place between different main shells, which severely limits the overlap, the polarization and the dipole moment (oscillator strength) of the independent-electron 3d-4f excitations. Nevertheless, the G^1(3d;4f) exchange integral is sufficiently large that there will be strong effects of polarization, most easily seen in the non-statistical ratio of the $3d_{5/2}$-4f to the $3d_{3/2}$-4f absorption lines.[14] For excitations from deeper levels (e.g. 2p-5d transitions; "white lines") polarization effects are negligible and and one can expect to observe statistical weighting of resonance peaks.

Giant Dipole Resonances and Collective Effects in Molecules and Solids

Inner-shell excitations in molecules, solids and adsorbates show giant dipole resonances and collective behaviour if the constituent atoms do so: Atomic-like core-level dynamics is then combined with diffraction of the emitted photoelectron by the neighbouring atoms, e.g. in solid Xe[15], in solid BaF_2 and $BaCl_2$[16], or in Te adsorbed on Ni metal[17].

To discuss collective effects in molecules[8,18] let us consider the I_2 molecule. The 4d-shell is deep and isolated and the giant dipole (shape) resonance lies fairly high in the continuum. This is basically an <u>atomic resonance</u>; however, the molecular potential determines the boundary conditions and imposes structure due to the <u>molecular</u> shape resonances, i.e. resonances in the <u>interatomic</u> scattering of the photoelectron. In addition, the chemical environment will of course affect the photoelectron spectrum (PE lines).

The 5p-shell of I_2 should be an interesting case. Atomic iodine must show strong collective behaviour in 5p-d transitions in the same way as Xe. However, the 5p-orbitals also form the <u>molecular valence levels</u>, with filled σ, π, π^* levels and empty σ^* levels. In this case, the molecular excitation properties are determined by the general collective behaviour of the 5p-bonding charge <u>and</u> atomic-core region. This means that the dynamics of the $\sigma \rightarrow \sigma^*$ transitions in Cl_2, Br_2, and I_2 should be different from F_2. In the same way, the dynamics of the $\sigma \rightarrow \sigma^*$ and $\pi \rightarrow \pi^*$ transitions in N_2 should be different from the heavier molecules with the same valence configuration (disregarding purely molecular differences).

Note that the above picture is similar to the previously discussed picture of dynamic core-valence interaction in metals, between collective excitations of the core (GDR) and of the electron gas (plasmons).

Finally, note that also for purely molecular shape resonances one may distinguish collective-like giant dipole resonances from "ordinary" shape resonances. The basis for collective valence excitations is a half-filled valence shell[8,18] (e.g. N_2, where the σ and π levels are filled and the σ^* and π^* levels are empty). This allows intravalence $\sigma \rightarrow \sigma^*$ and $\pi \rightarrow \pi^*$ transitions which, in the independent particle approximation, carry the bulk of the oscillator strength, and which therefore may be characterized as giant dipole resonances. Whether the resulting transitions are discrete or continuous is of secondary importance, just as in the atomic case.

However, if the transitions take place from core levels, e.g. $1s \rightarrow \sigma^*$ or $1s \rightarrow \pi^*$, there is relatively small overlap and oscillator strength. The $1s \rightarrow \pi^*$ transition shows up as a very prominent line in photoionization spectra of unsaturated hydrocarbons. However, strictly speaking this is not a giant dipole resonance, and is not associated with collective behaviour (cf. 3d-4f in La). In the same way, the $1s \rightarrow \sigma^*$ shape resonance is then essentially a scattering resonance in the final state one-electron (average) potential, and may be traced back to the corresponding shape resonance in e^--N_2 scattering.

2. MANY-ELECTRON LINEAR RESPONSE IN TERMS OF EFFECTIVE ONE-ELECTRON WAVE FUNCTIONS

The purpose of the present discussion is to outline how many-electron response may be accounted for by an effective driving field or by an effective one-electron wave function[19-21], and to discuss in some detail the effective-wave function formulation.

In general, the transition matrix element for core-level ionization may be writen as

$$M_{\varepsilon i}(\omega) = \langle \varepsilon_{eff} | v_{eff}(\mathbf{r}; \omega) | i \rangle \qquad (11)$$

The matrix element contains both an effective wave function $\langle \varepsilon_{eff} |$ and an effective driving potential $v_{eff}(\mathbf{r}; \omega)$. There is freedom in accounting for various many-electron effects through $\langle \varepsilon_{eff} |$ or $v_{eff}(\mathbf{r}; \omega)$ or both. The choice is a matter of personal taste and computational convenience. My personal opinion is that there is a "natural" division from a physical point of view, in which $\langle \varepsilon_{eff} |$ incorporates effects of the core-hole left behind (V^{N-1} potential; ladder interaction, including relaxation), while $v_{eff}(\mathbf{r}; \omega)$ incorporates screening of the external field (bubbles; RPA).

In more extreme divisions of many-electron effects one arrives at the following possibilities,

$$M_{\varepsilon i}(\omega) \begin{cases} \langle \varepsilon | v_{eff}(\mathbf{r}; \omega) | i \rangle & (a) \\ \langle \varepsilon_{eff} | v_{ext}(\mathbf{r}; \omega) | i \rangle & (b) \end{cases} \qquad (12)$$

In Eq. 12a all effects of the dynamics (including relaxation) have been transferred to an effective driving field, while in Eq. 12b they have been transferred to an effective wave function. This effective wave function describes final state term-level dependence, initial state (ground state) correlations, and final state relaxation effects. We shall concentrate on the effective wave function formulation, and the work to be presented[19-21] represents an extension of work by Wendin and Starace[5] and Wendin and Crljen[22].

In order to bridge the two formulations in Eq. 12, we write

the ionization matrix element according to

$$M_{\varepsilon i}(\omega) = \langle \varepsilon | v_{eff}(\mathbf{r};\omega) | i \rangle \qquad (a)$$

$$= \int \phi_\varepsilon^*(\mathbf{r}) \, [\int \varepsilon^{-1}(\mathbf{r},\mathbf{r}';\omega) v_{ext}(\mathbf{r}';\omega) d\mathbf{r}'] \phi_i(\mathbf{r}) d\mathbf{r} \qquad (b) \quad (13)$$

$$= \int [\int \phi_\varepsilon^*(\mathbf{r}) \phi_i(\mathbf{r}) \varepsilon^{-1}(\mathbf{r},\mathbf{r}';\omega) d\mathbf{r}] \, v_{ext}(\mathbf{r}';\omega) \, d\mathbf{r}' \qquad (c)$$

By defining

$$\int \phi_\varepsilon^*(\mathbf{r}) \phi_i(\mathbf{r}) \varepsilon^{-1}(\mathbf{r},\mathbf{r}';\omega) d\mathbf{r} = \phi_\varepsilon^{eff*}(\mathbf{r}') \phi_i(\mathbf{r}')$$

and interchanging the dummy indices $\mathbf{r}$ and $\mathbf{r}'$ we finally obtain

$$M_{\varepsilon i}(\omega) = \int \phi_\varepsilon^{eff*}(\mathbf{r}) \, v_{ext}(\mathbf{r};\omega) \, \phi_i(\mathbf{r}) d\mathbf{r} \qquad (14)$$

where the effective wave function $\phi_\varepsilon^{eff*}(r)$ is given by

$$\phi_\varepsilon^{eff*}(\mathbf{r}) = \phi_\varepsilon^*(\mathbf{r}) - \sum_{nj} \frac{\Gamma_{\varepsilon j n i}(\omega) \, \phi_n^*(\mathbf{r}) \phi_j(\mathbf{r}) \, \phi_i^{-1}(\mathbf{r})}{\omega_{nj} - \omega - i\delta \, \text{sgn}(\omega_{nj})} (f_n - f_j) \qquad (15)$$

f_i and f_n are Fermi factors, i.e. $f_i=1$ for (i) occupied, and $f_i=0$ for (i) empty. $\Gamma_{\varepsilon j n i}(\omega)$ is the effective electron-hole interaction described by the Bethe-Salpeter integral equation

$$\Gamma(\omega) = V + V \chi^0 \Gamma(\omega) \qquad (16)$$

in terms of the elementary (irreducible) electron-hole interaction V, i.e. the bubble interaction $\langle \varepsilon j | 1/r_{12} | in \rangle$ and the ladder interaction $\langle \varepsilon j | 1/r_{12} | ni \rangle$. The precise form of V depends on the choice of one-electron basis. Moreover, one may go beyond the RPAE by including screened ladder diagrams and other higher-order contributions.[6,19,20,22]

We now consider a single atomic subshell (i), and perform the angular integration. Eq. (15) then reduces to an equation for a complex effective radial one-electron wave function for an atomic subshell ($\omega_n \equiv \omega_{ni}$)

$$u_\varepsilon^{eff*}(r) = u_\varepsilon(r) - \sum_n [\frac{F_{\varepsilon n}(\omega) \, u_n(r)}{\omega_n - \omega - i\delta} + \frac{K_{\varepsilon n}(\omega) \, u_n(r)}{\omega_n + \omega}] \qquad (17)$$

$F_{\varepsilon n}(\omega)$ and $K_{\varepsilon n}(\omega)$ are complex effective electron-hole matrix elements describing final and initial state interaction. The signs have now been chosen so that the bubble interaction is positive.

The complex character of the effective wave function may be represented by a complex normalization N, giving a real wave function $P_\varepsilon^{eff}(r)$, and real interactions $V_{\varepsilon n}(\omega)$ and $G_{\varepsilon n}(\omega)$,

$$u_\varepsilon^{eff*}(r) = P_\varepsilon^{eff}(r)/N^* \qquad (a)$$

$$N^* = 1 + i\pi \, V_{\varepsilon \varepsilon}(\omega) \qquad (b)$$

$$F_{\varepsilon n}(\omega) = V_{\varepsilon n}(\omega)/N^* \qquad (c) \quad (18)$$

$$K_{\varepsilon n}(\omega) = G_{\varepsilon n}(\omega)/N^* \qquad (d)$$

The real effective one-electron wave function $P_\varepsilon^{eff}(r)$ is obtained from

$$P_\varepsilon^{eff}(r) = u_\varepsilon(r) - \sum_n \left[\frac{V_{\varepsilon n}(\omega) u_n(r)}{\omega_{nj} - \omega - i\delta} + \frac{G_{\varepsilon n}(\omega) u_n(r)}{\omega_{nj} + \omega}\right] \qquad (19)$$

$V_{\varepsilon n}(\omega)$ and $G_{\varepsilon n}(\omega)$ are given by the Bethe-Salpeter equation (16) keeping only the real (principle value) part of the one-electron susceptibility χ^0 (reaction matrix integral equation).

Finally, with the notation $u_\varepsilon^{eff}(r) = |u_\varepsilon^{eff*}(r)|$ and $N=|N|$ we obtain an expression for a normalized, real, effective one-electron wave function

$$u_\varepsilon^{eff}(r) = \text{<TDAE|} + \text{<CORR|} \qquad (a)$$

$$\text{<TDAE|} = N^{-1} \left[u_\varepsilon(r) - \sum_n \frac{V_{\varepsilon n}(\omega) u_n(r)}{\omega_{nj} - \omega - i\delta}\right] \qquad (b) \qquad (20)$$

$$\text{<CORR|} = N^{-1} \sum_n \frac{G_{\varepsilon n}(\omega) u_n(r)}{\omega_{nj} + \omega} \qquad (c)$$

<TDAE| and <CORR| represent effective wave functions describing final and initial state interactions resp. TDAE (Tamm-Dancoff approximation with exchange) includes only forward propagating bubble and ladder diagrams. With suitable choices of angular coefficients in the electron-hole interaction matrix element $V_{\varepsilon n}(\omega)$, the TDAE provides a reaction matrix formulation of term-dependent Hartree-Fock. In particular, in the case of dipole excitations in a closed-shell system, including bubble and ladder diagrams describes the Hartree-Fock ^1P approximation,

$$\text{<TDAE|} = \text{<HF } ^1P| \qquad (21)$$

Effective One-Electron Potentials

From Eq. (19) for the effective one-electron wave function we may extract an _effective one-electron potential_ by inverting the Schrödinger equation

$$V_\varepsilon^{eff}(r) \equiv V_\varepsilon^{eff}(r;\omega)$$
$$= [u_\varepsilon^{eff}(r)]^{-1} \left[\frac{d^2}{dr^2} + \varepsilon\right] u_\varepsilon^{eff}(r) \qquad (22)$$

This effective one-electron potential will contain term-dependence as well as effects of inital state interactions. Technically, the potential may be obtained on numerical form by applying Eq. (22) to numerical one-electron wave functions. This is what we have done so far.[19-21] Some of the results will be discussed below.

However, one may also apply Eq. (21) directly to the analytical expression in Eq. (19). Of particular interest is to calculate the modification of the zeroth-order one-electron potential $V_\varepsilon(r)$ (e.g. average-of-configuration Hartree-Fock, HF_{av}). In this case one obtains[23]

$$\Delta V_\varepsilon^{eff}(r) = V_\varepsilon^{eff}(r) - V_\varepsilon(r) \qquad (23)$$

$$\equiv [P_\varepsilon^{eff}(r)]^{-1} \sum_n [V_{\varepsilon n}(\omega) + G_{\varepsilon n}(\omega) \frac{\omega_n - \omega}{\omega_n + \omega}] u_n(r)$$

If $G_{\varepsilon n}(\omega)$ is set to zero, Eq. (23) describes the difference between the HF ^{1}P and HF$_{av}$ V^{N-1} potentials. Since the continuum function at kinetic energy ε enters both in the numerator (through $V_{\varepsilon n}(\omega)$ and $G_{\varepsilon n}(\omega)$) and in the denominator (through $P_\varepsilon^{eff}(r)$), this difference potential is quite insensitive to the kinetic energy, and varies very slowly when the kinetic energy is raised (we disregard here the problem of singularities, arising from defining a local potential from a non-local one). This is in acordance with previous numerical results.[22,24]

Including $G_{\varepsilon n}(\omega)$ gives the RPAE-length (^{1}P) potential. At high photon(=kinetic) energies the <u>inititial state</u> interaction $G_{\varepsilon n}(\omega)$ enters in a critical manner, cancelling the contribution from the <u>final state</u> interaction $V_{\varepsilon n}(\omega)$ (in the simplified RPAE, $V_{\varepsilon n}(\omega) = G_{\varepsilon n}(\omega)$, which guarantees cancellation; however, the cancellation can also be demonstrated in a general way[23]). In conclusion, initial (ground) state correlations within the RPAE are essential for establishing the correct high-energy dependence of effective one-electron potentials and wave functions and, consequently, photoionization cross sections.

Examples of effective one-electron potentials

In Fig. 1 we present a systematic comparison of different HF-based potentials in the case of a $4d^{-1}4f$ electron-hole pair excitation in atomic Ba. Since the radial 4f one-electron wave function has no nodes, there are no singularities in the one-electron potential [Eq. (22)]. The most important thing to be noted is the difference between the frozen ($4d^{-1}4f$) HF$_{av}$ and HF^{1}P V^{N-1} potentials, which directly reflects the importance of dipole polarization (exchange) effects[5]. A large repulsive contribution grows up in the 4d-region with two consequences. First, the inner well

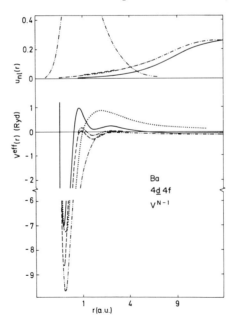

Fig. 1. Effective one-electron local potentials and effective wave functions for atomic Ba: ($\cdots\cdots$) εf HF V^N, ε=0.1 Ryd; ($-\cdot\cdot-\cdot\cdot-$) $4d^{-1}4f$ HF^{1}P V^{N-1}, frozen 4d-hole; ($-\cdot-\cdot-$) $4d^{-1}4f$ HF$_{av}$ V^{N-1}, frozen 4d-hole; ($-\cdot\cdot\cdot-\cdot\cdot\cdot-$) $4d^{-1}4f$ RPAE, frozen 4d-hole; ($-\cdot\cdot\cdot\cdot-\cdot\cdot\cdot\cdot-$) $4d^{-1}4f$ RPAE, relaxed 4d-hole.

becomes much more narrow, pushing the energy of the collapsed 4f orbital about 20 eV upwards into the continuum, turning it into a band of resonant continuum orbitals (shape resonance). Secondly, there arises a second barrier at around r=1 a.u. Introducing ground state correlations [RPAE; Eq. (19)] leads to small but significant changes in the barrier region, as was demonstrated by Wendin and Crljen[22].

A very important thing is that not even the RPAE is sufficiently good in the present case, as long as we do not include core relaxation effects.[6,13,25] As a reasonable approximation, static relaxation effects may be included through a relaxed-ion potential for the one-electron f-levels. The result for the relaxed ($4d^{-1}4f$) RPAE (length) V^{N-1} potential is shown in Fig. 1. There is a very pronounced an approximately rigid shift upwards of the inner-well and barrier regions due to the contraction of the atomic shells ("flow" of screening charge towards the hole). The role of this barrier is extremely important because it leads to a delayed onset of the photoionization cross section and therefore changes the shape of the spectral distribution, making it broader and less peaked.

The shape of the barrier does not change much for kinetic energies slightly above the 4d-ionization threshold, as shown in Fig. 2: The height of the barrier is slightly reduced, and this trend persists when the kinetic energy is increased. The presence of this barrier in the relaxed ($4d^{-1}\varepsilon f$) HF^1P V^{N-1} potential has been demonstrated by Griffin and Pindzola[24], who also investigated its energy dependence in the region within a few Rydbergs above threshold.

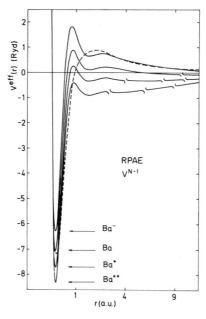

Fig. 2. Effective one-electron potentials for atomic Ba, $4d^{-1}\varepsilon f$ RPAE, relaxed 4d-hole, ε=0.1 Ry Ba$^-$ denotes the case where extra-atomic screening is simulated by adding an electron in a 5d-orbital. This removes the outer Coulomb well and Rydberg levels, resulting in a potential close to the HF V^N potential (----) in the outer region.

There are a few more things to be pointed out in connection with Fig. 1. First, the effect of relaxation on the HF_{av} V^{N-1} potential is directly reflected in the upwards shift of the second barrier in the RPAE potential. In the frozen case, this bump coincides with the peaking of the HF_{av} V^{N-1} potential, and the same is true in the relaxed case. This means that the importance of polarization effects is essentially independent of whether one uses a frozen or relaxed basis.

Secondly, one should note how the HF_{av} V^{N-1} potential turns into the HF(av) V^N potential when the 4d-hole is removed. The V^N potential also describes a broad, delayed onset 4d-f continuum cross section but the physics, reflected in the excitation potential is totally different from the $(4d^{-1}4f)$ HF^1P of RPAE V^{N-1} potentials.

Thirdly, it should be noted that the LDA potential (not shown) is very close to the relaxed HF_{av} V^{N-1} potential (not shown!) in the inner-well and barrier region. This explains why the LDRPA[6,14] and the TDLDA [26,27] give results similar to the RPAE on a relaxed basis.

Figure 2 describes how the effective one-electron potential (relaxed $4d^{-1}\varepsilon f$ RPAE V^{N-1}) changes when we add or remove valence electrons. In a first approximation, the shape of the inner-well and barrier region is rigid, and the general shape of the excitation cross section will therefore not vary much. Moreover, the energies of the core-levels will follow the bottom of the inner well in response to variations of the external charge. The position of the broad peak in the spectral distribution will then occur roughly at a fixed photon energy, independent of the position of the 4d-ionization threshold.

The inner-well f-resonance (shape resonance) essentially determines the overall distribution of oscillator strength, while the outer-well region determines the detailed structure (discrete or continuum density-of-states; atomic, molecular or solid state boundary conditions). If the resonance lies in the continuum (this is what is usually referred to as a shape resonance) there is not much difference between the free atom and the metal; however, in ionic crystals, such as BaF_2 or $BaCl_2$, there are very strong effects of the molecular field, leading to very pronounced structure while conserving the overall distribution of oscillator strength[16]. In Ba^{1+} and, in particular, Ba^{2+}, the the giant dipole resonance will straddle the 4d-thresholds. Below threshold it will be distributed over discrete levels which are eigenstates of the double well but largely localized in the outer well.

In conclusion, the giant dipole resonance should primarily be identified with the envelope of the oscillator strength distribution, and not with any particular line.

Finally, in Fig. 3 we show effective potentials for Xe using both a HF_{av} and an LDA basis. The kinetic energy is chosen in the region of the 4d-f absorption maximum (peak of the giant dipole resonance). First of all we note (Fig. 3a) the difference between the frozen-core HF^1P and RPAE: The way the ground state correlations influence the radial effective potential is not easy to anticipate. Secondly, we note again that relaxation effects shift the inner-well and barrier region upwards without much distortion of the shape.

In Fig. 3 we also show the effective one-electron potential corresponding to the LDRPA (Fig. 3b). The potential is quite similar to the relaxed RPAE, as could be expected. Again, the LDA is very close to the relaxed HF_{av} potential, and the role of polarization and ground state correlation is clearly seen.

It should be noted that Nuroh et al.[28] have presented an

effective one-electron potential for the 4f-electron (4d-f excitation) extracted from their TDLDA calculation for Ba^{2+} (the TDLDA is closely related to the LDRPA).[6,14]

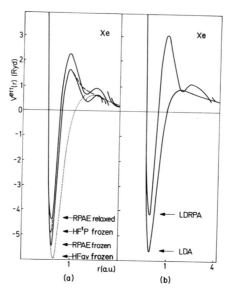

Fig. 3. Effective one-electron local potentials for 4d-εf transitions in Xe in the RPAE (a) and LDRPA (b); $\varepsilon = 2$ Ry, at the absorption maximum.

Average Length-Velocity Formulation

Equation (15) was derived under the assumption that the external perturbing potential $v_{ext}(\mathbf{r};\omega)$ is a local function of $\mathbf{r}$ (no derivatives). In particular, in the case of dipole excitations this means that we are working within the dipole-length formulation (dipole operator d^L). Equation (20a) may then be written in the following notation,

$$\langle RPAEL| = \langle HF^1P| + \langle CORR| \qquad (24)$$

Furthermore, since we are working in a one-electron basis where excited levels are calculated in a potential very similar (HF_{av}) or identical (LDA) to the one used for occupied levels, we have that

$$\langle n|d^V|i\rangle = -\langle i|d^V|n\rangle \qquad (25)$$

Equation (24) is approximate in the case of the HF_{av} potential. However, we shall assume the relation to be exact. In that case, starting from the dipole-velocity expression for the external perturbing potential ($v_{ext}(\mathbf{r};\omega) \propto \nabla$) we obtain an effective dipole-velocity one-electron wave function within the RPAE according to

$$\langle RPAEV| = \langle HF^1P| - \langle CORR| \qquad (26)$$

where $\langle HF^1P|$ and $\langle CORR|$ are <u>the same</u> as in Eq. (23). Consequently, in the RPAE, transferring effects of initial state correlations to the final state wave function leads to an operator dependence of the effective one-electron wave function.

However, since the RPAE ionization amplitude is the same in the length and velocity formulations[1], we have

$$d_{\varepsilon i}^{RPAE}(\omega) = <RPAEL|d^L|i> = <RPAEV|d^V|i> \quad (a)$$
$$= [<RPAEL|d^L|i> + <RPAEV|d^V|i>]/2 \quad (b) \quad (27)$$
$$= d_{\varepsilon i}^{av}(\omega) + \Delta d_{\varepsilon i}(\omega) \quad (c)$$

where

$$d_{\varepsilon i}^{av}(\omega) = <HF^1P|(d^L+d^V)/2|i> \quad (a)$$
$$\Delta d_{\varepsilon i}(\omega) = <CORR|(d^L-d^V)/2|i> \quad (b) \quad (28)$$

We can now define an <u>average length-velocity</u> photoionization cross section

$$\sigma^{av}(\omega) \propto \omega |d_{\varepsilon i}^{av}(\omega)|^2 \quad (29)$$

which approximately fulfills the f-sum rule (we could have defined the weighting in Eq. (28a) exactly to fulfill the f-sum rule).
It has been known for a long time that the average of HF^1P length and velocity photoionization <u>cross sections</u> give reasonable approximations to RPAE and experimental cross sections.[29,30] The advantage with the present formulation is that we have a systematic formulation of an average length-velocity <u>amplitude</u> and therefore a way to exactly calculate corrections in order to arrive at the RPAE (or beyond).

Figure 4 illustrates the case of 4p-d ionization of Kr, which is characterized by collective, giant dipole resonance behaviour. It seems to be a general feature[23] of the RPAE that the difference

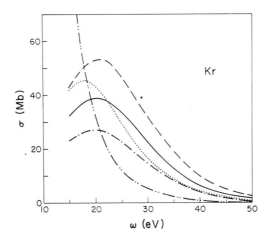

Fig. 4. Photoionization cross sections for 4p-εd transitions in Kr calculated in the simplified RPAE [$G_{\varepsilon n}(\omega) = V_{\varepsilon n}(\omega)$]; $\varepsilon = 0.1$ Ry.

(— — —) HF^1P (length)
(— · —) HF^1P (velocity)
(———) HF^1P (average length-velocity)
(— ·· —) HF_{av} (length or velocity)
(·······)   RPAE

term (28b) changes sign in the region of collective behaviour where the oscillator strength begins to fall. It follows that as the kinetic energy is increased through the giant dipole resonance region, the RPAE cross section will be closer to the HF ^{1}P length result from the threshold towards the maximum of the cross section, and then quite rapidly decrease and become closer to the HF ^{1}P velocity result. Note however that all of the HF ^{1}P length, velocity and average-length-velocity cross sections are incorrect at high energies (cross sections too high) due to omission of interference from the RPAE-type of initial state correlations.

Finally, in Fig. 5 we show the RPAE-length and -velocity effective one-electron potentials. The average is precisely the HF^{1}P potential and the difference represents twice the contribution from ground state correlations. Figure 5 also shows the RPAE-L and RPAE-V effective one-electron wave functions. Again, the average describes the $4p^{-1}\varepsilon p$ HF^{1}P wave function while the difference represents the effects of ground state (inital state) correlations. This difference is needed to compensate for the different actions of the dipole-length and dipole-velocity operators.

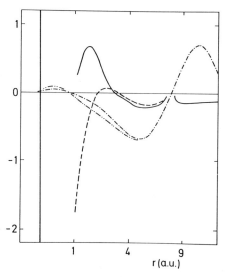

Fig. 5. Effective local (l=2) potentials for $4p$-εd transitions in Kr calculated in simplified RPAE.
(─────) length, potential
(── · ──) length, wave function
(── ── ──) velocity, potential
(── .. ──) velocity, wave function

3. GIANT RESONANCES AND SINGLE-PHOTON IONIZATION SPECTRA

The RPA represents <u>linear response</u>, involving single excitations. It therefore only describes <u>subshell</u> ionization cross sections, with no reference to main lines or satellites. In order to account for relaxation, shake-up and Auger processes one has to consider <u>non-linear response</u>, involving double or multiple excitations. This is not an easy task and very little work has been done along these lines.[6]

In the present section we shall consider resonance effects in photoemission of a single electron using the formulation where the dynamics is taken into account via an effective driving potential $v_{eff}(\mathbf{r};\omega)$. The partial photoionization cross section in channel $v=(i\to\varepsilon)$ is given by (α is the fine structure constant, a_0 is the Bohr radius)

$$\sigma^v(\omega) = 4\pi^2 \alpha a_0 \, \omega \, |\langle\varepsilon| \, M_{\varepsilon i}(\omega) \, |i\rangle|^2 \tag{30}$$

where

$$M_{\varepsilon i}(\omega) = \langle\varepsilon|v_{eff}(\mathbf{r};\omega)|i\rangle \tag{31}$$

is the effective amplitude for single-electron (1-electron) ionization. In the dipole approximation

$$\begin{aligned}
v_{eff}(\mathbf{r};\omega) &= - \mathbf{E}\cdot\mathbf{r}\, \varepsilon^{-1}(\mathbf{r};\omega) & \text{(a)} \\
&= - \mathbf{E}(\mathbf{r};\omega)\cdot\mathbf{r} & \text{(b)} \\
&= - \mathbf{E}\cdot\mathbf{r}(\omega) & \text{(c)} \\
&= - E\,\hat{\mathbf{e}}\cdot\mathbf{r}\,r(\omega) & \text{(d)}
\end{aligned} \tag{32}$$

$\hat{\mathbf{e}}$ is a unit vector along the direction of the external electric field. Eq. (32) defines a number of equivalent formulations to be used throughout this article. In particular, it illustrates the possibility to interpret the screened electron-photon coupling in Eq. (32a) in terms of an effective, space-dependent electric field (32b), or in terms of an effective dipole operator (32c).

In the following, we shall consider the RPA, evaluated using a LDA basis, the so called LDRPA. Its properties were discussed in some detail in the previous section.

A diagrammatic representation of the RPA integral equation (4) for the effective perturbing potential $v_{eff}(\mathbf{r};\omega)$ is shown in Fig. 6. The corresponding radial effective dipole operator $r(\omega)$ is given by the RPA integral equation

$$r(\omega) = r - \sum_{nj} c_{nj} \frac{\langle j|r_<\!/r_>^2|n\rangle \, \langle n|r(\omega)|j\rangle}{(\omega_{nj}^2-\omega^2)/2\omega_{nj}} \tag{33}$$

Taking the matrix elements of Eq. (33) gives back the usual matrix form of the RPA equation.

$$\langle\varepsilon|r(\omega)|i\rangle = \langle\varepsilon|r|i\rangle - \sum_{nj} c_{nj} \frac{\langle\varepsilon j|r_<\!/r_>^2|in\rangle \, \langle n|r(\omega)|j\rangle}{(\omega_{nj}^2-\omega^2)/2\omega_{nj}} \tag{34}$$

Fig. 6. Diagrammatic representation of the random phase approximation (RPA).

Extension of the RPA Method: Inclusion of spin-orbit splitting

Eq. (33) describes coupling of the atomic subshell excitations via the mean field. If one would like to take an extreme point of view one could say that each electron forms an excitation channel of its own. In this way one does not distinguish between intra-subshell and inter-subshell interaction. Moreover, if a given subshell is split by spin-orbit interaction, this simply means that we get two subshells instead of one, weighted according to their statistical weight and coupled via the mean field.

In the present approach, the spin-orbit energy splitting Δ_{so} of the core-level is introduced as an <u>external parameter</u>, while the strength of the induced potential, measured by the exchange splitting Δ_{ex}, is calculated. The ratio Δ_{so}/Δ_{ex} is a measure of the deviation from LS coupling: With this form of the RPA we are in fact describing an average form of <u>intermediate coupling</u>.

In the region of a giant dipole resonance, the exchange splitting Δ_{ex} is large, and the spin-orbit splitting Δ_{so} has to become very large in order to cause any major changes. Figure 7 shows the measured and calculated photoabsorption spectrum of uranium metal in the region of the 5d-threshold. In this case, the spin-orbit splitting $\Delta_{so} \approx 7$ eV while $\Delta_{ex} \approx 15$ eV. The main point is that the giant dipole resonance is only weakly modified by the spin-orbit splitting: The collective resonance is insensitive to whether the independent-particle excitations are degenerate or spread over a band of width considerably less than the exchange splitting Δ_{ex}.

On the other hand, the intensity (and position) of the low-energy resonance directly depends on the Δ_{so}/Δ_{ex} ratio: It vanishes in the limit $\Delta_{so}=0$ and it rapidly grows with increasing ratio. In a term-level dependent HF treatment, this resonance corresponds to a group of triplet levels (^3P, ^3D). In the present treatment we only calculate the amount of ^1P character without addressing the problem of detailed level structure. In a sense, the low-energy resonance is also a collective feature, but it is certainly not a giant dipole resonance.

The theoretical result in Fig. 7 concerns an "LDA-atom" with a spherically averaged 5f-shell and, obviously, it can only describe the metal spectrum in a limited way. In both the free atom and the

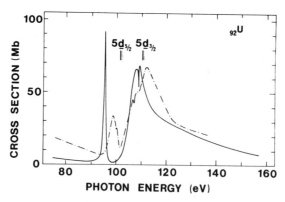

Fig. 7. Total LDRPA photoionization cross section of uranium ($7s^2 6d^1 5f^3$) in the 5d-region (solid line; Ref. 14) compared with the experimental photoabsorption cross section of uranium metal (Ref. 31).

metal, the open $5f^3$-shell may couple in a number of ways to the 5d-5f, 5d-nl and 5d-εl excitations, spreading the oscillator strength and the ionization thresholds over many levels. In the metal the 5f-electrons have an <u>extended (delocalized)</u> character: They form a narrow band and they are also hybridized with the highly itinerant 6s/5d conduction electrons.

The broadening and the structure of the photoabsorption spectrum of uranium metal in the 5d-region therefore has several origins, and different regions of the spectrum may be dominated by different decay processes. These points are essential for understanding and calculating photoemission spectra, i.e. the cross sections of main and satellite photoelectron lines.

<u>In the region of the giant dipole resonance</u> (around 110 eV in Fig. 7) the decay width is determined by the internal dynamics of the atomic core (5d-shell). The time scale for decay is very short and set by the width of the giant dipole resonance. The broadening due to the term-level structure of the 5d-εf ^{1}P transitions is clearly non-negligible but is nevertheless of secondary importance. Broadening due to 5d Auger decay is negligible. The concepts of 5f bandwidth and hybridization width are not particularly relevant in this region; if we think in terms of a 5f-level shifted up into the continuum, the inner-well 5f-electron is now coupled to a continuum of band states instead of a continuum of atomic f-states.

<u>In the region of the low-energy resonance</u> (around 95 eV in Fig. 7) the situation is quite different. There is now competition between decay and broadening involving 5d-5f recombination, 5d-Auger decay, 5f-hybridization and bandwidth, and 5d-5f term-level structure. In principle one should therefore include the 5d-Auger width and the 5d-5f term-level structure in the atomic calculation, and also the 5f-hybridization and bandwidth in the case of the metal. In the calculation for U (Fig. 7) we have not included anything of all this. The results are obtained with the basic LDRPA[14], which only accounts for electron-hole recombination (autoionization). Due to emission from the $5f^3$-shell, this results in a quite large autoionization width of $\approx$1 eV for the resonance at 95 eV photon energy. This width is not much larger than the $5p^{-1}5f^3 \rightarrow 5f\varepsilon g$ Auger (super Coster-Kronig) width of a 5d-hole, and the (LD)RPA therefore underestimates the width of the 95 eV resonance. In the case of Th, the 5d-width is probably less important (no occupied 5f-levels) and recombination should dominate.

In uranium metal, one can expect effects of 5f-hybridization and bandstructure to be competitive with intra-atomic electron-hole recombination. As a result, there will be direct 5d photoemission to continuum band levels below the vacuum level, which can be indirectly observed via resonant Auger emission. The solid state problem is clearly very difficult to treat. Moreover, the problem might have to be solved in a self-consistent manner: For instance, the 5d-5f term-level structure depends on the localization of the 5f-electrons.

In Fig. 8. we show the calculated[14] 3d-ionization cross section of La in order to illuminate some further points. In this case, the LDRPA has been <u>augmented</u> by inclusion of 3d-Auger decay. The essential effect is to introduce a complex 3d-energy ($\omega_{nj} \rightarrow \omega_{nj}+i\gamma$; j=3d) in Eq. (33), broadening the 3d-4f resonances and also reducing the probability with which the resonances decay to the single-hole channels ("main lines"). However, the sum of these channels no longer gives back the total photoionization cross section: One must add a resonant contribution describing the <u>total decay probability</u> via the <u>Auger mechanism</u> to two-hole-one-electron final states. In this way we include the total cross section from all the resonant satellites.

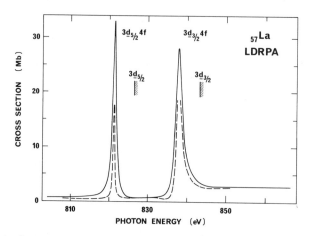

Fig. 8. Total LDRPA photoionization cross section of lanthanum ($6s^2 5d^1 4f^0$) in the 3d region. The solid line includes Auger broadening of the 3d core hole.

Figure. 8 illustrates a case where the spin-orbit splitting is five times larger than the exchange splitting, which does not prevent important coupling and intensity transfer from the low-energy ($3d_{5/2}$-4f) to the high-energy ($3d_{5/2}$-4f) peak. For the $3d_{5/2}$-4f peak the Auger width is much larger than the recombination width, while for the $3d_{3/2}$-4f peak they are comparable. This will influence the relative weights of the 3d-4f resonance structures in the various ionization channels. In particular, the $3d_{3/2}$-4f peak will appear much reduced in photoemission from outer shells (i.e. 4s and outwards).

Finally, Fig. 9 shows some preliminary results for the photoionization cross section of Yb in the LDRPA (spin-orbit splitting 6 eV ≈ exchange splitting). In this case, the <u>atomic</u> transitions (5p-d) are not characterized by large angular momentum barriers and delayed onset; the spectrum is then not dominated by continuum transitions and discrete 5p-5d transitions play an important role.

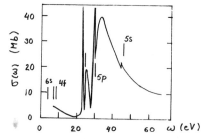

Fig. 9. Total LDRPA photoionization cross section of ytterbium ($4f^{14} 6s^2 5d^0$) in the 5p region (preliminary results).

As a result, there is a large difference between the atomic[32,33] and metallic[12,34] ionization spectra. Local-density based RPA schemes (LDRPA, TDLDA) will then not give any good description of the atomic absorption spectrum but may give a reasonable description of the broad 5p-d giant dipole resonance region in the metal. For accurate descriptions of the discrete resonance regions one must use the RRPA[33] or similar schemes. This is further illustrated by the recent application of the LDRPA to 5p- and 4d-ionization of laser excited atomic Ba($6s^2 \rightarrow 6s5d$) (Ref. 35).

<u>Resonant Photoemission Spectroscopy (RPES) for free atoms</u>

Let us now study the general behaviour of PE lines in a resonance region. For a free atom, the transition matrix element may be written as[6,36]

$$M_\varepsilon^v(\omega) = M_{o\varepsilon}^v - \frac{V_{\varepsilon r}^v \, d_r(\omega)}{\omega_r - \omega - i\gamma/2} \quad (35)$$

This form is closely related to the form of the RPA in Eq. (34): It may be obtained[6] by selecting a particular transition $r \equiv nj$ from the sum in Eq. (34). In this way, $M_{o\varepsilon}^v$ becomes an effective matrix element for direct ionization, while $d_r(\omega)$ and $V_{\varepsilon r}^v$ become effective excitation and decay matrix elements for the discrete intermediate level $r \equiv nj$. $\omega_r - i\gamma/2$ is the dressed, complex excitation energy. Within the RPA, γ would represent the recombination width. However, in general γ is the total width, including Auger decay.

Equation (34) is valid for any type of final ionic state, i.e. main lines as well as satellites. Moreover, it may be written on the Fano-like form[6,36-38]

$$M_\varepsilon^v(\omega) = M_{o\varepsilon}^v \left[1 + \eta^v \frac{q-i}{\varepsilon+i}\right] \quad (36)$$

where

$$q = -\frac{\text{Re } d_r(\omega)}{\text{Im } d_r(\omega)} \quad (a)$$

$$\varepsilon = 2(\omega-\omega_r)/\gamma \quad (b) \quad (37)$$

$$\eta^v = -\frac{V_{\varepsilon r}^v \, \text{Im } d_r(\omega)}{M_{o\varepsilon}^v \, (\gamma/2)} \quad (c)$$

and where

$$\gamma^v = 2\pi |V_{\varepsilon r}^v|^2 \quad (a)$$
$$\gamma = \sum_v \gamma^v \quad (b) \quad (38)$$

$$\text{Im } d_r^v(\omega) = \pi M_{o\varepsilon}^v V_{\varepsilon r}^v \quad (a)$$
$$\text{Im } d_r(\omega) = \sum_v \text{Im } d_r^v(\omega) \quad (b) \quad (39)$$

The Fano q-parameter characterizes the lineshape of the absorption resonance and is <u>common for all the decay channels</u>. η^v is a channel-specific lineshape parameter which describes the role of the continuum channels in excitation and emission. η^ν can be written as the ratio of two branching ratios,

$$\eta^\nu = B_e^\nu / B_x^\nu \qquad (40)$$

where

$$B_e^\nu = \gamma^\nu/\gamma \qquad (41)$$

describes branching in emission [(e); decay)], and where

$$B_x^\nu = \mathrm{Im}\, d_r^\nu(\omega)/\mathrm{Im}\, d_r(\omega) \qquad (42)$$

describes branching in excitation (x).

It should be noted that Eqs. (35)-(41) represent a general description and therefore also describe broad asymmetric giant dipole resonances. In that case, however, the lineshape parameters may become strongly energy-dependent and the interference minimum may lie below the ionization threshold (i.e. not observable).

Sufficiently close to a narrow resonance we may neglect the direct term in Eq. (35), obtaining

$$|M_\varepsilon^\nu(\omega)|^2 = |d_r(\omega)|^2 \; \frac{\gamma/2\pi}{(\omega_r-\omega)^2+(\gamma/2)^2} \; \frac{\gamma^\nu}{\gamma} \qquad (43)$$

or

$$\sigma^\nu(\omega) = \sigma_r(\omega)\, B_e^\nu(\omega) \qquad (44)$$

Eqs. (43),(44) describe the decay of a Lorentzian absorption line $\sigma_r(\omega)$ into the various emission channels (branching ratio $B_e^\nu(\omega)$).

This expression is useful for discussing the appearance of different resonance peaks in the partial cross sections. Of major importance is that the total width γ will increase while the partial width γ^ν will remain (approximately) constant when new ionization channels open up. This means that, for a given channel (ν), resonances will become relatively weaker as the photon energy is increased past successive ionization thresholds. The 4d cross section in the 3d-4f resonance region in La (Fig. 8) would be an excellent example of this phenomenon. The same is true for the 4f cross section in the 5p-5d resonance region in Yb (Fig. 9).

A particularly clean and nice example[39-40] is given by the $4s^{-1}$ main line and $4s^{-2}nl$ satellites in atomic Ca. There the progression of $3p^{-1}3d,nl\; ^1P$ absorption lines traverses a succession of $3p^{-1}$ shake-down thresholds, and it is very clear how the resonance intensity is lost from the 4s main line and satellite channels (Ca^{1+} channel) to other ($3p^{-1}$) channels (Ca^{2+} channel).

Qualitative discussion of RPES for Narrow f-band Metals: Effects of Hybridization

Let us return to the uranium 5d-photoionization cross section in Fig. 7.

We have already noted that we might have to consider 5f-hybridization with a continuum of extended (band) states for the 95 eV resonance.

For describing an experiment where one measures the photon energy dependence of the main PE lines (single-hole final states), one may use Eq. (35) [or Eqs. (33),(34)] by including the hybridization width γ^h in the total width γ, together with the recombination (autoionization) width γ^a and the Auger width γ^A. To lowest order it is the recombination matrix element $V_{\varepsilon r}{}^a$ that leads to single-hole final states. Therefore, at resonance [Eq. (43)] the relevant branching ratio is given by

$$B^a{}_e(\omega) = \gamma^a/\gamma \qquad (45)$$
$$= \gamma^a/(\gamma^a + \gamma^A + \gamma^h)$$

Beginning with the low-energy 5d-5f "^3L"(^1P) resonance around 95 eV (Fig. 7), the <u>atomic</u> ($\gamma^h=0$) decay goes via 5d-5f recombination with 5f-emission ($\gamma^a\approx 1$ eV), and Auger (super Coster-Kronig) decay of the 5d-hole ($\gamma^A\approx\gamma^a$), giving a possibility for resonant satellites. In atomic Th, on the other hand, one might expect recombination to dominate ($\gamma^A\ll\gamma^a$), in which case satellites should not resonate relative to main lines.[41]

In U <u>metal</u> ($\gamma^h\neq 0$) the 5f-electron may <u>escape</u>, leading to direct 5d emission (below the vacuum level) and to reduction of the resonance strength of the 5f,7s,6d,6p and 6s main PE lines. The 5f-orbital in the 5d-5f "^3L"(^1P) resonance will certainly be atomic-like; however, it must also be quite strongly coupled to the metallic continuum. The experimental proof is that there are no shake-down satellites in the 5d-XPS spectrum; the conduction electrons can easily flow in and fill the atomic-like 5f-screening orbitals (seems to be the case also for metallic Th in Fig. 10).

In the presence of 5f-hybridization, the electron emission spectrum will consist of four lines (we disregard any problems connected with the $5f^3$ open-shell structure): (i) a main valence PE line (weight γ^a/γ; Eq. (45)), (ii) a two-hole-one-electron PE satellite line (weight γ^A/γ), (iii) an AE line (weight γ^h/γ), and (iv) a main core-level PE line (weight γ^h/γ) (for a detailed discussion, see Ref. 36). In the 95 eV resonance, the 5f-hybridization width will most likely be comparable to the 5d-5f recombination width and to the 5d-Auger width ($\gamma^h\approx\gamma^a\approx\gamma^A$). The 5f-electron will then sometimes have escaped when the Auger decay occurs, and the result is a normal Auger line (<u>resonant Auger at fixed kinetic energy</u>) with weight γ^h/γ. There is also the possibility that the 5f electron is still around when the Auger decay occurs; the system is then left in a one-electron-two-hole configuration corresponding to a PE satellite line (<u>resonant satellite, moving with photon energy</u>) with weight γ^A/γ.

If $\gamma^h\approx\gamma^A$, then <u>both features</u> (<u>resonant satellite and resonant Auger</u>) <u>will exist at the same time</u>. They will both show resonance enhancement, and at resonance their kinetic energies will be separated by the Coulomb interaction. Both lines correspond to the same final state (two localized holes and a delocalized electron) and therefore interfere and <u>share intensity</u>.

In the region of the giant dipole resonance (around 110 eV in Fig. 7), it is convenient to think of the resonance as a 5d-5f ^1P excitation shifted to above the 5d-threshold by polarization effects and decaying into all the available continua (this is the way it appears in Eq. (35), which is exact). Formally, the main decay channel represents 5d-5f recombination with 5d-εf emission. However, this mixes the inner-well 5d-5f ^1P excitation with 5d-εf continuum states of the atom or solid and describes, in fact, "tunnelling" from inner to outer well region, i.e. the <u>shape resonance</u>. The time scale for decay is very short and set by the total width of the GDR. In this region, the GDR therefore mainly shows up in the 5d-main line and its satellites, and may be quite weak in the outer-shell cross sections relative to the low-energy 5d-5f resonance.

4. COLLECTIVE (AND OTHER) EFFECTS IN ELECTRON IMPACT EXCITATION OF CORE LEVELS IN LANTHANIDE AND ACTINIDE METALS

Electron energy loss spectroscopy (EELS) has for a long time been a very useful tool for investigating giant dipole resonances in solids. Work using high incident electron energies[42,43] (transmission EELS) has recently been followed by work at low incident electron energies[12,44-48] (reflection EELS). An example[44] is given in Fig. 10, showing thorium metal EELS for <u>varying final state electron energies</u> ε_f, from about ten times the 5d-binding energy ($\approx$90 eV) (ε_f=1200 eV) down to a value nearly equal to the 5d-binding energy (ε_f=100 eV).

Although the final state energy ε_f is not very high in the "high-energy" spectrum in Fig. 10, the loss spectrum is very similar to truly high-energy loss spectra[43] or to the photoabsorption spectrum[43]. However, at low final state energies ε_f, new structure grows up, reflecting the fact that exchange scattering with electron capture into empty 5f-levels is becoming important.

It is of major interest to be able to describe theoretically the development of the EELS spectra as ε_f tends to zero. The impact excitation mechanism is central to various experimental techniques where X-ray or Auger electron emission is studied as a function of the incident electron energy, in particular passing through core-level thresholds and resonances in the incident electron energy E.

For instance, in appearance potential spectroscopy (APS) the X-ray <u>yield</u> is studied as a function of incident electron energy in the regions of core-level thresholds and giant dipole resonances[49,50]. Studies of electron impact excitation functions of X-ray emission lines have revealed a number of extremely interesting resonance phenomena in lanthanide metals[51,52], well ahead of corresponding investigations using resonant photoemission. More recently, there have been studies of electron impact excitation functions of energy resolved Auger lines in lanthanide metals[53].

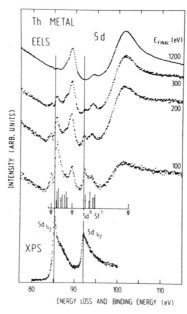

Fig. 10. EELS for thorium metal (from Ref. 44).

In this section I shall concentrate on a a discussion of resonance effects in bremsstrahlung isochromat spectra (BIS). An extensive experimental investigation of resonance effects in X-ray emission spectra in the 3d-region in metallic La and Ce was published by Liefeld and coworkers[51,52] in 1974. Related work in the BIS mode (resonant BIS, RBIS) was subsequently published by Riehle[54] and by Motais et al[55]. Moreover, the BIS results of Riehle[54] extend down towards the 4d-threshold.

Comparison of Resonant Bremsstrahlung Spectroscopy (RBIS) and Resonant Photoemission Spectroscopy (RPES)

The similarities between the two spectroscopies should be evident from Fig. 11. Off-resonance, PES (Fig. 11a) probes filled levels (removes a particle) while BIS (Fig. 11b) probes empty levels (adds a particle) (providing basic information about the single-particle Greens function propagator in many-body theory). BIS may be regarded as the inverse of PES and is therefore often referred to as inverse photoemission spectroscopy (IPES) in current solid state literature.

In the case of a core-level resonance, Fig. 11c describes resonant PES (discussed in Sec. 3). In the BIS case, resonant excitation implies capture with excitation (two-electron-one-hole intermediate level) followed by decay via X-ray emission (Fig. 11d). This process is also called dielectronic recombination. Comparing the processes in Figs. 11c and 11d, it is clear that RBIS may be regarded as the inverse of RPES.

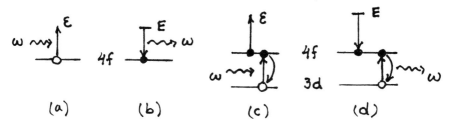

Fig. 11. Comparison of RBIS (a,c) and RPES (b,d) (see text).

Qualitative Discussion of Electron Impact Excitation of X-ray (and Electron) Emission in the 3d-threshold Region of La Metal

In Fig. 12 we show a series of pictures illustrating <u>radiative capture</u> (i.e. the BIS process) and related processes in metallic La.

Figure 12a shows <u>non-resonant</u> radiative capture into the 4f-level, including the possibility that the captured 4f-electron may hop into the metallic conduction band (hybridization). At <u>electron impact resonance</u> ($E=E_r$), there is formation of an $3d^{-1}4f^2$ intermediate level (Fig. 12b), which may decay in a number of ways: One or two 4f-electrons may hop into the conduction band (Fig. 12b), and there may be X-ray emission (and Auger emission) in the presence of two 4f-electrons (Fig. 12c), one 4f-electron (Fig. 12d), or perhaps even zero 4f-electrons.

At higher impact energies (off-resonance; $E>E_r$; $\varepsilon_f>0$), there is the possibility of inelastic scattering with 3d-4f excitation and reemission, either via the <u>direct</u> process in Fig. 12e, or via the <u>indirect, filling</u> process in Fig. 12f. The latter process depends directly on the 4f-hybridization. The two processes in Fig. 12 e,f cannot be separated experimentally without theoretical considerations: They should have different excitation cross sec-

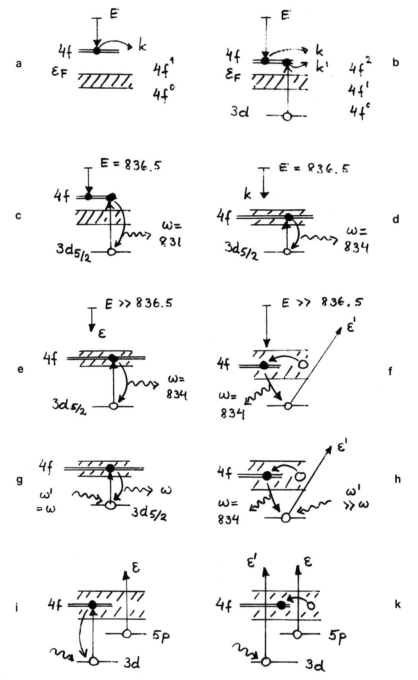

Fig. 12. Processess entering the description of radiative capture in metallic lanthanum

tions as a function of electron impact energy E. These types of inelastic excitation cross sections are presently being investigated.

There is another way of explicitly studying the filling processes, namely by photon excitation. By selectively exciting on resonance (Fig. 12g) one will observe the direct recombination process, while exciting off resonance (Fig. 12h) one will observe only the filling process.

Such experiments are difficult to perform at present. However, one could obtain the same information by studying photon excited electron spectra on (Fig. 12i) and off (Fig. 12k) resonance.

Incidentally, at least theoretically there is a possibility to separate the processes in Figs. 12e,f if one detects the emitted electrons, as in Figs. 12 i,k. The electron emission spectrum in Fig. 12k will have many more levels than in Fig. 12i, and this difference migh perhaps be observed in high-resolution electron impact spectra.

Theoretical Model of RBIS in the 3d-threshold Region of La Metal

Figure 13 shows the variation of the intensity of the BIS 4f-peak in metallic La as a function of electron impact energy[54] (normalized to the intensity of the 6s/5d band extending from the Fermi level and 2 eV upwards. The Fano-type resonance structure in the 3d-4f region is clearly seen. Structure is also seen in the 4d-region, possibly associated with the 4d-f giant dipole dipole resonance. There also seems to be structure in the 4p-region.

A more complete experimental picture[54] of the 3d-4f region is given in Fig. 14, showing BIS at a sequence of photon energies in the 800-860 eV range. In addition to the resonant BIS 4f-peak tracking the incident energy E, one observes the characteristic 4f→3d recombination emission at fixed photon energy, at resonance as well as off-resonance above the 3d-ionization thresholds. Furthermore, one observes resonant (in the incident energy) and non-resonant excitation of characteristic $3d^{-1} \to 5p^{-1}$ emission <u>in the presence of a 4f-electron</u> at around 812 eV and 829 eV. There is also $3d^{-1} \to 5p^{-1}$ emission in the <u>absence</u> of a 4f-electron at around 819 eV, more clearly seen in the work of Motais et al.[55]

In the region of the 3d-core excitations, direct and resonant BIS are described by the processes in Fig. 12a,c if hybridization is not included[56,57], and by Figs. 12a,c,d if 4f-hybridization is included[58]. We shall discuss the latter case here.

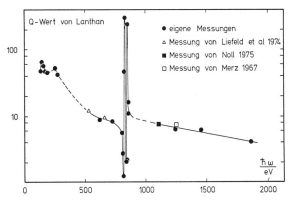

Fig. 13. Intensity variation of the 4f BIS peak in metallic La with incident electron energy (from Ref. 54).

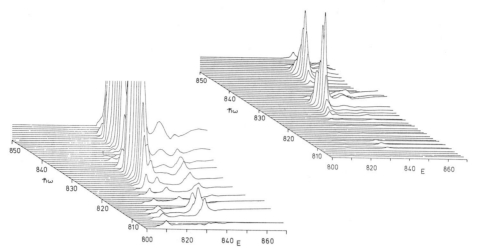

Fig. 14. Electron impact excited x-ray spectra, recorded in BIS mode for a number of different x-ray energies (from Ref. 54).

Figure 15 shows the basic RBIS X-ray emission amplitude $t_k(\omega;E)$, including a direct and a resonant part. The analogy with RPES is obvious, as previously discussed. The 3d-4f transition is excited by a screened interaction, leading to non-statistical weighting. The 4f-level is hybridized and degenerate with a continuum of extended states. The 3d-level is broadened by Auger decay, the 4f-3d excitation by recombination with electron emission, and the $3d^{-1}4f^2$ level by 1h-2p recombination (resonant scattering).

The $E \to k, 4f$ X-ray emission intensity is given by

$$I_k(\omega;E) \propto \int d\varepsilon_k |t_k(\omega;E)|^2 \delta(\varepsilon_k+\omega-E) \tag{46}$$

By straightforward application of the diagrammatic rules to the amplitude $t_k(\omega;E)$ in Fig. 15 we obtain[58]

$$I_{4f}(\omega;E) \propto \sigma_{4f} A_{4f}(E-\omega) R(\omega;E) \tag{47}$$

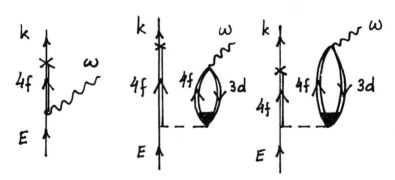

Fig. 15. Diagrammatic processes for (a) direct and (b,c) resonant BIS in metallic La. (b) and (c) describe 4f-3d x-ray emission in the presence resp. absence of the captured 4f-electron

where

$$\sigma_{4f} \propto \omega^3 \ |<4f|r|E>|^2 \qquad (48)$$

is the cross section for non-resonant radiative capture,

$$A_{4f}(E-\omega) = \frac{\gamma_h/2\pi}{(E-\omega-\varepsilon_{4f})^2+(\gamma_h/2)^2} \ \theta(E-\omega) \qquad (49)$$

is the lineshape function (spectral function) for the hybridized 4f-level ($\theta(E-\omega)$ is a step function), and

$$R(\omega;E) = \left| 1 + \sum_i \frac{a_i(E)}{E-E_{ri}-i\Gamma_{ri}/2} \left[1 + \frac{E-\omega-\varepsilon_{4f}+i\gamma_{4f}/2}{\omega-\omega_{ri}+i\gamma_{ri}/2} \right] \right|^2 \qquad (50)$$

describes the resonance enhancement of the radiative capture process.

In comparison with the previous description of RBIS[56,57] there are <u>two new features</u>: the <u>spectral function $A_{4f}(E-\omega)$</u> for the hybridized 4f level, and the last term in Eq. (50) describing <u>X-ray emission at fixed photon energy ω</u>.

We may summarize the most important properties of Eq. (50) for the emission intensity in the following manner: There is

(i) general resonance enhancement at $E = E_{ri}$ of the X-ray emission for any photon energy ω

(ii) a 4f-density-of-states peak at $\omega = E-\varepsilon_{4f}$, moving with incident electron energy (analogous to PE main line in PES)

(iii) a 4f→3d X-ray fluorescence peak at fixed photon energy $\omega = \omega_{ri}$ (analogous to resonant Auger)

Around resonance $E = E_{ri}$ there are therefore <u>two</u> X-ray emission peaks. The intensity of the 4f→3d X-ray resonance fluorescence peak depends directly on the 4f-hybridization width. More precisely, at resonance in the incident electron energy, $E = E_{ri}$, the fluorescence peak has the weight γ_{4f}/γ_{ri} relative the 4f-capture level (BIS-peak). γ_{ri} represents the total decay width of the $3d^{-1}4f$ excitation, i.e. the line width in e.g. absorption. This width is essentially given by by the non-radiative recombination width of the $3d^{-1}4f$ e-h pair plus the non-radiative (Auger) width of the 3d-hole.

From the above it is clear that the relative intensity of the resonance fluorescence peak is determined essentially by the ratio of the <u>non-radiative</u> lifetime of the electron impact excited $3d^{-1}4f$ pair to the lifetime for hopping of the captured 4f electron. This is easy to understand since the X-ray emission has to occur during the non-radiative lifetime of the excitation. In fact, if the core excitation could only decay radiatively, there would only be resonance fluorescence, because the captured 4f-electron would nearly always have time to escape.

As an illustration, we have evaluated Eq. (50), ignoring spin-orbit splitting and taking the coupling constant $a_i(E)$ to be real and ≈ 15 eV (see Ref. 56) and $\gamma_{4f}/\gamma_r = 0.2$ ($\gamma_{4f}=0.2$ eV; $\gamma_r=1.0$ eV). The result is shown in Fig. 16 and may be compared with the experimental results in Figs. 14 and 17.

There is reasonable qualitative agreement between the model and experiment. The essential new results are (i) that we obtain 4f-3d recombination x-ray emission at fixed photon energy, and (ii) that reasonable choices of 4f-hybridization with leads to the correct order of magnitude of the intensity of the 4f-3d characteristic emission relative to the resonant 4f-bremsstralung peak.

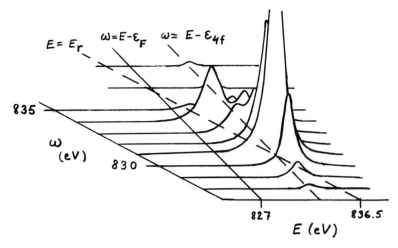

Fig. 16. Theoretical RBIS results, from Eq. (47) (see text). The spectrum has been broadened by 1 eV with respect to the incident energy (E) to account for the energy width of the incient electron beam.

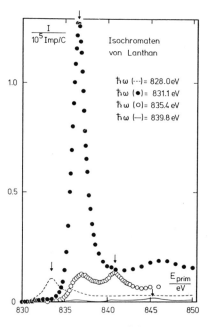

Fig. 17. Experimental RBIS results[54], at a few different X-ray energies. Resonances at $E=3d_{5/2}4f^2$ and $\omega=3d_{5/2}4f$.

Some Comments on RBIS in Regions of Giant Dipole Resonances

The great challenge, both theoretically and experimentally, is to study energy resolved x-ray emission (or Auger emission) as a function of incident electron energy (E) when E is scanned through the 4d-thresholds in the lanthanides or the 5d-thresholds in the actinides. We must then be able to account for excitation dynamics inherent in electron energy loss spectra (e.g. the uranium EELS in Fig. 10) and we must account for the associated radiative and non-radiative decay. In particular, we must be able to describe how EELS develops when the energy of the final state electron approaches threshold and exchange scattering with capture into localized 4f or 5f levels becomes important. Experimentally, very little work has been done.

The 4d-threshold region is dominated by the 4d-f giant dipole resonance, which must drastically change the excitation and x-ray emission properties in comparison with the 3d case. The approach in Eqs. (46)-(50) is still valid if the 4d-4f transitions are properly renormalized (cf. Sec. 3), but the relative importance of the various paramenters must be quite different. Let us therefore briefly discuss how to extend the previous discussion from the 3d-threshold (Fig. 15) to the 4d-threshold in metallic La[59].

Just like in the 3d-case, the captured 4f-electron will be localized in the inner well region. However, the 4d-4f excitation will now primarily belong to the ($4d^{-1}\varepsilon f\ ^1P$) channel, appearing as a <u>giant dipole resonance</u> in the continuum. In a first approximation, we may regard the captured 4f-electron and the $4d^{-1}\varepsilon f\ ^1P$ excitation as non-interacting. The resonance factor $R(\omega;E)$ in the excitation cross section for the BIS 4f-peak (Eq. 47) then is roughly proportional to the photoabsorption cross section of the target system (cf. Ref. 49), describing radiative decay of the electron-impact excited giant dipole resonance. In an improved treatment, the captured 4f-electron will experience a more attractive spherical potential due to the $4d^{-1}\varepsilon f\ ^1P$ excitations, while the $4d^{-1}\varepsilon f\ ^1P$ excitations will take place in a negative-ion system. However, the processes are coupled and the dynamics is closely related to that of relaxation in photoionization.

In addition, a small part of the dipole transition strength goes into triplet excitations in which both the captured and the excited electrons reside in localized (inner-well) 4f-orbitals. This part of the spectrum is excited for incident electron energies close to the 4d-threshold. Here, the hybridization, 4f-4d recombination, and 4d Auger decay widths probably are comparable, which could lead to a rather coplicated emission spectrum.

The discussion of the metal is based on the existence of localized, collapsed 4f-levels. The presence of such 4f affinity (N+1 electron) levels is proven experimentally by BIS (see e.g. refs. 51,52). The fact that electrons can be added in such localized levels is probably due to relaxation: In the metal, the 4f-electron will expel one unit of screening charge and dig a hole in the central cell, which may stabilize the localization (polaronic effect).

The present treatment of the resonantly enhanced radiative capture process can of course be generalized to the case where the final state electron ends up in the continuum with finite kinetic energy ε_f. This is bremsstrahlung in the conventional sense: Eq. (47) will then go over into a differential scattering cross section and in the bremsstrahlung continuum there will be a resonance structure reminiscent of (but not the same as) the photoabsorption spectrum. This problem has recently been studied theoretically by Amusia and coworkers[60] in the limit of high incident (and final) electron energy (50 keV). Experimentally, Zimkina et al.[61] have studied the x-ray emission spectrum in the 4d- and 5d-regions of lanthanum and thorium compounds at intermediate energies (around 1 keV).

It should be noted that under appropriate circumstances, the resonant bremsstrahlung ($\varepsilon_f > 0$) emission corresponds to <u>characteristic resonance fluorescence</u>. In Eq. (50), the summation normally contains a sum over discrete levels and integration over the continuum, and $R(\omega;E)$ represents the absolute square of a polarizability. However, in the neighbourhood of a prominent discrete resonance, like the 3d-4f excitation, Eq. (50) reduces to an expression similar to Eq. (43), i.e. the absorption process determines the emission (Fig. 12e). Note however that it could be that the experimentally observed emission essentially originates from an indirect process (Fig. 12f).

The above discussion can in principle be applied to free atoms; a limiting factor is then the absence of more or less collapsed affinity (N+1 electron) levels. However, if the target is an <u>ionic</u> system, the formulation in this section (without hybridization, $\gamma^h = 0$) may be used to study resonance enhancement of x-ray emission[62] and electron emission (recombination or Auger).

Very recently, Younger[63] has studied the problem of electron impact ionization of Cs^+ in the region of the 4d-threshold (i.e. threshold EELS). He makes the same division of the problem as outlined above: The "projectile" f-electron goes into a $4d^{10}\varepsilon f$ resonance of the static V^{N+1} potential, while the f-electron of the core transition sees the potential of the $4d^9 \varepsilon f\ {}^1P$ channel (giant dipole resonance).

In particular, Younger[63] finds a low-energy resonance in the V^{N+1} f-wave potential which he refers to as a <u>giant resonance</u>. As far as I understand, this resonance is a precursor of the bound 4f-levels, and the general situation then closely resembles the problem of impact ionization and BIS of metallic La above.

I would like to point out that one has to distinguish carefully between <u>giant scattering resonances</u> and the giant (dipole) resonances discussed in connection with excitation and ionization: The concept of giant resonance has so far only been applied to the <u>giant oscillator strength</u> of collective-like <u>transitions</u>. Certainly one could <u>define</u> valence-like shape resonances in elastic electron scattering (e.g. the 4f BIS peak in lanthanide metals or the σ^* resonance in molecular N_2) to be giant resonances. However, one should then note that this does not necessarily make dipole transitions to such levels giant dipole transitions: In the terminology of the present paper, 3d-4f transitions or 1s-σ^*, 1s-π^* transitions are <u>not</u> giant (dipole) transitions.

Finally, it should be noted that there is an interesting similarity between e.g. 4f-RBIS in the 4d-region and the radiative decay of 4p core-holes[64,65], influenced by giant Coster-Kronig processes[8,65] involving 4d-f giant dipole transitions.

5. GIANT RESONANCES AND COLLECTIVE EFFECTS IN MULTIPHOTON IONIZATION

Introduction

The dynamics of heavy atoms in intense laser fields has become a hot subject during the last five years due to experimental observations[66-70] of multielectron ionization, and even stripping, of valence shells of heavy atoms in cases where two or more photons are necessary even for single-electron ionization. The most spectacular effects have been observed for systems like Xe or U which are known to be characterized by giant dipole resonances and collective effects. This made Rhodes and coworkers[68,69,71,72] suggest that collective behaviour, in particular large amplitude oscillations and non-linear processes, should play an essential role for the processes leading to absorption of energy and emisson of many electrons.

One of the major issues has been whether the electrons come off sequentially (multi-step process) or all at the same time (single-step process). The present understanding[73] is that with the picosecond (or even subpicosecond) laser pulses used so far, ionization is sequential, proceeding mainly via multiphoton single-electron ionization from successive ion stages, e.g. $Xe^{1+} \to Xe^{2+} \to Xe^{3+}$ etc.

The purpose of the present work[74-76] is to study various aspects of multiphoton ionization of heavy atoms which are known to show collective behaviour in single-photon ionization. The ultimate goal is to treat the non-linear collective and single-particle dynamics when the field is strong enough to seriously perturb the atomic states (dynamic Stark shift and broadening; ponderomotive effects; harmonic generation, etc.). However, this is a rather formidable problem for a many-electron atom. Strong field effects have essentially only been seriously considered within one-electron formulations and model calculations, and many problems remain to be solved.

Here we shall consider the low-intensity limit and discuss 2-photon 1-electron ionization of the 5p-shell of Xe using a zero-field atomic basis. This will demonstrate important ways in which collective effects may enter, and it also represents the first calculation of its kind[76].

Non-Linear Response

Previously we discussed <u>linear response</u> and the random phase approximation with exchange (RPAE) in connection with 1-photon ionization. The RPAE is equivalent to <u>linearized</u> time-dependent Hartree-Fock (linearized TDHF) and only allows one real electron-hole pair at any time in the perturbation expansion (incidentally, this is why relaxation effects are not described by the RPAE).

However, if one <u>did not linearize</u> the TDHF equation, one could in principle obtain an exact description of the time-dependent deformations of the wave functions and charge density of the system under the influence of intense external fields.

To lowest order including non-linear response, the ground state many-electron wave function may be written as

$$|\Psi(t)\rangle = [1 + E \sum_{ni} a_1(t) c_n^+ c_i + c.c. + \qquad (51)$$
$$+ E^2 \sum_{nimj} a_2(t) c_n^+ c_i c_m^+ c_j + c.c. + \ldots]|0\rangle$$

2-photon 1-electron ionization

We begin by presenting the independent electron approximation. Figure 18 shows some elementary pictures of the 2-photon 1-electron ionization process. In particular, Fig. 18a,b describes two important situations, namely that the photon energy ω is (a) below or (b) above the 1-photon ionization threshold. The latter case requires laser photon energies above 12.8 eV, which are not yet

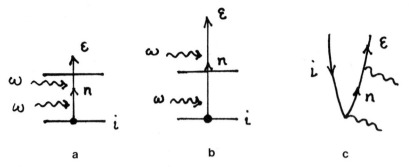

Fig. 18. Elementary pictures of 2-photon 1-electron ionization.

available for these applications. However, it is already possible to study the corresponding problems in, say, the 6s-shell of Ba or the $5p^6$ shell of I^-.

Figure 18c shows the digrammatic 2-photon 1-electron amplitude, given by the expression

$$t_{\varepsilon i}(\omega) = \sum_n \frac{\langle \varepsilon | \mathbf{E} \cdot \mathbf{r} | n \rangle \langle n | \mathbf{E} \cdot \mathbf{r} | i \rangle}{\omega_{ni} - \omega} \quad (52)$$

These independent-electron results cannot give any accurate descriptions of the true cross sections. As discussed in Sec. 2, screening of the external dipole perturbation $v_{ext}(r;\omega) = -\mathbf{E}\cdot\mathbf{r}$ is extremely important when describing 1-photon ionization of the 5p-shell in Xe. It may be represented in terms of a screened, effective perturbing potential

$$v_{eff}(r;\omega) = -\mathbf{E}\cdot\mathbf{r}\,\varepsilon^{-1}(r,\omega)$$
$$= -\mathbf{E}\cdot\mathbf{r}(\omega) \quad (53)$$

The question now is how this screening is going to enter the 2-photon amplitude and, in particular, what we mean by RPA in this context. If by RPA we mean linear response, we can only allow single-pair excitations (first order term in Eq. (51)). If in addition we only allow induced dipole polarization, only the first electron-photon operator will be screened at frequency ω (Fig.19c,d), and the 2-photon amplitude becomes

$$t_{\varepsilon i}(\omega) = \sum_n \frac{\langle \varepsilon | \mathbf{E} \cdot \mathbf{r} | n \rangle \langle n | \mathbf{E} \cdot \mathbf{r}(\omega) | i \rangle}{\omega_{ni} - \omega} \quad (54)$$

If we also allow induced monopole and quadrupole polarization (Fig. 19e), the second electron-photon operator will be screened at frequency 2ω

$$t_{\varepsilon i}(\omega) = \sum_n \frac{\langle \varepsilon | \mathbf{E} \cdot \mathbf{r}(2\omega) | n \rangle \langle n | \mathbf{E} \cdot \mathbf{r}(\omega) | i \rangle}{\omega_{ni} - \omega} \quad (55)$$

It is on this level of approximation that many-body photoionization calculations have been performed so far[77,78].

If we only consider forward propagating diagrams (TDA), this amplitude may be written in terms of intermediate- and final-state

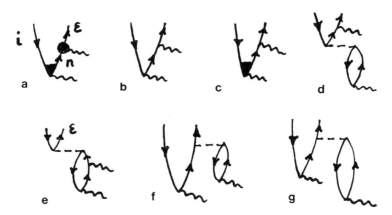

Fig. 19. Diagrammatic representation of 2-photon 1-electron ionization. Black dots represent RPA screening (for further explanations, see the text).

LS-coupled wave functions (and corresponding energies) and bare electron-photon operators

$$t_{\varepsilon i}(\omega) = \sum_n \frac{<\varepsilon_{eff}|\mathbf{E}\cdot\mathbf{r}|n_{eff}> <n_{eff}|\mathbf{E}\cdot\mathbf{r}|i>}{\omega_{ni} - \omega} \quad (56)$$

There is however no reason why the response should be limited to linear response. In fact, it would appear natural to consider the response to at least the same non-linear order as the multi-photon process itself. This means that each incident photon is allowed to polarize the system by exciting electron-hole pairs. To second order, this effect of non-linear response is digrammatically represented in Fig. 19f,g. It involves double excitations and provides additional contributions to the 2-photon 1-electron amplitude.

If the two e-h pairs in the double excitation in Fig. 19f, and 19g are independent, i.e. $\omega_{12}=\omega_1+\omega_2$ ("independent boson" approximation), then adding together the diagrams in Fig. 19f and 19g leads to the summations over (in) and (jm) becoming independent. It follows that the 2-photon 1-electron amplitude takes the form

$$t_{\varepsilon i}(\omega) = \sum_n \frac{<\varepsilon|\mathbf{E}\cdot\mathbf{r}(\omega)|n> <n|\mathbf{E}\cdot\mathbf{r}(\omega)|i>}{\omega_{ni} - \omega} \quad (57)$$

where both effective photon operators are screened in the same way (Eq. (33)). Note that this is an exact result within the "independent boson" approximation.

This form of the amplitude is very appealing from a physical point of view: It says that due to the atomic response there is a "macroscopic" effective, space-dependent, electromagnetic field driving the motion of the atomic shells. In this approximation, every photon operator is screened by the same dielectric function, given by Eq. (33). This includes photons being absorbed and reemitted (or the other way around), describing perturbation of the atomic wave functions due to the field (dynamic Stark effect to any order).

Finally, combining Eqs. (55) and (57) we obtain

$$t_{\varepsilon i}(\omega) = \sum_n \frac{<\varepsilon|\mathbf{E}\cdot\mathbf{r}(\omega,2\omega)|n><n|\mathbf{E}\cdot\mathbf{r}(\omega)|i>}{\omega_{ni} - \omega} \qquad (58)$$

where the last perturbation (n→ε) incorporates also screening at frequency 2ω, representing final state interactions as described in Fig. 19e. However, it should be noted that, as usual, there is considerable freedom in accounting for the screeening effects. The screening at frequency 2ω could e.g. be represented in terms of a dielectric function $\varepsilon^{-1}(r,2\omega)$ screening the 2-photon amplitude in Eq. (57)[75].

Application to 2-photon 1-electron ionization of the 5p-shell of Xe

In order to illustrate these concepts, I would like to give a brief presentation of some very recent numerical results.[76]

The 2-photon 1-electron ionization generalized cross-section σ_2 (in cm^4 s) can be written as

$$\sigma_2 = 2\pi \ (\alpha c/a_0)\,(\omega/F_0)^2 \sum_{1_\varepsilon} |t_{\varepsilon i}(\omega)|^2 \qquad (59)$$

where α is the fine-structure constant, c the speed of light, a_0 the Bohr radius, ω the photon energy. F_0 is equal to 3.22 10^{34} ph/cm^{2}s. The partial amplitudes $t_{\varepsilon i}$ are defined by

$$t_{\varepsilon i}(\omega) = \sum_n C(\varepsilon,n,i) \frac{<\varepsilon|r(\omega)|n><n|r(\omega)|i>}{\omega_{ni} - \omega} \qquad (60)$$

|i>, |n>, |ε> are one-electron wavefunctions (LDA or HF-average). C(ε,n,i) are spin-angular coefficients which depend on the laser polarization. The screening effects are given by the RPA integral equation for $r(\omega) = r\ \varepsilon^{-1}(r,\omega)$ in Eq. (33).

The 2-photon ionization cross section σ_2 has been calculated over a large energy region, from the 2-photon ionization threshold up to 1.4 Ry. When the photon energy is above the 1-photon ionization threshold, expression (60) describes a so-called ATI (above-threshold ionization) process. It requires a different numerical treatment, involving, in particular, calculation of free-free dipole matrix elements.[75]

Figure 20 shows the 2-photon ionization cross-section σ_2 calculated with a LDA basis, for linearly-polarized light, as a function of the photon energy. The solid line is the independent-electron (zeroth-order) approximation (Fig.19b), while the dashed line is the LDRPA result (Fig.19a). We have also calculated the 2-photon ionization cross-section by only screening the first electron-photon interaction (Fig.19c), an approximation which has been used in several previous calculations.[77,78] This result is indicated by the dotted line.

The LDA ionization threshold (0.62 Ry) is lower than the experimental threshold (≈0.90 Ry). In the region 2ω < 0.9 Ry, the LDA gives an <u>average description of the discrete 2-photon absorption resonances</u> of the real system. The structures in this region of Fig. 20 are associated with 5p→s→p and 5p→d→p transitions. Above ≈0.5 Ry the cross sections are dominated by 5p→d→f transitions. The pronounced peak in the independent-electron cross-section at 0.68 Ry is related to the 5p→εd LDA oscillator strength distribution which presents a pronounced maximum in this region. From the threshold (0.31 Ry) to ≈0.7 Ry, the RPA screening

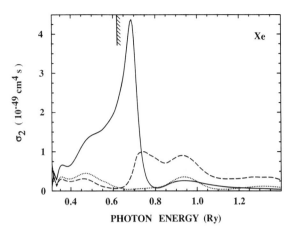

Fig. 20. 2-photon 1-electron ionization cross sections of Xe (linear polarization): (———) independent-electron approximation (LDA); (- - -) screened electron-photon interactions (LDRPA); (······) only the first interaction screened.

leads to a large reduction of the oscillator strength compared to the one-electron approximation.

We emphasize that <u>all</u> electron-photon operators should be dynamically screened. As seen in Fig. 20, omission of the screening of the second photon only leads to moderatly different results for ω below the LDA 5p-ionization limit but leads to dramatically different results for ω above that limit. The reason is that the present, fully screened, approach accounts for <u>double excitation (ionization) of the 5p-shell</u> (Fig.19f,g). Above 0.62 Ry, the spectrum is completely dominated by these processes. At photon energies ≥0.7 Ry, they induce an important enhancement of the 2-photon ionization cross-section compared to the one-electron approximation. This is not any anti-screening (field-enhancement) effect since the effective field is still reduced compared to the external field (up to ≈1Ry), but a purely <u>non-linear</u> effect: In the process, two photons are (virtually or really) absorbed by two <u>different</u> electrons.

The results obtained with the HF-basis are presented in Figs. 21,22. In Fig. 21, the photon energy is varied from 0.46 Ry (HF 2-photon ionization threshold) to 0.86 Ry. As in Fig. 20, the solid line is the independent-electron approximation, the dashed line represents the fully screened RPAE result, and the dotted line is the result obtained by screening only the first electron-photon interaction. We have also indicated the positions of the $5p^{-1}6s$ and $5p^{-1}5d$ ^1P resonances. Figure 22 shows the RPAE result over a wider energy region, together with the result of the LDRPA. As shown in the figure, we have included the $5p^{-1}ns$, n=6, 7,8 and $5p^{-1}nd$, n=5,6 resonances. Apart from the close neighbourhood of the $5p^{-1}ns$ resonances, the dominant channel leading to ionization is 5p→d→f.

Near the 2-photon-ionization threshold, our results in Fig. 21 can be compared to calculations of McGuire[79] and Gangopadhyay et al.[80] McGuire's result (Eq. (52)) evaluated with a Hartree-Fock-

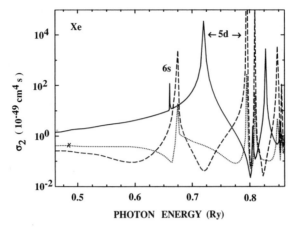

Fig. 21. 2-photon 1-electron ionization cross sections of Xe (linear polarization): (———) independent-electron approximation (HF); (- - -) screened electron-photon interactions (RPAE); (······) only the first interaction screened.

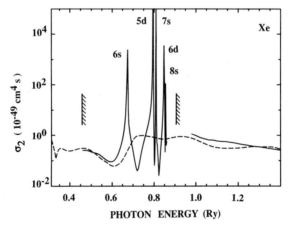

Fig. 22. 2-photon 1-electron ionization cross sections of Xe (linear polarization): (———) RPAE; (- - -) LDRPA. The 2-photon (0.46 Ry) and 1-photon (0.92 Ry) HF ionization thresholds are indicated by the hatched lines.

Slater type of basis and experimental excitation energies) is slightly below our independent-electron result[81] (full line). The multichannel quantum defect theory treatment of Gangopadhyay et al.[80] leads to good agreement with the present non-linear RPAE results in the 0.5-0.6 Ry range. Inclusion of spin-orbit interaction in the present scheme (Sec. 3) will probably give good agreement also in the vicinity of the $5p_{3/2}$-6s resonance. Finally, the cross in Fig. 21 indicates an experimental point[81] obtained at 0.47 Ry (193 nm). Although the good agreement may be somewhat fortuitous because this point lies in the region of autoionizing resonances (between the two 5p-ionization thresholds $P_{1/2}$, $P_{3/2}$), one can note that the experimental point is much closer to the RPAE (or LDRPA!) result than to the one-electron results[82].

Comparing the LDRPA and RPAE results we find that both results are in good agreement at low photon energy (0.46 Ry<ω<0.62 Ry) and in the continuum region (ω>0.98 Ry). This comparison enables us to point out the respective advantages of both methods. The RPAE result is, of course, much closer to the real spectrum in the region of the discrete resonances. However, the LDA greatly simplifies the numerical calculation (because of the absence of discrete states and the short range of the potential) and one gets a nice overview of the average oscillator strength distribution. We would even like to suggest that absence of discrete resonances in the LDA basis might provide results <u>simulating a situation of fairly high intensity</u> with very large ac Stark broadening.

Finally, in the present formulation[74-76] we neglect the electrostatic (Hartree-type) interaction between individual electron-hole pairs, in which case energies of multiple excitations equal the sum of single excitations. The continuum transition strength is dominated by collective (5p→d) excitations describing localized perturbations (small displacements) of the 5p-shell. These collective excitations are only weakly influenced by electrostatic monopole forces, and we feel that our "independent-boson" model could represent a fundamental starting point for describing the <u>gross features</u> of 2-photon ionization processes.

REFERENCES

1. M.Ya. Amusia, N.A. Cherepkov, and L.V. Chernysheva, Cross sections for the photoionization of noble-gas atoms with allowance for multielectron correlations, <u>Zh. Eksp. Teor. Fiz</u>. 60:160 (1971); (<u>Sov. Phys. JETP</u> 33:90 (1971))
2. G. Wendin, Collective effects in atomic photoabsorption spectra III. Collective resonance in the $4d^{10}$ shell in Xe, <u>J. Phys. B</u> 6:42 (1973)
3. H.P. Kelly, and R.L. Simons, Photoabsorption by neutral argon calculated by many-body perturbation theory, <u>Phys. Rev. Lett</u>. 30:529 (1973)
4. M.Ya. Amusia, and N.A. Cherepkov, <u>Case Studies in Atomic Physics</u> 5:47 (1975)
5. G. Wendin, and A.F. Starace, Perturbation theory in a strong-interaction regime with application to 4d-subshell spectra of Ba and La, <u>J. Phys. B</u> 11:4119 (1978)
6. G. Wendin, Application of many-body problems to atomic physics, <u>in</u>: New Trends in Atomic Physics, Les Houches Summer School, Session XXXVIII, 1982, eds. G. Grynberg, and R. Stora, (Elsevier Science Publishers B.V.,1984), p. 555
7. G. Wendin, Dynamic effects in core spectra, <u>in</u>: X-ray and Inner-Shell Physics, ed. B. Crasemann, AIP Conf. Proc. no. 94, 495 (1982)

8. G. Wendin, Breakdown of one-electron pictures in photoelectron spectra, in: Structure and Bonding Vol 45 (Springer Verlag, Berlin, 1981), p. 1
9. P. Feibelman, Surface electromagnetic fields, Progress in Surface Science 12: 287 (1982)
10. P. Apell, Å. Ljungbert, and S. Lundqvist, Non-local optical effects at metal surfaces, Phys. Scripta 30:367 (1984)
11. W. Ekardt, Size-dependent photoabsorption and photoemission of small metal particles, Phys. Rev. B 31: 6360 (1985)
12. F.P. Netzer, G. Strasser, G. Rosina, and J.A.D. Matthew, Core level excitations in the lanthanides by electron energy loss spectroscopy. II: 5p and 4p excitations, J.Phys. F 15:753 (1985)
13. G. Wendin, Giant dipole resonances in 4d photoabsorption of atomic barium, Phys. Lett. 46A:119 (1973)
14. G. Wendin, Photoionization of metallic lanthanum, thorium, and uranium in a local-density-based random phase approximation, Phys. Rev. Lett. 53:725 (1984)
15. R. Haensel, G. Keitel, and P. Schreiber, Optical absorption of solid krypton and xenon in the far ultraviolet, Phys. Rev. 188:1375 (1969)
16. P. Rabe, Ph.D. thesis, Universität Hamburg, 1974; DESY internal report F41-74/2 (1974); in German
17. G. Wendin, On the importance of atomic effects in photo-emission from Te adsorbed on Ni, Solid State Commun. 38:197 (1981)
18. G. Wendin, Collective effects, relaxation, and localization of hole levels in atoms, molecules, solids, and adsorbates, Int. J. Quant. Chem. S13:659 (1979)
19. Z. Crljen, and G. Wendin, Many-body theory of effective local potentials for electronic excitations. I, Phys. Scripta 32:359 (1986)
20. Z. Crljen, and G. Wendin, Many-body theory of effective local potentials for electronic excitations. II General theory, submitted to Phys. Rev. A
21. Z. Crljen, and G. Wendin, Many-body theory of effective local potentials for electronic excitations. III Application to giant dipole resonances, submitted to Phys. Rev. A
22. G. Wendin, Many-electron effects in photoionisation, in: Photoionization of Atoms and Molecules, ed. B. Buckley, Daresbury Laboratory report DL/SCI/R11 (1978), p. 1
23. G. Wendin, and Z. Crljen, Many-body theory of effective local potentials for electronic excitations. IV, to be published
24. D.C. Griffin, and M.S. Pindzola, LS term-dependent and con-figuration-average Hartree-Fock calculations for complex atoms and ions, Comments At. Mol. Phys. 13:1 (1983)
25. G. Wendin, Generalization of the RPAE: 4d-photoabsorption in atomic Ba including relaxation effects, Phys. Lett. 51A:291 (1975)
26. A. Zangwill and P. Soven, Density-functional approach to local-field effects in finite systems: Photoabsorption in the rare gases, Phys. Rev. A 21:1561 (1980)
27. A. Zangwill and P. Soven, Resonant photoemission in barium and cerium, Phys. Rev. Lett. 45:204 (1980)
28. K. Nuroh, M.J. Stott, and E. Zaremba, Calculation of the 4d-subshell photoabsorption spectra of Ba, Ba^+ and Ba^{++}, Phys. Rev.Lett.49:862 (1982)
29. A.F. Starace, Photoionization of argon and xenon including final-state correlation, Phys. Rev. A 2:118 (1970)
30. H.P. Kelly, S.L.Carter, and B.E. Norum, Calculations of photo-ionization of the 4d subshell of Ba and Ba^{2+}, Phys. Rev. A 25:2052 (1982)

31. M. Cukier, P. Dhez, B. Gauthé. P. Jaeglé, C. Wehenkel, and F. Combet Farnoux, J. Physique 39:L315 (1978)
32. D.H. Tracy, Photoabsorption structure in lanthanides: 5p subshell spectra of SmI, EuI, HoI, ErI, TmI and YbI, Proc. Roy. Soc. Lond. A357:485 (1977)
33. W.A. Svensson, M.O. Krause, T.A. Carlson, V. Radojevic, and W.R. Johnson, Photoelectron studies of the 4f and 6s subshells in atomic ytterbium, Phys. Rev. A 33:1024 (1986)
34. G. Rossi and A. Barski, Electron-emission processes following 5p photoexcitation in fcc Yb, Phys. Rev. B 32:5492 (1985)
35. J.M. Bizau, D. Cubaynes, P. Gérard, F.J. Wuilleumier, J.C. Keller, J.L. LeGouet, J.L. Picqué, D.L. Ederer, B. Carré, and G. Wendin, Experimental and theoretical determinations of the 5d photoionization cross section in laser-excited barium atoms between 15 and 150 eV photon energy, Phys. Rev. Lett. 57:306 (1986)
36. G. Wendin, Theoretical aspects of many-electron effects in atomic systems, Comments At. Mol. Phys. 17:115 (1986)
37. A.F. Starace, Behaviour of partial cross sections and branching ratios in the neighborhood of a resonance, Phys. Rev. A 16:231 (1977)
38. F. Combet Farnoux, Multichannel scattering theory of the resonant Auger effect in photoelectron spectroscopy, Phys. Rev. A 25:287 (1982)
39. J.M. Bizau, P. Gérard, F.J. Wuilleumier, and G. Wendin, Resonant photoemission in atomic calcium: A test case for atomic theory, Phys. Rev. Lett. 53:2083 (1984)
40. J.M. Bizau, P. Gérard, F.J. Wuilleumier, and G. Wendin, Photoionzation of atomic calcium in the 26-120 eV photon energy region, submitted to Phys. Rev. A
41. Cf. the discussion of satellite-to-main-line ratios in the 3p-resonance region in Ref. 40.
42. C. Colliex, M. Gasgnier, and P. Trebbia, J. Physique 37:397 (1976)
43. M. Cukier, B. Gauthé, and C. Wehenkel, J. Physique 41:603 (1980)
44. H.R. Moser, B. Delley, W.D. Schneider, and Y. Baer, Characterization of f electrons in the light lanthanides and actinides by electron-energy-loss and x-ray photoelectron spectroscopies, Phys. Rev. B 29:2947 (1984)
45. H.R. Moser, Ph.D. Thesis, ETH, Zürich (1983); in German
46. F.P. Netzer, G. Strasser, and J.A.D. Matthew, Selection rules in electron-excited 4d-4f transitions at intermediate incident energies, Phys. Rev. Lett. 51:211 (1983)
47. J. Kanski, and G. Wendin, Dynamic effects in electron-energy loss processes in barium, lanthanum, and cerium, Phys. Rev. B 24:4977 (1981)
48. J. Kanski (private communication, 1980)
49. P.O. Nilsson, J. Kanski, and G. Wendin, Many-body effects in the appearance potential spectrum of barium, Solid State Commun. 15:287 (1974)
50. M. Piacentini, R.J. Smith, and D.W. Lynch, Appearance potential spectra of La and Ce, in: Vacuum Ultraviolet Radiation Physics, eds. E.E. Koch, R. Haensel, and C. Kunz, (Vieweg-Pergamon, Braunschweig, 1974), p. 255
51. R.J. Liefeld, A.F. Burr, and M.B. Chamberlain, Electron-excitation energy-dependent features in the M-series x-ray spectra of lanthanides, Phys. Rev. A 9:316 (1974)
52. M.B. Chamberlain, A.F. Burr, and R.J. Liefeld, Excitation-

energy-dependent features in the continuum spectrum of cerium near the M_α and M_β x-ray emission lines, Phys. Rev. A 9:663 (1974)
53. J. Kanski and P.O. Nilsson, Appearance potential and characteristic-energy-loss spectroscopies of lanthanum: New interpretation of resonant x-ray emission, Phys. Rev. Lett. 43:1185 (1979)
54. F. Riehle, Ph.D thesis, Karlsruhe, 1977; in German
55. P. Motais, E. Belin, and C. Bonnelle, Electron-excited 3d-hole states of La^{3+} in solids, Phys. Rev. B 30:4399 (1984)
56. G. Wendin and K. Nuroh, Bremsstrahlung resonances and appearance-potential spectroscopy near the 3d thresholds in metallic Ba, La, and Ce, Phys. Rev. Lett. 39:48 (1977)
57. K. Nuroh and G. Wendin, Theoretical model for bremsstrahlung isochromat spectroscopy: Application to metallic lanthanum, Phys. Rev. B 24:5533 (1981)
58. G. Wendin, Resonance and threshold processes in bremsstrahlung isochromat spectra of metallic La in the 3d-region, 1983, unpublished
59. Some aspects of this problem have been discussed by K. Nuroh and G. Wendin, Resonant bremsstrahlung emission and APS near the 4d-thresholds of metallic Ba, La, and Ce, 1978, unpublished (in Ph.D. thesis of K. Nuroh)
60. M.Ya. Amusia, N.B. Avdonina, L.V. Chernysheva, and M.Yu. Kuichev, "Stripping" of the atom in bremsstrahlung, J. Phys. B 18:L791 (1985)
61. T.M. Zimkina, A.S. Lyakhovskaya Shulakov, V I Alaverdov, and M.V. Razuvaeva, Fiz. Tverd. Tela 25:26 (1983)
62. G. Wendin, On the influence of scattering resonances on radiative capture of electrons by atoms and ions, 1977, unpublished
63. S.M. Younger, Giant resonances in the electron-impact ionization of heavy atoms and ions, Phys. Rev. Lett. 24:2618 (1986)
64. M. Ohno and G. Wendin, Dynamics of screening and interference effects in x-ray and Auger emission spectra, Z. Phys. D, in press
65. M. Ohno and G. Wendin, Many-electron theory of x-ray photoelectron spectra: N-shell linewidths in the $_{46}Pd$ to $_{92}U$ range, Phys. Rev. A 31:2318 (1985)
66. A. L'Huillier, L.A. Lompré, G. Mainfray, and C. Manus, Multiply charged ions induced by multiphoton absorption in rare gases at 0.53 μm, Phys. Rev. A 27:2503 (1983)
67. P. Kruit, J. Kimman, H.G. Muller, and M. van der Wiel, Electron spectra from multiphoton ionization of xenon at 1064, 532, and 355 nm, Phys. Rev. A 28:248 (1983)
68. K. Boyer, H. Egger, T.S. Luk, H. Pummer, and C.K. Rhodes, Collision-free multiple photon ionization of atoms and molecules at 193 nm, Phys. Rev. A 32:214 (1984)
69. C.K. Rhodes, Multiphoton ionization of atoms, Science 229:1345 (1985)
70. S.L. Chin, F. Yergeau, and P. Lavigne, Tunnel ionization of Xe in an ultra-intense CO_2 laser field (10^{14} W/cm^2) with multiple charge creation, J. Phys. B 18:L213 (1985)
71. K. Boyer and C.K. Rhodes, Atomic inner-shell excitation induced by coherent motion of outer-shell electrons, Phys. Rev. Lett. 54:1490 (1985)
72. A. Szöke and C.K. Rhodes, Theoretical model of inner-shell excitation by outer-shell electrons, Phys. Rev. Lett. 56:720 (1986)
73. A.L. Robinson, Atoms in strong laser fields obey the rules, Science 232:1193 (1986)
74. G. Wendin, L. Jönsson, and A. L'Huillier, Screening effects in

multielectron ionization of heavy atoms in intense laser fields, <u>Phys. Rev. Lett</u>. 56:1241 (1986)
75. A. L'Huillier, L. Jönsson, and G. Wendin, Many-electron effects in multiphoton ionization: I. Screening effects in single-electron ionization, <u>Phys. Rev. A</u> 33:3938 (1986)
76. A. L'Huillier and G. Wendin, Two-photon one-electron ionization cross section of the 5p-shell of xenon including screening effects, submitted for publication
77. M.S. Pindzola and H.P. Kelly, Two-photon ionization of the neutral argon atom, <u>Phys. Rev. A</u> 11:1543 (1975)
78. R. Moccia, N.K. Rahman, and A. Rizzo, Two-photon ionization cross section calculations of noble gases: results for Ne and Ar, <u>J. Phys. B</u> 16:2734 (1983)
79. E.J. McGuire, Two- and three-photon ionization in the noble gases, <u>Phys. Rev. A</u> 24:834 (1981)
80. P. Gangopadhyay, X. Tang, P. Lambropoulos, and R. Shakeshaft, Theory of autoionization of Xe under 2- and 3-photon excitation, <u>Phys. Rev. A</u>, in press
81. The present independent-electron approximation includes all ladder diagrams, in which basis the RPA bubble expansion gives accurate RPAE results. The HF configuration-average results are slightly lower, and agree quite well with McGuire[79].
82. A.W. McCown, M.N. Ediger, and J.G. Eden, Two-photon ionization of xenon at 193 nm, <u>Phys. Rev. A</u> 26:3318 (1982)

THE STATE-SPECIFIC THEORY OF ATOMIC STRUCTURE AND ASPECTS OF THE DYNAMICS OF PHOTOABSORPTION

Cleanthes A. Nicolaides

Theoretical and Physical Chemistry Institute
The National Hellenic Research Foundation
48,Vas. Constantinou Ave., 116/35, Athens, Greece

ABSTRACT

A number of photoabsorption processes can be explained and calculated efficiently by adopting a **state-specific** point of view for the electronic structure of atomic and molecular ground and excited states. This view puts emphasis on the separate and appropriate choice and optimization of the function spaces describing the zeroth order and the remaining part of the N-electron wave-functions, especially in excited states. Analysis of the photoabsorption amplitude leads to a consistent understanding of the interplay between the atomic structure and the dynamics of photoabsorption. As applications, I discuss the cases of extraordinary photoabsorption phenomena, including the so-called "**giant resonances**", as well as the recently defined and computed Wannier two-electron ionization ladder (TEIL).

I. INTRODUCTION

According to the **state-specific** theory for the calculation of atomic and molecular wave-functions and properties, [1,2], given a property or a process which one wishes to understand and compute reliably, the aim is to obtain separately optimized zeroth order and virtual function spaces which are specific to the quantum mechanical state or states involved in the problem.

The theory is developed in terms of symmetry-adapted configurations (SAC) and not in terms of determinants, since this allows direct comparison with spectroscopy. (For atomic configurations with open shells of high angular momentum, one SAC is typically expanded in terms of hundreds of determinants). It is not restricted to treating only those cases which, to some approximation, can be computed by starting with a single determinantal zeroth order function (e.g. RPA applied to certain photoionization problems). It is implemented via state-specific numerical or analytic MCHF theory, which yields symmetry dependent, self-consistent zeroth order orbitals, via configuration-interaction procedures with fixed or iteratively optimized basis sets, which incorporate valence-Rydberg-continuum mixings and electron correlation, and via matrix-element calculational techniques capable of handling <u>nonorthonormal</u> basis sets. Nonorthonormality is a natural consequence of the state-specific nature of the theory and allows for the comprehension and easy evaluation with small, optimized wave-functions, of processes involving a variety of excited states [1-10] (E.g. multiple excitations, autoionization, charge-transfer, the Wannier two-electron ionization ladder-see below)

Finally, taking into account the computational experience with the discrete and quasi-discrete spectra [e.g. 1-10] and the multichannel reaction-matrix formalisms of Fano and Prats [11] and Ramaker and Schrader 12 for all channels being open a new, state-specific, configuration-interaction in the continuum (CIC) theory which incorporates the closed channels and derives the multichannel quantum defect has been developed 13 and applied to interesting cases of photoionization in regions of valence [14] or Rydberg-series resonances [4].

The above introductory remarks refer very briefly to the theoretical and computational foundations on which the short presentation which follows is based.

II. THE INTERPLAY BETWEEN ATOMIC STRUCTURE AND SPECTROSCOPY

The existing literature on the quantum theory of matter suggests that, in general, it is not necessary to obtain very accurate eigenvalues of the Schrödinger equation in order to have wavefunctions which can yield accurate (say within 2%-15% error) diagonal or off-diagonal properties. Inversely, even when an accurate energy solution is obtained it does not follow that all other properties can be calculated accurately in all cases. Examples are the hyperfine structure, dipole or transition moments of excited states exhibiting strong near-degeneracy effects, radiationless transition rates e.t.c. These facts are related to the details and to the mixing coefficients of the N-electron function spaces which constitute the wave-functions. A good understanding of their importance and contributions to the matrix element of a particular operator may allow the possibility of interpretation of observed spectra or of reliable predictions without even the application of exorbitant computational effort.

In the case of single photon absorption (emission), the relevant information is contained in the transition probability or cross-section

$$\sigma(E) \sim f(E) |<\Psi_i|\underset{\sim}{T}|\Psi_f>|^2 \qquad (1)$$

$\underset{\sim}{T}$ is the transition operator, $f(E)$ is a function of the transition energy which depends on the form of T, Ψ_i, and Ψ_f are the exact wave-functions for initial and final states.

Practical results on the choice of the appropriate form of T for accurate nonrelativistic calculations have been presented in [15,16]. Among other things, in [15] it was proposed that the fact that in RPA (or its equivalent, Time-dependent Hartree-Fock) the dipole length form is equal to the velocity one is due to its gauge invariance. In [16], it was shown that for electric quadrupole radiation, the corresponding velocity operator is always zero for many transitions in the Hartree-Fock approximation, a fact which makes such transitions exclusively electron correlation dependent.

Given that a large number of important phenomena depend on it, the accurate calculation of $\sigma(E)$ has been a central issue of quantum theory since its conception. Considerable computational simplification as well as reliability and physical transparency is achieved for systems with high symmetry, such as atoms and diatomics, when it is realized that the interelectronic interactions which affect $\sigma(E)$ the most are transition-specific [1,17]. This means that, by looking directly at the matrix element $<\Psi_i|\underset{\sim}{T}|\Psi_f>$, each T and each pair (Ψ_i,Ψ_f) are coupled in their own specific way, depending on the symmetry and the radial characteristics of the three quantum mechanical objects. A many-electron theory which suggests how to extract the major contribution to this coupling and compute it systematically for any type of transition was presented for discrete or quasi-discrete transition probabilities in the mid 70's under the name FOTOS [1,17,18] (First order theory of oscillator strengths). With appropriate formal and computational extensions to the many-body scattering problem [13], similar arguments apply for transitions involving the continuous spectrum [4]. The extension to some level of relativistic theory can be accomplished according to the procedure outlined in ref.1.

Apart from its quantitative capabilities, the FOTOS can also be used as a heuristic tool of analysis in determining roughly the relationship between the principal features of electronic structure (relaxation, electron correlation) and transition probabilities. Furthermore, within this framework, for atoms, a set of simple rules have been developed for the a priori identification of the most important correlation effects for the spectroscopic properties of doubly or inner hole excited states throughout the Periodic Table [1,6,7,19,20]. These are collected in Table 1.

III. EXTRAORDINARY PHOTOABSORPTION PHENOMENA

When the electron correlation effects of Table 1 are present, the deviations from the independent particle model predictions for $\sigma(E)$ can often be drastic. Nevertheless, the state-specific Hartree-Fock transition theory [3] can still account in a semi-quantitative way for phenomena such as multiple excitations or widths of shape resonances (often giving rise to the so-called <u>giant resonances</u> - the topic of this conference). This is possible because of the <u>nonorthonormality</u> between the orbitals of initial and final states. In ref.3 a number of demonstrative examples were given for the following cases: Bound-autoionizing state photoabsorption multielectron transitions, radiative autoionization, two-electron photoionization and photoionization into regions with shape resonances.

For the last case, the $He^- \ 1s2p^2 \ ^4P$ shape resonance was chosen. Given the topic of the conference, the following comments are relevant and, perhaps, useful:

Table 1

Rules for the easy identification of strong nonrelativistic correlation effects in atomic excited states throughout the Periodic Table. Their formulation is based on systematic information that has been extracted over the years, since the early 70s, from our electron correlation calculations (E.g., see refs. 1,2,6,7,14-20,29,30).

Type of Configuration-Interaction

I. Fermi-Sea configurations

II. Double jump of orbital momentum for single subshell excitations

$$l \leftrightarrow l + 2$$

Comments

Those strongly mixing configurations constituting the zeroth order MCHF wave-function. Their choice can be made either a priori[1,19],by considering the hydrogenic or the observed orbital near-degeneracies or the magnitude of off-diagonal matrix elements, or after a small CI calculation which yields information about the mixing coefficients.

In spin-orbital, determinantal excitation language, this mixing may correspond to the rearrangement of two spin-orbitals.

Examples

1) $s \leftrightarrow d$,
   e.g. optical spectra of metals with congiguration $nd^{\alpha}n's$.

   $nd^{\alpha}n'd \leftrightarrow nd^{\alpha}n's \leftrightarrow nd^{\alpha-1}n's^2$

2) $p \leftrightarrow f$
   e.g.photoelectron spectroscopy of Mn salts:
   $Mn^{++}\ 3p^63d^5\ 6s \leftrightarrow 3p^53d^5v_f\ 6S$

   $v_f = 4f + 5f + \ldots\ldots\ldots$

Table 1(cont.)

III. Hole-filling, "Symmetric Exchange of Orbital Symmetry" (SEOS)

$l^2 \leftrightarrow (l-1)(l+1)$

$(l \times l') \leftrightarrow (l-1)(l'+1)$

I.e., when a hole exists in the (l-1) subshell, two electrons in the neighboring subshells with l or with l and l' angular momentum symmetry will correlate strongly- if the overall symmetry allows it.

A very important correlation effect. Part of it may often fall in the Fermi-Sea (e.g. Mg $3p^2$ $^1D \leftrightarrow 3s3d$ ^1D, a mixing effect which was first noted by Bacher [31]). This was pointed out in ref.19. Its numerical importance has been demonstrated many-times (eg.14,20).

Examples

1) $p^2 \leftrightarrow sd$
   e.g. Photoionization of the ns shell below an np shell.

2) $df \leftrightarrow pg$
   e.g. 5p photoionization of the actinides. Even though g orbitals do not appear spectroscopically, our SEOS prediction is verified by computation [20].

3) $f^2 \leftrightarrow dg$
   e.g. nd photoabsorption in the rare earths.

IV. Brillouin-type excitations for certain inner hole excited states

$nl \leftrightarrow v_l$

v_l is a virtual orbital of l symmetry

This correlation mixing was determined to be significant recently [30], in connection with binding energy calculations of inner electrons. Its origin is related to angular momentum coupling peculiarities.

1. Because of symmetry, this state does not have any strong correlation effects. Although a many-electron treatment is required for the rigorous determination of its position [7], its Hartree-Fock wave-function constitutes a sufficiently accurate description of its structure and its photoabsorption properties.

2. The $He^- \, 1s2p^2 \, ^4P$ state is found just above the $He \, 1s2p \, ^3P^o$ threshold [7]. If Z is increased slightly, to a fictitious value $Z_{eff} = 2 + \varepsilon$, the ^4P state moves below the corresponding $1s2p \, ^3P^o$ threshold to become part of the discrete spectrum. If, on the other hand, Z is decreased to $Z_{eff}^- = 2 - \varepsilon$, the width of the resonance (computed as described below) increases. Thus, the effective Z and the Hartree-Fock theory determine the position and the character of this state (Compare with the extensively discussed Ba, Ba^+, Ba^{++} problem).

3. The $He^- \, 1s2p^2 \, ^4P$ state can be reached by photon absorption from the bound $He^- \, 1s2s2p \, ^4P^o$ ground state 21 or by photon emission from the bound $He^- \, 2p^3 \, ^4S^o$ excited state [22]. In both cases, the state-specific Hartree-Fock matrix element is large. (For $^4P^o \rightarrow \, ^4P$ it is given by

$<1s2p|\overline{1s2p}><2s|\underset{\sim}{d}|\overline{2p}> - <2s2p|\overline{1s2p}><1s|\underset{\sim}{d}|\overline{2p}>$ while for

$^4S^o \rightarrow \, ^4P$ by $<2p^2|\overline{2p}^2> <2p|\underset{\sim}{d}|\overline{1s}>)$.

4. The finite width of the shape resonance is calculated from the non-zero Hartree-Fock interaction matrix element

$<\overline{1s2p}^2|H-E|1s2p\varepsilon p>$, [7,22], where εp is a term-dependent, frozen core scattering Hartree-Fock orbital. This m$\underline{a}$trix element is essentially proportional to the overlap $<\overline{2p}|\varepsilon p>$. Thus, apart from its predictive value, the state-specific Hartree-Fock discrete-continuum configuration-interaction approach to core-excited shape resonances yields, in a natural way, the traditional picture of barrier penetration and decay, caused by the square of the overlap of the nonorthonormal HF orbitals.

Effects of Electron Correlation

Sometimes, the breakdown of Hartree-Fock transition theory is such that even qualitative deviations from its predicted spectrum occur. These can take place in the continuous as well as in the discrete spectrum, regardless of the number of electrons in the system. They are caused mainly by ground state Fermi-Sea correlations and by excited state valence-Rydberg (continuum) series interaction and presence of actual or virtual doubly excited states.

These deviations can be classified as follows [1,23,24]

1. Multiple excitations - especially those which are orbital symmetry forbidden in Hartree-Fock transition theory.

2. Elimination of Absorption

E.g. the Be $1s^22s2p\ ^1P^o \to {}^1D$ discrete spectrum [1]. In this case, the valence oscillator strength $2s2p \to 2p^2$ is <u>lost</u> completely to the Rydberg series.

3. Equalization of oscillator strength distribution

E.g. the B $1s^22s^22p\ ^2P^o \to {}^2S$ discrete spectrum [24]. In this case, the oscillator strength is distributed with the same order of magnitude over many Rydberg $1s^22s^2ns\ ^2S$ states and the heavily mixed valence $1s^22s2p^2\ ^2S$ state.

4. Localized exchange of oscillator strength

E.g. valence-Rydberg avoided crossings, in molecules as a function of geometry and in atoms as a function of nuclear charge [25,26].

5. Strong Enhancement of Absorption-"Collective Excitations"

Suppose that in an energy region relevant to a photoabsorption experiment, there is a very strong absorption peak over some interval and very little absorption elsewhere. Such situations have been observed in a number of atomic systems-to a good approximation- and the name "<u>collective giant dipole resonances</u>" has been employed by some researchers. Their "collective" nature has been associated with the interpretation of results of the RPA algorithm (e.g. see the many papers by Amusia, Wendin and coworkers). However, already in 1974, Fliflet et al. [27] pointed out that the "collective oscillations" of the spectrum of Ba 28] can be obtained by a term-dependent (i.e. state specific) Hartree-Fock calculation.

A couple of years later, the following analysis was published [23]. Assume that the phenomenon <u>cannot</u> be described by state-specific HF transition theory -as in the case of shape resonances discussed earlier. In other words, assume that the HF spectrum in this energy region distributes the oscillator strength over many states. Then, the phenomenon of strongly enhanced absorption emerges from a special case of heavy configurational mixing in the excited states. This special case occurs, (to a very good approximation), when each of the mixing coefficients, α_n, of the zeroth order HF functions, Φ_n, constituting the excited state Ψ_R which absorbs the oscillator strength, is related to the excitation amplitudes from the initial state Ψ_i to these wave-functions as follows [23]:

$$\alpha_n = \frac{\langle \Phi_n | D_z | \Psi_i \rangle}{\langle \Psi_R | D_z | \Psi_i \rangle} \quad (2)$$

where

$$\Psi_R = \sum_n \alpha_n \Phi_n \quad (3)$$

IV. IDENTIFICATION AND PHOTOABSORPTION TO THE WANNIER TWO-ELECTRON IONIZATION LADDER (TEIL)

More than thirty year ago, Wannier [32] treated the difficult problem of two-electron threshold ionization (TETI) using classical mechanics and a couple of meaningful approximations. Since then, there has been considerable effort to understand, this process and the related question of the properties of <u>doubly excited states</u> with large principal quantum numbers better, the aim being a fully quantummechanical approach capable of quantitative results [eg. 33-37]. Arguing that single particle coordinates are intrinsically unsuitable for constructing a theory of highly correlated quantum mechanical motions, most of the people involved in this effort have developed quantitatively approximate approaches based on the use of collective coordinates, particularly the hyperspherical coordinates used by Wannier. Yet, the progress made thus far has not been able to provide rigorous quantum mechanical, many-electron explanations and the concepts and information which have emerged are still vague and semiquantitative. It is thus evident that a many-electron quantitative treatment of aspects of this problem might provide a new view to it.

Since 1984, Komninos and the author [5,38,39] have been developing a theory for identifying a priori and computing with accuracy, **state-specific wave-functions** of a special class of doubly excited states (DES) which have the property that, as a function of excitation energy they lead smoothly to the Wannier state at threshold (E=0), characterized by $\hat{r}_1 = -\hat{r}_2$. Once they are known, it is possible to obtain observables such as the energy and the oscillator strength spectrum as well as the angular distribution of the ejected electrons at the corresponding threshold [5,38,39].

The theory is implemented in three steps:

a) We accept that one of Wannier's results from classical mechanics, i.e. that at $E=0, \hat{r}_1 = -\hat{r}_2$, must hold in the quantum mechanical case for the special class of DES mentioned above. This implies that, according to the state-specific theory of electronic structure, the zeroth-order <u>multi-configurational</u> vector much be constructed in terms of configurations with orbitals in the same shell ($n_1 = n_2$) and that, in order to minimize the (negative) electron correlation, it must correspond to the lowest root of the matrix formed by the possible configurations of each symmetry at each n. (E.g. consider the $^1P^o$ symmetry of He. For n=4, only the 4s4p, 4p4d and 4d4f configurations constitute the basis set).

b) The zeroth order vector for each shell, W_n^o, is computed selfconsistently via numerical (for low-lying) or analytic (for high-lying) MCHF theory with orthogonality contraints to core orbitals. Thus, relaxation and the concerted motion of the two electrons in the field of an <u>arbitrary</u> core (and not just a bare nucleus or zero angular momentum ion) is taken into account rigorously and systematically.

c) The remaining electron correlation, X_n, corresponding to each W_n^o, is added variationally by keeping the virtual arbitals orthogonal to the appropriate core orbitals. Thus, the total wave-function for the Wannier DES at each shell n is given by

$$W_n = W_n^o + X_n \tag{4}$$

The quantum mechanical version of the Wannier condition $r_1 = -r_2$ is in terms of the operators for the radial distance and the angle between the position vectors,

$$\langle r_1 \rangle = \langle r_2 \rangle \tag{5a}$$

$$\langle \vartheta_{12} \rangle = \pi \tag{5b}$$

In order to translate this into practical wave-mechanical language with many-electron, correlated wave-functions, we have conceptualized the problem in terms of <u>conditional probability distributions</u> and have introduced the definitions

$$\langle r_1 \rangle = \iint \rho(\vec{r}_1, \vec{r}_2) r_1 \, d\vec{r}_1 d\vec{r}_2 \tag{6a}$$

$$\langle r_2 \rangle = \frac{\int \rho(\langle r_1 \rangle, \vec{r}_2) r_2 d\vec{r}_2}{\int \rho(\langle r_1 \rangle, \vec{r}_2) d\vec{r}_2} \tag{6b}$$

where $\rho(\vec{r}_1, \vec{r}_2)$ is the exact electron density of the state of interest and $\rho(\langle r_1 \rangle, \vec{r}_2)$ is the density that results if r_1 is fixed at the average value $\langle r_1 \rangle$ and the angular dependence on $\vec{r}_1$ is integrated.

Our choice and calculation of the Wannier TEIL wave-functions lead to solutions for which the expectation values (6a) and (6b) indeed satisfy eq.5a and which, with increasing n tend to satisfy eq.5b. Furthermore, the MCHF zeroth order vectors become a better approximation to the exact wave-functions as n increases, a fact which justifies the use of Rydberg-like formulae for the TEIL (see eq.7) and which implies that the anticipated smoothness is setting-in already at about n=8-10 [5,38,39].

Among the new conceptual results which have been produced by our approach are the conditional probability plots of the wave-functions of the Wannier TEIL of $^1P^o$ and ^1S symmetries in H^-, He and Li^+. These show clearly how, along this ladder of DES, as the excitation energy increases toward the two-electron threshold, the two electrons move toward 180° apart while $\langle r_1 \rangle = \langle r_2 \rangle$ [5,38,39].

The numerical results which are given here can be tested by a photoabsorption experiment with high photonflux. Table 2 contains the total energies of the $^1P^o$ Wannier TEIL in H^-, He and Li^+. These were fitted to the Rydberg-like formula

$$E_n = -\frac{(Z-\sigma)^2}{(n+\mu)^2} \tag{7}$$

where for H^-, $\sigma=0.168$, $\mu=0.354$, for He, $\sigma=0.162$, $\mu=0.160$ and for Li^+ $\sigma=0.168$, $\mu=0.101$. Thus, σ is essentially Z-independent while μ goes like $\sim 0.3/Z$.

Table 3 contains the oscillator strengths from the ground state ^1S.

Table 2

Total energies of the Wannier TEIL of $^1P^o$ symmetry in H^-, He and Li^+. The calculations involve a Multi-configurational Hartree-Fock zeroth order vector W_n^o, with all the configurations of each shell n, to which the remaining correlation is added veriationally (in a.u.).

n	H^-	He	Li^+
3	-0.062386	-0.335694	-0.829753
4	-0.037351	-0.194871	-0.476859
5	-0.024624	-0.126743	-0.308552
6	-0.017434	-0.088984	-0.215760
7	-0.013017	-0.065871	-0.159051
8	-0.010086	-0.050714	-0.122208
9	-0.008045	-0.040247	-0.096812
10		-0.032718	-0.078597

Table 3

Oscillator strengths from the ground ^1S state to the $^1P^o$ Wannier TEIL in H^-, He and Li^+.

n	H^-	He	Li^+
3	2.3×10^{-5}	4.2×10^{-5}	2.3×10^{-5}
4	9.0×10^{-9}	2.1×10^{-6}	1.3×10^{-6}
5	2.7×10^{-10}	2.2×10^{-7}	1.4×10^{-7}
6	2.2×10^{-9}	3.4×10^{-8}	3.0×10^{-8}
7	1.1×10^{-9}	6.7×10^{-9}	6.8×10^{-9}
8	4.8×10^{-10}	1.7×10^{-9}	1.4×10^{-9}
9	1.5×10^{-10}	6.3×10^{-10}	4.8×10^{-10}
10		2.3×10^{-10}	1.8×10^{-10}

Finally, our results for the angular distribution of the ejected electrons, extracted from the asymptotic behavior of our TEIL wave-functions of ^1S symmetry, contradict the "Wannier theory" [33,36,37,40] in that they predict a Gaussian width for Li^+ where the previous theory predicts an isotropic distribution.

Acknowledgment

Most of the work described in this paper is a result of many years of collaboration with D.R.Beck and Y.Komninos.

REFERENCES

1. See articles by Beck and Nicolaides in "Excited States in Quantum Chemistry" eds. C.A.Nicolaides and D.R.Beck, Reidel (1978)
2. C.A.Nicolaides in "Advanced Theories and Computational Approaches to the Electronic Structure of Molecules" ed.C.E.Dykstra, Reidel (1984), p.161
3. C.A.Nicolaides, Chem.Phys.Lett. 101, 435 (1983)
4. Y.Komninos and C.A.Nicolaides, Phys.Rev. A33 (1986)

5. Y.Komninos and C.A.Nicolaides, J.Phys.B19, 1701(1986)
6. G.Aspromallis and C.A.Nicolaides, J.Phys. B19, L13(1986)
7. C.A.Nicolaides, Y.Komninos and D.R.Beck, Phys.Rev. A24, 1103(1981)
8. C.A.Nicolaides and Th.Mercouris, Phys.Rev. A32, 3247(1985)
9. I.D.Petsalakis, G.Theodorakopoulos,C.A.Nicolaides and R.J.Buenker, J.Chem.Phys.81, 3161(1984); 5952(1984)
10. G.Theodorakopoulos, I.D.Petsalakis, C.A.Nicolaides and R.J.Buenker, J.Chem.Phys. 82, 912(1985)
11. U.Fano and F.Prats, Proc.Natl.Acad.Sci.(India) A33, 553(1963)
12. D.E.Ramaker and D.M.Schrader, Phys.Rev. A9, 1980(1974)
13. Y.Komninos and C.A.Nicolaides, Z.Physik D, submitted
14. C.A.Nicolaides, Y.Komninos and D.R.Beck,Chim.Chron.10, 35(1981)
15. C.A.Nicolaides and D.R.Beck, Chem.Phys.Lett.35, 202(1975)
16. D.R.Beck , Phys.Rev. A23, 159(1981)
17. C.A.Nicolaides and D.R.Beck, Chem.Phys.Lett. 36,79(1975)
18. D.R.Beck and C.A.Nicolaides, J.Phys. B16, L627(1983)
19. D.R.Beck and C.A.Nicolaides, Int.J.Qu.Chem. S8, 17(1974); S10, 119(1976)
20. D.R.Beck and C.A.Nicolaides, Int.J.Qu.Chem.S14, 323(1980); Phys.Rev.A26, 857(1982)
21. A.U.Hazi and K.Reed, Phys.Rev. A24, 2269(1981)
22. C.A.Nicolaides and Y.Komninos, Chem.Phys.Lett. 80, 463(1981)
23. C.A.Nicolaides and D.R.Beck, J.Phys.B9, L259, (1976)
24. D.R.Beck and C.A.Nicolaides, Phys.Lett. 61A,227(1977)
25. C.A.Nicolaides and D.R.Beck, Chem.Phys.Lett.53, 87(1978)
26. D.R.Beck and C.A.Nicolaides, Chem.Phys.Lett.,53,91(1978)
27. A.W.Fliflet, R.L.Chase and H.P.Kelly, J.Phys.B7, L443(1974)
28. G.Wendin,Phys.Lett. 46A, 119(1973)
29. D.R.Beck and O.Sinanoglu,Phys.Rev.Lett.28,934(1972);C.A. Nicolaides and D.R.Beck,J.Phys.B6,535,(1973);C.A.Nicolaides,Chem.Phys.Lett.21,242(1973)
30. D.R.Beck,J.Chem.Phys.81,5002(1984)
31. R.F.Bacher,Phys.Rev.43,264(1933)
32. G.H.Wannier,Phys.Rev.90,817(1953)
33. A.R.P.Rau,Phys.Rev.A4,207(1971);J.Phys.Coll.43,211(1982)
34. U.Fano,Rep.Prog.Phys.46,97(1983)
35. J.Macek and J.M.Feagin,J.Phys.B18,2161(1985)
36. J.M.Feagin,J.Phys.B17,2433(1984)
37. P.L.Altick,J.Phys.B18,1841(1985)
38. Y.Komninos,M.Chrysos and C.A.Nicolaides,J.Phys.B,submitted
39. C.A.Nicolaides and Y.Komninos,Phys.Rev.A,submitted
40. A.R.P.Rau,J.Phys.B9, L283(1976)

INFLUENCE OF CENTRIFUGAL BARRIER EFFECTS

ON THE AUTOIONISING RESONANCES

M.A. Baig[1,3], K. Sommer[1], J.P. Connerade[1,2], and J. Hormes[1]

[1]Physikalisches Institut der Universität Bonn
[2]Blackett Laboratory Imperial College London
[3]Physics Department Quaid-i-Azam University Islamabad, Pakistan

The high resolution photoabsorption spectra of rare gases and alkali atoms are investigated in the p-subshell excitation region. It is demonstrated that centrifugal barrier plays a crucial role in determining the autoionisation widths and the intensities of the resonances.

1. INTRODUCTION

Potential barrier effects for d-electrons have been an explicit subject of study in recent years, following the detailed ab initio calculations by Griffin, Andrew and Cowan (1969), but the importance of the centrifugal term for d-electrons in the radial Schrödinger equation was implicitly recognised long before 1969 in a number of empirical statements such as the qualitative treatment of alkali atoms by penetrating and non-penetrating orbits (White 1934). The latter model leads to the statement that δ_l, the quantum defect, depends strongly on the orbital angular momentum l of the excited electron but is almost independent of the principal quantum number. To explain the detailed dependence on l is of course beyond this simple model and requires ab initio calculations.

2. A SIMPLE EXAMPLE OF CENTRIFUGAL BARRIER EFFECTS

To illustrate the influence of the centrifugal barrier on the spectra we are going to consider the principal series of the alkali atoms (Li through Cs). The fractional parts of the experimental quantum defects for ns- and nd-configurations as deduced from Moore (1971) are:

	Li	Na	K	Rb	Cs
δ_s	.400	.350	.182	.140	.059
δ_d	.003	.014	.259	.330	.475

The largest change in each sequence occurs between elements with empty d-subshells (Li and Na) and those with filled d-subshells (K, Rb and Cs)

To rationalize these findings, consider the potential for the d-electron. The effective radial potential is represented as:

$$V_{eff} = V(r) + l(l+1)\hbar^2/2mr^2$$

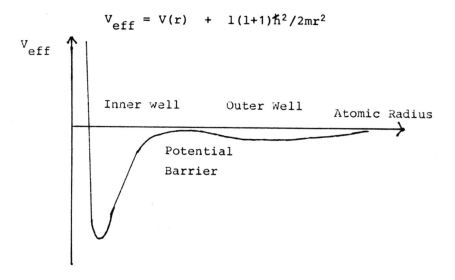

Fig. 1: Effective potential for d-electrons.

The electron moves in an effective potential arising from the attractive coulomb potential V(r) and the repulsive angular momentum dependent centrifugal potential $l(l+1)\hbar^2/2mr^2$.

For the s-electrons (l=o) the electron has no angular momentum and experiences only an attractive coulomb potential. With decreasing nuclear charge, the effective potential becomes more and more hydrogenic and its binding energy increases, born out by a decrease of the quantum defect δ.

For the d-electrons (l=2) there is a competition between the repulsive and the attractive parts of the potential. The repulsive part of the potential dominates in two regions (i) at very short radii, where it is stronger than the attraction of the unscreened nuclear charge and (ii) in the region of the barrier, where it is stronger than the coulombic potential in the outer reaches of the atom. The outer well is dominated by long range coulomb potential and behaves asymptotically as $-1/r^2$. It is shallow and can support an infinite number of bound states. The shape of the inner well depends on the nuclear charge. For light atoms, it is not deep enough to support any bound state and all the d-states are situated in the outer well, whereas, for heavy atoms, the inner well becomes deep enough to partially localize bound states.

The wavefunctions of the d-electrons for Li and Na are nearly hydrogenic, because the centrifugal barrier excludes them from the core, but become non-hydrogenic for K, Rb and Cs as a result of contraction into the attractive inner region of the potential. For s-electrons, there is no potential barrier but the filling of the d-subshells provides additional screening and the s-wavefunctions become more hydrogenic for K, Rb and Cs.

3. THE SPECTRA OF THE RARE GASES

Additional information on the spatial overlap between the core and ns- or nd-wavefunctions can be extracted by inner shell spectroscopy. The rare gases and alkali atoms are the best families of elements for a comparison in the autoionising region, taking into account the findings for the region below the first ionisation potential, where both systems are suitable candidates for the usual application of Multichannel Quantum Defect Theory for discrete levels. Thus we will discuss our spectra, at least for the rare gases, in both regimes. We have studied the photoabsorption spectra of the rare gases and the alkali atoms in the p-shell excitation region at high resolution. Synchrotron radiation emitted by the 5oo MeV electron accelerator at Bonn was used as the background continuum. The spectra were recorded in the first order of a 3 meter Eagle spectrograph equipped with a 5ooo line/mm holographic grating. The experimental details are described in our earlier papers (Baig et al. 1981, Connerade et al. 198o).

The high resolution photoabsorption spectra of rare gases have been extensively studied in connection with the Multi-channel Quantum Defect Theory (Seaton 1966, Lu and Fano 1970) to investigate the interchannel interactions among the overlapping Rydberg series. The series consist of two distinct sets of singly excited Rydberg series converging to two seperate ionisation limits. The Rydberg electron sees an inverted doublet mp^5 ($^2P_{1/2,3/2}$) core, the j=3/2 fine structure level being the ion ground state. The Rydberg series observed from the mp^6 $^1S_0$ ground state can be described in terms of five interacting channels, a Rydberg series together with its adjoining continuum being referred to as one channel. With J=1, these are labeled as :

$$mp^6 \rightarrow mp^5\ (^2P_{3/2})\ ns\ [3/2]$$
$$nd\ [1/2]\ ,\ [3/2]$$
$$mp^5\ (^2P_{1/2})\ ns\ [1/2]$$
$$nd\ [3/2]$$

The first three channels describe the bound Rydberg series converging to the $^2P_{3/2}$ limit and the remaining two describe those converging to the $^2P_{1/2}$ limit. We used the l [K] J coupling notation (Racah 1942, Cowan and Andrew 1965) for the level designation. The members of the series to the $^2P_{1/2}$ limit, which lie below the ionisation threshold, perturb the series to the $^2P_{3/2}$ limit, while those which lie above the ionisation limit autoionise into the mp^5 ($^2P_{3/2}$) ϵd, ϵs continua. Both effects are described within the framework of Multichannel Quantum Defect Theory.

Below the first ionisation threshold, a Lu Fano plot (Lu and Fano 1970) describes the interchannel interaction among the Rydberg series. In this two limit system, the energy levels are referred to both ionisation potentials and the two effective quantum numbers n_1^* and n_2^* are deduced for each level. The fractional parts of n_1^* and n_2^* are plotted as abscissa and ordinate for all experimentally determined levels and the resulting curve establishes the empirical function which describes the periodic nature of the series perturbation. The avoided crossing of this curve provides information about the width of the autoionising resonances observed above the first ionisation limit (Macek 1970). A strong interchannel interaction shows a larger avoided crossing and strongly autoionising resonances appear in the continuum.

This is easily understood considering the following arguments. At large distances the difference between the two ionisation limits dominates the electron-core interaction. If the different channels were not interacting one would expect each one to have a characteristic energy-independent quantum defect, just as though each channel could be considered a one channel case. At short range, the electrostatic interaction between the excited electron and the core is dominant and the difference in ionic thresholds becomes less important in comparison with the excited electron's large kinetic energy. This short range interaction results in series perturbations, i.e. quantum defects that vary dramatically along a Rydberg series due to the interaction between nearby levels of different Rydberg series.

Thus, the "perturbing interactions" take place as short range interactions of the Rydberg electron when penetrating the ion core. Consequently, both interacting states must have a significant overlap with the core state. In the same way as this effect causes the quantum defect to vary strongly if two discrete levels lie close together, the autoionisation width increases correspondingly as a sign for the strong interaction between the Rydberg state and the open-channel continuum belonging to the lower ionisation limit.

In figure 2 we reproduce the Lu-Fano plot for Neon (Baig and Connerade 1984) and Xenon (Hill et al. 1982). For Neon we find a small avoided crossing indicating a very weak interchannel interaction and the width of the autoionising resonances is also very small. For Xenon the interchannel interaction is strong leading to a larger avoided crossing and to strongly autoionising resonances in the continuum.

Comparing the autoionising resonances in the spectra of Ne, Ar (Yoshino 1970), Kr (Yoshino and Tanaka 1979) and Xe (Yoshino and Freeman 1985, Bonin et al. 1985), two striking differences are apparent:

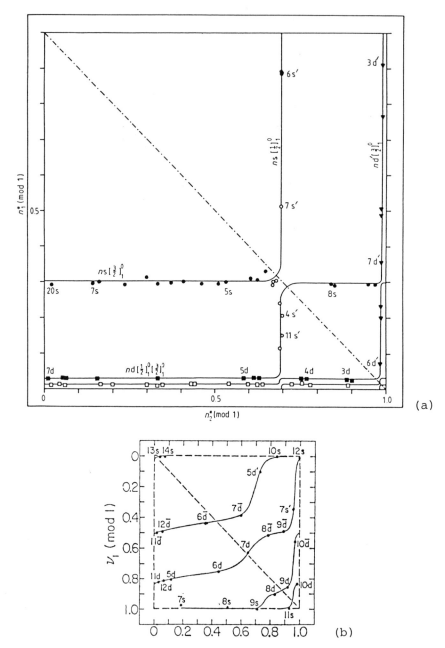

Fig. 2: Two-dimensional quantum defect plot of Neon (a) (Baig and Connerade 1984) and Xenon (b) (Hill et al. 1982).

(i) The nd-resonances in Ar, Kr and Xe show a typical Beutler-Fano asymmetric line shapes (Fano 1961) and the ns-resonances are very sharp. In Neon, the situation is different: the ns- and nd-resonances remain sharp and autoionisation is not prominent.

(ii) The ns-resonances are shifted towards higher energies as one proceeds from Xe to Ne. In Xenon, a pair of lines is formed by 11d and 13s with 11d lying on the lower energy side. In Kr a similar situation prevails for corresponding features, but the splitting decreases. In Ar, the pairs of lines are nearly degenerate while in Ne, the 13s level crosses over to higher energy of 11d and forms a pair with 1od. In fact, the ns-series members can be regarded as energy markers which give the potential of the atom without centrifugal effects, and the energies of the nd-series members relative to them yield information on centrifugal barrier penetration.

4. THE SPECTRA OF THE ALKALI ATOMS

In the case of the alkalis, the parent ion configuration p^5s provides four levels namely $^3P_{0,1,2}$ and $^1P_1$ which serve as the limits in the photoabsorption spectra. Additional complexity of the structure arises due to (s x d)-mixing in the ground state. However, the series converging to the $^1P_1$ limit can be picked out to compare them with the autoionising resonances in the rare gases.

In figure 3 we reproduce the high resolution absorption spectra of Na, K and Cs showing the autoionising resonances which are classified according to the following excitation scheme:

$$mp^6 \ (m+1)s \ ^2S_{1/2} \rightarrow mp^5 \ (m+1)s \ [^1P_1] \ ns \ [1,1/2]$$
$$nd \ [1,3/2]$$

The nd-resonances, which start out comparatively sharp in Na where the d-subshell is empty, rapidly become broad in going to K, Rb and Cs, where the d-subshell is filled. This is due to the large increase in spatial overlap between the nd-wavefunction and the core. Following the same sequence, ns- and nd-states which form the nearest pair in Na cross over in energy in going towards Cs, becoming energy degenerate around K, in precise analogy with the rare gases (see above). There is, however, a slight distinction between the situation in the

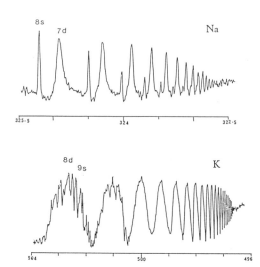

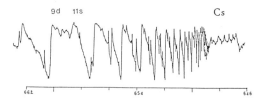

Fig. 3: High resolution absorption spectra of Na, K and Cs

alkalis and in the rare gases: for the rare gases, the ns-states can be considered as energy marker, allowing the contraction of the d-orbital to be followed. In the alkalis, the s-states also become more hydrogenic towards Cs and move oppositely in energy compared with the d-states.

The reason for the contraction of the d-orbitals with respect to the s-orbitals for increasing atomic number is the gradual deepening of the inner well in the effective radial potential. The effect described as "wavefunction collaps" (Connerade 1978) is present for values of l>1, but is less dramatic for d-electrons, for which there is only a shallow barrier between the inner and outer reaches of the potential, than for f-electrons, which possess a well characterised double valley potential.

5. DISCUSSION OF THE OBSERVED BARRIER EFFECTS

The facts described in chapters 3 and 4 can be explained through the behaviour of the nd- and ns-wavefunctions in the p →d and

p→s configurations. Therefore we have performed ab initio
Hartree-Fock calculations to study the penetration of the centrifugal barrier as a function of atomic number. In figure 4 the
p^5d- and the p^5s-wavefunctions for rare gases are reproduced.
It is clear from the figure that the d-wavefunctions move progressively inwards and the core penetration increases from Ne
to Xe. These calculatons were also performed for the alkalis
and in figure 5 the p sd and the p ss´wavefunctions of Na and
K are presented which show a similar trend of increased spatial
overlap between the core and the excited state. The consequences
of the increase in overlap are:

(i) The transition probability of the transitions from the core
to the d-states increases regularly from Ne to Xe and from
Na to Cs.

(ii) The autoionisation probability, which is determined by the
spatial overlap between the wavefunctions of the excited states
and the core, because there is no possibility of Auger broadening, increases regularly from Ar to Xe and from Na to Cs.

(iii) A departure from hydrogenicity which is apparent from the
quantum defects.

We compare the quantum defects for the rare gases below.

Identification		Ne	Ar	Kr	Xe
$(^2P_{3/2})$	ns [3/2]	1.303	2.134	3.097	4.009
	nd [3/2]	0.011	0.101	1.138	2.145
	nd [1/2]	0.028	0.391	1.462	2.519
$(^2P_{1/2})$	ns [1/2]	1.308	2.144	3.113	4.050
	nd [3/2]	0.013	0.181	1.195	2.279

The quantum defects of the ns-series grow by approximately
one unit from one rare gas to the next. For the nd-series, the
situation is slightly different: filling of the 3d-subshell begins only after Ar, beyond which the quantum defect increases
by roughly one unit from one rare gas to the next. Between Ne
and Ar, there is no change in the occupation of the d-subshell
and the quantum defect changes only by a small fraction of
unity. This behaviour is readily understood in terms of quantum
defect theory for the situation where a short-range well develops inside the atom and a phase change of π (equivalent to

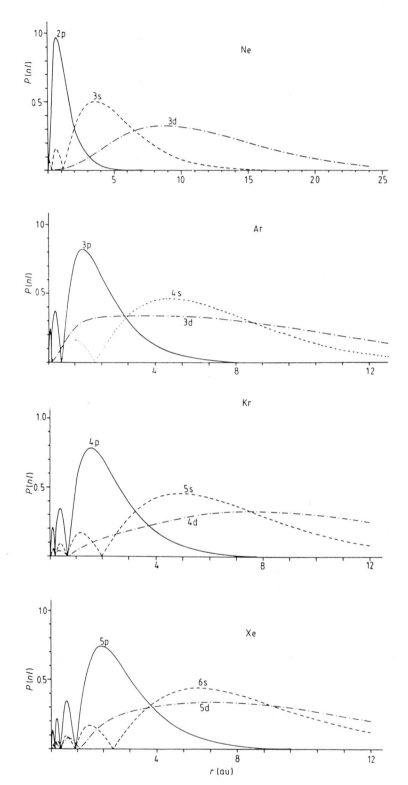

Fig. 4: Hartree-Fock radial wavefunctions for core excited Ne, Ar, Kr and Xe (Baig and Connerade 1984)

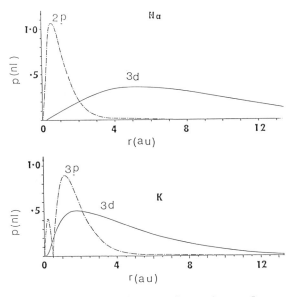

Fig. 5: Hartree-Fock radial wavefunctions for core-excited Na and K.

a change of unity in the quantum defect) occurs every time a new bound state is trapped in the short-range well.

The fractional part of the quantum defects for the ns- and nd-states of core excited Na to Cs are collected below.

	Na	K	Rb	Cs
δ_s	.72	.55	.40	.32
δ_d	.35	.55	.68	.75

As can be seen from this comparison, the quantum defects of the ns- and nd-series behave exactly in the same way as described above for the rare gases.

6. CONCLUSION

In conclusion, we are able to understand the changes in quantum defects and autoionisation widths by reference to multichannel quantum defect theory and to independent particle ab initio calculations. Centrifugal barrier effects are shown to play a dominant role in the p-shell excitation spectra of rare gases and the alkali atoms.

It is interesting to compare this situation with the recent study by Hill et al. (1982) of the evolution of the ns- and nd-series along the isoelectronic sequence Xe, Cs^+ and Ba^{++}. It was demonstrated that the widths of the ns-resonances are independent of nuclear charge, whereas the widths of the nd-resonances increases drastically from Ba^{++} to Xe and this fact was attributed to modification of the balance between the centrifugal and nuclear terms in the effective radial potential in the same way as discussed above for the rare gases and alkalis.

The present work was financially supported by SERC (UK) and BMFT (West Germany).

REFERENCES

Baig, M.A. and Connerade J.P., 1984, J.Phys.B:At.Mol.Phys. 17, 1785
Baig, M.A., Hormes J., Connerade J.P., and McGlynn S.P., 1981, J.Phys.B: At.Mol.Phys. 14 , L 725
Bonin K.D., McIlrath T.J. and Yoshino K., 1985, J.Opt.Soc.Am. B2, 8, 1275
Connerade J.P., 1978, Contemp.Phys. 19, 415
Connerade J.P., Baig M.A., Garton W.R.S. and McGlynn S.P. 1980, J.Phys.B:At.Mol.Phys. 13 , L 7o5
Cowan R.D. and Andrew K.L., 1965, J.Opt.Soc.Am. 55, 5o2
Fano U., 1961, Phys. Rev. , 124, 1866
Griffin D.C., Andrew K.L.and Cowan R.D., 1969,Phys.Rev. 177, 62
Hill W.T., Cheng K.T., Johnson W.R., Lucatorto T.B., McIlrath T.J. and Sugar J., 1982, Phys.Rev.Lett. 49, 1631
Lu K.T.and Fano U., 197o, Phys.Rev. A2, 81
Macek J., 197o, Phys.Rev. A2, 3, 11o1
Moore C.E., 1971, Atomic Energy Levels NBS Circular 35
Racah G., 1942, Phys.Rev. 61 , 536
Seaton M.J., 1966, Proc.Phys.Soc. 88, 8o1
Yoshino K., 197o, J.Opt.Soc.Am. 6o, 122o
Yoshino K. and Tanaka Y., 1979, J.Opt.Soc.Am. 69, 159
Yoshino K. and Freeman D.E., 1985, J.Opt.Soc.Am. B2, 1268
White H.E., 1934 Introduction to Atomic Spectra, (New York: McGraw-Hill Book Co.)

GIANT RESONANCE PHENOMENA IN THE ELECTRON IMPACT IONIZATION OF HEAVY ATOMS AND IONS

Stephen M. Younger

A-Division
Lawrence Livermore National Laboratory
Livermore, California 94550

INTRODUCTION

Heavy atoms and ions offer an interesting opportunity to study atomic physics in a region where the atomic structure is dominated by the interelectronic interactions. One illustration of this is the profound term dependence of atomic orbitals for certain configurations of heavy atoms and ions. This subject is extensively discussed elsewhere in the proceedings to this Institute. The present paper deals with another manifestation of resonance behavior, namely the appearance of giant scattering resonances in the cross sections for ionization of heavy atoms by electron impact. Such resonant structures arise from the double well nature of the scattering potential and have recently been identified in the cross sections for the electron impact ionization of several xenon-like ions.[1] In the present report, we summarize the results of calculations showing similar effects for a variety of other ions.

THEORETICAL METHOD

The electron impact ionization cross sections presented here were computed in a distorted wave Born exchange approximation. The target wavefunctions were represented by Hartree-Fock orbitals. The incident and scattered continuum electrons were computed in an approximation to the Hartree-Fock potential corresponding to the initial state of the atom or ion.[2] The $l=3$ ejected electron continuum orbitals were computed in a non-local term-dependent Hartree-Fock potential. For ionization of $4d^{10}$ subshells the $l=3$ ejected orbitals were computed in

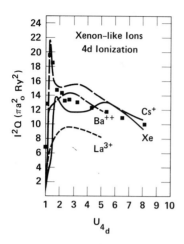
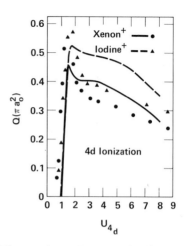

Figure 1. Scaled electron ionization cross section for xenon-like ions. Squares are the measurements of Hertling et. al.[6] for the double ionization of Cs^+ by a single electron impact.

Figure 2. Electron ionization cross sections for I^+ and Xe^+. The curves are distorted wave calculations and the points are the measurements of Achenbach et. al.[7].

term-independent approximate Hartree-Fock potentials. Sufficient partial waves were included in the triple partial wave expansion to ensure adequate convergence of the total cross section. Scattering exchange (exchange between the two final state continuum electrons) was included via the maximum interference theory of Peterkop[3]. In cases displaying very large resonance structures, the cross section was very sensitive to the method of integration over the final two-electron energy distribution. A 20 point Gauss integration was employed for this purpose. For ionization of $4d^{10}$ subshells, ground state correlation of the type $4d^{10}+4d^{8}4f^{2}$ is included.[4] Further details of the computational methods used can be found in previous publications.[5]

EXAMPLES OF GIANT RESONANCES IN HEAVY ATOMS AND IONS

Xenon-like Ions: 4d Subshell Ionization

Ionization of a 4d electron in xenon-like ions leaves the residual ion in a singly autoionizing state, so that the 4d cross section contributes to effective double ionization of the target by a single electron impact. Figure 1 shows the scaled 4d electron ionization cross

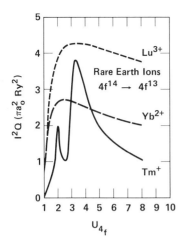
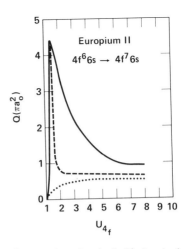

Figure 3. Scaled distorted wave electron ionization cross sections for the 4f ^{14} configuration in several rare earth ions.

Figure 4. Scaled distorted wave electron ionization cross section for the 4f subshell in Eu$^+$. Solid curve: distorted wave Born-exchange; dashed curve: distorted wave without scattering exchange; dotted curve: Coulomb-Born without scattering exchange.

sections, I^2Q, for Xe, Cs$^+$, Ba^{++}, and La^{3+} vs. incident electron energy. In this paper, the cross section, Q, is given in units of πa_0 the ionization energy, I, is given in Rydbergs, and the incident electron energy is expressed in ionization threshold units, i.e., u=E/I where E is the continuum energy and I is the ionization energy. In Cs$^+$, the resonance is narrower and much more pronounced than in Xe and occurs at lower incident electron energy, as does the normal maximum. For Ba^{++}, the resonance occurs just above threshold, resulting in a very rapid rise in the cross section at low electron energies. For La^{3+}, the resonance disappears from the continuum and only the "background" cross section remains. Figure 1 also compares the present results for Cs$^+$ with the measurement of Hertling et al. Very good agreement is obtained both for to the shape of the resonance and the amplitude of the total cross section. It has been shown[1] that the phase shift of the l=3 <u>scattered</u> partial wave undergoes a phase shift of more than pi at approximately 1 Rydberg, coincident with the appearance of the resonance in the cross section. The resonance structure thus represents a true scattering resonance rather than a manifestation of term-dependence in the ejected electron continuum.

Xenon⁺ and Iodine⁺

Figure 2 compares our theoretical results for Xe^+ and I^+ with the measurements of Achenbach et. al.[7] In this case, the theory underestimates the amplitude of the resonant enhancement of the cross section at low incident energies. We note, however, that calculations done using configuration-averaged ejected f-waves show a much larger resonant enhancement than is observed, so it is possible that the present method over-corrects for the effects of ejected orbital term-dependence and correlation.

Ionization of the $4f^{14}$ Configuration

Figure 3 plots the cross sections for ionization of the $4f^{14}$ closed subshell of Tm^+, Yb^{++}, and Lu^{3+}. In this configuration, significant resonant behavior is found only for Tm^+. Preliminary studies done with term-dependent ejected f-waves ($4f^{13}kf$ ^1S channel) indicate that the second maximum in Tm^+ is significantly reduced and broadened by ejected channel effects. The resonance at u=2 remains, however, and corresponds to a resonance in the scattered electron channel. Distorted wave calculations were done for 4f ionization from W^+ and Hg^+ to determine if the resonance reappeared when additional screening electrons were present. Neither of these elements was found to exhibit strong resonant behavior.

Europium⁺ $4f^76s - 4f^66s$ Configuration

Figure 4 presents cross sections for ionization of a 4f electrons from the ground configuration of Eu^+. A prominent resonant structure occurs at approximately 1.5 threshold units. The effect of scattering exchange is to broaden the resonance to higher energies, compared to the relatively narrow structure found when scattering exchange (exchange between the two final state continuum electrons) is neglected. Also shown in Figure 4 are the results of a Coulomb-Born calculation where the incident and scattered partial waves were computed in a Z=1 Coulomb potential and the ejected waves in a distorted wave potential. The Coulomb-Born approximation represents an approximate "background" cross section with which the full distorted wave cross section may be compared. Note that the effect of the resonance in the distorted wave calculation is to increase the low energy cross section by more than an order of magnitude over the "background" value.

SUMMARY

Distorted wave Born exchange calculations have revealed the presence of giant scattering resonances associated with the electron impact ionization of heavy atoms and ions. These resonances occur in the scattered electron channel, implying that they may serve as sensitive probes of the final state continuum-continuum electron interaction. Considering the complexity of such processes in heavy atoms the theoretical cross sections agree quite favorably with available experimental data. Additional calculations are in progress to further examine the nature of giant scattering resonances in electron ionization as well as other inelastic electron-ion processes.

REFERENCES

1. S. M. Younger, "Giant Resonances in the Electron Impact Ionization of Heavy Atoms and Ions," Phys. Rev. Lett. (1986).

2. S. M. Younger, "Photoionization and Electron Impact Ionization of the $4d^{10}$ Subshell in the Palladium isoelectronic Sequence," to appear in Phys. Rev. A.

3. R. K. Peterkop, "Theory of Ionization of Atoms by Electron Impact," Colorado Associated University Press, Boulder (1977).

4. M. S. Pindzola, D. C. Griffin, and C. Bottcher, "Electron Impact Double-Ionization of Xe II," J. Phys. B 16:L355 (1983).

5. S. M. Younger, "Distorted-Wave Electron-Impact Ionization Cross Sections for the Argon Isoelectronic Sequence," Phys. Rev. A, 26:3177 (1982), and references contained therein.

6. D. R. Hertling, R. K. Feeney, D. W. Hughes, and W. E. Sayle II, "Absolute Experimental Cross Sections for the Electron-Impact Single, Double, Triple, and Quadruple Ionization of Cs^+ Ions," J. Appl. Phys. 8:5427 (1982).

7. Ch. Achenbach, A. Muller, and E. Salzborn, "Electron-Impact Double Ionization of I^+ and $Xe^{\alpha+}$ ($\alpha=1,\ldots,4$) Ions: Role of 4d Electrons Like in Photoionization, Phys. Rev. Lett. 50:2070 (1983).

ACKNOWLEDGEMENT

Work performed under the auspices of the U.S. Department of Energy by Lawrence Livermore National Laboratory under contract #W-7405-Eng-48.

THE g-THOMAS-FERMI METHOD

Th. Millack and G. Weymans[*]

Physikalisches Institut der Universität Bonn
Nussallee 12
D-5300 Bonn 1
West Germany

INTRODUCTION

Density functional theories have proven to be a very powerful tool in many fields of physics, particularly in solid state and atomic physics. A major problem in the different approaches is the question of appropriately approximating the exchange-correlation functional. Here we outline a method to include many body correlations to arbitrary order and to construct a density functional which exactly describes the equilibrium properties of the inhomogeneous electron gas. Details of our calculations can be found in Ref. (1).

THE g-HARTREE ENERGY FUNCTIONAL

Starting point is the non-relativistic grand canonical partition function $Z[\mu(\vec{x}); \beta, \Omega]$, which is characterised by a space-dependent chemical potential $\mu(\vec{x})$ (β is the inverse temperature and Ω the volume of the system). In the framework of the g-Hartree method[2,3], Z is written as

$$Z[\mu(\vec{x}); \beta, \Omega] = Z_o[g; \mu(\vec{x}); \beta, \Omega] \, Z_{corr}[g; \mu(\vec{x}); \beta, \Omega] . \qquad (1.1)$$

Z_o is the partition in the g-Hartree mean field and Z_{corr} contains all fluctuations around the mean field. The real parameter g in (1.1) can be chosen ad libitum. By construction, Z is g-independent; it is the relative weight of Z_o and Z_{corr} which is controlled by the choice of g. Moreover, for a large class of potentials a g_o can be found such that

$$Z_{corr}\big|_{g=g_o} = 1 \qquad (1.2)$$

Hence, all equilibrium properties can be exactly calculated from $Z_o[g_o; \mu(\vec{x}); \beta, \Omega]$.

---

[*] present address: BAYER AG, Central Research, Polymerphysics, 5090 Leverkusen, West Germany

Fixing $g = g_0$ and using the heat kernel expansion

$$< x|e^{-tH}|y> \underset{t \to 0^+}{\sim} \frac{1}{(4\pi t)^{3/2}} e^{-\frac{|x-y|^2}{4t}} \sum_{\nu=0} \alpha_\nu(\vec{x},\vec{y}) t^{\nu/2} \qquad (1.3)$$

we construct the energy-density functional in the low-temperature limit[1] with the result

$$U[n] = \int d^3x \, \{ A_1 n^{5/3}(\vec{x}) + A_2 n^{4/3}(x) + A_3 n(\vec{x}) + \frac{g_0}{2} \int d^3x' \, \frac{n(\vec{x})n(\vec{x}')}{|\vec{x}-\vec{x}'|} \qquad (1.4)$$

$$+ \ldots \}$$

where

$$A_1 = \frac{(3\pi^2)^{5/3}}{5\pi^2}$$

$$A_2 = -(3\pi^2)^{4/3} \left(\frac{1-g_0}{4\pi^3} + \frac{\alpha_1(\vec{x},\vec{x})}{8\pi^{3/2}} \right)$$

$$A_3 = \frac{3}{16} \pi \alpha_1^2(\vec{x},\vec{x}) - \alpha_2(\vec{x},\vec{x}) + \frac{3(1-g_0)}{4\sqrt{\pi}} \alpha_1(\vec{x},\vec{x}) \qquad (1.5)$$

THE g-THOMAS-FERMI EQUATION

Using standard methods the g-Thomas-Fermi equation is then derived from (1.4)

$$\frac{d^2}{dx^2} \phi(x) = x \left(\alpha_g + \left(\frac{\phi(x)}{x}\right)^{1/2} \right)^3$$

$$\alpha_g = \frac{1-g_0}{g_0^{1/3}} Z^{-2/3} \frac{6^{1/3}}{4\pi^{2/3}} \qquad (2.1)$$

where $\phi(x)$ is the scaled charge density and

$$x = \frac{r}{a}$$

with

$$a = \left(\frac{3\pi}{4}\right)^{2/3} Z^{-1/3} g_0^{-2/3} \; .$$

The boundary conditions required to actually solve (2.1) are

$$\phi(0) = 1$$

$$-x_o \left(\frac{d\phi}{dx}\right)_{x=x_o} = 1 - g_o \frac{N}{Z} \qquad (2.2)$$

where

$$\phi(x_o) = 0 \ .$$

Inspecting (2.1) we see that taking $g_o = 1$ leads back to the Thomas-Fermi equation whereas dropping g_o and $(1-g_o)$ factors[3] results in the Thomas-Fermi-Dirac equation. Moreover, since g_o-values are always smaller than 1, solutions to eq. (2.1) exist – even for negative ions – as long as

$$N < \frac{Z}{g_o} \ . \qquad (2.3)$$

From (1.4) and (2.1) the total energy $E(N,Z;g_o)$ as a descending series in $Z^{1/3}$ can now be evaluated

$$E(N,Z;g_o) = a_1(N,g_o)Z^{7/3} + a_2(N,g_o)Z^2 + a_3(N,g_o)Z^{5/3} + \ldots \qquad (2.4)$$

where

$$a_1(N,g_o) = -\frac{16^{1/3}}{(3\pi)^{2/3}} g_o^{-1/3} \left(\frac{1}{5}\xi + \frac{1}{10} J_\varphi - \frac{1}{2} \frac{1 - \frac{N}{Z} g_o}{x_o}\right)$$

$$a_2(N,g_o) = \frac{1}{4}$$

$$a_3(N,g_o) = -\frac{6^{2/3}}{20\pi^{4/3}} g_o^{-2/3} (1-g_o) K_\varphi \qquad (2.5)$$

with

$$\xi = \phi'(0)$$

$$J_\varphi = \int_o^{x_o} dx \ \phi'^2(x)$$

$$K_\varphi = \int_o^{x_o} dx (x\phi(x))^{1/2} \phi''(x) \ .$$

RESULTS

We have performed numerical studies of eq. (2.1) for atoms using a semi-empirical determination of g_o; i.e. we compute "experimental g-values" by the requirement

$$E(N,Z;g_{exp}) = E_{exp}(N,Z) \quad . \tag{3.1}$$

In this way, we found optimised g-values as a function of Z, which can be used to calculate other observables[4]. The results are displayed in fig. 1. General trends, known from previous complete g-Hartree calculations[2,3,5], are reproduced. Moreover, the use of the g-Thomas-Fermi equation accounts for the effective screening near the boundary and automatically corrects for the self-energy contribution.

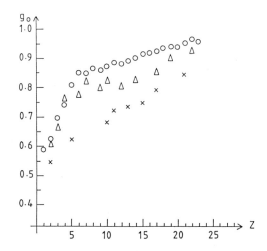

Fig. 1. Optimised g-values for neutral atoms, obtained by solving eq. (3.1), (o) for ground state configurations, (Δ) for outermost shell ionised atoms, (x) for inner shell (1s) ionised atoms

REFERENCES

1. Th. Millack, G. Weymans, to be published in J. Phys. B.
2. K. Dietz, these Proceedings;
   K. Dietz, G. Weymans, Physica A 131 (1985) 363;
   K. Dietz, O. Lechtenfeld, G. Weymans, J. Phys. B 15 (1982) 4301,4315.
3. K. Dietz, G. Weymans, J. Phys. B 17 (1984) 2987.
4. J.P. Connerade, K. Dietz, M.W.D. Mansfield, G. Weymans, J. Phys. B 17 (1984) 1211.
5. K. Dietz, M. Ohno, G. Weymans, to be published in J. Phys. B.

THE GIANT RESONANCES IN THE 3d TRANSITION ELEMENTS: SINGLE AND DOUBLE PHOTOIONIZATION IN THE REGION OF THE 3p EXCITATION

M. Schmidt and P. Zimmermann

Institut für Strahlungs- und Kernphysik
Technical University of Berlin
Hardenbergstrasse 12, D-1000 Berlin 12
Fed. Rep. of Germany

In the last years the process of double photoionization was studied experimentally by several groups especially in the alkaline earth and rare earths elements[1-4]. One reason is the strong interest of the theory for this process in the framework of many body effects[5,6].

We have used the monochromatized synchrotron radiation of the electron storage ring BESSY in Berlin, atomic beam technique and ion detection with a time-of-flight (TOF) mass spectrometer to measure the cross section of single and double photoionization of the 3d elements in the region of the 3p excitation between 30-100 eV. The photoabsorption and photoelectron spectra of these elements in this region were extensively investigated by the DESY group of B. Sonntag[7].

The atomic beams were produced by a conventional effusion oven which was operated in the temperature range of about 1300-2100 K, or by direct evaporation from an Ohmic heated wire (e.g. Ti). The mass and charge selective detection of the photoions was performed with a TOF spectrometer. Here the photoions were extracted from the interaction region of the atomic and the photon beam by a short electric pulse ($\approx$30 V) of about 0.8 µs duration and a repetition rate up to 0.1 MHz. The photoions were further accelerated by a constant voltage of about 200 V and detected by a channeltron after passing a drift tube of 15 cm. The time of flight was measured by a time-to-amplitude converter with the start signal from the pulse generator and the stop signal from the ion detector. After the identification of a specific mass and charge state X^+ and X^{++} in the TOF spectrum of a multichannel analyzer, appropriate gate pulses to a double discriminator were used as time windows to measure simultaneously the cross section of single and double photoionization in dependence of the photon energy.

The threshold for double photoionization of the 3d elements lies between about 19 eV(Sc) and 25 eV(Ni). The

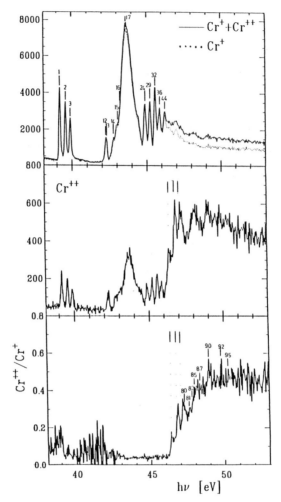

Fig. 1. Singe and double photoionization cross section of Cr in the region of the 3p excitation between 38-53 eV. The three vertical lines at about 47 eV show the threshold of 3p ionization $(3p^5 3d^5 4s\ ^8P_{9/2,7/2,3/2})$.

cross sections usually are very small just above these thresholds. However, they increase substantially if one reaches the region of the 3p excitation. There is a sharp increase of $\sigma(X^{++})$ after passing the threshold of 3p ionization with the subsequent Auger decay of these states. But there is also the possibility of 3p excitation to atomic states of configurations like $3p^5 3d^{N+1} 4s^2$ starting from the ground state configuration $3p^6 3d^N 4s^2$. These states, then, can decay to X^+ or X^{++}.

Fig. 1 shows as an experimental example the cross sections of Cr^+ and Cr^{++} and the ratio Cr^{++}/Cr^+ in the region of 38-53 eV. For the exceptional case of Cr as the only 3d element with a half filled 4s shell in its ground state configuration $3d^54s$ there is also the possibility of the 3p-4s excitation which gives rise to the three signals (1-3) belonging to the three ^7P-states of the configuration $3p^53d^54s^2$. The following signals (12-17) can be assigned to the 3p-3d excitation of levels of the configuration $3p^53d^64s$. Then there is a series of signals (24-44) due to Rydberg states of configurations $3p^53d^54snd$ converging to the thresholds of the 3p ionization ($3p^53d^54s\ ^8P_{9/2,7/2,5/2}$) at 46,36 eV, 46,72 eV and 47,05 eV (vertical lines).

It is interesting to see that all these sharp structures which are due to the excitation of autoionizing atomic levels are reproduced in the cross section of Cr^{++} which is reflected in a nearly constant ratio of Cr^{++}/Cr^+ up to the threshold of the 3p ionization. The sharp increase of the ratio Cr^{++}/Cr^+ after passing this threshold should be attributed to 3p ionization with the subsequent Auger decay.

We have measured the cross section of single and double photoionization for all 3d elements (exept Vanadium) in the region of 30-100 eV. There is good evidence for a nearly constant ratio of X^{++}/X^+ for Sc, Ti and Cr up to the threshold of 3p ionization whereas for Mn, Fe, Co and Ni (although with much larger error limits for Fe, Co and Ni because of much smaller counting rates) the ratio of $X^{++}/^+$ increases even before reaching this threshold. The mechanism of this process may be due to interference effects between 3p and Rydberg states but requests further investigations both theoretically and experimentally.

We acknowledge many helpful discussions with Prof. B. Sonntag and his coworkers at DESY (Hamburg) and with Dr. U. Becker (Berlin). Especially we are grateful for the photoelectron and photoabsorption results of the DESY group given to us prior to publication. This work was supported by the Bundesminister für Forschung und Technologie (BMFT).

REFERENCES

/1/ D.M.P.Holland and K.Codling, J.Phys.B: At.Mol.Phys. 14, 2345 (1981)
/2/ B.Lewandowski, J.Ganz, H.Hotop and M.-W. Ruf, J.Phys.B: At.Mol.Phys. 14, L803 (1981)
/3/ Y.Sato, T.H.Hayaishi, Y.Itikawa, Y.Itoh, J.Murakawi, T.Nagata, T.Sasaki, B.Sonntag, A.Yagishita and M.Yoshino, J.Phys.B: At.Mol.Phys. 18, 225 (1985)
/4/ D.M.P.Holland and K.Codling, J.Phys.B: At.Mol.Phys. 14, L359 (1981)
/5/ S.L.Carter and H.P.Kelly, Phys.Rev.A 24, 170 (1981)
/6/ G.Wendin, Les Houches Session 38, North Holland Physics Publishing, 1984
/7/ M.Meyer, Th.Prescher, E.von Raven, M.Richter, E.Schmidt, B.Sonntag and H.E.Wetzel, Z.Physik D (to be published, 1986)

DECAY CHANNELS OF CORE EXCITATION RESONANCES IN ATOMIC LANTHANIDES AND 3d
TRANSITION METALS

M. Meyer, Th. Prescher, E. v. Raven, M. Richter,
E. Schmidt, B. Sonntag and H.E. Wetzel

II. Institut für Experimentalphysik, Universität Hamburg,
Luruper Chaussee 149, D-2000 Hamburg 50, F.R. Germany

EXPERIMENT

Photoabsorption, photoelectron and ejected electron spectra of atomic lanthanides and 3d metals in the energy range of the strong 4d → (4f,εf) and 3p → 3d excitations have been investigated. The combination of three different experimental methods gives detailed information on the character of the giant resonances. In this context the study of the Z dependence turned out to be very important.

For the photoabsorption and photoelectron measurements, synchrotron radiation from the storage rings DORIS and BESSY was used in combination with a toroidal grating monochromator (TGM). In the absorption experiments the photon beam passed through a metal vapour column maintained inside a resistance heated tubular furnace. The spectra were recorded photoelectrically. For photoelectron spectroscopy the monochromatic photon beam was focused onto the interaction zone where it crossed an atomic beam. The atomic beam emanated from a high temperature furnace heated by electron bombardment. The kinetic energy of the photoelectrons was determined by a cylindrical mirror analyzer. Replacing the photon beam by a beam of 1.5 keV electrons, ejected electron spectra have been taken[1].

RESULTS

Lanthanides

The giant resonances displayed by the lanthanides in the energy range of the 4d ionization limits have been ascribed to $4d^{10}4f^n \rightarrow 4d^9(4f,\varepsilon f)^{n+1}$ excitations[8-11]. The shape and width of these resonances are strongly influenced by the interaction with the 4d, 4f, 5p, 5s and 6s ionization continua. Our recent photoelectron spectra of atomic Ce, Sm, Eu and Gd allow to assess the importance of the various decay channels. Figure 1 shows the partial cross sections accounting for most of the total cross section of these elements.

For Ce $(4d^{10}5s^25p^64f5d6s^2)$ the $4d^{10}4f \rightarrow 4d^9(4f,\varepsilon f)^2 \rightarrow 4d^94f,\varepsilon l$ channel dominates. The 4d partial cross section has been determined directly by detection of 4d electrons and indirectly by detection of

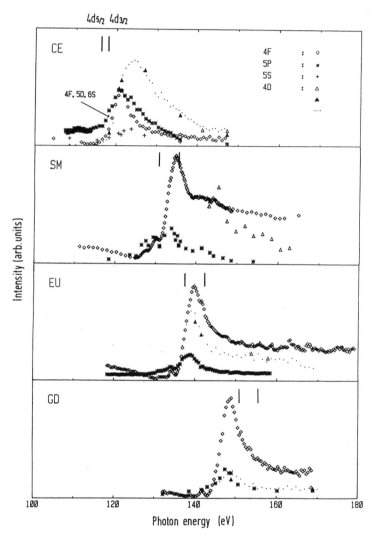

Fig. 1 Relative partial cross sections of the strongest photoemission lines for atomic Ce, Sm, Eu and Gd. The 4d cross section has been determined directly (△) and by the total Auger yield (▲). The dotted curve (...) displays the relative cross sections of a part of the Auger emission.

Auger electrons, which originate from the decay of the 4d hole state. The 4d cross section peaks well above ($\gtrsim$ 6 eV) the $4d_{3/2}$ ionization limit. The resonance profile closely resembles the 4d shape resonance of atomic Xe[3], Cs[2] and Ba[4,5]. Integrated over the resonance, the sum of the 4f, 5d and 6s partial cross sections is less than half of the 4d partial cross section. The 4f cross section accounts for ~ 70% of the sum. The asymmetric Fano type profile reaches its maximum $\gtrsim$ 3 eV below the 4d maximum, i.e. much closer to threshold. There is strong interference between the $4d^{10}4f \rightarrow 4d^9(4f,\epsilon f)^2 \rightarrow 4d^{10}, \epsilon l$ channel and the direct $4f \rightarrow \epsilon l$ ionization. The $4d^9(4f,\epsilon f)^2$ excited states change character with increasing energy. Close to the threshold approximately half of them decay via $4d^9(4f,\epsilon f)^2 \rightarrow 4d^{10}, \epsilon l$ autoionization. At higher energies the escape of the excited electron into the $4d^9 4f, \epsilon l$ continuum clearly is the main route of decay.

For higher Z and higher 4f occupation the probability for the emission of a 4d electron decreases. The 4f cross section gains in strength. The maximum of its Fano type profile shifts from a position above the $4d_{3/2}$ ionization threshold in Ce to a position below the $4d_{5/2}$ threshold in Gd. For Gd the 4d partial cross section only contributes less than 20% to the total cross section. Within the accuracy of our measurements the maxima of the 4f and 4d partial cross sections coincide for Eu and Gd. Transition to 4f-like states accounts for most of the giant resonances in Gd whereas for Ce the excited states are of dominant ϵf character. The ejected electron spectra and the absorption spectra confirm the photoelectron data.

3d transition metals

In Figure 2 the absorption spectra of atomic Ti, Cr, Mn, Fe, Co and Ni are presented[1]. The maxima are ascribed to $3p^6 3d^n \rightarrow 3p^5 3d^{n+1}$ excitations strongly coupled to the underlying $3d \rightarrow \epsilon l$ continua. The spectra have been recorded photoelectrically. Therefore the relative amplitudes are more reliable than those obtained from photographic data. Going along the series, the spectra change dramatically. Ti displays several almost symmetric, 0.7 eV wide lines distributed over 10 eV. The high energy lines ly above the lowest ionization limits. In contrast to this, Cr, Mn, Fe, Co and Ni display up to three broad asymmetric absorption maxima located below the ionization thresholds.

Our recent photoelectron spectra of atomic Ti prove that the absorption resonances are dominated by the

$$3p^6 3d^2 4s^2 \,\, ^3F \rightarrow 3p^5 3d^3 4s^2 \,\, ^3G \,\, ^3F \,\, ^3D \rightarrow \begin{array}{l} 3p^6 3d^2 4s \,\, ^2G \,\, ^{4,2}F \,\, ^2D \\ 3p^6 3d 4s^2 \,\, ^2D \\ 3p^6 3d^2 4p \,\, ^2D \end{array}$$

channels[6,7]. The symmetry of the resonance profiles can be attributed to the low oscillator strength of the $3d \rightarrow \epsilon l$ continuum, which renders interference effects negligible. For higher 3d occupation, i.e. higher Z, the direct $3d \rightarrow \epsilon l$ ionization and thus interference become stronger. For photon energies above 40 eV $3p \rightarrow \epsilon l$ ionization takes place in Ti leading to a considerable continuum in the absorption spectrum. The excited Ti II: $3p^5 3d^2 4s^2$ states undergo Auger decay to Ti III: $3p^6 3d^2$ and Ti III: $3p^6 3d4s$ states[7]. The corresponding Auger lines could be observed in the photoelectron spectra.

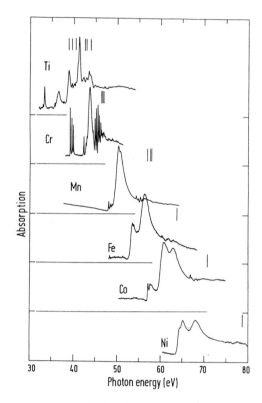

Fig. 2 Absorption spectra of atomic 3d transition metals. The spectra have not been normalized to each other. The vertical bars mark the positions of the 3p ionization limits.

1. M. Meyer, Th. Prescher, E. v. Raven, M. Richter, E. Schmidt, B. Sonntag, H.E. Wetzel, Zeitschrift für Physik D (1986), in press
2. Th. Prescher, M. Richter, E. Schmidt, B. Sonntag, H.E. Wetzel, J. Phys. B (1986), in press
3. U. Becker, Th. Prescher, E. Schmidt, B. Sonntag, H.E. Wetzel, Phys. Rev. A 33 (6) (1986), in press
4. F. J. Wuilleumier, in: X-Ray and Inner-Shell Processes in Atoms, Molecules and Solids, A. Meisel (ed.), p. 61. Karl-Marx-Universität Leipzig (1984)
5. U. Becker, R. Hölzel, H.G. Kerkhoff, B. Langer, D. Szostack, R. Wehlitz, in: Proc. XIV ICPEAK, Stanford (1985)
6. R. D. Cowan, The Theory of Atomic Structure and Spectra, Berkeley, University of California Press (1981) and references therein
7. Ch. Corliss, J. Sugar, J. Phys. Chem. Data 8:1 (1979)
8. A. F. Starace, Phys. Rev. B 5:1773 (1972)
9. J. Sugar, Phys. Rev. B 5:1985 (1972)
10. J. Dehmer, A.F. Starace, Phys. Rev. B 5:1792 (1972)
11. G. Wendin, A.F. Starace, J. Phys. B 11:4119 (1978)

II. MOLECULES — THEORY AND EXPERIMENTS

INNER SHELL RESONANCES IN MOLECULAR PHOTOIONIZATION

Irène Nenner

Lure, Lab. mixte CNRS, CEA et MEN, Bât. 209 D -
Centre Universitaire Paris-Sud, 91405 Orsay Cedex (France)
and
CEA/IRDI-DESICP - Département de Physico-Chimie
91191 GIF sur Yvette Cedex (France)

ABSTRACT

Discrete and shape resonances observed in photoabsorption spectra of molecules near core ionization edges are described using the SiH_4 (silane) and SiF_4 (Silicon tetrafluoride) molecules excited near the Si1s and Si2p edges as model systems. We illustrate the molecular specificity of such resonances by comparison of SiH_4 and the isoelectronic argon atom. Then we analyze the effect of the strength of the molecular field by a further comparison with SiF_4.

Electronic relaxation processes of such resonances identified primarily by photoelectron spectroscopy in SiH_4 and SiF_4 excited near the Si2p edge are presented and described in terms of autoionization and Auger effect. Competition with a primary fast dissociation process in SiH_4 is invoked.

Dissociation of the residual ions after a Si2p excitation of SiF_4 is evidenced with mass spectrometry in the normal and photoion-photoion coincidence modes. Multiple Auger are found to be other electronic decay channels. Limitations of the site selectivity of core excitation among final dissociation products in molecules are briefly discussed.

I - INTRODUCTION

Most resonances observed in inner shell photoabsorption spectra of an isolated molecule originate from transitions of a core electron into an antibonding valence, Rydberg orbitals or in the continuum. Their number, intensity and shape have been found (1) for a given absorbing atom to vary strongly with the nature of the surrounding ligands and therefore from one molecule to another. These observations show not only these resonances have little connection with resonances in the isolated atom case, but they are sensitive fingerprints of the chemical surroundings of the considered excited atom, i. e. geometrical arrangement, covalent or ionic nature of the bonds.

Theoretically, the concept of molecular shape resonances introduced by Dehmer (2) helped considerably to understand many intense features in molecular spectra. Basically, the core electron is trapped temporarily by a centrifugal barrier originating from the anisotropy of the molecular field through which it may eventually tunnel and escape. This approach known as the multiple scattering model (MSM) can be viewed as the scattering of a single electron by a positive ion. It has explained many important features of the discrete spectra and continuum parts of inner shell photoabsorption spectra of a number of molecules (1) in addition to the dynamics of the core electron election itself (3). This contrasts with the classical methods (e.g., 4) of quantum chemistry in which resonances are considered as core excited states i.e. valence or Rydberg and therefore account only for the discrete spectrum. Other methods (5,6) combine the e^-- ion collisional approach with quantum chemistry methods and have successfully calculated various resonances in the innershell continuum of some simple molecules.

Notice that some of the core excited states can be called shape resonances when they lie in the inner well region, this is often the case for transitions into antibonding valence orbitals. In contrast, Rydberg states which lie in the outer well region are not shape resonances and are found sometimes with very low oscillator strength in those cases where the potential barrier is high and favors all inner well transitions at the expense of outer well ones.

Another useful approach to analyze core resonances is to use the core equivalent model (7-9) in which a core excited molecular state can be compared with the state of another molecule, in which an atom with a nucleus Z and a core vacancy can be replaced by another atom with a nucleus Z + 1. This model has been found successful not only to explain line positions (8,10) but also to account for large Franck-Condon envelopes when the core excited state is repulsive (10-13). Notice that limitations of this model have been found when one considers vibrational frequencies and potential energy surfaces (14, 15).

A different method has been developed by Robin (16) for interpreting discrete resonances in an extensive number of simple or complex molecules. This approach takes advantage of the regularity of term values (i.e. the resonance energy measured with respect to the ionization threshold) independently of the inner hole. This has been verified in many occasions for Rydberg and also for valence state as stated by Robin (16) (for example, a good transferability of term values, has been found in SO_2 and CS_2 (17) when resonances are compared from the sulfur K edge to the sulfur L edge). However, some limitations of this statement exist : i) Term values generally are found to increase as the original orbital is deeper because the shielding of the ionic core by the valence electron decreases (16) or because of large changes in Coulomb and exchange integrals (6, 18). ii) A clear assignment to Rydberg or valence states is not always possible especially when both have the same symmetry. A comprehensive analysis of this problem in connection with the shape resonance concept has been performed by Keller and Lefebvre-Brion (19).

In the present paper, we will classify for the sake of clarity resonances in the discrete part of the spectrum (<u>discrete resonances</u>) separately from resonances in the continuum (<u>shape resonances</u>), primarily because they interact with totally different continua and have drastically different lifetimes.

When an electon is bound in a valence orbital or in a Rydberg orbital in the presence of the core hole, the resulting states may decay via electroic autoionization. If one restricts the discussion to a single "outermost" core orbital, ignoring all deeper ones, we have the following sequence:

$$(\text{core})^2 (\text{valence})^4 (\text{unoccupied})^0 \xrightarrow{h\nu} (\text{core})^1 (\text{valence})^4 (\text{unoccupied})^1$$

$$(\text{core})^1 (\text{valence})^4 (\text{unocc})^1 \rightarrow (\text{core})^2 (\text{valence})^3 (\text{unocc})^0 + e^-$$

or by resonant Auger

$$(\text{core})^1 (\text{valence})^4 (\text{unocc})^1 \rightarrow (\text{core})^2 (\text{valence})^2 (\text{unocc})^1 + e^-$$

Notice that autoionization involves the participation of the excited electron whereas in the Auger, it remains a spectator. Notice also that the final ionic state is either a one-hole state or a 2 hole-1 particle state. Higher order processes involving the ejection of two electrons are possible, we call them relaxation shake off or double resonant Auger or double autoionization. In certains situations, direct dissociation of the core excited state may compete with autoionization and resonant Auger. This aspect is reviewed in details by Morin (12).

When the core electron is ejected in the continuum, either directly or via a shape resonance one can consider this process as the following primary step

$$(core)^2 \; (valence)^4 \; (unoccupied)^0 \xrightarrow{h\nu} (core)^1 \; (valence)^4 \; (unoccupied)^1 \rightarrow$$

$$(core)^1 \; (valence)^4 \; (unoccupied)^0 + e^-_{core}.$$

Then the core hole decays via a normal Auger process as a second step.

$$(core)^1 \; (valence)^4 \; (unoccupied)^0 \rightarrow (core)^2 \; (valence)^2 \; (unocc)^0 + e^-_{Auger}.$$

Notice that the final ionic state has a 2 hole configuration. Higher order processes, i.e. multiple Auger may well occur, leading to the ejection of more than one electron.

Generally, the state of charge and configuration of the residual ion is intimately linked to the nature of the relaxation processes of core resonances. Furthermore if there are selective relaxation pathways, leading to the production of specific ionic states one may expect in certain circumstances to obtain specific ionic fragments. This particular point of strong photochemistry interest is only relevant to molecules with a significant size as we will see with the chosen examples.

In the present paper, we use the silane, SiH_4 and silicon tetrafluoride molecules as model systems photoexcited near the Si1s and 2p edges. Firstly, we analyze the specific molecular effects in photoabsorption spectra by comparizon of SiH_4 with the isoelectronic argon atom. Then we illustrate by choosing the silicon tetrafluoride (SiF_4) molecule, a stronger molecular field case compared to SiH_4. Secondly, we describe the many-electrons nature of the electronic relaxation processes of such resonances and the multicenter character of the electronic excited configuration of the residual (singly or multiply charged) ions, on the basis of photoelectron and photoion spectrometry experiments.

II - NEAR EDGE RESONANCES IN SiX_4 (X = H, F) PHOTOABSORPTION SPECTRA

A - The silane case : comparison with argon.

The silane molecule is a close shell species which has a tetrahedral geometry (T_d symmetry group).

The A_1 ground state configuration of SiH_4 can be described as follows :

$$1a_1^2 \quad 2a_1^2 \quad 1t_2^6 \qquad \underbrace{3a_1^2 \quad 2t_2^6}$$

Si1s   Si2s   Si2p   Si sp^3, H1s (four Si-H bonds)
    core orbitals   valence orbitals

In comparison the argon atom configuration is $1s^2 \; 2s^2 \; 2p^6 \; 3s^2 \; 3p^6$. We show in Fig. 1 the comparison of the Ar1s photoabsorption spectrum obtained by Parratt (20) with the Si1s spectrum of SiH_4 as measured by Bodeur and Nenner (21). Similar comparison is made near the Ar2p and Si2p edge edge for which we have used the photoabsorption spectrum of Hayes el al., so (22) (also measured by Friedrich et al., (4)) and the electron energy loss (23), of King et al., (23), as seen in Fig. 2.

Several observations can be made having in mind that the instrumental broadening is generally negligible compared to the natural line width.

i) The important broadening of resonances in the 1s compared to the 2p spectra (Ar and SiH_4) orginates from the core hole lifetime which is about ten times shorter for 1s than for 2p.

ii) Rydberg series can be identified in both Argon and SiH_4 as np for the 1s spectra and ns and nd for the 2p spectra. The latter are split into two series by spin-orbit coupling. Notice that the interpretation of the Si2p spectrum is of Schwarz (10) and is at varience with the analysis of Robin (16) who invokes uniquely np series. Following Schwarz (10) we observe that the tetrahedral molecular field splits the d series into two components d_e and d_{t_2}.

iii) The SiH_4 1s and 2p spectra reveals additional strong and wide resonances which have no equivalent in the atomic case. They result from transitions into antibonding valence unoccupied orbital (σ^*). In SiH_4, two low lying σ^* orbitals a_1 and t_2 are expected (4). In the T_d symmetry group, electric dipole selection rules, allow transitions a_1 (1s like) $\rightarrow t_2$, therefore the σ^* resonance in the Si1s spectrum has the t_2 symmetry. In the S12p spectrum, these rules allows t_2 (2p like) $\rightarrow a_1$, t_2, t_1, e. Therefore the transitions into σ^* with a_1 and t_2 symmetry calculated by Friedrich et al., (4) are both allowed. However the presence of close-by Rydberg transitions complicate the assignment of the SiH_4 spectra. The CI calculations of Friedrich

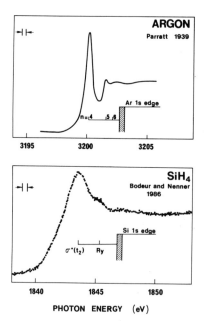

Fig 1 - Comparison of the Ar1s photoabsorption spectrum of ref 20 with Si1s spectrum of SiH_4 (ref 21)

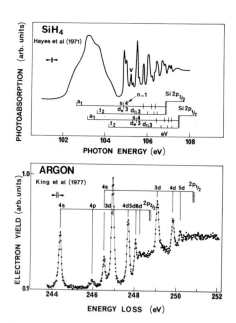

Fig 2 - Comparison of the Si2p photoabsorption spectrum (ref 22) (interpretation of ref 10) with the electron energy loss spectrum near the Ar2p edge (ref 23)

TABLE I - Experimental and theoretical term values (eV) of Rydberg and/or valence transition in core photoabsorption spectra of argon and silane.

Rydberg	valence σ*	Argon core excitation			SiH_4 core excitation			
		Ar1s ref 20	Ar2p ref 23	theory ref 10	Si1s ref 21	Si2p ref 4	theory ref 4	theory ref 10
4s		4.4	4.4	3.99	-	-	)	
	a_1^*	-	-	-	4.4	)4.04	4.2	
	t_2^*				3.4±0.5	3.4	2.85	
4p		2.7	2.73	2.60	)		2.14	2.5
5p		1.25	-		)1.7±0.5			
5s		-	-	-	-	2.2	1.63	
3d	3d(e)	-	1.75	1.51	-		1.56	)1.9
	$3d(t_2)$					1.6	0.99	)
Core ionization limit (eV) 1s or 2p (3/2, 1/2)		3202.9 ref 24	248.7 250.8 ref 23		1847.0 ref 26	107.2 107.8 ref 25	108.44 ref 4	

et al., (4) on the Si2p spectrum shows that the σ* (t_2) and 4p orbital are energetically separated whereas the σ* (a_1) and 4s are not. The latter states are probably strongly mixed (4, 10). This is known as the Rydberdization of the valence orbital (19) which probably vanishes at large internuclear distance. Notice that this assignment of the Si2p spectrum is found at variance with the interpretation of Robin (16) who suggests a unique σ* and a distinct 4s Rydberg line. Nevertheless we have reported in Table I experimental and theoretical term values of SiH_4 measured near the Si1s and Si2p edges, favoring the interpretation of ref 4 for the basis of the discussion.

iii) We observe in SiH_4 a good transferability of term values from one edge to another. This is true also for Argon. This shows that resonances depend primarily upon the final orbital, as stated by Robin (16) and that the orbital relaxation is independent of the core orbital. However, there is no clear correspondance of term values between argon and SiH_4, except for the totally symmetric orbital 4s.

iv) The calculations of Schwarz (10) based on the core equivalent species i.e. potassium or Argon and PH_4 for SiH_4 are found to be in satisfactory agreement with experiment (Table I).

v) All σ* resonances in SiH_4 are extremely broad showing probably a large Franck-Condon envelope which characterizes repulsive states (see below) and possibly other effect such as Jahn-Teller as analyzed in details in the SiF_4 case. Notice that this broadening (especially the Si2p spectrum) is not due to the rate of autoionization or Auger because of the sharpness of Rydberg lines implies, at the contrary, a rather long core hole lifetime. This point will be discussed in the next section.

vi) There is no features observed in the Si1s or Si2p continua of the SiH_4 spectra except for the Si1s a weak doubly excited state non represented in Fig 1. The Ar1s spectrum shows also a doubly excited state as reported by Wuilleumier (27), but at higher relative energy. It is probable that excitation of an additional electron of the outer shell requires less energy in SiH_4 than in Argon.

The absence of any other shape resonances in SiH_4 (Si1s or Si2p) shows that the molecular field brought by the hydrogen atoms around the silicon atom is very weak. This behaviour resembles basically the argon or silicon atom photoabsorption spectra (1s or 2p), despite their difference in core ionization energies. The same conclusions have been obtained by Pavlychev et al (28) when calculating the Si2p photoabsorption spectrum of silane.

B - The silicon tetrafluoride molecule.

In contrast to SiH_4, the silicon ligands i.e. the four fluorine atoms dominate the parentage of the valence shells primarily because of the three 2p lone pairs attached to each F atoms.

The X^1A_1 ground state has the following configuration :

$1a_1^2 \ \underbrace{1t_2^6 2a_1^2}_{} \ 3a_1^2 \ 2t_2^6 \ \underbrace{4a_1^2 \ 3t_2^6}_{} \ \underbrace{4a_1^2 \ 4t_2^6 \ 1e^4 \ 5t_2^6 \ 1t_1^6}_{}$

Si1s   F1s   Si2s Si2p   Si-F bond   F lone pairs and Si-F bond

We show in Fig 3, the Si1s photoabsorption spectrum of Bodeur et al (18) as well as the Si2p photoabsorption spectrum of Friedrich et al (29). From the comparison of Fig 3 several observations can be made.

i) The discrete spectrum

Having in mind that the uncertainty on the Si1s edge is much less than 1 eV (18), the Si1s spectrum does not show any evidence of Rydberg lines. In contrast, one sees a weak feature (line 1 in Fig 3a) followed by a strong doublet (lines 2 and 3). Friedrich et al., (29) have predicted as for SiH_4, two low lying antibonding Si-F unoccupied orbitals with the a_1^* and t_2^* symmetries.

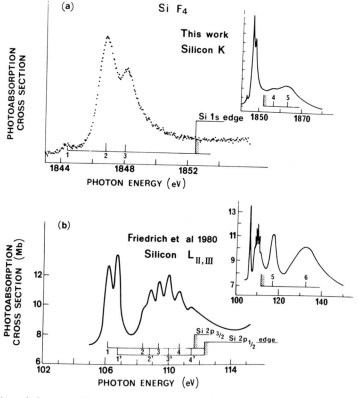

Fig 3 - Si1s (a) and Si2p (b) photoabsorption spectra of SiF_4 from ref 18 and 29 respectively.
inserts : extended spectra.

The observation of a weak line is certainly consistent with the electric dipole forbidden σ* (a_1). However the existence of a unique σ* (t_2), which is electric dipole allowed, is not sufficient to explain the observed doublet. We have offered (18) an interpretation based on the core equivalent species PF_4. The PF_4 ground state is known (30) to be stable in distorted geometries[4] (Fig. 4) in one of the trigonal bipyramidal (TBP) arrangements C_{2v} or C_{3v}. One expects a very large potential energy barrier between tetrahedral and TBP geometries. The existence of this barrier combined with the degenerate character of σ* (t_2) states have led us to analyze the spectrum in terms of a Jahn-Teller effect (31, 32). This results from a vibronic coupling which departs from the Born-Oppenheimer approximation. This implies that vibrational modes normally forbidden by summetry rules are involved. They lead to a distorsion of the molecule and are called the active Jahn-Teller mode ν_{JT}. Relevant potential energy curves are schematically represented in Fig. 5.

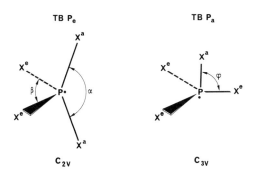

Fig. 4 - Equatorial (C_{2v}) and axial (C_{3v}) trigonal bipyramid after ref 30.

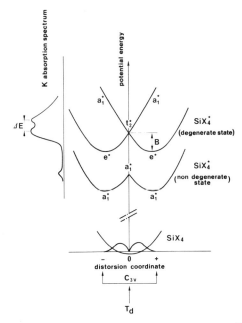

Fig. 5 - Schematics of the Jahn-Teller effect.

265

In the strong coupling case, a large broadening and even a splitting ΔE of the line may be observed. This splitting ΔE is related to the potential barrier B by $\Delta E = 2\nu_{JT}\sqrt{D}$ for which $D = B/\nu_{JT}$. It is likely that barriers of a few eV are necessary to explain a 1.4 eV splitting of the t_2 line seen in the Si1s SiF$_4$ spectrum and the extra broadening of the t_2 line in the Si1s spectrum. However this interpretation must be supported by suitable theoretical calculations.

The former interpretation (16, 29) of the Si2p spectrum (Fig. 3b) may be then revised. The spin orbit coupling splits all features in two series composed of σ^* (a_1) and σ^* (t_2) defined within the T_d symmetry group. These two transitions are both electric dipole allowed and are expected to be intense. However, assuming that Rydberg transitions are absent like in the Si1s spectrum, we analyze line 1 as σ^* (a_1) and lines 2, 3, 4 as σ^* (t_2). We have suggested (18) that the latter is affected also by a Jahn-Teller effect but there is a total removal of the three-fold degeneracy of the σ^* (t_2) orbital. The former interpretation of Friedrich et al. (29) does invoke σ^* (a_1) and σ^* (t_2) but also d-like Rydbergs. Notice that our interpretation (18) implies that the transferability of term values for $\sigma(a_1)^*$ and $\sigma^*(t_2)$ is not valid ($\sim$ 2 eV difference). This is not explained at the moment.

ii) in the continuum

One observes (fig 3) strong resonances above both the Si1s and Si2p edges. In the former we believe that the two features (lines 4, 5) are shape resonances in the t_2 continuum originating from the strong molecular field brough by the fluorine atoms. In the latter, Pavlychev (33) has performed MSMXα calculations and has shown that the two strong shape resonances line 5 and 6 in Fig 3 b (insert) have a d character with two components e and t_2 because of the effect of the tetrahedral field.

Notice that the presence of strong shape resonances is consistent with the absence of Rydberg states. Following the description of Dehmer et al. (3) a large potential barrier favors all inner well bound (σ^*) and continuum states (shape resonance) at the expense of outer well states (Rydberg).

III - ELECTRONIC RELAXATION PROCESSES

The large amount of internal energy stored in the molecule, through inner shell excitation, as compared to binding energies of valence electrons favors generally a strong coupling with various types of ionization continua. In this picture one assumes that any ionization process is much faster than dissociation. We will see that this is not always true. The most important relaxation channel of a core hole is the Auger and autoionization decay. Notice that this statement is true for molecules built with light atoms for which radiative decay rates of core holes are vanishingly small. As stated in the introduction, the nature and probability of such electronic relaxation processes are intimately connected to the location of the core hole and the resonances below and above threshold.

A - Discrete resonances

When a core electron is excited into a bound orbital such as an antibonding molecular orbital (σ^*) or a Rydberg orbital, the decay of the core hole through resonant Auger or autoinization produces the residual molecular ion respectively with 2h-1p or 1h configurations. This is illustrated nicely by the SiH$_4$ molecule (34) photoexcited near the $\sigma^*(a_1)$ resonance at 103.2 eV already analyzed in section II A (see Fig. 2). We show in Fig. 6 two photoelectron spectra of silane obtained on and off resonances. We also present in Fig. 7 the partial cross section measurements of the two valence bands 2 t_2 (Fig. 7a), 3 a_1 (Fig. 7 c), the 24 eV satellite band (Fig. 7 d) and the background around 15 eV binding energy (Fig 7 b), along the resonance region (see Fig. 7 top). The dramatic differences seen between Fig. 6 a and 6 b, also emphasized by the strong resonance behavior of Fig. 7 are analyzed in details in ref 34 and can be summarized as follows.

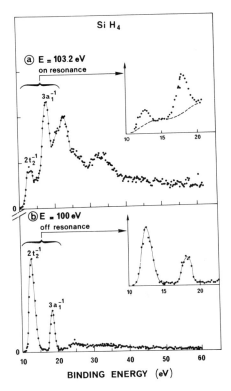

Fig. 6 - Photoelectron spectra of silane measured on and off the σ^* (a_1) resonance from ref (34). The inserts show an enlargement of the 10-20 eV binding energy region.

i) The intensity of the main $3a_1$ valence band increases drastically on resonance. This can be interpreted in terms of autoionization in which the σ^* electron participates to the transition leaving SiH_4 in a $3a_1$ hole state (mainly Si3s).

ii) There is an enormous intensity of satellite bands above 20 eV photon energy. This can be explained through a resonant Auger decay in which the σ^* remains a spectator leaving the SiH_4^+ ion with an excited configuration. However it is not clear, in the absence of precise theoretical calculations, to claim that such a process occurs within the parent molecule as a primary electronic relaxation process rather than a secondary event as seen now.

iii) There is a surprising increase of the "background" extending from 10 to 20 eV binding energy (Fig 6 a), which is found to be strongly associated to the σ^* resonance. These electrons are found to have a two high kinetic energy to be associated to an Auger process. We have offered (34) an interpretation based on a fast dissociation followed by autoionization (and may be Auger) of the core excited fragment. As seen on the schematic diagram of Fig 8, we have :

$$SiH_4^* \xrightarrow{(1)} Si^*H_3 + H$$

$$Si^*H_3 \xrightarrow{(2)} SiH_3^+ + e^-$$

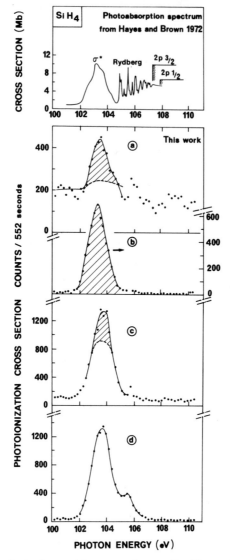

Fig. 7 - top, photoabsorption spectrum of silane from ref. 22. Bottom, partial photoionization cross sections from ref 34, for the a) the $2t_2$ valence note, b) the continuum at 15 eV binding energy, c) the $3a_1$ valence state, d) the 24 eV satellite line. Dashed areas represent the contribution of SiH_3^+ formed through dissociation of SiH_4^*.

This hypothesis is also based on the core equivalent species PH_4 which is known to be an unstable radical giving the PH_3 + H fragments. In other words, the core excited σ^* state is assumed to be purely repulsive as demonstrated previously by Morin and Nenner (11) in HBr and verified later (12) in other diatomic-like cases. The lifetime of the Si2p core hole estimated from the natural Rydberg linewidth (22) which amounts to 10^{-14} s allows the H atom to separate from the SiH_3 radical at a distance double from the normal Si-H bond length. This estimation is made under the assumption that

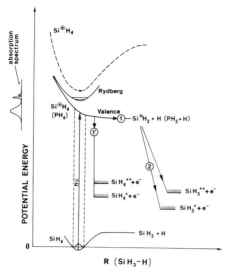

Fig. 8 - Relevant potential energy curves for SiH_4 and SiH_4^+ along the SiH_3-H coordinate, from ref. 34.

the energy released in the dissociation is totally converted into kinetic energy of the H atom. Notice that this fast dissociation which should proceed on a femto second time scale is very probably in competition with normal resonant Auger and autoionization of the core excited original molecule, see ii) above. Consequently, the satellite lines seen above 20 eV in the photoelectron spectrum of Fig 6 a may well be an admixture of SiH_4 and SiH_3 lines but this has not been established yet.

Let us analyze now the SiF_4 molecule and the effect of hydrogen atom substitution by fluorine ligands. The discrete part of the Si2p photoabsorption spectrum (Fig 3) is basically dominated by transitions into two distincts $\sigma^*(a_1$ and $t_2)$ as analyzed in section II B. We show in Fig 9, the photoelectron spectra of SiF_4 measured on and off resonances by de Souza et al (35). The outer valence bands appear as four lines below 23 eV binding energy and corresponds mostly to F2p lone pairs with some contribution of Si-F bonding orbitals. At higher energy, the 39 eV line is primarily associated to the F2s inner valence orbitals. All other lines are satellite lines. The dramatic changes found upon resonance excitation are summarized as follows

i) Satellite lines are strongly enhanced like for SiH_4.

ii) The symmetry of the resonances has a strong influence on the nature of the final satellite ionic state. This is fully consistent with the electron spectator transition model in which SiF_4^+ states with 2 h - 1p configurations with different (a_1 or t_2) additional electrons are likely found at different binding energies.

iii) The satellites line positions are found to be in equal number compared to normal Auger lines but shifted by aconstant amount of a kinetic energy scale. This is illustrated in Fig. 10. For the a_1 resonance (Fig. 10 b)

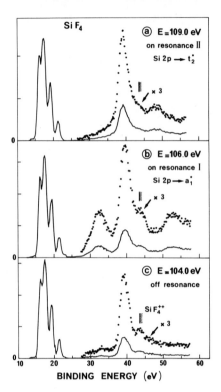

Fig. 9 - Photoelectron spectra of SiF_4 measured a) on the t_2 resonances, b) on the a_1 resonance, c) off resonance - From ref. 35.

the shift is found to be $\sim$ 7.5 eV and for the t_2 one, (Fig 10), it amounts to 5 eV. This comparison reveals that a satellite is superimposed on the F2s main line around 76 eV kinetic energy (39 eV binding energy). The increase of this main band observed in partial cross section measurements (not shown here) must not be interpreted in terms of autoionization but as a normal resonant Auger transition. This experimental observation already made in other core excited molecules (36,37) favors the electron spectator transition picture.

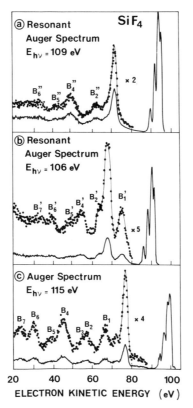

Fig. 10 - Comparison of photoelectron spectra of SiF_4 (ref 35) measured on resonances (a and b) with the Auger spectrum obtained at 115 eV. The Auger line are labelled B_1 to B_7 (10 c) and resonant Auger lines are labelled $B'_1 - B'_7$ (10 b) and $B''_1 - B''_7$ (10 a).

iv) The non resonant behavior of the outer valence or the F2s bands contrasts with SiH_4. This is because intratomic processes are largely favored by many orders of magnitures over interatomic ones (38). Indeed, in SiF_4, any process leading to one or two valence bands would have a clear interatomic character, because the original hole is evidently located on the silicon atom while the final holes would be found mostly on the fluorine ones.

v) There is no evidence of fast dissociation processes of the two σ^* resonances in contrast to SiH_4.

These two states results from transitions into Si-F antibonding orbitals and have also a strong repulsive character. Even if the Si2p core hole has a reasonably long life time as for SiH_4, the unfavorable mass ratio of the F over SiF_3 fragments (for example) probably reduces drastically the dissociation rates.

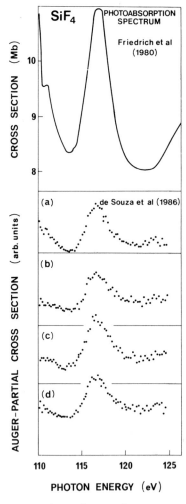

Fig. 11 - Top - Photoabsorption spectrum of SiF_4 above the Si2p edge. Bottom-Auger partial cross sections for the B_1, B_2, B_4 and B_6 lines defined in Fig. 10 c - From ref.35.

B - <u>Continuum resonances</u>

As stated in the introduction, we may consider the ejection of this electron via a molecular shape resonance to be faster than the decay (Auger) of the core itself. This is supported by the original shape resonance picture (2, 3) which involves a single electron trapped temporarily by a potential barrier. This trapping is faster in photoionization than in electron scattering experiments and occurs in a 10^{-16}-10^{-15} s time scale. In contrast, the Auger decay of the core hole involves the interaction of two electrons and occurs with smaller rates i.e. typically 10^{14} s^{-1}. Consequently, the existence of a shape resonance will be reflected also in the partial cross section for Auger lines.

This is illustrated in the SiF_4 case. The Si2p continuum is dominated around 118 eV by a e^* shape resonance (33) as reproduced on top of Fig 11 (see also section II B). The Auger spectrum has been measured in this energy region and shows 7 lines labelled B_1 to B_7 in fig. 10 c. The partial cross section for four of this lines, B_1, B_2, B_4 and B_6 have been measured (35) in this energy range and are reproduced in Fig 11 a, 11 b, 11 c and 11 d respectively. All four lines are found to resonate at 118 eV showing that the simple two step process is valid. Actually, this effect found for other molecules photoionized at L edges (39) or K edges (40) is likely to be very general.

IV - DISSOCIATION OF IONS : The SiF_4 case

The study of the charge, mass and kinetic energy of photoion following inner shell excitation must be viewed as complementary to photoelectron studies. The following points might be elucidated depending upon the sophistication of the ion detection method :

i) the identification of other electronic relaxation pathways of resonances especially those associated to the formation of double and multiple charged ions;

ii) the question of possible competition between dissociation and ionization during relaxation events of the core hole;

iii) the amount of damage (bond breaking) around the atomic site under consideration;

vi) the mechanism of multiple bond breaking (direct or in cascades)

v) the identification of the internal energy of the final fragmentation products.

We illustrate our present understanding of these different questions using recent results on SiF_4 photoexcited near the Si2p edge which have been described in details elsewhere (41). Fragment ions are detected by a time of flight (TOF) mass spectrometer operated in two modes.

i) a normal TOF mode in which ions are extracted by a pulsed field into the drift region. The ions are selected according to their mass over charge ratio by measuring their arrival time on the detector.

ii) The photoion-photoion coincidence (PIPICO) mode in which ions are extracted continuously by a constant field into the drift region. A single detector is used to measure TOF differences Δt of fragment ions issued from the same dissociation event. Only positive ion pair produced from $m^{2+} \rightarrow m_1^+ + m_2^+$ appears at constant Δt in contrast to all other ions issued from different dissociation events.

We present in Fig. 12, a typical mass spectrum of SiF_4 obtained at 115 eV photon energy. One readily recognizes f^{2+} anf f^+ types of fragments. SiF_3^+ is found to be the most intense fragment and the parent ion SiF_4^+ has a negligible intensity.

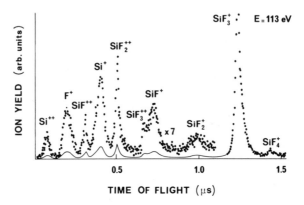

Fig. 12 - TOF. Mass spectrum of SiF_4 and reproduced from ref 41.

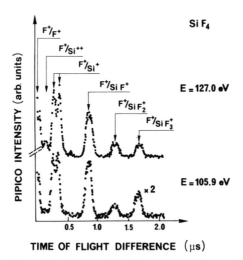

Fig. 13 - Photoion-photoion coincidence spectra of SiF_4 measured below (105.9 eV) and above (127 eV) the Si2p edge from ref 41.

The significant intensity of SiF_3^{2+}, SiF_2^{2+}, SiF^{++} and Si^{2+} shows that they have SiF_4^{2+} as a precursor and are associated with neutral F fragment. Notice the absence of F^{n+} (n ⩾ 2) shows that there is a site selectivity during the electronic relaxation processes. The observation of F^+ as well as SiF_n^+ fragments cannot be interpreted in a unique fashion because they may have SiF_4^+ or SiF_4^{2+} or SiF_4^{3+} as precursors.

The PIPICO spectra of SiF_4 obtained at 105.9 and 127 eV reported in Fig 13 shows nicely that some fragments are associated in paris. The SiF_3^+/F^+ pair has SiF_4^{2+} as a precursor. The unstability of such SiF_4^{2+} ions leads to a single bond (Si-F breaking in contrast to all other ion pairs which are evidently associated to multiple bond breading. The SiF_n^{2+}/F^+ pairs may result from a coulomb explosion process in which two or more bonds are broken instantaneously. They also can result from a series of sequential events. At present this question is unanswered because even if the large width of peaks in Fig 12 reveals equally large kinetic energy release in the process, there is no information on the undetected fragments. Along those lines, a triple coincidence between $F^+/F^+/Si^+$ (and SiF^+) has been performed (41) showing that SiF_4^{3+} is a precursor for the atomization ob the molecule and that the large kinetic energy release in the process favors the description of the dissociation in terms of a colomb explosion.

A - Discrete resonances

The photon energy dependence of intensity of individual fragments or pairs, as seen in Fig 14 in the region of the Si2p discrete spectrum of SiF_4, provides additional important informations. Qualitatively, one can identify the resonating dissociation channels and the possible precursors. Quantitatively, knowing the ion detection efficiency one can extract multiple ionization yield relative to single ionization one. Then a meaningfull comparison with photoelectron spectroscopy can be made because the latter only provides detailed information on events associated to the ejection of electron with a well defined energy. Using the results shown in Fig 14, we can make the following comments :

i) The most intense fragment SiF_3^+ a seen in Fig 12 does not resonate whereas all other fragments do. The relative abundance of these fragments compared to SiF_3^+ is compatible with the ratio "resonance over continuum" extracted from the photoabsorption of Fig 3b.

ii) The fragmentation branching ratio depends upon the symmetry of the resonance showing the significant influence of the excited electron on electronic relaxation pathways.

iii) The large amount of SiF_4^{2+} and SiF_4^{3+} ions evidenced through their fragmentation pattern along resonances compared to SiF_4^+ ions shows that multiple resonant Auger processes are efficient. Resonant double Auger has been also evidenced in abundance in other systems such as tetramethylsilane (42) or methyliodide (43). A further question arises : is this process considered as a direct one in which two electrons are ejected at once or is it a two step one where the electrons are ejected sequentially. Actually the second picture prevails because SiF_4^+ ions with excited configurations have been already evidenced by satellite lines in photoelectron spectra (Section III A). Moreover many of these lines are embedded in the double ionization continuum : consequently the ejection of a secondary electron is extremely probable, even if it has not been observed yet because of its low kinetic energy.

iv) No detailed information on the internal energy content of the fragments can be extracted from these experiments because the ion kinetic energy resolution achieved in TOF experiments is insufficient and also because multicoincidence detection techniques are required (44) to study a single dissociative potential energy surface.

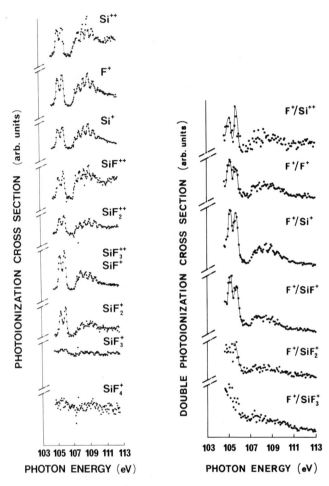

Fig. 14 - Left - partial photoionization cross section of singly charged positive fragments in SiF_4 measured below the Si2p edge, from ref. 41.

Right : partial photoionization cross sections for fragments pairs measured in the PIPICO mode, and below the Si2p edge from ref. 41.

B - <u>Continuum resonances</u>

The photon energy dependence of individual fragments or pairs in the region of the e* shape resonance is seen in Fig 15. Our photoelectron studies (39) have shown clearly the importance of Auger processes leading to doubly charged ions with two-holes configurations. The spectra of Fig 15 shows also that :

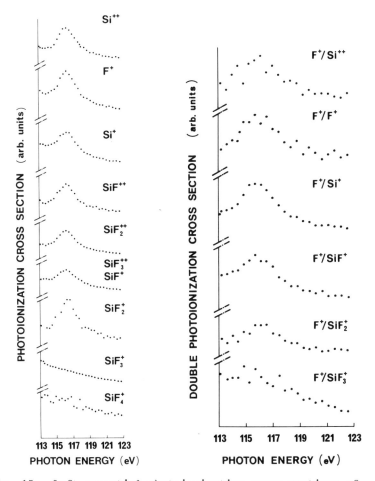

Fig. 15 - Left : partial photoionization cross sections of singly charged positive fragments in SiF_4 measured above the Si2p edge, from ref 41.

Right : partial photoionization cross sections of ion pairs measured in the PIPICO mode and above the Si2p edge, from ref 41.

i) SiF_4^{++} (2 holes) ions are highly unstable and produce multibond breaking. Indeed the F^+/SiF_3^+ pair does not resonate while all other pairs do. Generally, SiF_4^{++} is found to dissociate into SiF_n^{2+} + (4-n)F as well as into SiF_n^+ (n = 0 to 2) + F^+ + ...

This shows that either the two holes are located on the silicon atom or on Si-F bonds. In the latter case it is probable that two distinct fluorine sites are involved because of the large amount of multiple bond breaking. This is in qualitative agreement with the conclusions drawn from the analyzis of the Auger spectrum mode by Rye and Houston (45) and by Aksela et al (46).

Lastly, the SiF_4^{2+} (2 holes) ionic states produced through Auger are different from those produced through direct shake off. Indeed the PIPICO spectrum obtained around 50 eV (38) shows that the SiF_3^+/F^+ pair dominates the spectrum (not shown here). This is because only certain SiF_4^{++} states are produced through Auger.

ii) The core hole is found to decay also by double Auger because SiF_4^{3+} is found to be produced. This is seen through the F^+/Si^{++} pair and through triple coincidence measurements (not shown here) which have been reported in ref 38. This shows that two Auger electrons are produced with significant probability.

However, the question of interpreting this double Auger process in terms of a direct event or a cascade is not solved at the moment.

V - CONCLUDING REMARKS

Intense resonances are found in SiX_4 (X = H or F) in the vicinity of Si1s and Si2p core ionization edges. In the discrete spectrum, transitions into antibonding unoccupied orbitals (σ^*) are found to dominate the spectrum at the expense of transitions into Rydberg orbitals. This effect which is specific of molecules is found to increase from SiH_4 to SiF_4. It is also found to follow the importance of shape resonances in the continuum.

The substitution of hydrogen atoms by fluorine ones around silicon perturbs the σ^* bound states. Unexpectedly large band splittings in SiF_4 are observed and tentatively interpreted in terms of a Jahn Teller effect.

The combination of photoelectron spectroscopy and mass spectrometry methods has shown that

i) Discrete resonances decay equally through autoionization and Auger processes into ions with 1h, 2h-1p 2h or 3h-1p... configurations, in the valence shells. However, some pathways are found to vanish in SiF_4 in cases when the final holes are localized on fluorine atoms rather than silicon.

The relatively large amount of SiF_4^{2+} ions produced on resonances is probably due to cascade of resonant Auger processes. Fragment ions are interpreted in terms of dissociation of SiF_4^+, SiF_4^{2+} or SiF_4^{3+} through multiple bond breaking after complete electronic relaxation processes. An exception is found for SiH_4 in which direct dissociation of the core excited state is found fast enough to compete with autoionization or resonant Auger.

ii) Shape resonances enhance the formation of SiF_4^{2+} (2 h configurations in the valence shells) and its fragmentation products SiF_n^{2+} (n = 0 to 2) + (4 - n) F or SiF_n^+ + F^+ + ... because of Auger decay of the core hole - SiF_4^{3+} is also found with significant intensity showing that double Auger also occurs.

Lastly, selectivity among electronic relaxation processes is found to depend upon the nature, of the resonance i.e. discrete versus continuum or symmetry of the excited electron. All energetically allowed process occur unless they have a definite interatomic character. However, this selectivity is not found of importance among the dissociation products. because any residual ions produced with one or multiple valence excitations in the valence shells produce more or less the same ionic fragments. A much larger difference among dissociation products must be observed after a fluorine 1s excitation. Indeed Rye and Houston (45) have measured an Auger spectrum F (KVV) which is drastically different from Si (LVV). Nevertheless, the search of site selective dissociation pathways must be more fruitful for large size molecules as already evidenced in acetone (47), in trifluorethylene (48) or in tetramethylsilane (42, 49).

ACKNOWLEDGEMENTS

It is a great pleasure for me to thank my coworkers S. Bodeur, P. Lablanquie, P. Millié, P. Morin, G.G.B. de Souza and A.C.A. Souza for their essential contributions to the results presented here and for numerous stimulating discussions.

REFERENCES

1. A bibliography of inner-shell spectra of molecules is given by A.P. Hitchcock, J. Electron Spectrosc. 25, 245 (1982)
2. J.L. Dehmer, J. Chem. Phys. 56, 4496 (1972)
3. J.L. Dehmer, D. Dill, and A.C. Parr, in Photophysics and Photochemistry in the Vacuum Ultraviolet (Reidel, Dordrecht, 1985), p. 341. J.L. Dehmer, A.C. Parr and S.H. Southworth to appear in Handbook on Synchrotron Radiation, Vol. II, Ed. by G.V. Marr (North-Holland, Amsterdam, 1986).
4. H. Friedrich, B. Sonntag, P. Rabe, W. Burscher, and W.H.E. Schwarz, Chem. Phys. Lett. 64, 360 (1979).
5. D.L. Lynch, V. McKoy, and R.R. Lucchese, in Resonances in Electron-Molecule Scattering, Van der Waals Complexes, and Reactive Chemical Dynamics, ACS Symposium Series, n° 263, Ed., D.G. Truhlar (American Chemical Society, Washington, D.C., 1984) Chapter 6, p. 89.
6. P.W. Langhoff, in Resonances in Electron-Molecule Scattering, Van der Waals Complexes, and Reactive Chemical Dynamics, ACS Symposium Series. n° 263, Ed. D.G. Truhlar (American Chemical Society, Washington D.C., 1984), Chapter 7, p. 113.
7. M. Nakamura, M. Sasanuma, S. Sato, M. Watanabe, H. Yamashita, Y. Iguchi, A. Ejiri, S. Nakai, S. Yamaguchi, T. Sogawa, Y. Nokai and T. Oshio, Phys. Rev. 178, 80 (1969).
8. G.R. Wight and C.E. Brion, J. Electron Spectry 1, 457 (1973); 4, 313 (1974)
9. W.H.E. Schwarz, Angew. Chem. Intern. Ed 13, 454 (1974)
10. W.H.E. Schwarz, Chem. Phys. 11, 217 (1975); 9, 157 (1975)
11. P. Morin and I. Nenner, Phys. Rev. Lett. 56, 1913 (1986)
12. P. Morin, Proceedings of the "Giant resonances in atoms molecules and solids", Nato-ASI Series, R. Karnatak, J.M. Esteva and J.P. Connerade Eds. (This volume).
13. G.C. King and F.H. Read, "Atomic Inner-Shell Physics", Edited by B. Crasemann (Plenum, New-York, 1985), Chap. 8, p. 317.
14. H. Agren, L. Selander, J. Nordgren, C. Nordling and K. Siegbahn, Chem. Phys. 37, 161 (1979).
15. K. Radler, B. Sonntag, T.C. Chang, W.H.E. Schwarz, Chem. Phys. 13, 363 (1976).
16. M.B. Robin, Higher Excited States of Polyatomic Molecules, Vol. 3, Academic Press, Orlando, 1985.
17. S. Bodeur and J.M. Esteva, Chem. Phys. 100, 415 (1985) S. Bodeur and J.M. Esteva, Proceedings of the EGAS Conference, London (July 1984).

18. S. Bodeur, I. Nenner, P. Millié, Phys. Rev. A (in press).
19. F. Keller and H. Lefebvre-Brion, Z. für Physik D. Atoms. Molec. and Clusters (in press).
20. L.G. Parratt, Rev. Mod. Phys. 31, 616 (1959).
21. S. Bodeur and I. Nenner, J. Physique (France) Colloques "EXAFS and near edge structures IV" (in press).
22. W. Hayes, F.C. Brown and A.B. Kunz, Phys. Rev. Lett. 27, 774 (1971).
23. G.C. King, M. Tronc, F.H. Read and R.C. Bradford, J. Phys. B Atom. Molec. Phys. 10, 2479 (1977).
24. J.A. Burden and A.F. Burr, Rev. Mod. Phys. 39, 125 (1967).
25. W.B. Perry and W.L. Jolly, Chem. Phys. Rev. 17, 611 (1972).
26. R.G. Cavell and R.N.S. Sodhi, J. Elect. Spect. Rel. Phenom. 15, 145 (1979).
27. Willeumier, J. Physique (France) 26, 776 (1965).
28. A.A. Pavlychev, A.S. Vinogradov, T.M. Zimkina, D.E. Onopko and R. Stsargan, Opt. Spectrosc. (USSR) 48, 109 (1980).
29. H. Friedrich, B. Pittel, P. Rabe, W.H.E. Schwarz and B. Sonntag, J. Phys. B Atom. Molec. Phys. 13, 25 (1980).
30. R.A.J. Janssen, G.J. Visser and H.M. Buck, J. Am. Chem. Soc. 106, 3429 (1984).
31. P. Habitz and W.H.E. Schwarz, Theoret. Chim. Acta (Berlin) 28, 267 (1973).
32. C. Cossart-Magos, Proceedings of the XIXth Jerusalem Symposium in Quantum Chem. and Biochem. "Tunneling", 5-8 Mars 1986 (to be published in Reidel Pub. Co, Dordrecht, Holland).
    C. Cossart-Magos, D. Cossart and S. Leach, Chem. Phys. 41, 345 (1979).
33. A.A. Pavlychev, A.S. Vinogradov and T.M. Zimkina, Opt. (USSR), 52, 139 (1982).
34. G.G.B. de Souza, P. Morin and I. Nenner, Phys. Rev. A (in press)
35. G.G.B. de Souza, P. Morin and I. Nenner (to be published)
36. L. Ungier and T.D. Thomas, J. Chem. Phys. 82, 3146 (1985)
37. T.A. Ferrett, D.W. Lindle, P.A. Heimann, H.G. Kerkhoff, U.E. Becker and D.A. Shirley, Phys. Rev. A 34, 1916 (1986)
38. T.D. Thomas and P. Weightman, Chem. Phys. Lett. 81, 325 (1981)
39. G.G.B. de Souza, P. Morin and I. Nenner, J. Chem. Phys. 83, 492 (1985)
40. C. Truesdale, D. Lindle, P. Kobrin, U. Becker, H. Kerkhoff, P. Heimann, T. Ferrett and D. Shirley, J. Chem. Phys. 80, 2319 (1984)
41. A.C.A. Souza, P. Lablanquie, G.G.B. de Souza, P. Morin and I. Nenner (to be published)
42. P. Morin, G.G.B. de Souza, I. Nenner and P. Lablanquie, Phys. Rev. Lett. 56, 131 (1986).
43. G. Dujardin, L. Hellner, D. Winkoun and M. J. Bernard, Chem. Phys. 105, 291 (1986).
44. J.H.D. Eland, F.S. Wort, P. Lablanquie and I. Nenner, Z. für Physik D. Atoms. Molec. and Clusters (in press)
45. R.R. Rye and J.L. Houston, J. Chem. Phys. 78, 4321 (1983).
46. S. Aksela, K.H. Tan, N. Aksela and G.M. Bancroft, Phys. Rev. A 33, 258 (1986).
47. W. Eberhardt, T.K. Sham, R. Carr, S. Krummacher, M. Strongin, S.L. Weng and D. Wesner, Phys. Rev. Lett. 50, 1038 (1983)
48. K. Muller-Dethlefs, M. Sander, L.A. Chewter and E.W. Schlag, J. Phys. Chem. 88, 6098 (1984)
49. P. Morin in Photophysics and Photochemistry above 6 eV, Edited by F. Lahmani (Elsevier, Amsterdam, 1985), p. 1.

RESONANCES IN THE K-SHELL SPECTRA OF FLUORINATED ORGANIC MOLECULES

A.P. Hitchcock, P. Fischer and R. McLaren

Dept. of Chemistry and Institute for
Materials Research
McMaster University, Hamilton, Canada L8S 4M1

ABSTRACT

The inner shell electron energy loss spectra of a series of fluoroalkanes (C_nF_{2n+2}, n=1-6), fluoroethenes ($C_2H_nF_{4-n}$, n=0-4) and fluorobenzenes ($C_6H_nF_{6-n}$, n=0-6) are presented. C 1s ⟶ σ* (C-F) transitions are identified in each spectrum 3 to 5 eV below the ionization threshold(s), consistent with the correlation between positions of σ* resonances and molecular bond lengths. The analogies between spectral signatures of molecular shape resonances and atomic giant resonances are discussed briefly. The growth of potential barrier effects with the development of a fluorine cage around a core excited atom are demonstrated with the absolute cross-sections for C 1s ⟶ σ* (C-F) in the spectra of, C_6F_6, C_2F_4, C_2F_6 and CF_4, derived by sum rule normalization from the energy loss spectra.

INTRODUCTION

Centrifugal potential barriers are a characteristic component of the atomic giant resonance phenomenon. Similar potential barriers appear to occur in the anisotropic fields of low Z molecules. Such barriers produce spectroscopic effects analogous to those identified in transition metal and rare earth spectroscopies as being of giant resonance nature. These effects include : intensity enhancement of transitions to selected states of high angular momentum ; a decrease in the intensity of transitions to diffuse Rydberg states and to the direct ionization continuum; and, in some cases, a marked sensitivity of the spectrum to perturbations of the potential barrier by changes in the chemical environment of the core excited atom. In the field of molecular spectroscopy such phenomena are regularly encountered in the valence and, in particular, the core excitation spectra of highly fluorinated molecules (sometimes referred to as 'fluorine cage' species). Dehmer[1], along with others, has noted that potential barrier enhancements are greatest in the core spectra of an atom surrounded by a symmetrical distribution of fluorines, as in SF_6, SiF_4 and CF_4. In a molecular orbital model the core exitation spectra of these species can be interpreted in terms of enhancement of all symmetry allowed transitions to molecular, virtual valence orbitals accompanied by a reduction of the intensity of the direct ionization continuum and Rydberg transitions. The enhanced transitions can occur both above the core ionization threshold, where the quasi-bound states decay rapidly into the direct core ionization continuum and thus are best denoted as shape resonances ; and below the ionization threshold, where the excitation and subsequent decay of the core hole can be treated as separate processes such that the resulting states can be thought of as 'bound' although they are sometimes referred to as 'discrete' shape resonances. Further aspects of the relationship between atomic giant resonances and molecular shape resonances have been discussed recently[2,3].

In molecules some of these enhanced transitions (shape resonances) are excitations to antibonding σ^* orbitals. Recently[4] it has been shown that there is a linear relationship between the position of σ^* transitions in B, C, N, O and F K-shell spectra and the lengths of bonds to adjacent atoms. The sensitivity of resonance position to bond length derives from the strong geometry dependence of the potential barrier. The interplay between the resonance intensity enhancement effects in highly fluorinated molecules and the bond length sensitivity of σ^* transitions is of particular interest. In order to investigate this the inner shell spectra of a variety of fluorinated organic molecules have been recorded by electron energy loss spectroscopy (ISEELS). The contributions of shape resonance to the core excitation spectra of fluorocarbons is well illustrated by three series : fluoroalkanes ($C_n F_{2n+2}$, n=1-6); fluoroethenes ($C_2 H_n F_{4-n}$, n=0-4) and fluorobenzenes ($C_6 H_n F_{6-n}$, n=0-6). The ISEEL spectra of each of these series is presented and the principal features assigned in the context of the bond length correlation. The spectra of CF_4, $C_2 F_6$, $C_2 F_4$ and $C_2 F_4$ and $C_6 F_6$ are then compared on an absolute scale in order to demonstrate the strong dependence of the C 1s $\longrightarrow \sigma^*$ (C-F) resonance intensity on the spatial distribution of fluorines around a carbon atom.

The inner shell electron energy loss spectrometer was operated with 2.5 keV final electron energy, a 2° average scattering angle, a target gas pressure of 10^{-4} torr and a resolution of 0.6 eV FWHM, limited by the energy spread of the unmonochromated incident electron beam. The samples studied were obtained commercially and used without further purification, aside from degassing liquids. The spectra were calibrated from the energies of sharp, well-known core excitation features of CO or CO_2. The energies of sharp features are known to an accuracy of 0.1 eV.

RESULTS AND DISCUSSION

A. Fluoroalkanes

The manifestations of potential barriers in molecules are clearly illustrated by a comparison of the C 1s spectra of CF_4 and CH_4 (Fig. 1). The spectrum of methane can be satisfactorily interpreted[5,6] in terms of a Rydberg series converging onto a continuously declining C 1s continuum. In contrast, the C 1s oscillator strength in CF_4 is concentrated into the intense, broad t_2 σ* (C-F) resonance. The Rydberg transitions appear only as very weak features on the high energy side of the σ* (C-F) resonance and there is almost complete suppression of the direct ionization continuum at threshold.

Similar effects dominate the C 1s (and F1s) spectra of the fluoroalkanes. The C 1s spectra of C_2F_6 and $CF_3(CF_2)_nCF_3$, n=1,4 are presented in Fig. 2. The most intense feature occurs 4.4 eV below the C 1s(CF_3) ionization threshold and is assigned to a σ* (C-F) resonance in which the core hole is on the CF_3 carbon, analogous to the t_2 resonance in CF_4. In comparison to CF_4 two additional strong core→valence resonances are observed in C_2F_6. The higher energy one at 300.5 eV (0.7 eV above the IP) is attributed to C 1s→σ* (C-C) transitions, based on the bond length correlation. The sharp feature at 297.7 eV, between the σ* (C-F) and σ* (C-C) resonances is attributed to a mixed 3p/π* (CF_3) state. This is analogous to the π* (CH_3) and π* (CH_2) states recently identified in the ISEEL spectra of a large number of saturated hydrocarbons[7-9]. Excitations from two C 1s levels can be distinguished in the spectra of the longer alkanes, since the C 1s (CF_3) and C 1s (CF_2) IPs differ by 2 eV. Thus the C 1s (CF_2)→σ* (C-F) resonance appears at 293 eV in C_3F_8 and its intensity

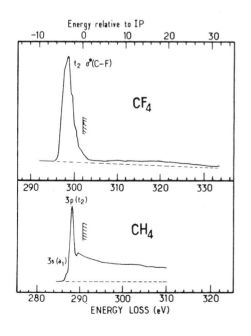

Fig. 1: The C 1s spectra of CH_4 and CF_4 recorded by electron energy loss. The absolute energy scales are indicated for each species, along with a scale indicating the energy relative to the respective ionization threshold, at which the spectra are aligned.

relative to the 295.3 eV feature increases as the number of CF_2 units increases. C 1s(CF_2) excitations to σ^* (C-C) and $3p/\pi^*$ (CF_2) also occur and are observed as additional shoulders or a broadening of the C 1s(CF3)$\longrightarrow\pi^*$ (CF_2) and C 1s(CF_3)$\longrightarrow\sigma^*$ (C-C) features. In addition to the sharp resonances a broad feature occurs in each continuum around 309 eV. The exact origin of this feature is unclear. Possibilities include double excitation; an additional σ^* (C-F) or σ^* (C-C) level; or the first EXAFS wiggle associated with the C-C bond. The spectra of the linear alkanes also contain a similar continuum resonance which has been interpreted as a second σ^* (C-C) state[7].

The F 1s ISEEL spectra of these fluoroalkanes (Fig. 3) can be interpreted in a manner similar to the C 1s spectra although the features are less well defined because of increasing linewidths associated with a shorter core hole lifetime and steeper, dissociative potential curves, and because the F 1s(CF_2) and F 1s(CF_3) IPs are separated by only 1.5 eV. The F 1s continuum is appreciably more intense relative to the resonance features, indicating a smaller potential barrier in the region of the F 1s core.

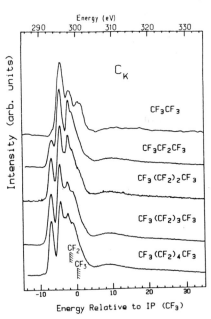
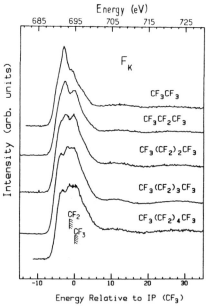

Fig. 2: The C 1s spectra of the linear perfluoroalkanes. The hatched lines indicate the locations of the C 1s (CF_3) IP of C_2F_6 and C_3F_8 and the C 1s(CF_2) IP of C_3F_8, as measured by XPS.

Fig. 3: The F 1s spectra of the linear perfluoroalkanes. The hatched lines indicate the locations of the F 1s (CF_3) IP of C_2F_6 and C_3F_8 and the F 1s(CF_2) IP of C_3F_8, as measured by XPS.

B. Fluoroethenes

The C 1s spectra of $C_2H_nF_{4-n}$, n=1-4 are presented in Fig. 4. The lower energy regions of these spectra are dominated by promotions from each distinct C 1s orbital (CH_2, CHF or CF_2) to the π^* (C=C) or σ^* (C-F) orbitals. Thus transitions to π^* (C=C) occur around 285 eV from C 1s(CH_2), 287.5 eV from C 1s(CHF), and 290 eV from C 1s(CF_2) while transitions to σ^* (C-F) occur around 288 eV (CH_2), 290 eV (CHF) and 292 eV (CF_2) respectively. In C_2H_4, where there is no C-F bond, Rydberg features (2-5) occur with appreciable intensity. Very weak Rydberg transitions can be identifed in C_2F_4. In the intermediate species these are difficult to identify because of lower intensity and spectral overlap.

The C 1s continuum shape of these spectra is difficult to interpret. In C_2H_4 the σ^* (C-C) resonance is attributed to the weak, broad peak at 302 eV. This assignment has been substantiated by comparison of gas and condensed phase spectra of various alkenes and recently by the polarization enhanced C 1s spectrum of C_2H_4 on Cu(100) at low temperature[10], an adsorbate situation where the molecular geometry is essentially unchanged. If the height of the potential barrier in C_2F_4 was as much as

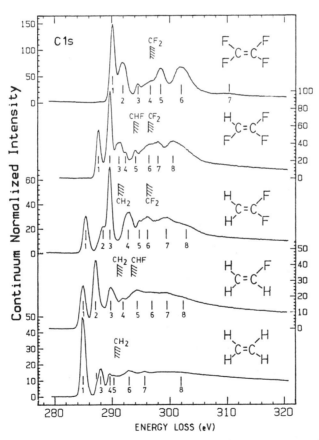

Fig. 4: The C 1s spectra of C_2H_4, CH_2CHF, CH_2CF_2, $CHFCF_2$ and CF_2CF_2. The hatched lines indicate the IPs as measured by XPS.

12 eV above the vacuum level then one might expect a systematic enhancement of the σ*(C-C) resonance. However there is no evidence for intensity enhancement of σ* (C-C). In particular the σ* (C-C) feature in C_2F_4 (7) is no more prominent than that in C_2H_4. On the other hand there is a systematic increase in the intensity of two features around 3(±1) and 7(±2) eV above the respective C 1s ionization thresholds, suggesting that the height of the potential barrier at carbon is between 6 and 10 eV above the vacuum level. The specific assignment of these two resonances is unclear. One possibility in C_2F_4 is transitions to additional σ* (C-F) orbitals since a minimal basis set description predicts 6 unoccupied orbitals with four having σ* (C-F) character. The presence of two continuum features (#6,7) at similar relative energies in C_2H_4 suggests that the previous assignment of these features to double excitation[5] may be incorrect and that they are one-electron transitions. A more detailed discussion of the C 1s as well as the F 1s spectra of these molecules is in preparation.

C. Fluorobenzenes

The C 1s and F 1s spectra of the higher symmetry, successively fluorinated benzenes, $C_6H_nF_{6-n}$, n=0-6 have recently been recorded and analysed[11]. The C 1s spectra (Fig. 5) are interpreted in terms of overlapping excitations from C 1s(CF) and C 1s(CH) to a common set of unoccupied valence levels which include 1π* (e_u in C_6H_6), 2π* (b_{2g} in C_6H_6), σ* (C-F) and 3 other levels in the continuum which appear to be predominantly σ*(C-C) in character (corresponding to orbitals of e_{1u}, e_{2g} and a_{2g} symmetry in C_6H_6, according to recent calculations[12]). The excitation of a C 1s (CH) or C 1s(CF) electron in the fluorobenzenes appears to be very independent of its environment (i.e. the number of CF or CH neighbours). This is analogous to our interpretation of the fluoroethene spectra outlined in the previous section, where each spectrum was viewed as the sum of C 1s(CH_2), C 1s(CHF) or C 1s(CF_2) excitations at fixed energies, with the components of the sum given by the molecular structure. However the variation of the energies and intensities of the individual transitions in the fluoroethenes is greater than in the fluorobenzenes.

The extremely local character of the C 1s excitations in the fluorobenzenes has been used to identify the C 1s(CF)⟶σ* (C-F) transitions through a spectral stripping procedure. This is necessary since the σ* (C-F) features are much weaker in the fluorobenzenes than in the other fluorocarbons. The C 1s(CH) and C 1s(CF) excitations were separated by subtracting the spectrum of C_6H_6, scaled to eliminate the C 1s(CH)⟶1π* transition at 285.3 eV in each case (left hand panel of Fig. 5). Following this subtraction the C 1s(CF)⟶σ* (C-F) transitions can be observed between the 1π* and 2π* transitions in all six molecules (right hand panel of Fig. 5). The position of the σ* (C-F) resonances in the fluorobenzenes are generally similar to that in other fluorocarbons (3-5 eV below threshold), consistent with the bond length correlation. There does appear to be a small shift to lower energy with increasing fluorination which is consistent with a 'perfluoro effect' on antibonding σ* (C-F) levels. However the shift of 2.6 eV between C₆H₅F and C₆F₆ is a smaller effect than the shift in σ* resonance positions with gross bond length changes since C-F resonances shift by 30 eV per angstrom according to the Z=15 correlation line[4].

D. <u>Variation of C 1s ⟶ σ*(C-F) intensity with spatial distribution of fluorines</u>

Qualitatively, the results presented in the previous three sections show that σ* (C-F) resonances dominate the fluoroalkane spectra, are well-defined features in the fluoroethenes and are a relatively minor

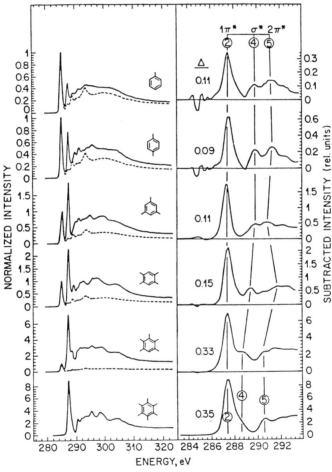

Fig. 5: The C 1s spectra of the fluorobenzenes.
The left hand panels present the spectra normalized to unit intensity at the first feature along with the spectrum of C₆H₆ adjusted for best match at this feature. The right hand panels give the subtracted spectra which approximate the contributions of C 1s(CF) excitation in each molecule. The Δ parameter is the energy in eV which must be added to the indicated scale to recover the absolute energy loss. This shift aligns both the C 1s⟶1π* and C 1s(CF)⟶1π* transitions in all molecules.

component of the fluorobenzene spectra. This suggests that the strength of the molecular potential barrier is a strong function of the spatial arrangement of the fluorines around the core excited atom rather than simply being a function of the total number of fluorines in the molecule. In order to demonstrate this effect clearly it is desireable to place the C 1s spectra on an absolute intensity scale and correct for degeneracy (differing numbers of C 1s⟶σ* (C-F) transitions). This allows a comparison of the strength of an individual C 1s⟶σ* (C-F) transition as a function of the environment of the C-F bond.

Absolute oscillator strength scales for the C 1s spectra of C_6F_6, C_2F_4, C_2F_6 and CF_4 have been obtained by a sum rule normalization. Each spectrum was extrapolated to 700 eV, its area evaluated and then normalized to the total number of C 1s electrons in the molecule (12, 4, 4 and 2 respectively). This procedure assumes that the spectrometer sensitivity is independent of energy loss, which is a good approximation since the ISEEL spectrometer operates in a fixed final energy mode providing constant analyser response and only a slight variation in the incident electron energy through the recorded spectrum. It does ignore an E^{-n}, $2<n<3$ factor which relates the relative intensities of photoabsorption and electron impact near the dipole limit. However this effect is constant for all spectra and the error associated with it appears to be relatively small since the integrated cross section of the C 1s⟶σ* (C-F) transition in CF_4 is 13.8 Mb.eV by this method, in reasonable agreement with that of 12.0 Mb.eV determined by a direct, quantitative photoabsorption measurement[13].

The absolute C 1s spectra of C_6F_6, C_2F_4, C_2F_6 and CF_4 obtained in this manner are presented in Fig. 6 with a vertical scaling which corrects

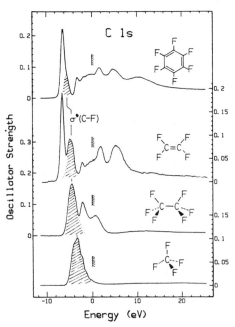

Fig. 6: The absolute oscillator strength for C 1s excitation of C_6F_6, C_2F_4, C_2F_6 and CF_4 derived by extrapolation and sum rule normalization. The spectra are aligned at their C 1s IPs and the vertical plot scales are in the ratio 4:6:4:6 to correct for the different number of C 1s⟶σ*(C-F) transitions contributing to each spectrum. The shaded area is thus proportional to intensity of C 1s⟶σ*(C-F) excitation in each molecule on a per transition basis.

for degeneracy such that the shaded area in the figure is proportional to the intensity in each molecule on a per C 1s⟶σ* (C-F) transition basis. The absolute integrated cross sections of the σ*(C-F) resonances in each molecule are summarized in table 1, which also gives the degeneracy corrected intensity. It is clear from both the graphical and numerical presentations that the intensity on a per C 1s⟶σ* (C-F) transition basis increases rapidly as the fluorine cage is completed. Thus this comparison provides a dramatic demonstration of the enhancement of so-called discrete shape resonances, presumeably because of the strengthened potential barrier associated with a more complete fluorine cage.

Table 1

Cross-Sections of C1s-->σ*(C-F) and F 1s-->σ*(C-F) Transitions

molecule	#C-F	I(C 1s) Mb.eV	I/σ*(C-F)	I(F 1s) Mb.eV	I/σ*(C-F)
C_6F_6	6	7.6	1.3	24.9	4.2
C_2F_4	4	15.9	4.0	38.3	9.6
C_2F_6	6	35.8	6.0	58.7	9.8
CF_4	4	25.9	6.5	68.7	17.2

Acknowledgements : Financial support for this research has been provided by NSERC. The donation of samples from AT&T Bell Laboratories and helpful discussions with M.B. Robin are gratefully acknowledged.

References

1. J.L. Dehmer, J. Chem. Phys. 56: 4496 (1972).
2. M.B. Robin, Chem. Phys. Lett. 119: 33 (1985).
3. F. Keller and H. Lefebvre-Brion, Z. fur Physik D in press.
4. F. Sette, J. Stöhr and A.P. Hitchcock, J. Chem. Phys. 81: 4906 (1984).
5. A.P. Hitchcock and C.E. Brion, J. Electron Spectrosc. 10: 317 (1977); A.P. Hitchcock, M. Pocock and C.E. Brion, Chem. Phys. Lett. 49: 1259 (1977).
6. M. Tronc, G.C. King, R.C. Bradford and F.H. Read, J. Phys. B 9: L555 (1976);   M. Tronc, G.C.King and F.H. Read, J. Phys. B 12: 137 (1979).
7. A.P. Hitchcock and I. Ishii, J. Electron Spectrosc. in press.
8. A.P. Hitchcock, D.C. Newbury, I. Ishii, J. Stöhr, J.A. Horsley, R.D. Redwing, A.L. Johnson and F. Sette, J. Chem. Phys. in press.
9. D.C. Newbury, I. Ishii and A.P. Hitchcock, Can. J. Chem. 64: 1145 (1986).
10. J. Stöhr, unpublished results.
11. A.P. Hitchcock, P. Fischer, A. Gedanken and M.B. Robin, J. Phys. Chem. in press.
12. J.A. Horsley, J. Stöhr, A.P. Hitchcock, D.C. Newbury, A.L. Johnson and F. Sette, J. Chem. Phys. 83: 6099 (1985).
13. F.C. Brown, R.Z. Bachrach and A. Bianconi, Chem. Phys. Lett. 54: 425 (1978).

ATOMIC AUTOIONIZATION OBSERVED IN nd $\to \sigma*$

MOLECULAR RESONANCES

P. Morin

Département de Physicochimie, CEA, CEN Saclay
91191 GIF SUR Yvette Cedex -France;
and
LURE Laboratoire commun CNRS, CEA et MEN
Université de Paris Sud - 91405 Orsay Cedex

ABSTRACT

The excitation of a core electron into an antibonding empty valence orbital of a molecule leads to a quite intense resonance compared to Rydberg transitions in the photoabsorption spectrum. We show the direct experimental evidence of its repulsive character by photoelectron spectroscopy in the case of a d-shell excitation of halogen compounds such as HBr, CH_3Br, HI and CH_3I. The first step of the resonance decay is a very fast dissociation process followed by autoionization of the core excited fragment as a second step. Competition with usual molecular autoionization is found when the mass of the radical linked to the excited atom is increased.

I - INTRODUCTION

Inner shell excitation and ionization of molecules reveal several kinds of resonances. Above threshold shape resonances may occur, which can be of atomic or molecular type according to the nature of the centrifugal barrier of the potential. In the "atomic" type, the outgoing is not appreciably affected by the molecular field and the barrier originates mainly from the high "atomic" angular momentum of the electron. In the molecular type the barrier originates from the multicenter nature of the molecular field which is responsible of the strong anisotropic potential. These two types of shape resonances have been discussed in detail by F. Keller and H. Lefèbvre-Brion[1] and observed experimentally in various molecules[2].

Below threshold, we may now distinguish between Rydberg series converging to inner shell ionization and transition from inner shell to molecular empty orbitals. These orbitals, built from antisymmetric linear combination of atomic orbitals, are specific of molecules and do not exist in atoms.

As an example of these various resonances, we show in fig 1 a threshold electron spectrum (TES) of HI which reflects the absorption spectrum as discussed in the appendix. and we compare it to the isoelectronic Xe absorption spectrum from ref 3.

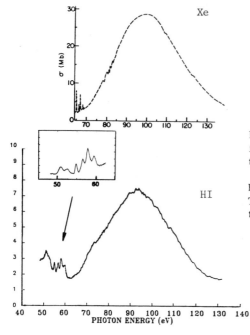

FIG.1 top: Xe absorption spectrum from Haensel et al in the region of the giant $4d \to \varepsilon f$ resonance
bottom: TES (see appendix) of HI. Note the similarity with Xe. The insert exhibits the below threshold resonances (σ^*, Rydberg).

The prominent feature of the 4d absorption spectrum of Xe is the well known $4d \to \varepsilon f$ giant resonance, discussed by many authors[3]. Below threshold sharp resonances are observable which correspond to the $4d \to np$ Rydberg series. The HI spectrum, presented in fig 1, exhibits an intense resonance extending from 70 to 120 eV (max 92.9 eV, FWHM 35 eV) which corresponds to the giant $4d \to \varepsilon f$ resonance. Note that this resonance is very similar to the isoelectronic Xe one. This points out the very atomic nature of the resonance. Another probe of that is the comparison of similar resonances observed in CHI_3, I_2 absorption spectra[2]. The similar position and width of the resonances indicate obviously that the giant resonance is not appreciably affected by changing the chemical environment.

Below threshold (see insert in fig 1), in addition to Rydberg states, two broad resonances are seen at lower photon energy corresponding to the transitions $4d_{5/2}$, $4d_{3/2} \to 5p\,\sigma^*$. Attention is now focused on the very nature of these resonances, and their relaxation pathways.

We will first describe briefly the experimental set up and then present the very specific decay channel of the $3d \to \sigma^*$ transition in HBr. In the next section the CH_3Br results are discussed in terms of the influence of the radical mass. Finally results on HI and CH_3I are presented and the role of the inner hole lifetime will be emphasized.

II - EXPERIMENTAL

The experimental set up is schematized in fig 2 and described in detail elsewhere[4]. Two kinds of experiments have been done detecting either photoelectrons (with kinetic energy analysis) or photoions (with mass/charge analysis). The photoelectron spectroscopy (PES) experiments were performed by use of a 127° cylindrical electrostatic analyzer[4] and the mass spectrometry studies by use of a time of flight mass spectrometer[5]. Photoemission measurements from a gold mesh allowed a normalization of the results with respect to the incident photon flux.

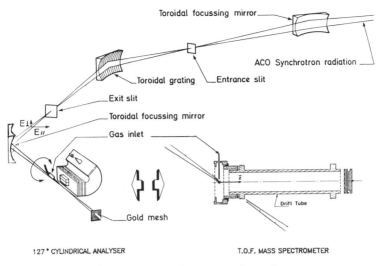

FIG.2 Experimental set up for electron (left) or ion (right) analysis.

III - 3d → σ* in HBr

Fig 3a is a PES of HBr taken just before the σ* resonance (hν = 68.2 eV). The peaks X, A, B correspond to the ionization of the 4pπ, 4pσ and 4sσ M.O's respectively in agreement with previous PES[6]. Fig 3b is a PES taken on the $3d_{5/2}$ → σ* resonance at 70.6 eV. The most striking observation is that numerous sharp lines appear along the spectrum, especially at low binding energy (18.4 eV). They do not correspond to any known states of HBr⁺ and some of them are located too low in binding energy to be due to doubly excited states of HBr⁺. In order to explain their existence, we have invoked a two step relaxation process as described in greater details in our recent publication[7].

FIG.3 PES recorded off (a) and on (b) the HBr resonance. On the left side is reported the electron energy loss spectrum of SHAW et al.[13]

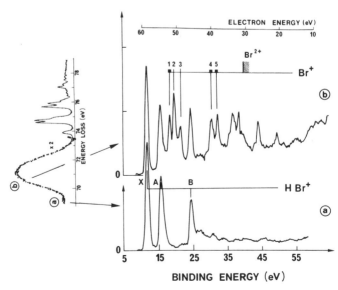

Let's assume the σ* resonance to be a strongly repulsive state and the 3d hole lifetime to be long enough : we can imagine the neutral dissociation of the resonance to occur before any electronic relaxation process take place. This two step process (neutral dissociation followed by the autoionization of the excited Br*($3d^9 4s^2 4p^6 \, ^2D$) atom is schematized in fig 4. Note that such a two step process was invoked by Cermak[8] in his pionneer work in O_2 to explain the decay of some long lived Rydberg states in the valence excitation region using metastable He impact.

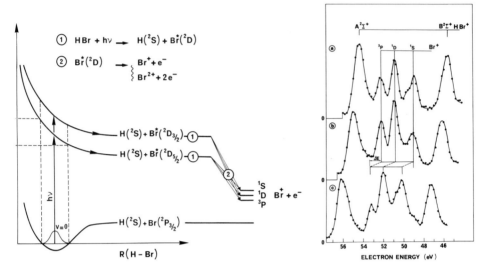

FIG.4 Schematic of the two step relaxation process.

FIG.5 PES recorded in HBr at 70 (a), 70.6 (b), 71.6 eV (c) photon energy, plotted on a kinetic energy scale.

To test the validity of our description, the following points have been examined :

1. position of the autoionizing lines of Br* (^2D)
2. dependence on photon energy
3. real time scale analysis

1. position of the lines

From available atomic data on Br (namely the energy levels of Br* states, and the energy of the Br($3d^{10} 4s^2 4p^5$ .. $3d^9 4s^2 4p^6$ transition) it was possible to calculate the position of the autoionizing lines of Br*(^2D) (see ref.7 for more details). They have been reported on fig 3b. The perfect coincidence between the observed peaks and lines 1, 2, 3, 4, 5 fully confirms our two step description and allows the assignment of the various lines : i. e. 1, 2, 3 correspond to the $^3P_{2,1,0}$, ^1D and ^1S states of Br$^+$($4s^2 4p^4$).

2. dependence on photon energy

According to the two step relaxation process schematized on fig 4 it is clear that the energy of the electrons originating from Br*(^2D) autoionization, doesn't depend upon photon energy. Excitation of the σ* resonance along its repulsive potential curve, changes only the kinetic energy of the fragments but not the electron ones. This is exactly what shows fig 5. PES a and b are recorded at 70 and 70.6 eV photon energy and correspond to the $3d_{5/2}$ excitation. On a kinetic energy scale, peaks A and B from HBr$^+$ are correctly shifted by .6 eV (i. e. the change in photon energy). But the Br$^+$ lines appear exactly at the same energy as expected from the previously described two step process. PES c has been recorded at 71.6 eV which corresponds to the $3d_{3/2}$ excitation. The Br$^+$ lines appear now shifted by ΔE (the $3d_{5/2} - 3d_{3/2}$ spin orbit splitting) which corresponds exaclty to the scheme proposed in fig 4.

3. real time scale analysis

The lifetime of the 3d hole can be estimated from the width of the Rydberg series converging to $3d^{-1}$ (see fig 3). The 90 meV thus determined corresponds to $\tau \leq 7 \cdot 10^{-15}$ s. Now what would be the internuclear distance at this time ? The kinetic energy released in the dissociation is 2.47 eV (see ref 7) which is entirely transferred to the hydrogen atom (due to the Br/H = 80 mass ratio). We deduce (ref.7) a 3 Å internuclear distance at time τ, which is large enough to justify a posteriori the relaxation process as two independent steps.

We will now replace the light hydrogen atom by the heavier methyl radical and discuss the extension of our model.

IV - $\underline{3d \rightarrow \sigma^* \text{ in } CH_3Br : \text{ROLE OF THE MASS}}$

The left side of fig 6 exhibits the TES of CH_3Br. The two $3d_{5/2}$, $3d_{3/2} \rightarrow \sigma^*$ resonances are seen around 70 eV and various PES recorded along them have been reported on the right side of the figure. Again the Br* autoionization lines appear in the valence region. We have reported the position of the $^3P_{2,1,0}$ ^1D and ^1S lines of Br$^+$ as calculated in the previous section. Undoubtly the peaks correspond to the two step process:

$$CH_3Br + h\nu \rightarrow CH_3 + Br^*(^2D)$$
$$Br^*(^2D) \rightarrow Br^+ + e^-$$

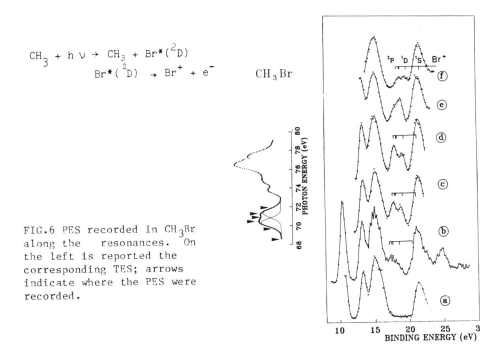

FIG.6 PES recorded in CH_3Br along the resonances. On the left is reported the corresponding TES; arrows indicate where the PES were recorded.

Nevertheless the situation is somewhat different from HBr. Namely :

1/ the intensity distribution of the lines is different : ^3P peak is greater than ^1D as opposed to HBr (unfortunately ^1S peak falls in the $2a_1$ band of CH_3Br^+ and cannot be precisely analyzed),

2/ the general intensity of the lines is weaker than in HBr, suggesting that the two step process is may be not the unique relaxation pathway of the σ* resonance. Examinating now the time scale of the phenomenon we obtain[9] that the dissociation would be achieved in $7.10^{-15}s$ which corresponds nearly to the inner hole lifetime. Thus it is not surprising that any competing process may occur. In order to confirm this point let us now analyze our photoion study of CH_3Br.

By use of a time of flight mass spectrometer (see fig 2) operated in the PIPICO mode[5], we were able to determine the partial cross section of ion pairs formation in the 68-78 eV region. The results are reported in fig 7. The existence of strongly resonant channels as for example $CH_1^+ + Br^+$ points out a competing process to the neutral dissociation which would lead only to Br^+ and Br^{2+} ion formation. Our observations show that an other decay of the σ* resonance occurs which is a molecular autoionization :

$$CH_3Br + h\nu \rightarrow CH_3Br^* \rightarrow CH_3Br^{2+} + 2e^-$$
$$CH_3Br^{2+} \rightarrow CH_3^+ + Br^+$$
$$H^+ + CH_2 + Br^+$$
$$etc...$$

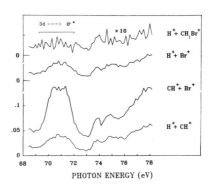

FIG.7 Ion pair formation in CH_3Br as a function of photon energy; the measurements are the result of a coincidence technique.

V - $4d \rightarrow \sigma^*$ in HI, CH_3I : EXTENSION TO OTHER SYSTEMS

We have extended our study to hydrogen and methyl iodide in order to check the validity of our description. On the left side of fig 8 is represented the TES of HI. It is interesting to note that the $4d_{5/2}$, $4d_{3/2}$ spinorbit splitting of iodine is so large ($\simeq 1.75$ eV) that the two σ^* resonances are not overlapping. Thus it was possible to excite very selectively the $4d_{5/2}$ and $4d_{3/2} \rightarrow \sigma^*$ resonances.

PES_2 "d" of fig 8 exhibits well defined lines, corresponding again to the $I^*(^2D_{3/2})$ autoionization into $^3P_{2,1,0}$, ^1D and ^1S states of I^+ as they are reported on the figure. We conclude then that the neutral dissociation followed by atomic autoionization is very present as a decay channel of the σ* resonance.

An interesting point is now to compare PES "c" and "d" which have been recorded at top of $4d_{5/2}$ and $4d_{3/2} \rightarrow \sigma^*$ resonance. It is striking to observe that the I^* lines intensity distribution is very different in both spectra. This reflects that the two 5/2, 3/2 components interact together and cannot be considered independently. Note that Mazoni et al (ref 10). looking at the 4d → 5p transition in atomic iodine. observed very different absorption features exciting the 5/2 or 3/2 component. Particularly they found a four times larger width for the 5/2 than 3/2 resonance. This will lead to an inverse ratio for the inner hole life times. This could be the reason why. in the HI case, the autoionizing lines seem to be much more perturbed in PES "b" and "c" ($4d_{5/2}$ excitation) than in PES "d" and "e" ($4d_{3/2}$ excitation). This point could be elucidated by measuring the $H^+ + I^+$ ion pair formation along the two resonances.

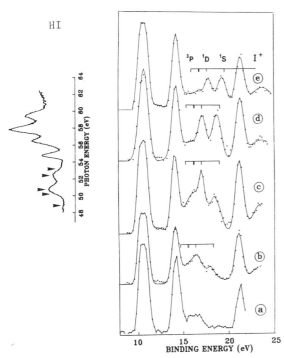
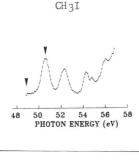
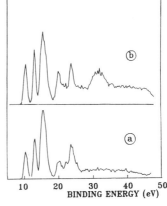

FIG.8 On the left is reported the TES of HI. On the right, PES recorded along the resonances (corresponding to the arrows).

FIG.9 PES of CH_3I recorded off (a) and on (b) the resonance. On top, the corresponding TES.

In fig 9 are shown two PES of CH_3I recorded "off" and "on" the $4d_{5/2} \rightarrow \sigma^*$ resonance. It is likely to observe that now the atomic autoionization lines have desappeared and that they have been replaced by a molecular autoionization band located around 31 eV binding energy. A high background of electron is also extending in the double ionization region : this may be due to a shake off process were two electrons are ejected simultaneously giving rise to very broad bands instead of sharp lines.

VI - CONCLUSIONS

We have given evidence of a new relaxation process of a core excited molecule : a fast neutral dissociation followed by the autoionization of the core fragment. It is interesting to note that our observations (specially in HBr, CH$_3$Br, HI) are a <u>direct probe</u> of the repulsive nature of the σ^* resonance in these halogeno compounds. We have shown that the existence of such a relaxation process was the result of a crucial balance between the inner hole life time and the dissociation rate of the excited molecule. Now the question is to know if we can find this process in other more complicated systems. It seems that hydrides are good candidates as the H atom is light enough to escape before electronic relaxation of the core hole occurs. For instance recent observations on SiH$_4$ by de Souza et al[11] have shown that the Si(2p) $\rightarrow \sigma^*$ (Si - H) resonance also decay via a two step mechanism. When dealing with heavier atoms (or radicals) than H, the phenomenon desappear more or less, but even if the dissociation is not achieved, it is possible for the system to keep some memory of the repulsive nature of the resonating state. This could be reflected as high vibrationnal excitation (even dissociation) of the residual ion.

APPENDIX :

Why can a threshold electron spectrum (TES) reflects the absorption spectrum of an inner hole ?

Electronic relaxation of an inner hole (after excitation or ionization) gives rise to various phenomena. Let us describe them on an example : the 3d shell of Krypton.

<u>above threshold</u> :

ionization : Kr ($3d^{10}$ $4s^2$ $4p^6$) + hν $\rightarrow$ Kr$^+$ ($3d^9$ $4s^2$ $4p^6$) + e$^-$

- The first type of relaxation is the ejection of an Auger electron of a well defined energy (an other electron filling the hole). For example :

Kr$^+$ ($3d^9$ $4s^2$ $4p^6$) $\rightarrow$ Kr^{2+} ($3d^0$ $4s^2$ $4p^4$) + e$_A^-$

- But, as the 3d threshold is above the Kr^{3+} appearance potential, it is also possible to have a shake off relaxation, namely the simultaneous ejection of two electrons (another electron filling the hole), leading to Kr^{3+} formation :

Kr$^+$ ($3d^9$ $4s^2$ $4p^6$) $\rightarrow$ Kr^{3+} ($3d^{10}$ $4s^2$ $4p^3$) + e$_1^-$ + e$_2^-$

In this case, only the sum E_{max} of the kinetic energy of the departing electrons is determined and the energy of one electron can have any value between 0 and E_{max}.

<u>below threshold</u> :

excitation : Kr ($3d^{10}$ $4s^2$ $4p^6$) + hν $\rightarrow$ Kr* ($3d^9$ $4s^2$ $4p^6$ 5p)

- The first type of relaxation is the ejection of a "resonant" Auger electron of a well defined energy. For example :

$$Kr^* (3d^9 4s^2 4p^6 5p) \rightarrow Kr^+ (3d^{10} 4s^2 4p^4 5p) + e^-_A$$

in the case where the Kr^+ ion formed is in a very excited state, it can eject a second electron as for example :

$$Kr^* (3d^9 4s^2 4p^6 5p) \rightarrow Kr^+ (3d^{10} 4s 4p^5 5p) + e^-_{A_1}$$

followed by :

$$Kr^{2+} (3d^{10} 4s^2 4p^4) + e^-_{A_2}$$

In this process the two Auger electrons have well defined energies[12]

- Also it is possible to have shake off relaxation

$$Kr^* (3d^9 4s^2 4p^6 5p) \rightarrow Kr^{2+} (3d^{10} 4s^2 4p^4) + e^-_1 + e^-_2$$

with ejection of two electrons, producing again zero energy electrons.

In conclusion, either below and above threshold, it is possible to have a shake off relaxation of the inner hole producing zero energy electrons. Now by detecting these zero (or near zero) energy electons as a function of photon energy we get a TES which reflects the formation of the core excited (ionized) species, namely its absorption spectrum.

REFERENCE

1 : F. KELLER and H. LEFEBVRE-BRION, Zeischrift für Physik D, in press

2 : - Molecular type : see for ex. A.P. HITCHCOCK and C.E. BRION
J. El. Spec. 13 (1978) 193

- Atomic type : F.J. COMES, U. NIELSEN and W.H.E. SCHWARZ
J. Chem. Phys. 58 (1973) 2230

G. O'SULLIVAN, J. Phys. B 15 (1982) L327

3 : see for ex. M. BEN AMAR and F. COMBET-FARNOUX
J. Phys. B 16 (1983) 2339

4 : P. MORIN, M.Y. ADAM, I. NENNER, S. DELWICHE, M.J. HUBIN-FRANSKIN and P. LABLANQUIE, Nucl. Inst. Meth. 208 (1983) 761

5 : P. LABLANQUIE
Thèse 3ème cycle ORSAY (1984)

6 : J. DELWICHE, P. NATALIS, J. MOMIGNY and J.E. COLIN
J. El. Spec. 1 (72/73) 219

W. von NIESSEN, L. ASBRINK and G. BIERRI
J. El. Spec. 26, (1982) 173

7 : P. MORIN and I. NENNER
Phys. Rev. Lett. 56, (1986) 1913

8 : V. CERMAK, J. El. Spec. 3 (1974) 329

9 : P. MORIN, M. SIMON, P. LABLANQUIE and I. NENNER (to be published)

10 : M. MAZZONI and P. PETTINI, Phys. Lett. 85A (1981) 331

11 : G.G.B de SOUZA, P. MORIN and I. NENNER
submitted to Phys. Rev. A (1986)

12 : P. MORIN (to be published)

13 : D.A.SHAW, D.C.CVEJANOVIC, G.C.KING and F.H.READ
J.Phys.B 17(1984) 1173

ARE THE d→f SHAPE RESONANCES

IN DIATOMIC MOLECULES 'GIANT RESONANCES' ?

       H. Lefebvre-Brion

       Laboratoire de Photophysique Moléculaire
       Bâtiment 213, Université de Paris-Sud
       91405 Orsay, France

INTRODUCTION

    In molecular photoionization, many shape resonances have been observed these last years (for a review, see for example[1]). They appeared mainly in non-hydride molecules built on atoms of the first or second rows of the periodic table. These resonances have been called molecular shape resonances[2]. Another class of shape resonances, named atomic-like shape resonances, is expected in heavy hydrides composed of atoms for which 'giant shape resonances' have been observed in their 4d inner shell photoionization spectra for example Xe, Ba and Eu[3]. A preliminary calculation of the 4d inner shell photoionization of HI has been made recently and gives a first theoretical example of such atomic like shape resonances[2]. A question could be asked : Are these two different types of shape resonances in molecules comparable to 'giant resonances' in atoms ? In this paper, we will try to give an answer to this question.

MOLECULAR SHAPE RESONANCES

    The majority of the molecular shape resonances observed and calculated until now, correspond to $\sigma \rightarrow \sigma^*$ (or $\sigma_g \rightarrow \sigma_u$ for homonuclear molecules) transitions in non hydride molecules. We restrict ourselves here to a review on the calculations performed in diatomic molecules. For such transitions from a bonding σ orbital to the corresponding antibonding σ^* orbital, namely a charge transfer transition, the intensity is especially strong. This transition can be described approximately as a $d\sigma \rightarrow \varepsilon(p+f)\sigma$ transition.

    In Table I, we list the results of calculations made for shape resonances of some diatomic molecules which appear in the near ultraviolet spectrum. The width, Γ, is defined as the energy interval between two points of half maximum cross section. The resonance energy, ε_r, is the energy at the maximum of the cross section measured from the ionization threshold. In theoretical calculations, the widths and the heights of the resonances are very dependent on the method of calculation. We have chosen to report in Table I only one center static exchange calculations, using the dipole length approximation. The other approximations, namely the MSXα method exaggerates the strength of the

Table 1. Characteristics of some molecular shape resonances

Molecule	Ion State	R (a.u.)	Maximum Height (Mb)	Width (eV)	ε_r (eV)
N_2 [a]	$X^2\Sigma_g^+$	2.068 [b]	10.4	21.5	15.6
		2.268	14.8	12.6	11.7
CO [c]	$X^2\Sigma^+$	2.132	7.1	~21.2	11.8
NO [d]	$b^3\Pi$	2.173	11.0	7.7	8.8
	$A^1\Pi$		4.6	5.5	5.2
O_2 [e]	Mean of $b^4\Sigma_g^-$ +$B^2\Sigma_g^-$	2.182	9.9	3.3	6.0
		2.282 [b]	13.4	2.4	3.7
		2.382	17.7	1.9	1.9
		2.482	21.4	~1.2 [f]	0.2

a. G. Raseev, H. Le Rouzo and H. Lefebvre-Brion, J. Chem. Phys. **72**, 5701 (1980)
b. $R = R_e$
c. R.R Lucchese and B.V. McKoy, Phys. Rev. **A28**, 1382 (1983). The width has been evaluated assuming the cross section symmetric.
d. Ref. 5
e. Ref. 6
f. This value has been estimated as twice the half-width for $\varepsilon > \varepsilon_r$.

resonances and gives too narrow peaks and the Stieltjes moment theory gives peaks, in general, too large lying at too low energy. The theoretical values are given for the equilibrium internuclear distance of each molecule. Consequently the experimental values have not been reported in this Table, because they correspond to cross sections averaged on the vibrational motion.

In Fig. 1, a plot of the width against the energy of the cross section maximum, ε_r, is given. In order to show that Γ decreases with ε_r, a very rough straight line has been drawn through the points corresponding to the equilibrium internuclear distance of each molecule. This relation can be explained in terms of a simple picture of a one dimensional potential for the photoelectron. The coordinate here is the distance of the photoelectron to the center of mass of the molecular ion. The shape resonances originate from a barrier in this potential. The closer to the ionization threshold the resonance appears, i.e. the smaller ε_r is, the larger the potential barrier is and the more stable the resonance i.e. the longer its lifetime (proportional to the inverse of the width). The potential becomes deeper and the resonance narrower, when Z increases. The extreme case is for F_2 (Z=18, R_e = 2.68 a.u.) where the $\sigma_g \to \sigma_u$ transition holds in the discrete spectrum[4]. Note that for Ne_2, σ_u is occupied in the ground state. For identical Z and R_e values, the relation Γ against ε_r depends on the details of the molecular potential. For example, in the NO molecule, the potentials corresponding to the $b^3\Pi$ and $A^1\Pi$ states respectively are slightly different[5]. The potential for the $A^1\Pi$ state contributes a little more attractive exchange energy than that for the $b^3\Pi$ state of same configuration. Consequently $\varepsilon_r(A^1\Pi)$ is smaller than $\varepsilon_r(b^3\Pi)$. The calculations for O_2 [6] have been made for a multiplet-averaged potential.

Fig. 1 is comparable to the universal curve plotted by Connerade[7] for 'giant resonances' in the $4d \to \varepsilon f$ photoionization of Xe, Ba and Eu.

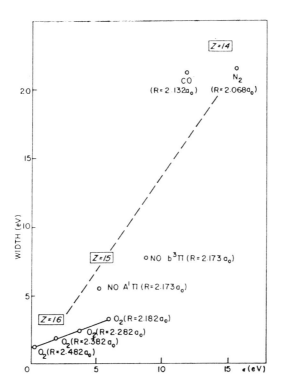

Fig. 1. Relation between the width Γ and the resonance energy ε_r referred to the threshold energy for static exchange calculations.

It is clear[8,9] that molecular shape resonances correspond well to $\sigma \rightarrow \sigma^*$ non Rydberg states and are strongly localized at the resonance energy value near the molecular core (r < 2 a.u., where r is the electron distance to the center of mass) as it has been shown by Loomba et al. [10].

Nevertheless, the description by one potential is difficult in the case of molecular shape resonances because we have shown[6] that the resonance disappears when the equations giving the continuum wave functions are uncoupled. This means that these resonances are induced by a typical molecular effect. This molecular character is emphasized by comparison with the united atom cases for which no shape resonance appears.

However, in molecules, a supplementary variable, R, the internuclear distance complicates the situation. For a given molecule, the characteristics of the shape resonances vary strongly with the R value for which the calculations have been made. When R increases, the resonance becomes more and more narrow and approaches the threshold energy. This is clearly seen for the O_2 molecule in Fig. 1. This behaviour, which is responsible for strong observed vibrational effects, has been explained in terms of Rydbergization of the continuum valence orbital [2,11] which acquires more and more f character when R increases. For different molecules, similar relation, namely ε_r decreases as R increases, has been noted for K-shell resonances of polyatomic molecules [12]. Finally, as stated by Dehmer et al.[1], 'the continuum resonant functions are substantially uncoupled from the internal environment of the molecule. This means that the shape resonant phenomena often persist in going from the gas phase to the condensed phase'. But it could be predicted that the vibrational effects, due to valence-Rydberg mixing, would be different in the condensed phase compared to those known in the gas phase.

All these characteristics of the molecular shape resonances are very similar to those of 'giant resonances' described by Connerade [3] as non Rydberg states, localized inside the inner potential well and surviving in the solid. Thus, the question can be asked if the molecular shape resonances can be called also 'giant resonances.' This has been discussed by Robin[13,14], which has classified shape resonances observed in cyclopropane and cyclohexane as 'giant resonances.' But, because these resonances have observed in the absorption spectra, it is not clear if their width corresponds to a photoionization rathan than a photodissociation process. A question ased by Robin concerns the absolute value of the 'grain resonances.' The values of about 20 Mb for the O_2 molecule seem perhaps small compared to that for the Xenon atom ($\sim$30 Mb). Also we can conclude that these molecular resonsnces are shape resonances but it is perhaps difficult to call them 'giant resonances.'

ATOMIC-LIKE SHAPE RESONANCES

In hydrides, the $\sigma \to \sigma^*$ transition lies in general below the ionization limit. This is particularly true for the hydrogen halides where the transition 'd'$\sigma \to \sigma^*$ has been observed below the 'd'σ ionization threshold. Consequently when the photoionization corresponding to ionization of the d inner shell gives rise to a shape resonance, this resonance is not of molecular type. It corresponds to a 'd'$\to \varepsilon$'f' resonance where d and f are quasi atomic like functions. Such a case has been studied recently[2] in the HI molecule. In this paper, the ion core molecular states have been described in terms of Hund's case (a) wave functions for which Λ, Σ, the projection of the total electronic angular and spin momentum on the internuclear axis and S the spin momentum, are well defined. As the spin-orbit interaction is large in this heavy hydride, a more physical description must be made in terms of Hund's case (c) wave functions. This description is directly related to the (J, M_J) atomic functions. For a d hole in atom, giving $^2D_{5/2}$ and $^2D_{3/2}$ sublevels, we have the correspondences, given in Table 2, for simplicity for the d configuration, with the Hund's case (c) wave functions with Ω defined, expressed in the Hund's case (a) wave functions $^2\Sigma^+_{1/2}$, $^2\Pi_{3/2, 1/2}$ and $^2\Delta_{5/2, 3/2}$ functions used in Ref. 2. The limit case (c) functions is convenient, because the electrostatic differences between the $^2\Sigma^+$, $^2\Pi$ and $^2\Delta$ states can be neglected (< 0.5 eV) in comparison with the spin orbit splitting (1.76 eV [15]) of the two groups of molecular peaks related to the two atomic limits ^2D_J with $J = 5/2$ and $3/2$. It is easy to show that the atomic relations subsist in the case (c) Hund representation, namely for the partial cross sections

$$\sigma_{5/2} = \frac{3}{2} \sigma_{3/2} \quad \text{or} \quad \sigma_{\ell+1/2} = \frac{\ell+1}{\ell} \sigma_{\ell-1/2} \quad (1)$$

The total cross section corresponding to a d hole is given by

$$\sigma_d = \sigma_{5/2} + \sigma_{3/2} = \sum_\Omega \sigma_\Omega = \sum_\Lambda \sigma_\Lambda \quad (2)$$

Table 2. Projection of the (J, M_J) atomic wave functions on the molecular wave functions expressed in Hund case (c) wave functions

	J	M_J	Atomic functions	Ω	Molecular functions
$^2D_{5/2}$	$\frac{5}{2}$	$\frac{5}{2}$	$\|d_{+2}\|$	$\frac{5}{2}$	$^2\Delta_{5/2}$
	$\frac{5}{2}$	$\frac{3}{2}$	$\frac{1}{\sqrt{5}}\{2\|d_{+1}\|+\|\overline{d_{+2}}\|\}$	$\frac{3}{2}$	$\frac{1}{\sqrt{5}}\{2(^2\Pi_{3/2})+{}^2\Delta_{3/2}\}$
	$\frac{5}{2}$	$\frac{1}{2}$	$\frac{1}{\sqrt{10}}\{\sqrt{6}\|d_{+0}\|+2\|\overline{d_{+1}}\|\}$	$\frac{1}{2}$	$\frac{1}{\sqrt{10}}\{\sqrt{6}(^2\Sigma^+_{1/2})+2(^2\Pi_{1/2})\}$
$^2D_{3/2}$	$\frac{3}{2}$	$\frac{3}{2}$	$\frac{1}{\sqrt{5}}\{\|d_{+1}\|-2\|\overline{d_{+2}}\|\}$	$\frac{3}{2}$	$\frac{1}{\sqrt{5}}\{{}^2\Pi_{3/2} - 2(^2\Delta_{3/2})\}$
	$\frac{3}{2}$	$\frac{1}{2}$	$\frac{1}{\sqrt{10}}\{\sqrt{6}\|\overline{d_{+1}}\|+2\|d_{+0}\|\}$	$\frac{1}{2}$	$\frac{1}{\sqrt{10}}\{2(^2\Sigma^+_{1/2}) - \sqrt{6}(^2\Pi_{1/2})\}$

For the total cross section, the expression (2) is identical to that used in Ref. 2. The calculated σ_d cross section is given in Fig. 2. Very recently, Morin has observed [15] the $4d^{-1}$ ionization in HI. The averaged ionization potential is 57.96 ± 0.05 eV. The maximum of the cross section is at about 35 eV or 1.28 a.u. from the threshold (ε_r(calc) = 0.5 a.u.) and the width is about 35 eV (width (calc) = 19.6 eV). Its absolute value is not known but is probably near from that obtained for CH_3I, namely 11 Mb [16] (calc. 51 Mb). The model used for HI is here not satisfying to represent accurately the resonance. We have suggested [2] that it would be necessary to introduce here interchannel interactions between channels of same symmetry but with different electronic ionic cores.

We have shown previously[2] that this resonance does not vary with R, contrarily to the molecular shape resonances treated in the preceding paragraph. Furthermore, we know that the molecular shape resonances disappear if the continuum equations are uncoupled. To study this point for the HI molecule, we have performed two types of comparisons. At first, the static potential is included with the exchange potential but the equations are uncoupled. The resultant cross section is nearly unchanged. However, this type of uncoupled calculation can be criticized because the exchange equations and the orthogonalization with the bound orbitals produces a weak residual ℓ-mixing. Also, we have performed another comparison in excluding the exchange potential. When the equations are coupled, because the sum of the exchange terms between the continuum electron and all the core electrons is a positive quantity, the potential without these terms is less attractive and the resonance lies at higher energy (ε_r = 1 a.u. with the static potential instead of ε_r = 0.5 a.u. with the static-exchange potential), see Fig. 2. If now furthermore the continuum equations are uncoupled, the resonance is nearly unchanged. In the uncoupled cases, the selection rule $\Delta\ell = \pm 1$ is practically satisfied. In the coupled cases, the transition moments have non zero values for any $\Delta\ell$, but the cross section, proportional to the

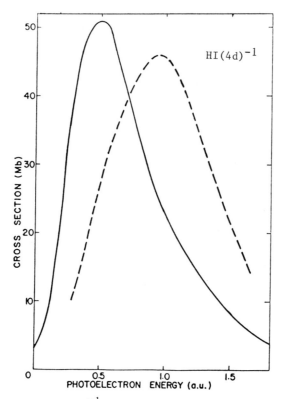

Fig. 2. Calculated $4d^{-1}$ photoionization cross sections for the HI molecule. The full curve is for the static exchange results. The dotted curve for the static potential without exchange. The results for the uncoupled equations are not reported because they differ from these curves of less than two per cent.

sum of the square of the transition moments, is about the same in the coupled and uncoupled calculations for a given approximation for the exchange potential. We see that the molecular character of this resonance is very weak compared to the case of the molecular shape resonances in light molecules. This is the reason why this resonance can be called 'atomic-like shape resonance'.

In conclusion, since the character of the resonance in the HI molecule is nearly atomic, as we have shown it, this resonance could be compared to 'giant resonances' observed in atoms such as the Xenon atom or better the iodine atom. Experiments and calculations on other heavy hydrides would be interesting to confirm this result.

CONCLUSION

Many properties of shape resonances in molecules seem comparable to those of 'giant resonances' in atoms. Unfortunately, it is difficult to prove better the similarity without detailed study of the localization of the resonances in the wells of the molecular potential.

Consequently, we think that if all these molecular resonances are well characterized shape resonances, the name 'giant resonance' cannot be used for them without caution.

ACKNOWLEDGEMENTS

The author thanks Dr G. Raseev and Dr F. Combet-Farnoux for fruitful discussions.

REFERENCES

1. J.L. Dehmer, A.C. Parr and S.H. Southworth, to appear in Handbook on Synchrotron Radiation, vol. II, edited by G.V. Marr (North-Holland, Amsterdam, 1986).
2. F. Keller and H. Lefebvre-Brion, Zeits. für Physik D, to be published.
3. J.P. Connerade, J. Less Common Metals, 93, 171 (1983).
4. A.E. Orel, T.N. Rescigno, B.V. McKoy and P.W. Langhoff J. Chem. Phys., 72, 1265 (1980).
5. M.E. Smith, B.V. McKoy and R. Lucchese, J. Chem. Phys., 82, 4147 (1985).
6. G. Raseev, H. Lefebvre-Brion, H. Le Rouzo and A.L. Roche, J. Chem. Phys., 74, 6686 (1981).
7. J.P. Connerade, J. Phys. B 17, L165 (1984).
8. W. Thiel, J. Electron. Spect., 31, 151 (1983).
9. P.W. Langhoff, in 'Resonances in electron molecule scattering', edited by D.G. Truhlar (American Chemical Society, Washington, 1984), p. 113.
10. D. Loomba, S. Wallace, Dan Dill and J.L. Dehmer, J. Chem. Phys. 75, 4546 (1981).
11. P. Morin, I. Nenner, M.Y. Adam, M.J. Hubin-Franskin, J. Delwiche, H. Lefebvre-Brion and A. Giusti-Suzor, Chem. Phys. Letters, 92, 609 (1982).
12. F. Sette, J. Stöhr and A.P. Hitchcock, J. Chem. Phys., 81, 4906 (1984) and references therein.
13. M.B. Robin, Chem. Phys. Letters, 119, 33 (1985).
14. M.B. Robin, 'Higher excited states of polyatomic molecules', vol. III, Academic Press, Orlando, Florida, 1985.
15. P. Morin, private communication.
16. D.W. Lindle, P.H. Kobrin, C.M. Truesdale, T.A. Ferrett, P.A. Heimann, H.G. Kerkhoff, U. Becker and D.A. Shirley, Phys. Rev. A 30, 239 (1984).

III. SOLIDS – THEORY AND EXPERIMENTS

THE QUASI-PERIODIC TABLE AND ANOMALOUS METALLIC PROPERTIES

A. M. Boring and J. L. Smith

Los Alamos National Laboratory
Los Alamos, NM 87545

ABSTRACT

We compare the electron states in atoms and solids in an attempt to gain some insight into the quasi-localized behavior of the heavy-electron materials. These materials occur in compounds containing elements that have uncertainties in their electronic configurations both in the free atoms and in the metals. We find that unlike the long range phenomena of conventional superconductivity in the transition metals and local-moment magnetism in the rare-earth systems, the site quantum numbers in the heavy-electron systems have not yet been determined. These uncertainties lead to the anomalous properties of these systems.

INTRODUCTION

Because they defy many of our preconceptions about possible condensed matter ground states the newly discovered heavy-fermion (or heavy-electron) systems are intriguing.[1] Superconductivity and local-moment magnetism are relatively well-understood states of long-range order for the electronic ground states of metals. The periodic table of the elements can be arranged to show that between these two phenomena there exists a region where the properties of the ground states are uncertain.[2] The valence d or f electron states of the atoms seem uncertain as to whether to form bands (leading most likely to superconductivity) or to remain localized (forming local-moment magnets) upon forming a solid. This uncertainty or instability on the part of the electrons leads to metallic ground states that are more challenging than either conventional superconductivity or local-moment magnetism. While different, ground states such as itinerant magnetism, spin fluctuations, charge density waves, or spin density waves remain in the conceptual framework of itinerant behavior. The heavy-electron systems, on the other hand, yield so many anomalous ground state properties as to question our understanding of localized versus itinerant behavior of metallic systems in a fundamental way.

These materials are ordered intermetallic compounds in which one of the constituents is either a rare earth or actinide atom with partially filled 4f or 5f electron shells.[3] Another characteristic of these systems is that the f-atom components are well separated spatially - a fact conventionally thought to lead to local-moment magnetism and a decoupling of the f electrons from the conduction bands. However, heat capacity measurements indicate that an enormous density of states develops at the Fermi energy (E_F) at low temperature, which is usually taken as a signature of a very narrow energy band pinned at E_F. This peak in the density of states corresponds to electron states with an energy width of 1-10 meV (~10-100 K), and it is these states with small dispersion that are hard to understand and lead to the strange ground state properties observed. This narrow f band at low temperatures still indicates rather localized states and can still lead to magnetism. However, it is found that some of these compounds become superconducting. Hence it is not surprising that the heavy electron superconductors have anomalous properties and that the pairing interaction may well be electron-electron rather than the conventional electron-phonon interaction found in all other superconducting materials.

These heavy-electron systems seem to show new behavior for nearly every experiment that is performed. They are clearly a new electronic state in solids. We are a long way from a proper theoretical description[4] of these systems, but in this paper we will focus on our general understanding of forming electron states in solids with the thought of obtaining a clearer idea of how this heavy-electron state develops. To this end we will first discuss the electronic states of atoms and their associated quantum labels. Next we describe how these quantum labels change in forming solids and how this leads to new quantum states and new phenomena. However, we discover that the heavy-electron systems are just those for which the atomic-like quantum numbers are important and uncertain.

Electrons in Atoms

As we go from light to heavy atoms the overall feature is an increase in the complexity of the valence states. This is seen, for example, in the complex spectra of the uranium atom as opposed to the simple doublet of the sodium atom. The important quantum numbers for light atoms are of course the total orbital angular total momentum L and the total spin S of the electrons of the partially filled shells. (We remind the reader that there are two parts to the calculation of atomic states: one determines the orbitals and average of configuration energies of the central field problem, and then one calculates the lowest term of a multiplet by the appropriate coupling of the open shell states.) For partially occupied shells with <L> and <S> $\neq$ 0 we have a magnetic moment with orbital and spin contributions. For the open shell, Hund's rules apply for the occupation numbers and Russel-Saunders (or L-S) coupling is appropriate. As Z increases the coupling between L and S (or ℓ and s of each electron) becomes important, and the total angular momentum $J(\vec{J} = \vec{L} + \vec{S})$ is then the good quantum number. However, even for uranium (Z = 92) while the core electrons obey J-J coupling, the valence electrons (configuration: $5f^3 6d^1 7s^2$) are intermediately coupled and are actually closer to L-S than J-J coupling. This preference for L-S coupling of the valence d and f electrons can be seen in the stability of the half filled shells as the 3d, 4d, 4f, and 5f shells have either d^5 or f^7 configurations and only in the 5d elements do we find a d^4 ground state. This raises the question of the importance of other relativistic effects (mass-velocity and Darwin correction) for valence electrons. (One goes from solving the Schrödinger equation to solving the Dirac equation for high Z atoms because relativistic effects are obviously large for the core electrons.) It has been shown that even to get the average of configuration energy right for the valence electron of uranium all electrons must be treated relativistically.[5] Another feature of increasing Z is that the energies of the valence electrons for different quantum numbers (different ℓ values) are closer together. This leads to

an uncertainty in the occupation numbers of the ground state. This is
seen in the late 3d transition metal atoms (Fe, Co, and Ni) where the 4s
and 3d orbital energies are so close that earlier workers had a hard time
determining the correct ground state configuration. This uncertainty is
also seen in Ce and the early actinides in that they contain more than
one partially occupied ℓ shell (f and d).

As Z is increased the centrifugal term $\ell(\ell+1)/r^2$ of the Hamiltonian
also becomes important. One effect of this is to decrease the spin-orbit
interaction of the higher angular momentum states in a central field and
in part explains the preference for L-S coupling of valence electrons of
high Z number. (The spin-orbit coupling goes as $\sim 1/r^3$, averaged over the
wavefunction, and so if the wavefunction is small at small r due to the
centrifugal barrier term the spin-orbit coupling will be reduced.)
Another effect of the centrifugal term is that it leads to the so-called
double-well in the one-electron potential for the excited states of some
atoms.[6] This phenomenon occurs when the orbital of the next higher ℓ
value becomes bound, a remnant of the excited Rydberg states of the Z-1
atom that leaves the one-electron potential with two wells. This effect
has been discussed for Ce in a solid state context but it also occurs in
Sc, Y, La, Ce, and Pa atoms.[7] This effect occurs in atoms because the
potential of the last electron in the atom must be hydrogenic at large r
and fit onto the fully screened potential at small r, and the $\ell(\ell+1)/r^2$
term creates a barrier between the large and small r behavior. An argument against this effect being important in metals is that the large r
behavior is not hydrogenic.

Electrons in Metals

We use Figure 1 to discuss trends in the electron states of the
periodic table and the development of the atomic quantum numbers for
metals. Figure 1, besides showing the regions of superconductivity and
local-moment magnetism, indicates the itinerant versus localized behavior
of d and f electron systems. As the atoms come together to form a periodic
array this change yields a new quantum number (the $\vec{k}$ vector) which reflects
the translational symmetry. The valence electron wavefunctions of the
atoms overlap and this leads to a quasi-continuum of energy levels. If
the overlap is large enough the electrons spread out into the solid and
are no longer associated with any one site. Delocalized states formed
from valence s and p levels tend to have very large energy dispersion in
metals (bandwidths of 5-10 eV wide). For these materials a free electron
model seems to describe much of the metallic behavior (roughly). However,
the electron states of the solid generated from valence d and f levels do
not reflect free electron behavior. In any event, the calculated energy
bands of most s, p, and d electron systems do give an accurate account of
much of the behavior of metals. For f electron systems we may not yet
have a proper electron theory.

One can ask however, how the atomic quantum numbers such as L or S
(or rather the average of L and S over the crystal wavefunction; <L> and
<S>) have changed at a given atomic site in the condensed phase. The
experimental data expressed in Figure 1 give the answer. For the superconductors the values of <L> and <S> around a given site must be zero
otherwise these terms would give rise to magnetism which would destroy
the superconductivity. (This argument would be modified if these were
not conventional superconductors.) The fact that <L> and <S> are zero in
these solids is also the reason why one only needs to determine the
average of configuration energies to obtain good average properties
(cohesions, bulk moduli, shear constants, etc.). Of course the dynamical
spin of each itinerant electron is still important as superconductivity
is a correlated electron motion of bound spin pairs. On the other hand,

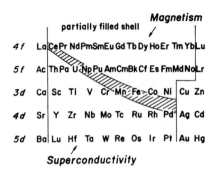

Fig. 1. Nearly periodic table of the f- and d-electron elements where the hatched regions shows the cross-over between itinerant (bonding) and local-moment magnetism (non-bonding) behavior of the valence d and f electrons.

in the magnetic regime of Figure 1 we know <L> and <S> are not zero and are in fact equal to their atomic values for the f shell in question. This is known from two observations: the experimental solid state magnetic moments are equal to the calculated atomic values, and the atomic multiplet structure is observed in the photoemission spectra of these solids. These local moments are coupled indirectly through interactions with the conduction band states (RKKY interaction) to form the complicated magnetic structures observed. Thus the two long range ordered states in condensed matter are determined by whether or not the atomic angular moments are quenched.

Now we consider the systems in the cross over region of Figure 1. Looking at Fe, Co, and Ni we find systems for which <S> ≠ 0 and <L> = 0. Again this is based on two observations: the experimental moments of these itinerant ferromagnets agree with the L = 0 atomic values, and the calculated electron structure leads to a <S> ≠ 0 ground state (spin-polarized calculations). Obtaining the T = 0 ground state for these systems has, however, not been an easy task for modern band theory as evidenced by the fact that a recent calculation yielded ferromagnetic fcc Fe stable relative the correct ferromagnetic bcc Fe crystal. The theory of the itinerant magnetism of these systems at finite temperatures is still being developed. For the other systems in the region (Ce, Pa, U, Np, and Pu) the experimental data shows no clear trend in the values of <L> and <S>. Some of these systems (Ce and U) are both superconducting (<L> and <S> = 0) in one phase and magnetic (<L> and/or <S> ≠ 0) in another. Also it has been argued that itinerant magnetism in these systems is more consistent with <S> and <L> ≠ 0 (but not equal to their atomic values).[8] The magnetism in these systems seems to be spin driven as in the case of Fe, Co, and Ni. However, due to spin-orbit coupling (a relativistic effect) an orbital (L) moment is generated in the itinerant states.[8] For a theoretical treatment this creates a problem. For the non-relativistic case (Fe, Co, and Ni) one can perform a spin-polarized calculation ($<S> \neq 0$), but for higher Z systems cannot obtain a relativistic (Dirac equation) solution for the spin-polarized case. It is only recently that such a theory has been worked out and it has not become a standard tool of modern electron theory.[9]

The correlated motion of electrons in atoms and solids, which is left out of mean field calculations such as the Hartree-Fock or the Local Density Functional approximation, can be important. Although we will not consider this topic in any detail, a few comments seem appropriate. For

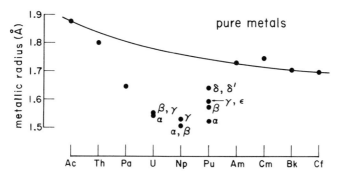

Fig. 2. Lattice spacings of the actinide metals showing the deviation from rare earth behavior (solid line).

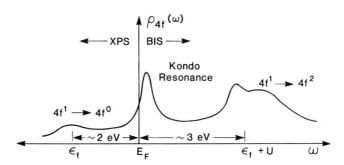

Fig. 3. The "Kondo" spectrum of a Ce impurity in a metal. The parameters were chosen to emphasize three features: the occupied peak at ε_f (f level), unoccupied peak at $\varepsilon_f + U$ (where U is the intrasite coulomb integral) and the Kondo resonance near E_F.

atoms one performs Configuration Interaction calculations to determine correlation effects. This process can be viewed as including fluctuations or increasing the calculational basis set. For removal energies these are 1 or 2% effects. Also it is usually found that there are only 1 or 2 other important configurations and the dominant ground state can always be determined. For solids where one has a quasi-continuum of levels in the ground state, such calculations (if they could be performed) are probably meaningless. One can envision having in the solid a large number of small (in energy) charge and spin fluctuations near E_F such that there is no dominant configuration. In such cases the one-electron band theory methods, which determine properties for an average configuration, must fail to properly describe the properties that are mainly a function of the states near E_F. Although the success of the band theory methods in determining properties of solids indicates this state of affairs does not occur frequently, it may be the rule and not the exception in the heavy-electron systems.

Besides giving a clear separation between the different long range behaviors in metals Figure 1 also separates the bonding (3d, 4d, and 5d) valence electrons from the non-bonding (4f and late 5f) valence electrons. In Figure 2 the lattice constants of the rare earth (4f) and actinide (5f) metals are compared. The departure (smaller values) of the earlier actinides (Pa through Pu) from the rare earth contraction curve is taken as evidence that in these sytems the 5f electrons are bonding. Based on a comparison with the transition metals one might expect $<L> = 0$ and even maybe $<S> = 0$, however as mentioned above some data indicates $<S> \neq 0$ and maybe even $<L> \neq 0$. Because electronic structure calculations have given very good predictions of the bonding in the transition metals we look at what those calculations predict for the 5f states. Unfortunately these calculations are inconclusive at the present time: the 5f electrons treated as itinerant and bonding leads to overbinding whereas if they are treated as non-bonding one obtains too little binding for the solid.[10] These results indicate that for these systems something may be lacking in the standard electronic structure calculations for the average properties of the ground state. (Since cohesion depends on all the bonding electrons and not just the electron states at E_F this failure is more crucial to the standard electronic structure calculations than the neglect of fluctuations mentioned earlier.)

Returning to atomic electronic states for clues, we see that Hund's first rule shows the double occupancy of a spatial orbital is energetically unfavorable, indicating that the intracoulomb integral is very repulsive. Hubbard argued more than 20 years ago that the proper treatment of this integral in the solid state would yield improved results.[11] The basic effect this should have on the electronic structure is to decrease the hopping of electrons from site to site leading to less disperion or even localization of some states in the solid. A development related to this effect is the Kondo problem.[12] The Kondo problem is the observed low temperature minimum in resistivity of metals with magnetic impurities. The development of the Kondo model or the related Anderson model[13] for the impurity problem has been a major achievement of solid state theory. It yields a localized state for the impurity with an energy level that overlaps the conduction bands of the host and the hybridization between these two states leads to spin fluctuations of conduction bands at E_F and modifies the one-electron picture in important ways.[14] In Figure 3 we show the energy dispersion of this model for both occupied and unoccupied states.[15] This "Kondo" spectrum is seen in the combined XPS/BIS spectra of Ce compounds and is the most compelling evidence to date for this model for more than the impurity problem. The development of a lattice Kondo model for intermetallic compounds is currently an active research topic. Having shown the development of

Table 1. The heaviest electron compounds, and their ground states.

Compound	Ground State:
$CeCu_2Si_2$	Superconducting at 0.5 K
UBe_{13}	Superconducting at 0.9 K
UPt_3	Superconducting at 0.5 K
U_2Zn_{17}	Antiferromagnetic at 10 K
$NpBe_{13}$	Antiferromagnetic at 3 K
UCd_{11}	Antiferromagnetic at 5 K
$CeAl_3$	no ordering
$CeCu_6$	no ordering

electron states in solids and why the systems in the cross-over region are the most difficult to describe we turn to some of the specific features of Ce and the light actinide heavy-electron systems.

Heavy-Electron Systems

For the heavy-electron systems discovered to date the "active" (f electron) atom component has been an element from the cross-over region of Figure 1 (Ce or a light actinide). In Table I we give the eight most actively studied systems and indicate their ground state behavior. As we have seen it is not clear what the good quantum numbers are in these systems, and this is one reason that a correct theoretical description is still lacking. What we wish to discuss here are some of the specific features of the heavy electrons that underlie the problems. As mentioned earlier the anomalous behavior in these systems occurs at low temperatures and so we will discuss the high and low temperature properties separately. A good reason for discussing the high temperature state is that knowing what state the highly correlated low temperature state is condensing from may indicate how to construct a proper theory. Also we mention that most of the low temperature experimental data only samples the energy states very near E_F, so that the global electronic structure is not obtained and this may be important.

All of these systems at high temperature have Curie-Weiss magnetic behavior and this has been taken as evidence for a local-moment description at these temperatures. There are three arguments that indicate that this may not be a forgone conclusion. First, it has now been shown that at high temperatures the itinerant states may give rise to a Curie-Weiss-like law (independent of the band dispersion).[16] Secondly, part of the argument against itinerant states giving Curie-Weiss behavior is that the Fermi temperature T_F for band states is orders of magnitude greater than the Curie temperature, and so in normal metals one cannot access enough states. However, if T_F is less than the Curie temperature as it is in these systems then the whole f band can respond at room temperature. The third problem arises because the high temperature (above 20 K) XPS/BIS spectra for uranium compounds does not show the "Kondo" spectrum of the Ce compounds which has been taken as a signature of localized f states. However, the high temperature resistivity in both Ce and U heavy-electron systems tends to be high ($\rho \sim 100$ μΩcm whereas $\rho \sim 1\text{-}10$ μΩcm is typical for metals) which could indicate scattering of the conduction electrons from a localized state as in the Kondo problem. All of this leads us to

believe that for now it is not obvious whether or not the low temperature
state is condensing out of itinerant or localized states.

Turning to the low temperature behavior of these materials where
most of the experimental data has been taken we find the situation is no
better. Of course, the high specific heat coefficient (γ) indicates very
narrow bands pinned at E_F. That this is not a conventional band is
evidenced by the fact that all the standard band calculations performed
on these systems give f bandwidths which are one to two orders of magnitude wider than the data indicates.[17,18,19] This leads to calculated
densities of states at E_F which maybe as much as 20 times too small.
Another indication of itinerant states is the deviation of the magnetic
susceptibilities from precise Curie-Weiss behavior. This coupled with
failure of conventional local density band theory to describe this highly
correlated itinerant state has lead to what has been labeled renormalized
band theory.[20] The basic idea is a renormalization of the f-band states
by including the effects of an intrasite coulomb term U (decreased
hopping) and modifying the hybridization of this f band states with other
bands. (Here the argument is being made that U is strong enough to
narrow the f band by decreasing the coupling to other states but does not
localized these f states as in the Anderson or Kondo model.) To date
such theories have been applied mainly to Ce compounds which have an f^1
configuration and the realistic treatments have not yet been made self-consistent. These theories and calculations look very different from
atomic theory, but one must remember they are a remnant of Hund's first
rule.

The most intriguing behavior of the low temperature phase is the
existence of magnetism and superconductivity with the same electron shell
(f electrons). This phenomena is very different from the coexistence of
superconductivity and magnetism in the rare-earth-transition metal borides
in which where one sublattice is superconducting (with transition metal
atoms) and the other sublattice is magnetic (with rare-earth atoms).
Here the f electrons show spin-fluctuation behavior (UPt_3 and UBe_{13})
before going superconducting. Because it has been argued that spin-fluctuations will destroy conventional superconducting pairing ($k\uparrow$, $-k\downarrow$)
with S = 0, this has lead speculation about the pairing in these systems.
Because the only known system to have triplet (S = 1) pairing is ^{3}He
there have been attempts to extend theoretical models of liquid ^{3}He to
the heavy-electron systems. However, it was soon realized that in heavy-electron metals the spin-orbit coupling is large and a correct symmetry
analysis does not lead to a singlet-triplet classification as in the ^{3}He
case.[21] This comparison between liquid ^{3}He and heavy-electron systems
has generated efforts to apply Fermi liquid theory to these sytems.[22]
Such calculations are conceptually far removed from atomic theory - the
model was put forth by Landau to explain why the strongly interacting
electrons in a solid have a free electron energy spectrum. The weak
residual interactions left between the quasiparticles are then used to
describe systems that do not behave like a free electron gas. Since we
do not see any real connections between this model and atomic concepts we
will not discuss it further, but end by saying that this model has not to
date been very successful at describing heavy-electron behavior either.

CONCLUSIONS

In this paper we have shown that the conventional long range ordered
states of superconductivity and local-moment magnetism in most d and f
electron metals in the periodic table are consistent with either both <L>
and <S> (around a given site) being zero or both having their atomic
values. For the late 3d transition metals (Fe, Co, Ni) which fall
between localized and itinerant behavior the properties are consistent

with $\langle S \rangle \neq 0$ (and not equal to the atomic value) but $\langle L \rangle = 0$. However, for the other systems in the cross-over region (Ce and the earlier actinide) may have $\langle S \rangle$ and $\langle L \rangle \neq 0$ and also not equal to their atomic values. It is from these sytems that the "active" f atom component of the heavy-electron systems arises. It is also for these systems that the intrasite coulomb integral appears to be important. However, at the present time it is not clear why this leads to the narrow f bands found in the heavy-electron system at low temperatures.

ACKNOWLEDGMENTS

We thank R. D. Cowan for discussions of some of the aspects of atomic problems considered in this paper. This work was performed under the auspices of the U.S. Department of Energy, Office of Basic Energy Sciences, Division of Materials Science.

REFERENCES

1. Z. Fisk, H. R. Ott, T. M. Rice, and J. L. Smith Nature 320, 124 (1986).
2. J. L. Smith and E. A. Kmetko, J. Less-Common Metals 90, 83 (1983).
3. G. R. Stewart, Rev. Mod. Phys. 56, 755 (1984).
4. P. A. Lee, T. M. Rice, J. W. Serene, L. J. Sham and J. W. Wilkins, Comments Cond. Matter Phys. 12, 99 (1986).
5. J. H. Wood and A. Michael Boring, Phys. Rev. B 18, 2701 (1978).
6. J. P. Connerade, J. Less-Common Metals 93, 171 (1983).
7. R. D. Cowan, The Theory of Atomic Structure and Spectra (Univ. of California Press, 1981).
8. M. S. S. Brooks and P. J. Kelly, Phys. Rev. Lett. 51, 1708, (1983).
9. A. H. MacDonald and S. H. Vosko, J. Phys. C 12, 2977 (1979); BuXing Xu, A. K. Rajagopal, J. Phys. C 16, 1339 (1984); P. Strange, J. Staunton, and B. L. Gyorffy, J. Phys. C 17, 3355 (1984); G. Schadler, P. Weinberger, A. M. Boring, and R. C. Albers, Phys. Rev. B 34, xxxx (1986).
10. M. S. S. Brooks, J. Phys. F14, 1157 (1984); also unpublished calculations by A. M. Boring and R. C. Albers.
11. J. Hubbard, Proc. Roy. Soc. A 276, 238 (1963).
12. J. Kondo, Prog. Theoret. Phys. 32, 37 (1964); Solid State Physics 23, 183, (1969) (F. Seitz and D. Turnbull eds. Academic Press, N.Y.).
13. P. W. Anderson, Phys. Rev. 124, 41 (1961).
14. N. Andrei, K. Funya, and J. H. Lowenstern, Rev. Mod. Phys. 55, 331 (1983); O. Gunnarsson and K. Schönhammer, Phys. Rev. Lett. 50, 604 (1983); N. E. Bickers, D. L. Cox, and J. W. Wilkins, Phys. Rev. Lett. 54, 230 (1985); F. Patthey, B. Delley, W.-D. Schneider, and Y. Baer, Phys. Rev. Lett. 55, 1518 (1985).
15. O. Gunnarsson and K. Schönhammer, Phys. Rev. B 28, 4315 (1983).
16. B. L. Gyorffy, A. J. Pindor, J. Staunton, G. M. Stocks, and H. Winter, J. Phys. F 15, 1337 (1985).
17. T. Jarlborg, H. F. Braun and M. Peter, Z. Phys. B 52, 295 (1983).
18. K. Takegahara, H. Harima, and T. Kasuya, J. Magn. Magn. Mats. 47 & 48, 263 (1985) and A. M. Boring, R. C. Albers, F. M. Mueller, and D. D. Koelling, Physica 130B, 171 (1985).
19. R. C. Albers, Phys. Rev. B 32, 7646, (1985); R. C. Albers, A. M. Boring, and N. E. Christensen, Phys. Rev. B 33, 8116 (1986); P. Strange and B. L. Gyorffy, Physica 130B, 41 (1985); T. Oguchi and A. J. Freeman, Physica 135B, 46 (1985); J. Sticht and J. Kubler, Solid State Commun. 54, 389 (1985).
20. P. Strange and D. M. Newns J. Phys. F 16, 335 (1986); N. d'Ambrumenil and P. Fulde, J. Magn. Magn. Mat. 47 & 48, 1 (1985); B. H. Brandow, Phys. Rev. B 33, 215 (1986).

21. G. E. Volovik and L. P. Gor'kov, Pis'ma Zh. Eksp. Teor. Fiz. 39, 550 (1984); E. I. Blount, Phys. Rev. B 32, 2935 (1985).
22. O. T. Valls and Z. Tesanovic, Phys. Rev. Lett. 53, 1497 (1984); T. M. Rice, K. Ueda, H. R. Ott, and H. Rudigier, Phys. Rev. B 31, 594 (1985).

A CONDENSED MATTER VIEW OF GIANT RESONANCE PHENOMENA

Andrew Zangwill

School of Physics
Georgia Institute of Technology
Atlanta, Georgia 30332

INTRODUCTION

It is my intent in this article to present a view of giant resonance phenomena (an essentially atomic phenomenon) from the perspective of a condensed matter physicist with an interest in the optical properties of matter. As we shall see, this amounts to a particular prejudice about how one should think about many-body effects in a system of interacting electrons. Some of these effects are special to condensed matter systems and will be dealt with in the second half of this paper. However, it turns out that my view of the main ingredient to a giant resonance differs significantly from that normally taken by scientists trained in the traditional methods of atomic physics. Therefore, in the first section that follows, I will take advantage of the fact that my contribution to this volume was composed and delivered to the publishers somewhat <u>after</u> the conclusion of the School (rather than before as requested by the organizers) and try to clearly distinguish the differences of opinion presented by the lecturers from the unalterable experimental facts.

Let us take a purely pragmatic definition of a giant resonance. It is a prominent "bump" that appears in the optical absorption spectrum of an atom, molecule or solid whose leading edge is not far from a specific near core level threshold and whose integrated oscillator strength is approximately equal to the number of electrons in the atomic shell associated with that threshold. Although many different spectral features might be shoehorned into this definition I will restrict my attention only to the most dramatic examples, i.e., those associated with intra-subshell dipole transitions from an initial state to an unfilled bound or virtual bound state characterized by the same principal quantum number. Hence, I associate "giant resonances" primarily with np → nd optical excitations in the atoms of the 3d, 4d, and 5d transition metal rows and the elements immediately preceeding them and nd → nf optical excitations in the atoms of the lanthanide and actinide rows and the elements immediately preceeding them. This places the excitation energies of interest in the vacuum ultraviolet part of the electromagnetic spectrum. As an illustration, Figure 1 shows the total absorption from barium metal (it has been <u>de rigueur</u> to discuss barium at this Summer School) just above the 4d threshold. Our task is to formulate a convenient conceptual framework within which to understand why absorption features like this one appear and why they might be interesting and/or useful.

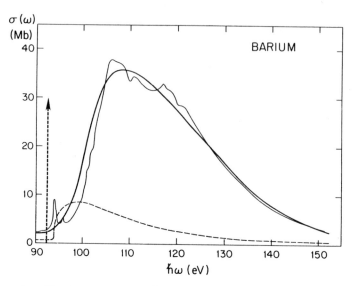

Fig. 1. Giant resonance absorption in barium metal above the 4d threshold. Data from Ref. 1; calculations discussed in the text.

ONE-BODY EFFECT OR MANY-BODY EFFECTS?

The experimental observation that giant resonances carry significant fractions of atomic shell oscillator strength immediately brings to mind the analogy with γ-ray absorption resonances in nuclei[2] and plasmon excitations in metals.[3] Both of these phenomena commonly are regarded as "collective effects" entirely produced by many-body interactions within the nucleon or electron system, respectively. To what extent is this true in the present context? Consider the Golden Rule expression for the total photoabsorption cross section in the long wavelength limit

$$\sigma(\omega) = 4\pi^2\alpha\hbar\omega \sum_F |\langle F|\varepsilon \cdot U_{ext}(x)|0\rangle|^2 \delta(\hbar\omega - E_F + E_0) \qquad (1)$$

In this formula, $U_{ext}(x)$ is the dipole operator, α is the fine structure constant, and ε is the polarization vector of the incident radiation. At fixed excitation frequency ω the absorption occurs between the exact ground state $|0\rangle$ and one of the exact excited states $|F\rangle$. Of course, only approximate evaluation of the requisite matrix elements is possible since the exact many-electron wave functions are unknown. The simplest <u>independent particle</u> approximation writes these quantities as single Slater determinants constructed from one-particle orbitals that are the eigenstates of an appropriate effective potential function. The Golden Rule then takes the form

$$\sigma(\omega) = 4\pi^2\alpha\hbar\omega \sum_{i,f} |\langle f|\varepsilon \cdot U_{ext}(x)|i\rangle|^2 \delta(\hbar\omega - \varepsilon_f + \varepsilon_i) \qquad (2)$$

By definition, "many-body effects" appear when the independent particle model fails to describe the experimental data and one resorts to corrective perturbation theory of some sort. Evidently, different choices of a one-body basis set will lead to differing conclusions: one person's many-body effects might be another person's one-body effect.

This ambiguity does not arise in the nuclear and plasma resonance cases. There is broad agreement that the eigenstates of a Woods-Saxon (or related) potential and free electron plane waves, respectively, are the most natural one-particle states for these problems. The corresponding independent particle absorption cross sections fail badly and a perturbation theory sufficient to create a many-body collective resonance is required. Unfortunately, the situation is not quite so clear cut in the present case. From the point of view of traditional atomic and molecular physics, long experience suggests that Hartree-Fock orbitals ought to constitute the independent particle basis set of choice. Indeed, if one uses the most correct "term-dependent Hartree-Fock" eigenstates, many giant resonances (like the barium example[4]) are reasonably reproduced by straightforward application of Equation (2). The resonant structure simply reflects transitions to a virtual bound state in the continuum with some natural distribution of oscillator strength. One would say that one is dealing with a "shape" resonance and collective effects are insignificant. Many-body refinements ("ground-state correlations", "channel mixing", etc.) can be handled either via low-order diagrammatic perturbation theory[5] or configuration interaction.[6]

A rather different picture emerges if one uses the most "natural" set of one-electron orbitals from the point of view of a condensed matter physicist. The Hartree-Fock approximation is notoriously poor for study of the electronic structure of extended solid state systems. For example, it yields zero density of states at the Fermi energy of a metal. By contrast, the local density approximation (LDA) to the density functional formalism of Hohenberg, Kohn and Sham[7] has proved very effective for these problems.[8] This approach yields a complete one-electron basis set of so-called Kohn-Sham orbitals that can be used (somewhat cavalierly) to compute optical matrix elements as in Equation (2). The results are quite reasonable for the fundamental optical spectra ($h\nu < 12$ eV) of solids traditionally studied by condensed matter physicists.[9] However, similar independent particle LDA calculations of the photoabsorption cross section of atoms at higher excitation energy never reproduce the giant resonances of interest here. Instead, one finds (see below) that a time-dependent generalization of this theory that includes polarization-type many-body effects yields remarkably good agreement for the total cross section. In this approach the giant resonances are quite usefully viewed as strongly damped collective oscillations of entire atomic shells quite akin to the nuclear and plasmon examples. Hence, one is led to the point of view lucidly propounded many years ago by Roulet and Nozieres[10] and implemented by Amusia and co-workers[11] and Wendin[12] through RPAE calculations using <u>configuration average</u> Hartree-Fock orbital basis sets.

The main point of this section is to emphasize that the assignment of "many-body effects" in the context of giant resonances in atoms, molecules and solids is very much a question of definition. There is no "correct" answer. Nevertheless, I shall argue in the next section that the interpretation of these phenomena in accordance with the dielectric response and density functional formulations common to condensed matter physics provides a particularly economical and physically appealing picture. Succeeding sections will deal with specific features that are unique to the problem of giant resonances in solid state environments.

IELECTRIC RESPONSE

The optical properties of solids commonly are discussed in terms of the frequency dependent dielectric function $\varepsilon(\omega)$ whose imaginary part is proportional to photoabsorption cross section[13]:

$$\sigma(\omega) \propto \mathrm{Im}\, \varepsilon(\omega) \tag{3}$$

At the elementary macroscopic level, we think of $\varepsilon(\omega)$ as that function that describes the reduction (screening) of an external field by a polarizable medium. This notion remains correct at the microscopic level[14] as long as we recognize the non-local spatial dependence of this phenomenon, viz.,

$$U_{eff}(x|\omega) = \int \varepsilon^{-1}(x,x'|\omega)\, U_{ext}(x')\,dx' \tag{4}$$

In a translationally invariant crystal we Fourier transform the spatial variables and discover that the relationship between the microscopic and macroscopic dielectric functions is non-trivial:

$$\varepsilon(\omega) = \lim_{q \to 0} \frac{1}{\{\varepsilon^{-1}(q+G, q+G')\}_{G=G'=0}} \tag{5}$$

For fixed frequency ω and wavelength q of the external perturbation, the dielectric function is a matrix in the reciprocal lattice vectors (G) of the crystalline lattice. Optical properties are recovered in the long wavelength limit ($q \to 0$) only after inversion of a large matrix.

We now require some expression for this dielectric matrix. In the most systematic approach, one expresses[3,10] this quantity in terms of the density-density correlation function $\chi(x,x'|\omega)$,

$$\varepsilon^{-1}(x,x'|\omega) = \delta(x-x') + \int dy\, \frac{1}{|x-y|}\, \chi(y,x'|\omega) \tag{6}$$

The latter is simply related to a two-particle Green function for which a systematic perturbation theory is available. This type of scheme has been carried out with some success to describe the low energy optical spectra of covalent semiconductors by Hanke and collaborators[15]. More commonly, one adopts the result that emerges from the lowest order in perturbation theory — the so-called random phase approximation (RPA):

$$\varepsilon(q+G, q+G'|\omega) = \delta_{G,G'} + \frac{8\pi e^2}{N\Omega |q+G|^2} \sum_{k,v,c}$$

$$[\langle vk|e^{-i(q+G)\cdot r}|c\, k+q\rangle \langle c\, k+q|e^{+i(q+G')\cdot \vec{r}}|v\vec{k}\rangle] \tag{7}$$

$$\left[\frac{1}{E_c(k+q)-E_v(k)-\hbar\omega-i\delta} + \frac{1}{E_c(k+q)-E_v(k)+\hbar\omega+i\delta} \right]$$

Here, the sums over k, v and c refer to sums over the wave vectors in the first Brillouin zone of the crystal and the one-particle states (e.g., LDA Kohn-Sham eigenfunctions) of the valence and conduction bands, respectively. The structure of this equation makes the relationship between Equations (2) and (3) more evident. Unfortunately, the sums indicated above are very difficult to perform reliably and results are available only at zero frequency. Nevertheless, even in the static limit, it is very appealing physically to actually "see" the internal effective field given by Equation (4). Figure 2 is a plot of $U_{eff}(x|0)$ (actually its gradient – the electric field) for one unit cell of a zinc selenide crystal[16]. We see wild variations that reflect nonuniform microscopic screening of the external field by the polarizable cation and anion. The static dielectric constant of this crystal – about 5 – can be read off directly from the average value of the internal field.

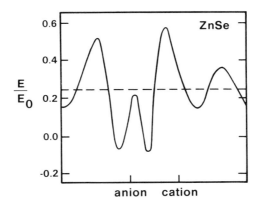

Fig. 2. Microscopic internal field $E(x)/E_{ext}$ along a cut through a ZnSe crystal. Horizontal line denotes the average microscopic field. From Ref. 16.

The RPA calculation discussed above evidently describes the polarizability of the electronic system. At zero frequency, the principal source of this polarization may be regarded as virtual transitions between occupied and unoccupied one-electron states induced by the external field[17]. These particle-hole pairs (relative to the ground state Slater determinant) describe a <u>distortion</u> of the original ground state charge density. The induced charge generates an induced dipolar potential (via Poisson's equation) and the total screened effective field is the sum of the external field and this induced field. The screening is largest in systems for which the virtual transitions have large overlap and hence large oscillator strength. With this in mind, it is quite reasonable to suppose that polarization effects will be important in photoabsorption at excitation frequencies where strongly overlapping orbitals are connected by real - rather than virtual - transitions. But, as the catalogue given in the first section indicates, this is precisely where the giant resonances occur.

TIME-DEPENDENT LOCAL DENSITY APPROXIMATION

The time-dependent local density approximation (TDLDA) was conceived as a means to include polarization-type many-body effects in a dielectric[18] response formulation of photoabsorption within the context of the density functional formalism. I have reviewed this subject in some detail elsewhere[19] and Wendin[20] has provided an account of the relationship between this method and earlier RPAE calculations. Therefore, here I shall focus on the economy and interpretative advantage offered by the method and note some recent results that point the way to further improvement.

Density functional theory[8] provides an explicit prescription for calculation of the <u>exact</u> ground state charge density $n(x)$ of an N-electron system by solution of the following Hartree-like equations:

$$[-\nabla^2 + v_{eff}(x)]\psi_i(x) = \varepsilon_i \psi_i(x)$$

$$v_{eff}(x) = v_{ext}(x) + e^2 \int dx' \frac{n(x')}{|x-x'|} + v_{xc}[n(x)] \quad (8)$$

$$n(x) = \sum_{i=1}^{N} |\psi_i(x)|^2$$

In this expression, $v_{ext}(x)$ is a fixed external potential (the nuclear Coulomb potential for the atomic case) and the functional $v_{xc}[n(x)]$ is called the exchange-correlation potential. The latter quantity contains all of the many-body quantum mechanics of the problem and is generally unknown. However, there exist a number of excellent approximations to the exact $v_{xc}[n(x)]$ beginning with the remarkable LDA[8] and proceeding to more recent and significantly improved functionals[21]. For present purposes, I wish to emphasize that $v_{xc}[n(x)]$ is not to be confused with any approximation to the non-local exchange operator of Hartree-Fock theory such as the X-α scheme of Slater. Its conceptual and theoretical foundations are completely different although the frequently employed local density approximation to it turns out to be numerically similar to one choice of an X-α potential. In what follows, we will suppose that an adequate form of $v_{xc}[n(x)]$ is available and regard the $\psi_i(x)$ obtained from Equations (8) as basis functions for use in, say, Equation (2).

As noted earlier, the resulting independent particle LDA calculations of atomic photoabsorption fail badly in the regime of observed giant resonances. This is illustrated by the dashed curve in Figure 1 for barium. The calculated oscillator strength resides almost entirely in a delta function that corresponds to a 4d → 4f bound-bound transition. By contrast, the measured resonance is significantly shifted to higher energy and broadened relative to the LDA result. Indeed, independent particle calculations of giant resonances quite generally yield strong, narrow absorption features that exhibit a characteristic transition energy ω_0 and width γ that are both smaller than observed. This fact alone already signals the importance of polarization effects as may be seen by use of a schematic driven oscillator model[22].

Consider the equation motion for the change in density δn that is induced in an atom in the presence of an external field $E_o \cos\omega t$ in the independent particle approximation:

$$\ddot{\delta n} + \gamma \dot{\delta n} + \omega_o^2 \delta n = E_o \cos\omega t \quad (9)$$

The absorption cross section is related to the charge distortion through the complex, frequency-dependent polarizability $\alpha(\omega)$:

$$\sigma(\omega) = 4\pi \frac{\omega}{c} \operatorname{Im} \alpha(\omega) \quad (10)$$

$$\alpha(\omega) = c \int dx \, z \, \delta n(x|\omega) \quad (11)$$

As noted earlier, optical transitions that occurs in the neighborhood of ω_o contribute significantly to the polarizability if there is large overlap between the initial and final states, i.e., if δn is large. In that case, we are compelled to compute the induced Coulomb field,

$$E_{induced} \sim e \int \frac{\delta n(x')}{|x-x'|} dx' \sim -\kappa \delta n \quad (12)$$

written here for simplicity as if the integral kernel were a delta function. The constant κ is taken to be complex to represent the fact that the induced density generally oscillates out of phase with the external field. Equation (12) now must be added to the right hand side of Equation (9) since the electrons cannot distinguish the external field from internal fields. Hence, the effective equation of motion is

$$\ddot{\delta n} + (\gamma + \mathrm{Im}\kappa)\dot{\delta n} + (\omega_o^2 + \mathrm{Re}\kappa)\delta n = E_o \cos\omega t \tag{13}$$

which clearly demonstrates the shifting and broadening required to recover agreement with experiment.

The essence of the TDLDA is precisely that of the oscillator model sketched above – made a bit more quantitative by inclusion of the quantum-mechanical dynamical response of the atom. The most transparent statement of the defining equations is:

$$U_{ind}(x|\omega) = \int dx' \frac{\delta n(x'|\omega)}{|x-x'|} + \frac{\partial V_{xc}(x)}{\partial n(x)} \delta n(x|\omega)$$

$$\delta n(x|\omega) = \int dx' \chi_o(x,x'|\omega)[U_{ext}(x') + U_{ind}(x'|\omega)] \tag{14}$$

The first line takes account of the classical induced potential considered above and, in addition, includes an induced exchange-correlation potential. The latter can be significant ($\sim$ 30% of the total induced field for low density outer shells) and should not be neglected. The second line describes precisely how the electronic charge density distorts in the presence of an effective driving field $U_{eff}(x|\omega)$. This response is dictated by a density-density correlation function χ_o appropriate to the independent particle system and is exactly calculable for an atom within the LDA. Solution of the integral equation (14) is equivalent to generating the local density analog of the random phase approximation to the exact χ from χ_o for use in Equation (6). The cross section may be found <u>either</u> from the converged value of δn and Equations (10) and (11) <u>or</u> by <u>replacing</u> the external field by the effective field in the Golden Rule[18], viz.,

$$\sigma(\omega) = 4\pi^2 \alpha \hbar\omega \sum_{i,f} |\langle f|\varepsilon \cdot U_{eff}(x|\omega)|i\rangle|^2 \delta(\hbar\omega - \varepsilon_f + \varepsilon_i) \tag{15}$$

This formula is particularly valuable as an aid in the interpretation of calculations.

The heavy solid curve in Figure 1 is the result of a TDLDA calculation of the photoabsorption of barium near the 4d threshold. The agreement with the experimental "giant resonance" is quite good and is typical of what one finds with this method. In its present form, the TDLDA averages over any structure due to multi-electron excitation and/or multiplet effects. However, in my view, the principal utility of this method is that direct visual inspection of the space and frequency dependent effective driving potential $U_{eff}(x|\omega)$ (or its gradient – the internal "electric" field) leads to substantial physical insight into the excitation process. For example, the idea that the 4d shell may be thought of as a single damped driven "collective" oscillator in the giant resonance regime is borne out in the following manner. If this notion is correct, the oscillations induce a dipolar charge separation within the 4d shell (see Figure 3). Below resonance, i.e., below the peak in the absorption curve, we expect the charge to oscillate in phase with the external field. The resulting induced field then screens the external field within the 4d charge radius and antiscreens the external field at radii in excess of the 4d shell mean radius. The situation is reversed at frequencies above the resonant absorption maximum. The shell now oscillates $180°$ <u>out of phase</u> with the external field and the screening/antiscreening characteristics switch accordingly. Gratifyingly, this simple

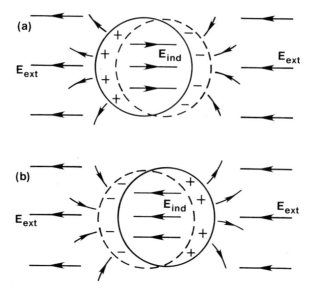

Fig. 3. Schematic model of an atom in an time-dependent electric field: (a) induced charge oscillates in phase with the field; (b) induced charge oscillates out of phase with the field.

picture is borne out precisely by the calculated effective fields[19].

It is possible to cast these results in a fashion even more akin to the traditional dielectric response approach of condensed matter physics. We define an effective local screening function by the relation

$$\varepsilon^{-1}(x|\omega) = \frac{U_{eff}(x|\omega)}{U_{ext}(x)} \quad (16)$$

Combining Equations (15) and (16) it is clear that the TDLDA method ascribes all of the "many-body effects" to a dielectric function that renormalizes the transition operator while retaining the original independent particle orbitals. Hence, we can retain our intuition about what orbital wave functions "look like". This contrasts with the configuration-interaction approach of atomic and molecular physics whereby one systematically improves the initial and final state wave functions. Of course, only the squared matrix element is an observable so that any partitioning of many-body effects among the initial state wave function, the transition operator, and the final state wave function is possible[23].

Now let us return to the barium example and study the behavior of the function $\varepsilon^{-1}(\omega)$[19]. In particular, Figure 4 illustrates the frequency dependence of both the real and imaginary parts of this quantity at two different radial positions in the atom - an "inner" region where the 4d wave function resides and an "outer" region where the valence orbitals, e.g., 5p reside. We see immediately from Equation (15) that the <u>partial</u> photoionization cross section from these two orbitals are expected to be radically different. Indeed, the valence orbitals are mere spectators to the wild oscillations of the 4d shell and are "blown out" of the atom (variously termed "resonant photoemission", "direct recombination", etc.) in accordance with their spatial sampling of the effective driving field. In particular, we expect two types of resonant line shapes: one for all orbitals that occupy the same region of space as the polarizable shell and a different one for all orbitals that reside at greater radial distances[18].

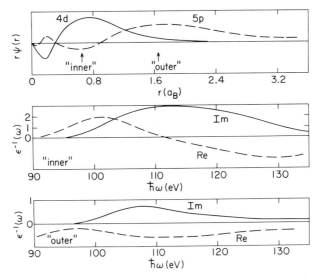

Fig. 4. Barium 4d and 5p radial wave functions and frequency dependence of the effective dielectric function at two radial distances from the nucleus. From Ref. 19.

Another advantage of the TDLDA scheme is computational economy. It is not necessary to recompute potentials anew as one enters different energy regimes. This means that many calculations can be performed rapidly in order to extract trends. For example, Figure 5 is a plot of LDA (independent particle) and TDLDA total photoabsorption cross sections over a wide energy scan for atomic ruthenium. These results were obtained by Gary Doolen of the Los Alamos National Laboratory using a relativistic version of the TDLDA[24]. Some type of polarization effect is operative whenever these two curves differ. I have chosen this example because it illustrates an amusing case of what might be called a "double giant resonance" in the 10-100 eV range. The very sharp resonant structure around 40 eV is a giant resonance that corresponds to 4p → 4d transitions analogous to the 3p → 3d resonances in the 3d transition metal row discussed below. The interesting point is that this sharp structure is superimposed on a very broad absorption bump that occurs both in the independent particle cross section and in the TDLDA result. We would describe this feature to a dull shape resonance if it were not for the fact that the TDLDA resonance is conspicuously shifted to higher energy and broadened relative to the LDA curve by a significant amount (note the double logarithmic scale). In fact, this many-body effect arises from a polarization of the ruthenium 4d shell by "4f-like" continuum states well above threshold. If we trace this feature across the 4d transition metal row we discover that it evolves continuously into the barium giant resonance!

In closing this section it is worth remarking that the original construction of the TDLDA for photoabsorption[18] was not sanctioned within formal density functional theory. The latter is a ground state theory and says nothing about excitations; one could only show that Equations (14) were valid at zero frequency[25]. Since that time, the existence of a formal time-dependent density functional theory has been proven[26] and Gross and Kohn[27] have investigated the quality of the TDLDA in this context. Their analysis provides an a posteriori justification for the quality of the results obtained to date and suggests that future improvements in the theory, such as the inclusion of multiple excitations ("shake-up", "double-photoionization",

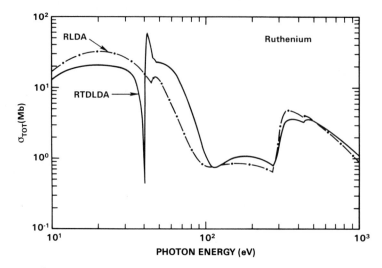

Fig. 5. Relativistic LDA and TDLDA total absorption cross sections for a ruthenium atom.

etc.) must proceed in tandem with improvements in the $v_{xc}[n(x)]$ functional itself beyond the LDA.

RESONANT PHOTOEMISSION IN SOLIDS

By far the most common application of giant resonance phenomena in solid state physics is the use of resonant photoemission to "project out" angular-momentum resolved components of the valence band density of states. At its root, this technique relies on a simple quantum mechanical effect. Suppose that two distinct one-electron excitation processes can occur at precisely the same photon energy, e.g., a bound-bound transition and a bound-free transition (Figure 6). The total absorption exhibits a characteristic line shape that reflects the interference between these two

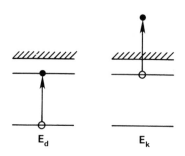

Fig. 6. Two degenerate excitation processes that interfere to produce a Fano absorption profile.

channels. This problem was first analyzed by Fano[28] using configuration interaction theory. Here I will sketch a derivation that uses a method more common to condensed matter physics – a model Hamiltonian.

First, let us rewrite Equation (1) in a slightly different form using the symbol T to stand for the transition operator.

$$\sigma(\omega) = \sum |\langle F|T|0\rangle|^2 \delta(\hbar\omega - E_F + E_0)$$

$$= \frac{1}{\pi} \sum \langle 0|T^\dagger|F\rangle \, \text{Im} \, \frac{1}{\hbar\omega - E_F + E_0 - i\eta} \langle F|T|0\rangle \quad (17)$$

$$= \frac{1}{\pi} \, \text{Im} \, \langle 0|T^\dagger \frac{1}{\hbar\omega + E_0 - H} T|0\rangle$$

$$= \frac{1}{\pi} \, \text{Im} \, \langle 0|T^\dagger G T|0\rangle$$

Since T connects the ground state to the excited states we see that the photoabsorption depends only on the matrix elements of the Hamiltonian within the manifold of excited states. One way to proceed is to write down explicitly a model Hamiltonian, calculate the excited states (usually approximately) and compute the matrix elements (cf. the contribution of Gunnarsson to this volume). Alternatively, one can simply assign the matrix elements. This is the practical approach for our simple problem:

$$\langle d|H|d\rangle = E_d$$

$$\langle k|H|k\rangle = E_k \quad (18)$$

$$\langle d|H|k\rangle = V_k$$

E_d and E_k are the energies of the bound-bound and bound-free transitions (the latter is degenerate with the former for one particular value of k) and V_k is a coupling matrix element that describes the dipole-dipole interaction between the two types of particle-hole pairs or, less precisely, describes the decay of one excitation into the other. All that remains is algebra[29] and one finds the familiar Fano result that the pure absorption cross section to the continuum is multiplied by a modulation factor that reflects the interaction with the discrete excitation:

$$\sigma(\omega) = \sum |T_{ok}|^2 \delta(\hbar\omega - E_k) \frac{(q+\varepsilon)^2}{1+\varepsilon^2}$$

$$q = \frac{T_{od} + \sum T_{ok} \frac{V_k}{\hbar\omega - E_k}}{\pi \sum T_{ok} V_k \delta(\hbar\omega - E_k)} \quad (19)$$

$$\varepsilon = \frac{\hbar\omega - E_d - \sum \frac{|V_k|^2}{\hbar\omega - E_k}}{\pi \sum |V_k|^2 \delta(\hbar\omega - E_k)}$$

What happens if more than one continuum channel is open, i.e., there exist one-hole final states where the hole can occur in different subshells? A (rather more complicated) multi-channel Fano-type calculation is possible[30] and one finds resonances of the form of Equation (19) with different

q-values for each of these partial photoionization cross sections. The
resonances are particularly large and distinctive if one of the degenerate
channels exhibits a large one-particle transition moment - much of this
oscillator strength can be transferred to resonant photoemission in a different channel. This is precisely the situation at a giant resonance. One
finds core-to-valence absorption strength transferred to valence band photoemission intensity. The reader will recognize that the TDLDA approach discussed above provides a ready means to gauge the <u>relative</u> partitioning of
strength amongst the open partial photoionization channels: unless direct
emission from the polarizable core level itself predominates, the strongest
resonance will occur from the state with maximal overlap with the core level
and, hence, with the effective driving field. One reaches the same conclusion by asking which of the V_k Coulomb matrix elements is largest. This
idea has been used particularly extensively to accentuate the 4f and 5f partial densities of states in lanthanide[31] and actinide[32] materials near their
respective 4d and 5d giant resonances*.

The explicit application of these ideas to the barium example (and more
generally to the whole lanthanide row) provides another opportunity to
emphasize that "many-body" effects can be partitioned in different ways to
suit different purposes. Previously we imagined a one-step process whereby
the polarization effects built into the TDLDA transferred spectral oscillator strength from a bound-bound transition below the 4d threshold to a broad
giant resonance above threshold. An alternative and equally valid point of
view first was propounded by Dehmer et al.[33]. They broke the problem into
two parts. First, the large overlap-driven Coulomb exchange integral between the 4d initial state and 4f final state was imagined to drive the
oscillator strength up in energy through the vacuum threshold and into the
f continuum. This no-longer "discrete" excitation then interferes with,
i.e., decays into, the continuum and the width and line shape have the Fano
form. I wish to emphasize that the physics that enters this discussion is
identical to that included in the TDLDA. One can use both languages interchangeably to illuminate different aspects of the problem.

An interesting example of giant resonance photoemission used as a tool
to study an unrelated condensed matter problem occurs for samarium metal.
In this case, one finds[34] that as the photon energy is tuned above the 4d
threshold two distinct parts of the electron energy distribution curve
exhibit giant resonance intensity enhancements. However, the energy dependence of these features, i.e., the 4f partial cross sections, are not identical (Figure 7). This is interesting because it is known (from other considerations) that one curve is characteristic of the bulk of the sample
whereas the other is characteristic of just the uppermost atomic layer at
the surface of the sample. Why might these cross sections be so different?
The answer may be found by comparison of the experimental curves with relativistic TDLDA 4f partial photoionization cross sections for samarium atom
in two different valence configurations: $[Xe]4f^66s^2$ and $[Xe]4f^56s^25d$. Bulk
samarium metal is known to exhibit the trivalent configuration and the $4f^5$
cross section does indeed match nicely with the major peak in the data.
However, the principal peak in the surface spectrum matches well with the
$4f^6$ calculation - the configuration known to exist for a free samarium atom!
We conclude that the surface atoms are in a different valence state than the
bulk. This result is not outlandishly surprisingly if we note that the
bonding state of a surface atom is in some sense midway between that of a
normal bulk atom and a free atom. Of course, this argument applies to any

---

*It is worth remarking that the subtraction techniques ccommonly used in
experimental papers to extract "f" densities of states are not without
problems. Resonant photoemission from other parts of the valence band and
resonant Auger processes can considerably obscure the situation.

solid. It turns out[35] that the energy balance is quite delicate and it appears that samarium has the singular distinction of being the only elemental solid whose surface atoms maintain their free atom configuration when the solid is assembled.

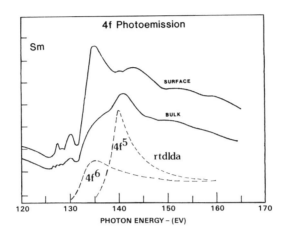

Fig. 7. 4f partial photoionization cross sections from the bulk and surface of samarium metal along with TDLDA calculations (see text). Data from Ref. 34. The experimental curves are not normalized to one another or to the theory.

CASE HISTORY: 3D TRANSITION METALS

Up till now, we have taken the attitude that giant resonance phenomena in condensed matter systems is essentially an "atoms-in-solids" effect. This is certainly true for the lanthanides where both the initial and final states in the 4d → 4f excitation are undoubtedly atomic-like. A major (and difficult) physics problem has been to infer the magnitude of solid state hybridization with the narrow f-states (see e.g., Ref. 31; a different approach to this problem is discussed by J. Fuggle in this volume). However, I now wish to argue that a new richness enters the problem if we study cases where explicit, solid state, one-body (finite bandwidth) and many-body (conduction electron response) effects come into play. To make this point I draw two examples from the 3d transition metal row: nickel and chromium.

Suppose many body effects did not exist. Then, for a metal, optical absorption from a core level is described by a Golden Rule expression that factorizes into the product of an atomic-like transition moment and the bulk unoccupied density of states[36]. This simple scheme is well borne out, for example, by $L_{2,3}$ absorption into the empty 3d states of nickel[37]. The simplest guess would be that $M_{2,3}$ absorption in the same system would reflect exactly the same 3d partial density of states. Unfortunately, it does not. Instead, the total absorption in nickel metal above the 3p threshold is almost identical to the corresponding spectrum for nickel vapor. (Figure 8). The unoccupied 3d-density of states of the metal is completely distorted by the large polarizability of the nickel 3p shell (as reflected by a large 3p/3d exchange Coulomb integral). The oscillator strength is pushed into the continuum just as in the atom. Indeed, an atomic Fano-type calculation including multiplet effects[41] reproduces the data quite well.

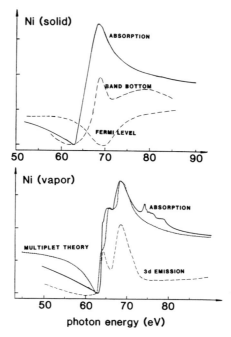

Fig. 8. Giant resonance phenomena in nickel. Top panel: total absorption (Ref. 38) and 3d partial cross sections (Ref. 39). Bottom panel: total absorption and 3d partial cross section (Ref. 40). Theory from Ref. 41.

Now focus attention on the metallic valence band photoionization cross sections. Notice that emission from occupied states near the bottom of the 3d band tracks the total absorption and is atomic-like whereas emission from 3d states near the Fermi level shows a distinct anti-resonance. I have surveyed the literature and discovered that this peculiar effect occurs in all the 3d transition metals to the right of chromium. The Fermi level emission (which corresponds to initial states near the middle and top of the band for Mn and Ni, respectively) may be explicable using a result obtained by Davis and Feldkamp[30]. They showed that a Fano effect occurs in a purely solid state model consisting of an unfilled s-band and a single deep core level. The role of the "discrete" excitation is played by absorption from the core

334

level to the Fermi level. The band emission shows an anti-resonance.

Unfortunately, this model shows an anti-resonance for emission from all parts of the occupied band - including the band bottom. Hence, the question remains: why do extended 3d-states at the band bottom act like free nickel atom states? A possible solution goes back (at least) to the work of Combescot and Nozieres[42] who were concerned with the response of a free electron solid to the sudden creation of a core hole. This is a many-body problem although the answer is easy to guess. The core hole screens itself by drawing electrons towards itself, both spatially and energetically. This tends to distort the density of states and piles up spectral weight at the bottom of the band. If the core hole potential is strong enough it "splits off" a discrete atomic-like state below the band edge. We infer from this argument that valence band wave functions at the bottom of the band are almost condensed into atomic states themselves in the presence of a core hole. Although no detailed calculations have been performed it appears that one might be in the interesting situation where the <u>direct</u> 3p/3d Coulomb interaction collapses extended states into localized states while the corresponding exchange interaction pushes the oscillator strength up into the giant resonance region.

The story is rather different for the transition metal elements to the left of manganese. In these materials the exchange interaction of the valence band with the 3p core hole is still large but the Coulomb interaction among the valence electrons is smaller than for the later members of the series. The latter statement means that the 3d electrons in, say, vanadium are rather less atomic like in the ground state than the 3d electrons in nickel. The best available data is for chromium (Figure 9) for which the atomic giant resonance is very much in evidence. However, neither the relevant atomic Fano calculation[41] nor the simple solid state model calculation[30] (both broadened to account for the unoccupied chromium d-band

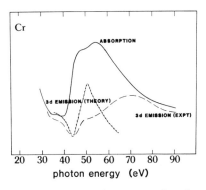

Fig. 9. Giant resonance phenomena in chromium: total absorption (Ref. 38) and 3d partial photoionization cross section (Ref. 43). Theory is described in text.

width) adequately describe the peculiar "delayed onset" of the 3d emission from this metal.

At present, I can offer only a (barely credible) suggestion that might be applicable to this situation. Oh and Doniach[44] studied the problem of resonant photoemission in the presence of core hole screening processes. However, instead of screening by a quasi-localized orbital as discussed above for nickel, they considered electron gas screening mechanisms such as plasmon and particle-hole excitation. They find that under certain circumstances oscillator strength in the main emission line indeed is shifted from the nominal peak maximum at energy ε to energy $\varepsilon + \hbar\omega_p$ where $\hbar\omega_p$ is the plasmon energy. Unfortunately, chromium is far from a free electron metal with a well-developed collective plasma excitation. Instead, its "plasmon" is formed from weakly coupled low frequency interband transitions and an analysis along these lines is problematical. Nonetheless, my main point is that giant resonance processes persist in the early 3d transition metals (and very likely also in the nominally band-dominated 4d and 5d transition rows) but are significantly modified from a simple atomic picture. A systematic study of precisely how this dramatic effect washes away as solid state effects become increasingly important would be quite interesting - both theoretically and experimentally.

CONCLUSION

In this report I have tried to analyze the problem of giant resonance absorption and photoemission from the point of view of condensed matter physics. The basic phenomenon may be understood in a very appealing manner by the use of density functional methods in a real-space approach to the dielectric response of inhomogeneous electron systems. The resonance effect now is a standard tool in the condensed matter electron spectroscopic community and is used to probe many aspects of the electronic structure of solids that are inaccesible by other means. Here, I have focused only on metallic solids. The reader is invited to explore related work in compound semiconductors[45] and small metal particles[46] and, indeed, invent new applications himself!

I gratefully acknowledge the support of the US Department of Energy for my research on this subject under Grant DE-FG05-86ER45243.

REFERENCES

1. P. Rabe, K. Radler and H. W. Wolff in Vacuum UV Radiation Physics, edited by E. E. Koch et al. (Vieweg-Pergamon, Berlin, 1974), p. 247.
2. D. J. Rowe, Nuclear Collective Motion (Metheun, London, 1970).
3. D. Pines and P. Nozieres, The Theory of Quantum Liquids (W. A. Benjamin, New York, 1966).
4. H. P. Kelly, S. L. Carter and B. E. Norum, Phys. Rev. A25, 2052 (1982). See also the contribution of C. Clark to this volume.
5. H. P. Kelly, Phys. Scr. 21, 448 (1980) and this volume.
6. C. A. Nicolaides in Advanced Theories and Computational Approaches to the Electronic Structure of Molecules, edited by C. E. Dykstra (Reidel, Dordrecht, 1984), pp. 161-184 and this volume.
7. P. Hohenberg and W. Kohn, Phys. Rev. 136, B864 (1964); W. Kohn and L. J. Sham, Phys. Rev. 140, A1133 (1965).
8. For a review, see Theory of Inhomogeneous Electron Gas, edited by S. Lundqvist and N. H. March (Plenum, New York, 1983).
9. C. S. Wang and B. M. Klein, Phys. Rev. B24, 3417 (1981) and references therein.
10. P. B. Roulet and P. Nozieres, J. Phys. (Paris) 29, 167 (1968).

11. M. Ya. Amusia and N. A. Cherepkov, Case Studies in Atomic Physics 5, 47 (1975).
12. G. Wendin in Photoionization and Other Probes of Many-Electron Interactions, edited by F. J. Wuilleumier (Plenum, New York, 1976).
13. F. Bassani and G. Pastori Parravicini, Electronic States and Optical Transitions in Solids (Pergamon, Oxford, 1975).
14. A. Zangwill and Z. Levine, Am. J. Phys. 53, 1177 (1985).
15. W. Hanke, H. J. Mattausch and G. Strinati in Electron Correlations in Solids, Molecules and Atoms, edited by J. T. Devreese and F. Brosens (Plenum, New York, 1983), p. 289.
16. R. Resta and A. Baldereschi, Phys. Rev. B23, 6615 (1981).
17. L. I. Schiff, Quantum Mechanics (McGraw-Hill, New York, 1968), 3rd edition, p. 265. See also W. Brandt and S. Lundqvist, Arkiv Physik 28, 399 (1964).
18. A. Zangwill and P. Soven, Phys. Rev. Lett. 45, 204 (1980); A. Zangwill and P. Soven, Phys. Rev. A21, 1561 (1980).
19. A. Zangwill in Atomic Physics 8, edited by I. Lindgren, A. Rosen and S. Svanberg (Plenum, New York, 1983), p. 339; A. Zangwill in EXAFS and Near Edge Structure III, edited by K. O. Hodgson, B. Hedman and J. E. Penner-Hanh (Springer, Berlin, 1984), p. 13.
20. G. Wendin in New Trends in Atomic Physics, edited by G. Grynberg and R. Stora (North-Holland, Amsterdam, 1984), p. 555.
21. D. C. Langreth and M. J. Mehl, Phys. Rev. B28, 1809 (1983).
22. A. Zangwill and P. Soven, J. Vac. Sci. Tech. 17, 159 (1980).
23. Z. Crljen and G. Wendin, Phys. Scr. 32, 359 (1985) and to be published.
24. D. A. Liberman and A. Zangwill, Comp. Phys. Commun. 32, 75 (1984).
25. M. J. Stott and E. Zaremba, Phys. Rev. A21, 12 (1980).
26. H. Kohl and R. M. Dreizler, Phys. Rev. Lett. 56, 1993 (1986).
27. E. K. U. Gross and W. Kohn, Phys. Rev. Lett. 55, 2850 (1985).
28. U. Fano, Phys. Rev. 124, 1866 (1961).
29. A. Shibatai and Y. Toyozawa, J. Phys. Soc. Jpn. 25, 335 (1968).
30. L. C. Davis and L. A. Feldkamp, Phys. Rev. B23, 6239 (1981).
31. D. Wieliczka, J. H. Weaver, D. W. Lynch and C. G. Olson, Phys. Rev. B26, 7056 (1982).
32. A. Fujimori and J. H. Weaver, Phys. Rev. B31, 6411 (1985).
33. J. L. Dehmer, A. F. Starace, U. Fano, J. Sugar and J. W. Cooper, Phys. Rev. Lett. 26, 1521 (1971).
34. J. W. Allen, L. I. Johansson, I. Lindau and S. B. Hagstrom, Phys. Rev. B21, 1335 (1980).
35. B. Johansson, Phys. Rev. B19, 6615 (1979).
36. J. E. Muller and J. W. Wilkins, Phys. Rev. B29, 4331 (1984).
37. R. D. Leapman, L. A. Grunes and P. L. Fejes, Phys. Rev. B26, 614 (1982).
38. B. Sonntag, R. Haensel and C. Kunz, Solid State Commun. 7, 597 (1969).
39. J. Barth, G. Kalkoffen and C. Kunz, Phys. Lett. 74A, 360 (1979).
40. E. Schmidt, H. Schroder, B. Sonntag, H. Voss and H. E. Wetzel, J. Phys. B16, 2961 (1983).
41. L. C. Davis and L. A. Feldkamp, Solid State Commun. 19, 412 (1976).
42. M. Combescot and P. Nozieres, J. Phys. (Paris) 32, 913 (1971).
43. J. Barth, F. Gerken and C. Kunz, Phys. Rev. B31, 2022 (1985).
44. S. J. Oh and S. Doniach, Phys. Rev. B26, 1859 (1982).
45. A. Franciosi, S. Chang, R. Reifenberger, U. Debska and R. Riedel, Phys. Rev. B32, 6682 (1985).
46. M. J. Puska, R. M. Nieminen and M. Manninen, Phys. Rev. B31, 3486 (1985); W. Ekhardt, Phys. Rev. B31, 6360 (1985).

PHOTON AND ELECTRON - STIMULATED DESORPTION FROM RARE EARTH OXIDES

G. M. Loubriel

Sandia National Laboratories
Albuquerque, New Mexico 87185

ABSTRACT

Resonances in the photon-stimulated desorption (PSD) spectra of La, Ce, Pr, Sm, Er, Tm and Yb oxides are reported at photon energies near their 4d edges and at the Er(3d) edge. These resonances, also seen in Soft X-ray Absorption (SXA), arise from excitations of d electrons to the 4f shell. Comparisons of SXA and PSD show how PSD can be used to determine surface valency. ESD data from Ce and Er oxides is used to establish the importance of the 5p level in the electronic pathways which lead to desorption.

INTRODUCTION

The electronic structure of rare earth elements and their compounds has been recently studied with renewed vigor.[1-6] These materials have extremely varied physical properties and are used in the automobile, petroleum, photographic, nuclear and vacuum tube industries. The recent interest in them has stemmed from their uses as hydrides and oxygen storage devices and from their mixed valency properties.[3] The majority of recent studies of rare earth materials has focused on resonant processes seen in various spectroscopies such as photoemission,[2,3] Appearance Potential Spectroscopy[4] and Soft X-Ray Absorption (SXA).[7,8] Our goal in this paper is to show how these resonances appear in Photon[9] and Electron - Stimulated[10] Desorption (PSD and ESD) experiments. We conclude that resonant PSD is an effective monitor of surface 4f level occupancy: it not only shows that the desorbed ion (in our case H+) bonds to a given rare earth atom, but it also gives the atom's valency. For the problems of cerium and erbium oxidation we show that while Ce may have (under certain oxidation conditions) a +3 surface layer with subsurface layers of +4 valency, erbium does not show such a strong surface effect. While this study does not claim to be a definitive oxidation study on erbium, the results suggest that the surface of erbium oxide is stable with Er atoms of 3+ valency.

EXPERIMENTS

The PSD experiments were carried out using beam lines I-1 (4 degree), I-2 (8 degree) and III-3 (Jumbo) at the Stanford Synchrotron Radiation Laboratory (SSRL), which cover the energy ranges of 64-800 eV, 5-34 eV and 500-4000 eV respectively. The ESD and PSD ion measurement apparatus are those previously described[9,10] but with modifications to the PSD system necessary to insure electron and photon rejection. The total photo-electron yield (PEY) was measured with a spiraltron whose front end was biased positive. This PEY is taken since it should be proportional to the bulk SXA. Both PSD and PEY measurements were normalized to the incident photon flux which was continuously monitored.

Polycrystalline Ce and Er films were used as samples. Both were cleaned by prolonged Ar+ ion bombardment and subsequent oxidation to produce thick oxide layers. For erbium, 90 Langmuirs of O_2 were passed through a hot filament to provide active oxygen and increase the oxidation rate. This process is probably the cause of our trace F+ desorption. On the other hand, the large amounts of H+ desorption were probably due to bulk hydrogen which segregated to the surface. The oxides of several rare earths were prepared starting with the powder oxides of La, Pr, Sm, Er, Th and Yb which were mixed in a methanol slurry and deposited on a tantalum film. For Sm we used a foil which was exposed to air to produce a thick oxide layer.

The possibility that photoemitted electrons (mainly composed of inelastic electrons) themselves cause desorption processes, as they cross the surface, cannot be ruled out a priori. This mechanism would cause desorption yields of all ions to be identical and the desorption spectrum merely to reflect the absorption of photons deep inside the bulk of the material studied. We tested this possibility experimentally, as follows: First we measure the total current escaping the sample if no field is present. This current turns out to be (given reasonable cross section estimates) insufficient to cause the desorption signals observed unless the negative bias voltage (usually -700 V) used for measuring the positive ions deflects the electrons, causing a multiplicative effect on the current (by a factor of ~ 1000). If such were the case, lowering the bias voltage to ~ 50 V would allow most of the electrons to escape and would reduce the ion yields. We observe an ion yield at ~ 50 V bias which was within 10% of that at higher voltages. Other authors[11] have reached the conclusion that secondary electrons of energies of ~ 5 eV are responsible for the desorption as opposed to direct processes. The major evidence for this conclusion is that when oxidized Yb is deposited on Sm the H+ PSD yield shows desorption peaks at energies corresponding to the 4f levels of the Sm. As we will see the thresholds for ESD from these samples are of the order of 20 eV and this energy is well above those for secondary electrons. A possible interpretation of their data is that the Yb does not completely cover the Sm, as can be seen in their photoemission spectrum of their sample.

RESULTS - PSD

The SXA of various lanthanides have been measured for photon energies around the 4d threshold. The peaks observed are due to transitions of the type:

$$4d^{10}\ 4f^n \longrightarrow 4d^9\ 4f^{n+1}$$

The multiplicity of peaks is the result of large spin orbit splittings due to the exchange interaction between the 4f electrons themselves and between these and the 4d hole.[12] The peak energies and intensities have been calculated for many of the lanthanides using intermediate coupling and show that the number of peaks and their splittings are a strong function of the assumed 4f shell occupation (and valency).[12] The best example of this change in absorption with valency is afforded by the SXA of Ce (Ref. 13), which can exist as CeO_2 ($4f^0$) or Ce metal and Ce_2O_3 ($4f^1$). A large change in the SXA spectrum in going from CeO_2 to Ce metal is evident in the number of peaks and energy splittings. The SXA of CeO_2 ($4f^0$) differs less from that of La_2O_3 ($4f^0$) than that of Ce metal. All rare earths with the same 4f occupation show similar SXA spectra.[14-16] Thus SXA is an efficient technique to monitor bulk rare earth chemistry. Shown below are examples of PSD spectra which show these resonances for La, Ce, Pr, Sm, Er and Yb compounds and illustrate how PSD can be used to study rare earth surface chemistry in much the same way as SXA is used to study bulk rare earth chemistry.

Previous work[1] shows that the PSD H+ ion yield spectra from cerium oxides has features reminiscent of the SXA spectrum of Ce (in the 100 to 113 eV photon energy range). The PSD H+ ion yield spectrum after 750 L O_2 exposure at 300 K and subsequent heating to approximately 475 K is virtually identical to the Soft X-ray Absorption spectrum (SXA) for Ce metal ($4f^1$) (Ref. 13). If the surface is oxidized without heating a new peak shows up in the PSD H+ ion yield in the region of the dominant SXA peak in CeO2 ($4f^0$) at 108.7 eV. The conclusions of Ref. 1 are that since CeO_2 is a highly efficient ionic conductor of oxygen (Ref. 17), heating results in oxygen being transported away from the surface as it is reacted at the metal/oxide interface resulting in a reduction of the surface to pre- dominantly Ce_2O_3 and that even at room temperature the surface has a mixture of +3 and +4 Ce atoms.

Figure 1 shows the SXA of oxidized Tm (Ref 15) and the PEY and PSD of H+ ions. All three curves exhibit three peaks of similar intensity ratios and energy spacings. This agreement shows that the initial process which leads to H+ desorption at the surface and electron emission from the bulk is photon absorption by Tm atoms in a +3 state. Specifically, H+ atoms at the surface were bonded to Tm atoms in a $4f^{12}$ configuration. Small disagreements in peak intensities and peak energies can probably be traced to different photon energy resolution and to problems in monochromator calibration which, in our case, can only amount to an error of .5 eV. A third possible cause for shifts in energy is surface core level shifts. Indeed, our peak energies for Tm: 171.1, 174.2, and 178.5 differ by at most 0.5 eV from those reported using SXA spectroscopy (171.6, 174.5 and 178.5 eV).[15]

Comparison of Tm_2O_3 spectra and those of Yb_2O_3 and Er_2O_3 show that neither a $4f^{11}$ or $4f^{13}$ configuration is possible for the Tm surface.

As in the case of Tm, the spectra from Er and Yb oxides (Fig. 2) show excellent agreement between PSD and PEY which implies the same valency in the surface as in the bulk ($4f^{11}$ for Er and $4f^{13}$ for Yb). The peak energies for Er occur at 163.0, 164.8, 167.0, 170.7 and 174.2 eV; only slightly shifted from those reported using SXA. For Er we also observe a sharp edge (0.5 eV total width) at 162.8 eV. The Yb spectrum shows a main peak at 180.6 with a small peak at 172.4. SXA spectrum of Yb shows peaks at 179.6 and a peak, slightly larger than ours, at 171.3 eV.

Figure 3 shows the PEY and PSD of H+ ions from Pr and Sm oxides. These spectra differ from those previously shown in that small pre-edge resonances can be observed. These resonances are expanded in Fig. 3c and occur in Sm at photon energies of 147 to 132 eV and in Pr at 106 to 120 eV. The pre-edge structure is an excellent indicator of the valency of the rare earth atom being probed. The low ion signal in our PSD experiments did not allow comparison of pre-edge structure with that observed in PEY, although the PEY agrees well with SXA experiments.[13,15] For Pr we find pre-edge structures at 106.2, 107.2, 108.1, 108.4, 109.7, 110.2, 110.6, 111.1, 112.4, 113.6, 114.9, 115.3, 116.1, 116.7, 116.9, 117.5 and 118.5 eV and large peaks at 123.6 and 131.0 eV. Due to our good photon energy resolution, 0.12 eV, we are the first to show some of these structures (see Fig. 3c). For Sm we find pre-edge structures at 126.1, 127.1, 127.6, 128.6, 129.9, 130.5 and 131.7 eV and large peaks at 136.7, 140.0 and 147.9 eV which have shoulders at 133.0, 133.4, 135.0 and 136.7 eV. Again, agreement between PSD and PEY show similar oxidation states in the surface as in the bulk for Sm and Pr oxides.

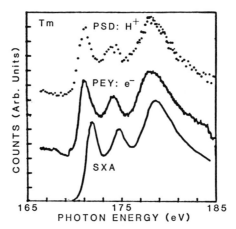

Figure 1   Comparison of the PSD H+ ion yield, the PEY from Tm_2O_3 and the SXA of Thulium oxide (solid line, ref. 15). These show the typical absorption spectrum of $4f^{12}$ compounds.

Figure 4 shows the SXA of oxidized lanthanum (Ref. 15) and the PSD H+ ion yield from La_2O_3. Note first that the SXA from La_2O_3 ($4f^0$) has a peak at similar energy as the CeO_2 ($4f^0$) SXA spectrum. This similarity shows the sensitivity of SXA to 4f occupancy. Also, note that the SXA and PSD spectra in Fig. 4 are similar except that the smaller peak at ~ 96.5 eV is missing (the 0.2 eV photon energy increments of this region of the PSD spectrum may have missed the peak whose FWHM is expected to also be

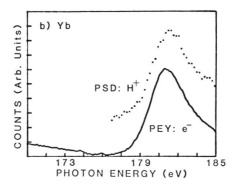

Figure 2a   The PEY and PSD - H+ ion spectra from Er_2O_3.

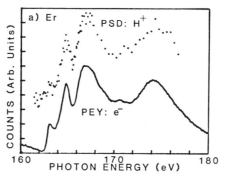

Figure 2b   The PEY and PSD - H+ ion spectra from Yb_2O_3.

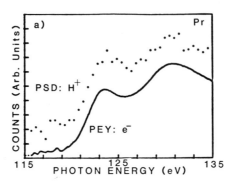

Figure 3a   The PEY and PSD - H+ ion spectra from Pr_2O_3.

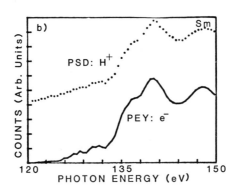

Figure 3b   The PEY and PSD - H+ ion spectra from Sm_2O_3.

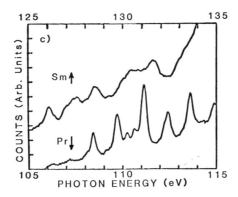

Figure 3c   Pre-edge structures in the PEY of Pr_2O_3 and Sm_2O_3.

0.2 eV) and for the small signal to noise ratio in PSD which was caused by a large H+ desorption background. Also a 0.7 eV difference in peak positions is seen between PSD and SXA. This shift is not considered significant, since either monochromator error or surface core level shifts might be the cause of such a small shift.

Fig. 5 shows the H+ and F+ PSD yields and the photo electron yield (PEY) from erbium oxide in the M5(3d5/2) and M4(3d3/2) spectral region. The curves are normalized to have the same intensity at the M5 peak; experimentally the H+ to F+ count rate ratio was 10/1. There were no other peaks in the mass spectra. Any OH+, O+ or Er+ emission present was at least two orders of magnitude lower than the H+ yield.

The spectra in Fig. 5 are almost identical. The M5 region includes peaks at 1402.4 and 1404.5 with a shoulder at 1407.5 eV while the M4 region has a single maximum at 1445.5 eV. The similarity of the spectra implies that the erbium atoms sampled in all cases (bulk atoms for PEY and, neglecting desorption due to secondary electrons, surface atoms for PSD) have the same 4f occupancy and perhaps the same valency. Possible surface oxide states of Er metal ($4f^{11}$), are ErO($4f^{11}$), Er_2O_3($4f^{11}$) and ErO_2($4f^{10}$). Only the 3+ valency Er_2O_3 is known to form in the bulk and thus, we would expect our PEY results to be similar to the absorption spectrum of Er_2O_3. Such a comparison is carried out in Fig. 6 for the M5 edge. Note that the SXA and PEY spectra are similar in numbers of peaks, peak splittings and peak intensities, with the only discrepancy being the difference in experimental photon energy resolution. To illustrate this point, we also show in Fig. 6 the calculated absorption line spectrum of Er3+ (Ref. 13) where the lines have been broadened using gaussians of 2.0 eV full width at half maximum to yield reasonable agreement with our PEY data. In similar fashion, a width of 1.5 eV agrees with the SXA spectrum, the major discrepancy being that the theory underestimates the height of the first peak relative to the most intense feature. Nonetheless, the similarity of SXA for Er_2O_3 and our PEY implies that the bulk of our sample is Er_2O_3 with Er in a 3+ valency state.

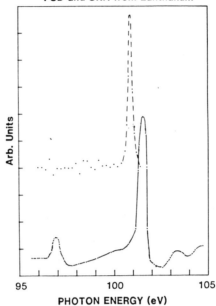

Figure 4   Comparison of the PSD H+ ion yield from La (dotted line) and the SXA of lanthanum (solid line, ref. 15). These show the typical absorption spectrum of $4f^0$ compounds visible here and in the absorption spectrum of CeO_2: two peaks separated by ~ 5.0 eV and with an intensity ratio of ~ 4.

Since the H+ and F+ PSD spectra are almost identical to the PEY results we must conclude that the surface Er atoms from which H+ and F+ desorb have 3+ valency. This valency could be produced by surface species of different stoichiometries, for example, ErOF, Er_2O_3...H and ErOF...H (where ... implies a weak-covalent bond and not much perturbation of 4f occupancy) all have Er(3+) valency. We also do not know the state of the surface prior to hydrogen and fluorine attack since all the surface molecules mentioned could be formed from either Er_2O_3 or ErO surfaces (by substitutional adsorption or not). We favor the hypothesis of an Er_2O_3-like surface since an ErO surface implies that all of the hydrogen on the system would be bonded to fluorinated ErO (as in ErOF...H) and never to ErO (as in ErO...H), and since it is known that fluorine readily replaces oxygen atoms in Er_2O_3 to produce ErF_3 (Ref. 18). Also, the observation of negligible O+ desorption yield is consistent with an Er_2O_3-like surface in that previous PSD studies[4] show no O+ yield from Ce_2O_3.

Not shown in Fig. 6 is the M4 region of the PEY and SXA. It is important to note that the ratio of the total intensities of the M5 to M4

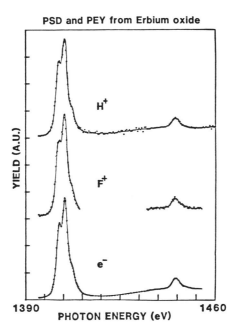

Figure 5   The PSD spectra of H+ and F+ from oxidized
            erbium is shown together with the PEY
            of the same sample across the erbium M5 and M4
            edges. Close agreement of all curves shows
            that bulk and surface Er atoms have the same
            valency.

regions of the spectra is 8/1 in PSD and perhaps larger in SXA. This result disagrees with the ratio (6/4) of the 2j+1 degeneracies of the 3d5/2 and 3d3/2 levels but agrees more closely with the theoretical (Ref. 13) ratio of 3.64. The fact that there is disagreement with the degeneracy ratio shows that the edge spectrum is not due to 3d to continuum transitions (as in x-ray photoemission) but due to a line absorption spectrum (the calculations in Ref. 13 are carried out assuming 3d to 4f transitions only).

RESULTS - ESD

The ESD spectra of Ce and Er oxides were measured in the 10 - 200 eV energy range. In both cases we were able to measure H+ and O+ desorption while for Er we also see H2+, OH+ and F+ ions. Typical spectra for Ce and Er are shown in Fig. 7 where the thresholds for desorption are compared to those obtained in PSD. The first point to notice is that the PSD spectrum exhibits a sharp peak and decays quickly thereafter while all the ESD spectra increase constantly. This difference is due to the inherent cross section dependence of electrons and photons at an edge. Due to this large background at excitation energies > 100 eV the giant resonances observed in SXA at 120 eV (Ce) and ~ 160 eV (Er) were not distinguishable in ESD.

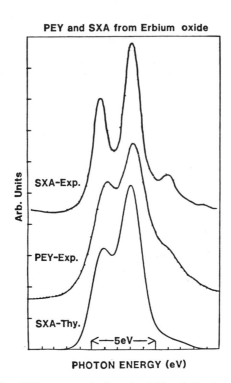

Figure 6. The PEY spectrum from oxidized Er is shown together with the SXA spectrum[14] of Er_2O_3 and a theoretical calculation[13] of the 3d to 4f absorption line spectrum of Er+3. The SXA data has been shifted by 3.75 eV to align major features in the spectrum, for this reason no energy scale is shown. The agreement seen shows that the bulk of our sample is Er_2O_3.

The second point to be made from the data shown in Fig. 7 is that the ESD H+ ion threshold from Ce is ~ 18.5 eV (after expansion of the yield scale by a factor of 10 a very sharp threshold is indeed observed). This threshold agrees with the PSD threshold of H+ ions from Ce and with the ionization threshold of the Ce 5p, or O2s levels in that sample[1]. Thus, we assign the initial step in the desorption process (at low excitation energies) to ionization of the Ce 5p or O2s levels. At higher excitation energies (> 100 eV) we have also argued previously[1] that ionization of 4d levels results in Auger cascades that create holes in the low lying core levels (Ce 5p or O2s) and that desorption then proceeds as at lower energies.

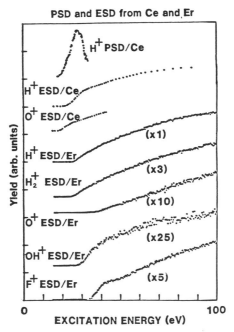

Figure 7   Low excitation-energy desorption spectra from Ce and Er oxides.

At 100 eV incident electron energy the relative yields of these species desorbed from oxidized Er, species are 100 for H+, 36 for H2+, 8 for O+, 4 for OH+ and 24 for F+ desorption. These ratios are not equal to those observed in PSD due to sample and sample treatment differences. The thresholds are at ~ 25 eV for H+ and H2+, 28 eV for OH+, 31 eV for F+ and between 33 eV and 37 eV for O+. The thresholds correlate with specific electronic transitions. The easiest one to explain is the F+ desorption threshold at 31 eV which is undoubtedly due to ionization of the F2s level (at ~ 31 eV). The H+ and H2+ thresholds at ~ 24 - 25 eV can be explained as being caused by ionization of the O2s level (at ~ 24 eV) or as due to Er 5p to 5d excitations[19] with a threshold at ~ 25 eV. The O+ ESD threshold occurs at an excitation energy between 34 and 37 eV. Compared to the O+ ESD threshold from Ce (taken under ~ 10^{-6} Torr O2 pressure) of ~ 18 eV the high threshold for O+ desorption from Er is remarkable. This threshold cannot be correlated with any of the features of the optical absorption spectrum of Er (Ref. 19) or with photoemission peaks of oxidized Er (Ref. 20 shows the binding energies of the Er 5p3/2, 5p1/2 and 5s levels in a Er surface oxide layer to be 26.2, 32.2 and 53.2 eV respectively). The electron energy loss spectrum of oxidized Er (103 L O2) shows[21] a peak at 39.5 loss energy which is attributed to excitations from the 5p level. The peak starts at ~ 33 eV loss energy and could be the cause for the O+ desorption. The concave upwards shape of this curve is reminescent of desorption of H+ from sites that afford efficient screening of the initial excitation[22] a fact which is known to occur in rare earth oxides.[23]

We must point out that the desorption thresholds for Er oxide, shown above, cannot involve the 5s level since the binding energy of the latter is 53.2 eV.[20]

CONCLUSION

We have shown the usefulness of comparisons between PSD and SXA spectra to gain chemical insight into specific local bonds. It has been shown[1] that for Ce oxidized at 300 K the Ce valency of the surface oxide is 3+ with subsurface layers of 4+ valency. In contrast to Ce this study demonstrates that the oxides of Pr, Sm Er, Tm, and Yb have the same surface and bulk valency. These resonances, which give valency information in PSD, are difficult to see in ESD. ESD has been used to determine that the 5p level in the rare earths is most likely involved in the electronic steps which give rise to desorption of surface species.

ACKNOWLEDGEMENTS

We wish to acknowledge useful discussions with C. C. Parks, B. E. Koel and M. L. Knotek and the technical assistance of J. M. Borders. This work was supported by the U. S. Department of Energy (DOE) under contract No. DE-AC04-00789. The PSD experiments were conducted at the Stanford Synchrotron Radiation Laboratory which is supported by NSF grant number DMR-77-27489.

REFERENCES

1. B. E. Koel, G. M. Loubriel, M. L. Knotek, R. H. Stulen, R. A. Rosenberg and C. C. Parks, Phys. Rev. B25, 5551 (1982).

2. A. Zangwill and P. Soven, Phys. Rev. Lett. 45 , 204 (1980).

3. J. W. Allen, L. I. Johansson, I. Lindau and S. B. Hagstrom, Phys. Rev. B21, 1335 (1980).

4. D. Chopra, G. Martin, H. Naraghi and L. Martinez, J. Vac. Sci. Technol. 18, 44 (1981).

5. G. Loubriel, M. L. Knotek, R. H. Stulen, B. E. Koel and C. C. Parks, J. Vac. Sci. and Technology A1, 1145 (1983).

6. G. Loubriel and C. C. Parks, J. Less Common Met. 93, 213 (1983).

7. W. Gudat and C. Kunz, Phys. Rev. Lett. 29, 169 (1972).

8. H. W. Wolff, R. Bruhn, K. Radler and B. Sonntag, Phys. Lett. 59A, 67 (1976).

9. M. L. Knotek, V. O. Jones and V. Rehn, Phys. Rev. Lett., 43, 300 (1979) and Surf. Sci. 102, 566 (1981).

10. M. L. Knotek, Surf. Sci. 91, L17 (1980) and Surf. Sci. 101, 334 (1980).
    M. L. Knotek and P. J. Feibelman, Phys. Rev. Lett. 40, 964 (1978).

11. J. Schmidt-May, F. Senf, J. Voss, C. Kunz, A. Flodstrom, and R. Stockbauer, p94. Desorption Induced by Electronic Transitions, DIET II, W. Brenig and D. Menzel Eds., Springer-Verlag, New York, 1984.

12. R. Haensel, P. Rasbe and B. Sonntag, Solid State Commun. 8, 1845 (1970).

13. J. Sugar, Phys. Rev. B5, 1785 (1972) and Phys. Rev. A6, 1764 (1972).

14. E. A. Stewardson and J. E. Wilson, Proc. Phys. Soc., London, A69, 7 (1956).

15. V. A. Fomichev, T. M. Zimkina, S. A. Bribovskii and I. I. Zhukova, Sov. Phys. Solid State 9, 1163 (1967).

16. D. W. Fischer and W. L. Baun, Advances in X-Ray Anal., 11 230 (1968) and J. Appl. Phys. 38, 4830 (1967).

17. See for example: Superionic Conductors, edited by G. D. Mahan and W. L. Roth (Plenum Press, New York, 1976).

18. The Rare Earths, F. H. Spedding and A. H. Daane (R. E. Krieger, Publ Co., Inc., Huntington, NY, 1961).

19. D. H. Tracy, Proc. Royal Soc., London A357, 485 (1977).

20. B. D. Padalia, W. C. Lang, P. R. Norris, L. M. Watson and D. J. Fabian, Proc. Royal Soc., London A354, 269 (1977).

21. E. Bertel, F. P. Netzer and J. A. D. Matthew, Surf. Sci. 103, 1 (1981).

22. J. A. Kelber, M. L. Knotek, Surf. Sci. (to be published).

23. M. Cardona and L. Ley, Editors, Photoemission in Solids, Vol. II, (Springer, Berlin, 1978) p. 252.

APPLICATIONS OF ELECTRON ENERGY LOSS SPECTROSCOPY TO GIANT RESONANCES IN RARE EARTH AND TRANSITION METAL SYSTEMS

J. A. D. Matthew

Department of Physics
University of York
York   YO1 5DD   UK

INTRODUCTION

The electron gun is sometimes termed the "poor man's synchrotron" - it is a relatively cheap high flux particle source capable both of high energy resolution and high spatial resolution. Its disadvantage relative to monochromated synchrotron sources is that it is not as selective in its excitation. In principle a monoenergetic beam of energy Ep creates all excitations and ionisations requiring an energy loss ΔE less than Ep; it is in some ways analogous to a white radiation source, but in addition it can reach excited states that photons cannot reach, and it is that aspect which will be the main topic of this review.

Here we consider electron energy loss spectroscopy (EELS) in reflection mode of giant resonances excited at the surfaces of rare earth and transition metal systems. Firstly we review briefly the interactions of electrons with matter and typical experimental set ups for such studies. After examining new insights into the properties of 4d $\rightarrow$ 4f giant resonances in the rare earths we examine evidence for corresponding 4p $\rightarrow$ 4f resonances. Finally we briefly assess recent work on giant resonances at transition metal surfaces.

THE INTERACTION OF ELECTRONS WITH SURFACES

The nature of inelastic scattering of electrons by atoms, solids or surfaces depends on whether their velocities are "fast" compared to the orbital speeds of the target electrons or merely comparable to them. In the former case the Born approximation leads to a direct dependence of differential scattering cross section on momentum transfer, which for a one electron excitation in an atom is given in atomic units by

$$d\sigma = \frac{4}{v^2} \frac{1}{\theta^2 + \theta_E^2} \frac{1}{q^2} |\langle f|\exp(i\underline{q}\cdot\underline{r})|i\rangle|^2 \, d\Omega \quad - (1),$$

where θ is the scattering angle, $\theta_E = \Delta E/(2Ep)$, v the incident electron speed, and the scattering vector $\underline{q} = \underline{k}-\underline{k}'$ with $\underline{k}(\underline{k}')$ the incident (scattered) wave vector; $\underline{r}$ is the position vector of the electron and the matrix element connects the initial and final states of the system. The inner core transitions from solid surfaces that we are concerned with here are quasi-atomic, and equation (1) will be approximately valid, although the many body character of the resonances is important in determining the details. Expanding $\exp(i\underline{q}.\underline{r})$ gives a leading dipole term at low momentum transfer (small θ), so that $d\sigma/d\Omega$ depends to a good approximation on the same matrix elements as X-ray absorption cross sections; however, if $|\underline{q}|$ is comparable to $1/r_c$, where r_c is the mean core orbital radius, higher order transitions become important, and the electron energy loss spectrum contains transitions additional to those observed in photon absorption. Once the incident energy Ep is only slightly larger than the excitation energy ΔE, the Born approximation ceases to be valid, selection rules are relaxed, and an excitation free for all sets in. Inelastic electron scattering therefore directly complements photon absorption investigations of giant resonances by systematically extending the range of final excitation states.

Here we are concerned with electron energy loss spectroscopy (EELS) in reflection mode, where monoenergetic electrons incident on clean solid surfaces under ultra high vacuum conditions are scattered into electrostatic analysers (Figure 1). The electron energy distribution N(E) is measured as a function of loss energy ΔE, ie with E = Ep-ΔE. For the primary energies used in this work (200 < Ep < 3 kev) the inelastic mean free path λ varies from about two atomic layers to around 2 nm, so that experiments are highly surface sensitive. In addition the typical scattering angle is less than θ_E so that forward elastic scattering is dominant, and there must be a high angle elastic scattering before or after loss (Figure 1). For some of the systems discussed here this surface sensitivity is not of importance because giant resonances are such parochial excitations that the location of the excited atom is irrelevant, but in some systems we shall see that the environment is of some significance. Surprisingly EELS in reflection geometry has not been comprehensively reviewed, but EELS in the rare earths has been recently discussed by Netzer and Matthew (1).

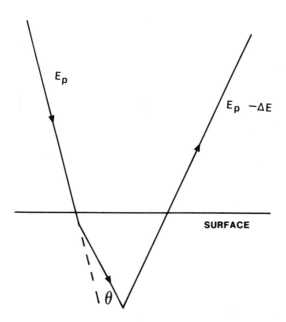

Figure 1   Schematic diagram of electron energy loss in reflection geometry showing a small angle energy loss ΔE followed by an elastic reflection.

WHAT'S NEW?

What additional insights to giant resonance phenomena does EELS provide? As discussed above rich new seams of excited states are revealed, and progress in their spectroscopic classification is discussed here. In principle this extra experimental information should provide new tests of the nature of giant resonances, and might shed light on the basic controversy over whether resonance profiles are controlled mainly by the potential well experienced by the excited electron - see Connerade (2), or by the mode of de-excitation through Fano-like coupling to other channels. Such tests prove more difficult than it might first seem, but possible future experiments are considered.

RARE EARTHS

In assessing giant resonance losses in the rare earths it is important to relate the spatial extent of the key orbitals to typical momentum transfers in electron energy loss. Table 1 compares $<r_{n\ell}>$ in au in the ground state of free Ce with the minimum momentum transfer $q_{min} = \Delta E/\sqrt{(2Ep)}$ for typical scattering conditions; $q_{min}<r_{n\ell}>$ values of order unity give indications of substantial non-dipole excitation.

(1) 4d -> 4f TRANSITIONS

$4d^{10}4f^{n} \rightarrow 4d^{9}4f^{n+1}$ transitions in the rare earths are obvious starting points for EELS studies of giant resonances (3). The simplest case is that of trivalent Yb as in Yb_2O_3, where dipolar $4d^{10}4f^{13}(^2F_{7/2}) \rightarrow 4d^94f^{14}(^2D_{5/2})$ transitions are totally dominant at high Ep, while $4d^94d^{14}\ ^2D_{3/2}$ final states become accessible as Ep is lowered (Figure 2); in contrast divalent Yb metal is

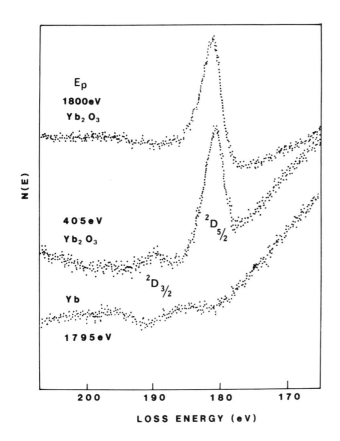

Figure 2

4d-4f electron energy loss spectra of Yb_2O_3 and Yb metal characterised only by ionisation thresholds. From a giant resonance viewpoint the results are of limited interest because the excitation energies lie below the relavent ionisation thresholds.

At the other end of the rare earth series things are spectroscopically more interesting. In La metal or La_2O_3 virtually identical $4d^{10} \rightarrow 4d^9 4f^1$ or $4d^9 \epsilon f$ transitions are observed (Figure 3). At higher Ep EELS results might be expected to correspond to the X-ray absorption. However, even at Ep = 2920 ev significant non-dipole excitation to ^1F and ^1H states is already occurring in addition to the main ^1P resonance and the sharp $^3P_1$ and $^3D_1$ transitions that derive their oscillator strength from spin-orbit interaction (4). As the primary energy is lowered a feast of new structure appears with the overall profile accounted for by the full range of $^{1,3}P$, $^{1,3}D$, $^{1,3}F$, $^{1,3}G$, $^{1,3}H$ multiplets accompanied by spin-orbit splitting. As the spectra themselves suggest and as Hartree-Fock calculations confirm the new galaxy of states all lie below or close

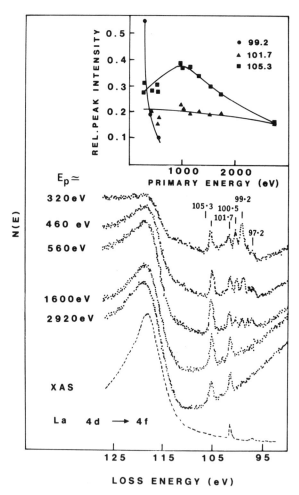

Figure 3   4d-4f electron energy loss spectra of clean metallic La at various primary energies Ep - inset plots the variation in intensity of various features as a function of primary energy.

to the 4d ionisation thresholds, and are very strongly inner well in character; little changes in the main ^1P resonance region with energy variation. This pattern has been confirmed by parallel studies of Moser *et al* (5).

Pr (Figure 4a) shows a similar pattern to La and it is recognised that the main resonance is now split into a number of components, each of which carries ^1P parentage from the basic 4d-4f interaction (6). Again the non-dipole components are on the low energy side of the main dipole resonance, and correspond to highly localised final 4f states.

In contrast, at the other end of the rare earth series, the dipole allowed transitions lie at energies lower than the principal non-dipole components; in Tm (Figure 4b) $4d^{10}4f^{12}(^3H_6) \rightarrow 4d^94f^{13}(^3G_5, ^3H_5$ and $^3H_6)$ transitions between 170 and 180 ev are allowed optically; the $^1H_5$ final state requires change of spin and is not clearly isolated. The calculations of Larkins (7) for the similar $4d^94f^{13}$ state of Yb^{++} indicate that many additional multiplets should appear between 180 and 190 ev for low Ep. None are in fact observed. This may be due to difficulties with the secondary electron background and valence losses accompanying the main dipole components, but it should be noted that these non-dipole allowed final states lie well above the $4d_{3/2}$ ionisation threshold. State dependent Hartree-Fock calculations for this system would be of interest, while by systematic variation of Ep it may be possible to see whether any broad resonances lie above 180 ev.

(2) 4p LOSSES

In the case of 4d losses EELS has built on a well defined dipolar base determined by X-ray absorption. In contrast EELS data on losses around the 4p thresholds (8,9) preceded investigation by X-ray absorption. The case of 4p excitation is complicated by breakdown of the one electron approximation through $4p^{-1} \leftrightarrow 4d^{-2}\epsilon f$ giant Coster-Kronig coupling (10). This is strongest for $4p_{\frac{1}{2}}$ holes, which are characterised by broad ill defined peaks in photoemission; $4p_{3/2}$ shows a conventional peak, but at an energy well shifted from Hartree-Fock predictions. Two possibilities arise: dipole allowed 4p-5d transitions may occur or quadrupolar 4p-4f transitions. It ought to be possible to distinguish these processes by Ep variation, but the changes that are then observed are not

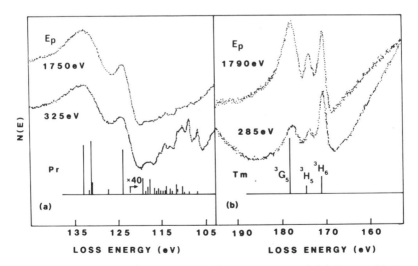

Figure 4   4d-4f electron energy loss spectrum of (a) Pr   (b) Tm

TABLE 1

Ce

(a) Orbital size(17)		(b) Momentum transfer		
Orbital	$<r_{n\ell}>$ au	Ep(ev)	ΔE(ev)	q_{min}(au)
4s	0.67	400	120	0.8
4p	0.69		210	1.4
4d	0.74	2000	120	0.4
4f	0.95		210	0.6
5d	2.64			

amenable to unambiguous determination. The reason for the dilemma is that 4p and 4f wave functions may overlap very strongly, while 4p-5d overlap is likely to be much weaker (Table 1). This implies that quadrupole excitation may dominate dipolar excitation throughout the primary range available in reflection EELS experiments. Recently Morar et al (11) have measured the 4p absorption spectrum of weakly oxidised Ce, and their results are compared with EELS and photoemission data in Figure 5. EELS yields a large peak that shows primary energy dependence, and lies below the $4p_{3/2}$ binding energy, as well as a smaller feature approximately coinciding with a satellite photoemission peak that may form part of the $4p_{\frac{1}{2}}$ spectrum. In contrast X-ray absorption has no structure at the energy of the main EELS feature, but does peak in the region of 218 ev where the other techniques have a subsidiary peak. This provides conclusive evidence that the main EELS peak is non-dipole in character - a giant 4p-4f quadrupole resonance. Such an interpretation is supported by Hartree-Fock calculations of $4p^6 \rightarrow 4p^5 4f$ transitions in La (11), which confirm that the 4f state will be collapsed, that there will be strong 4p-4f exchange coupling accompanied by a large quadrupole matrix element. Calculations also suggest that the excited 5d state will not be collapsed, so that 4p-5d exchange splitting is weak. Empirically it appears that the 4p-5d oscillator strength is concentrated around the energy of the subsidiary 4p peak in photoemission at 218 ev, an effect that can only be understood by going beyond the single particle approximation. A systematic comparison of 4p EELS and X-ray absorption throughout the lanthanide series is now required, accompanied by calculations that account for at least some of the many body character of these systems. However, the work of Morar et al (11) confirms the dominance of quadrupole 4p excitation in EELS.

The properties of 3d, 5p and 4s losses in the rare earths are discussed by Netzer et al (9) and Strasser et al (3). From the insights into the 4p excitation discussed here it is likely that 4s-4f transitions will dominate the 4s losses with large exchange interaction arising from the strong 4s-4f overlap. Although these losses are much weaker than either the 4p or 4d, they form an interesting class of resonances that deserve further study.

TRANSITION METAL SYSTEMS

3p -> 3d resonances in elements of the first transition series (12) are analagous to 4d -> 4f giant resonances in the rare earths. In the lower part of the series (Sc,Ti,V) the dipole resonances are, as in La and Pr, well above the ionisation threshold and non-dipole transitions are to be expected at lower loss energies. These have been observed by Erickson and Powell (13), but the additional multiplets are less well resolved than in the rare earths. Spin dependent electron loss spectroscopy has, however, been applied to the 3p -> 3d losses by Mauri et al (14) using an unpolarised electron beam scattered from a ferromagnetic crystal, the asymmetry in spin polarisation of the scattered electrons being observed using a Mott detector. As the primary energy decreases towards threshold, the spin asymmetry rises due to the increase in importance of exchange scattering. The need for a ferromagnetic or ferrimagnetic phase to define a preferred spin direction in the solid implies that cooling below room temperature is essential in most rare earth metals and compounds. Spin polarised electron sources may also be used with or without spin detection (15), and such techniques shall undoubtedly enhance our understanding of giant resonances over the next decade. Such techniques have great potential for investigating the rich non-dipole resonance structure in the rare earths, but thus far have only been applied to spin dependent valence excitations in ferromagnetic Gd (16).

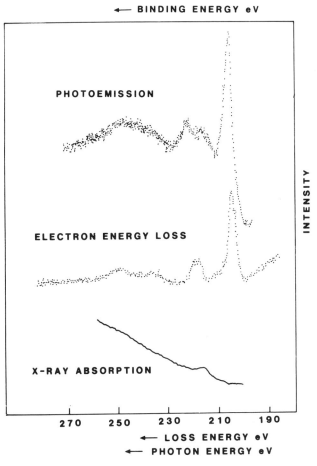

Figure 5   Comparison of the X-ray absorption spectrum, the electron energy loss spectrum, and the photoelectron spectrum of lightly oxidised Ce in the region of the 4p ionisation thresholds.

CONCLUSIONS

Electron energy loss spectroscopy is an important complement to photon absorption in studying giant resonance phenomena. It has been shown that the well known dipolar giant resonances are accompanied by a range of non-dipole transitions, while strong overlap between 4p and 4f wave functions leads to a giant quadrupole resonance to which synchrotrons can only turn a blind eye. Variation of spectral structure with primary energy helps to disentangle dipole and non-dipole effects, but spin dependent loss phenomena will hold the key to probing transitions involving change of spin.

ACKNOWLEDGMENTS

The author particularly wishes to thank Falko Netzer, Erminald Bertel, Gregor Strasser and John Rivière for helping to formulate the view of EELS and giant resonances presented here. The SERC of UK is gratefully acknowledged for supporting the work.

REFERENCES

1. F P Netzer and J A D Matthew   Handbook of Physics and Chemistry of the Rare Earths   Vol 10 (Ed K A Gschneidner Jr and L Eyring (Amsterdam:North Holland))

2. J P Connerade   J.Less.Common.Metals 93, 171 (1983)

3. G Strasser, G Rosina, J A D Matthew and F P Netzer   J.Phys.F Met.Phys. 15, 739 (1985)

4. F P Netzer, G Strasser and J A D Matthew   Phys.Rev.Lett. 51, 211 (1983)

5. H R Moser, B Delley, W D Schneider and Y Baer   Phys.Rev. B29, 2947 (1984)

6. J Sugar   Phys.Rev. B5, 1785 (1972)

7. F P Larkins   Atomic Data and Nuclear Data Tables 20, 311 (1977)

8. G Strasser, F P Netzer and J A D Matthew   Solid State Commun. 49, 817 (1984)

9. F P Netzer, G Strasser, G Rosina and J A D Matthew   J.Phys.F Met. Phys. 15, 753 (1985)

10. G Wendin   Photoelectron Spectra Structure and Bonding 45 (Springer Verlag, Berlin) 1981

11. J Morar, F P Netzer, J A D Matthew and C W Clark   (To be presented)

12. M W D Mansfield and G H Newsom   Proc.Roy.Soc. A357, 77 (1977)

13. N E Erickson and C J Powell   Phys.Rev.Lett. 51, 61 (1983)

14. D Mauri, R Allenspach and M Landolt   Phys.Rev.Lett. 52, 152 (1984)

15. J Kirschner   Phys.Rev.Lett. 55, 973 (1985)

16. D Weller and S F Alvarado   Z.Phys.B 58, 261 (1985)

17. A D McLean and R S McLean   Atomic Data and Nuclear Data Tables 26, 197 (1981)

HIGH RESOLUTION SOFT X-RAY SPECTRA OF D-F TRANSITIONS:

NEW SPECIFIC FEATURES IN THE 3d RESONANCES OF CeO_2, PrO_2 AND TbO_2

J.-M.Esteva, R.C.Karnatak

Laboratoire de Spectroscopie Atomique et Ionique
Unité Associée 040775, Bât. 350, Université Paris-Sud
91405 Orsay   and   Lure, Bât. 209d, Université Paris-Sud
91405 Orsay, France

INTRODUCTION

The 4d-f transitions in Ba and the rare earth (RE) elements exemplify the so-called "Giant resonances" and play an important role in the theory of atomic spectra[1]. As opposed to 4d-f transitions, the 3d-f transitions which cover 1 keV range have simpler features but retain certain characteristics related to the wave function collapse in the l=3 radial functions. Due to the dominant 3d spin-orbit interaction the spectral features are separated into two distinct groups. In the case of Ba, La, Tm and Yb3+ single absorption lines are clearly distinguished. The other RE show multiplet structures.

The nd-4f (n=3,4) transitions in RE vapours and solids generally show quasi-atomic character. Their spectra present spectral features which are characteristic of the number of 4f shell electrons: (i) The systematic investigation of the 4d-4f absorption spectra[2] revealed that for La, Eu and to certain extent Ce, there is an excellent agreement between the spectra in vapour phase and the solid phase and no spectral shift is observed. Similar behaviour is expected for Gd. In these cases the number of the 4f electrons in both phases remains the same. (ii) The Pr, Tb, Dy, Ho and Er vapour spectra were found to be similar to those from the corresponding Z+1 trivalent metals. For Sm, Tm and Nd vapour spectra such similarity with Z+1 metals was not observed because the corresponding Eu and Yb metals are divalent and Pm is unstable. (iii) The RE compounds showing valence change give different spectra. This is due to change in the number of 4f shell electrons. Here again the valence states in different elements leading to an iso-electronic sequence give similar spectra. For example the divalent Eu, trivalent Gd and the tetravalent Tb give similar 3d-4f spectra. Similarly some of the spectral features of the tetravalent Ce correspond well with those of the trivalent La. For RE elements near the begining, middle and end of the RE series, in certain chemical environment some modifications in the spectral features due to the instability of the 4f shell are observed. The temperature and pressure are sometime the cause of such instability in certain cases. The ground state properties of such solid related to the instability of the 4f shell, at present covers a wide field in the solid state Physics and Chemistry.

The localized nature of the 3d-4f transitions in some normal RE was also demonstrated in the past by comparing the $3d^9 4f^{n+1}$ multiplets obtained by atomic calculations with the observed spectra[3-8]. Recently the 3d-4f multiplet calculations were performed for whole of the RE series[9] and the results were found to be in good agreement with the experimental spectra observed for RE metals under ultra high vacuum conditions. These calculations yield data on the positions and the strengths of the lines corresponding to the $3d^{10} 4f^n - 3d^9 4f^{n+1}$ dipole allowed transitions. In order to compare them with the experimental spectra one needs to broaden the line strengths by taking into account the life time width of the upper state and the instrumental broadening function. Useful data on the total integrated intensities of the $3d_{3/2}$ and $3d_{5/2}$ peaks and the absorption length (λ) were obtained from the theoretical spectra by using the estimated values of the life time widths. Under these circumstances it becomes crucial to know the instrumental resolution precisely. If the instrumental resolution and other broadening functions are known, one can, by trials optimise the agreement between the theoretical prevision and experiment by choosing different life time widths.

The new data published[8,9] on the corrected 3d-4f line widths of RE were obtained by photo yield method. It is interesting to note that the observed spectra reported in ref. 8 and 9 were obtained with similar spectrometers and are quite comparable to those obtained earlier[7] with the same resolution by transmission through thin films. In this case[7] the instrumental broadening was estimated to be small and was neglected in the determination of the natural width of the 3d-4f lines. The corrected line widths in ref. 8 were obtained by taking into account the width of the rocking curve of the double crystal spectrometer as the instrumental broadening function and the width show a better agreement with the theoretical estimation of the line width by McGuire[10]. In ref. 9 the line saturation effects due to comparable photon penetration depth (absorption length) and the electron sampling depths were invoked to be the cause of the broadening of the 3d-4f lines. The estimated values for the $3d_{5/2}$ line width of La was found to be 0.4 eV, a value much below the theoretical estimate[10].

Natural line widths are fundamental in the studies of fluorescence yield and line shapes in various core level spectroscopies. In the context of the 3d core level life times in RE, Crecelius et al[11] determined the 3d level widths from the X-ray photo electron spectra (XPS) of La-Pr metals and compared them with the theoretical life time widths obtained by McGuire[10]. In the case of La the appreciable difference observed between the $3d_{5/2}$ XPS line width (1.8 eV) and the theoretical 3d level width (0.7 eV) were attributed to be due to the neglect in the theory the decay paths in which an empty short-lived 4f level is filled. For La, similar mechanism which involves the decay paths connected with the capture of two electrons in the short-lived 4f states was suggested[7] to explain the residual difference between the $3d_{5/2}$ XAS line width (1.1 eV) of La and the theoretically obtained 3d level width. These experiments pointed out the importance of the available decay paths due to the presence of the short-lived 4f states in RE.

Thus we see that the values of 3d-4f line widths of normal RE metals and oxides obtained (i) by neglecting the instrumental broadening effects or (ii) by over estimating these and (iii) invoking the line saturation effects differ considerably. Then a doubt persists whether the 3d-4f line widths of RE are really influenced by the existence in solids of further decay paths for the 3d hole in presence of an excited 4f electron. From the experimental stand point this situation is further complicated if we consider the observation with the same resolution of the narrower 3d-4f lines in RE fluorides[7] and unusually broad lines or multiplets in RE dioxides[12,13]. So interest attaches to know whether in certain chemical environment the change observed in the 3d-4f lines or multiplet widths is real. For this reason a quantitative estimation of the instrumental and other broadening effects seems to be necessary.

In this paper we will revist the problem of the resolution of double crystal (beryl) monochromator[14]. We will briefly discuss the line saturation effects usually encountered in the absorption measurements by transmission and the photo yield methods. We will give here, for the first time, a semi quantitative estimate of line saturation by comparing the 3d line width of La obtained by photo yield and the resonant photon scattering. Finally the line narrowing mechanism in the RE fluorides and the recently observed unusual broadening in the 3d-4f lines in RE dioxides[12,13] will be discussed.

Monochromator resolution

A description of the theory of double crystal spectrometer is out of scope of the present paper. We will give here some essential information related to the resolution which can be obtained with the beryl crystal in the 1 keV region. The flat polished slabs of beryl (1010) crystals are mounted as diffracting elements in the double crystal monochromator placed behind the synchrotron radiation source. A precise estimation of the resolution of a double crystal spectrometer raises certain problems. In fact, the resolution of the spectrometer can be defined as the observed broadening of an infinitely narrow line of given intensity. This in fact is the measure of angular interval within which one can not distinguish between two nearby wavelengths after Bragg reflection. In such ideal case the recorded spectrum $f_c(\theta)$ from a single crystal involves miscellaneous uncertainties in defining the Bragg angle θ. The distribution $f_c(\theta)$ is called the crystal diffraction pattern. W_c is the full width at half maximum (FWHM) of this distribution. In the case of Bragg reflections from two parallel crystals the width W_{2c} of the final diffraction pattern depends upon the shapes and the widths of the individual $f_c(\theta)$. Then, there exists a certain relationship between W_{2c} and W_c. For two parallel and identical crystals the final intensity after two reflections follows a curve which is the product of two single crystal patterns. In the case of the gaussian shape of $f_c(\theta)$ we have:

$$W_{2c} = W_c/\sqrt{2} \quad \ldots\ldots\ldots\ldots\ldots\ldots(1)$$

A quick test of the alignement of the monochromator is obtained by rotating the second crystal around an axis perpendicular to the beam direction. For a particular Bragg incidence θ on the first crystal, the intensity of the reflected beam from the second crystal as a function of the second crystal rotation passes through a maximum. This maximum corresponds to the position when the reflecting planes of the second crystal are parallel to those of the first crystal. The intensity versus crystal rotation curve is called the "rocking curve". The width W of the rocking curve gives an indication of the mesure of the resolution. In such double crystal arrangement the intensity distribution from the second crystal is in fact, given by the convolution of two individual single crystal diffraction patterns. Then the relation between W and Wc is found to be : $W=\sqrt{2}W_c$ and from (1) it becomes

$$W = 2 W_{2c} \quad \ldots\ldots\ldots\ldots\ldots\ldots (2)$$

The rocking curves for beryl crystals were obtained for 5 different values of the Bragg angle θ by rotating the second crystal around the parallel position. In Table 1 we give the relevant data on the position (in E and θ) and the width of the rocking curves (in sec., mrad and eV). The observed widths of the 1s-3p lines of Ne and Na are also included in Table 1 for comparison. In fig. 1 we plot the W(sec) against the energy positions obtained at LURE comparatively those obtained at the Standford Synchrotron Radiation Laboratory (SSRL) for beryl crystals[15]. Thus we see that the widths of the rocking curves obtained at two different laboratories are quite comparable.

TABLE 1

E (eV)	θ (deg)	tg θ	W measured (mrd)	(sec.)	(eV)	Observed 1s-3p line width (eV)
822	70.92	2.89	1.7	350	0.48	
865	63.92	2.04	1.2	247	0.51	0.50 (Ne)[a]
972	53.06	1.33	0.66	137	0.48	
1162	41.96	0.90	0.50	103	0.64	0.55 (Na)[b]
1491	31.40	0.61	0.28	58	0.68	

(a) ref.16
(b) ref.17

From Table 1 it appears clearly that the rocking curve widths are somewhat greater than the Ne and Na line widths in the same range of energy. These observations clearly demonstrate that the FWHM of the rocking curves is an approximate indication of the resolution. The reason for this comes out of the fact that when the second crystal is rotated, the rocking curve thus obtained is the convolution of the incident beam angular profile and the actual rocking curves of the first and second crystals. In our earlier experiments instead of the second crystal , only

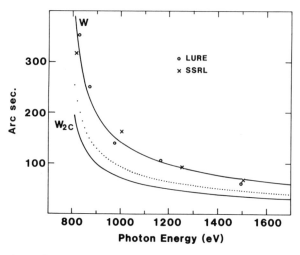

Fig.1. Resolution of two beryl crystals monochromator.
_____ ; W- experimental FWHM of the crystal rocking curves.
_____ ; W_{2c} calculated ultimate resolution (W_{2c}=W/2).
.... ; estimated resolution (from the observed 1s-3p line widths of Ne and Na).

first crystal was rotated to obtain the rocking curve. In such set up, at 865 eV the actual width of the observed rocking curve was found to be 0.76 eV. This is much higher than the 0.51 eV width of the rocking curve at 865 eV (cf Table 1) obtained by rotating the second crystal. In all these experiments, it appears that it is difficult to extract the contribution of the beam line profile and hence know the true width W_{2c} of the rocking curve which may correspond to the actual resolution of the monochromator.

We find in the literature that there is certain evidence of the radiation damage to the first crystal by intense incident beam[15]. This damage causes an appreciable broadening of the rocking curve. Recently[8] the widths of the rocking curve for the crystals similar to our beryl crystal were mesured to be varying from 0.72 eV at 800 eV to 0.96 eV at 1400 eV. The use of the Be window in our experiments considerably improved the life of our Beryl crystals. We have been using the same pair of crystals for several years without any appreciable change in the rocking curve widths.

In actual photoabsorption experiments with synchrotron (continuous) radiation, the spectral scans are obtained by rotating both crystals in such a way that the direction of the outgoing beam always remains fixed[14]. During the scan, due to availability of a continuous range of wavelengths in the incident beam, one obtains a continuous flow of indisitinguishable crystal diffraction patterns after Bragg reflections from two crystals. Under these circumstances it is rather impossible to obtain by double rotation the width W_{2c} of the rocking curve which may correspond to the actual resolution of the double crystal monochromator. In such case the observed widths of the narrowest possible lines become a guide for the estimation of the resolution W_{2c}. In the 1 keV region, among the various elements studied, the 1s-3p lines of Ne and Na were found to be narrowest so far. The widths observed with the use of a double beryl crystal monochromator were found to be 0.50+ 0.05[16] and 0.55+0.05[17] eV respectively for Ne and Na 1s-3p lines in 900-1100 eV region. These observed widths are in fact the convolution of the natural line widths and the instrumental resolution function. If for simplicity, we neglect the natural width of the 1s hole, these values can be taken as an upper limit of the resolution. The first conclusion which we draw from these mesurements is that the resolution of the monochromator is certainly better than 0.5 eV in this range of energy.

So, for an estimate of the resolution W_{2c} one has to rely either (i) upon the values of W_{2c} obtained by the relation $W_{2c}=W/2$ by supposing that the single crystal diffraction patterns are identical and have a gaussian shape or (ii) on the theoretical 1s level widths of Ne and Na. The theoretical 1s level widths given in the literature[18] are 0.24 and 0.3 eV respectively for Ne and Na. The fold of the lorentzian line of 0.3 eV width (theoretical) with a suitable gaussian line whose width is W_{2c} would yield a curve which may correspond to the actual 1s-3p observed line of Ne and Na. This procedure yields a value of about 0.35 eV for W_{2c}. It is interesting to mention that the values of W_{2c} obtained by using theoretical line widths are much larger than those obtained from the observed W. This may be due to the asymmetry of the single crystal patterns which may also not be identical. The W_{2c} curves obtained by using the theoretical level widths (dotted curve) and that obtained by using the experimental rocking curve widths ($W_{2c}=W/2$) are given in Fig. 1.

One should bear in mind that an observed line is the fold of a lorentzian line with the gaussian shape instrumental function. The observed line widths and shapes depend upon the width of the broadening function. In the case of the RE the instrumental broadening function has a width much smaller than the true line width. So the distortion of the observed lines is very small. Nearly lorentzian and Fano shape 3d-4f lines are observed for these elements.

The shapes and the relative widths of the instrumental broadening and the spectral line are some of the important factors which govern the profiles and the widths of the observed lines. The other factors which may influence the shapes and the widths are connected with the sample thickness, response of the detector and the penetration and escape depths of the incident and the outgoing particles respectively. We will describe them in the following paragraph.

Line saturation effects in photoabsorption by transmission and photoyield methods

The saturation effects due to the sample thickness in the X-ray absorption spectra have been known since long. In fact, these effects arise due to the insensitivity of the dectector to the variation of the transmitted intensity I after being attenuated by an exponential factor $e^{-\mu x}$ in passing through a sample of thickness x.

An estimate of the variation of I with μ can be obtained by differentiating :

$$I = I_0 e^{-\mu x} \quad \ldots\ldots\ldots\ldots(3)$$

partially with respect to μ, which gives:

$$\frac{dx}{d\mu} = -I_0 \left(x\, e^{-\mu x} \right) \quad \ldots\ldots\ldots\ldots(4)$$

where Io is the intensity incident on the sample.

For increasing values of x the quantity inside the parenthesis in eq. (4) passes through a maximum at $x=1/\mu$, and for a given value of μ the sensitivity dI is maximum at this value. Thus for smaller values of μ, the larger x are needed to obtain the desired maximum sensitivity in dI. For values of x greater than the optimum the detector is not able to discriminate the variations in I. This saturation effect leads to the flattening of the absorption peaks. This effect can be minimized by the choice of thinner samples. Thus the constraint due to the tedious sample preparation in transmission method leads one to use the easier photoyield method for absorption measurements[19].

The general theory of photo yield related to the migration of electrons to the surface in solid after photoabsorption is complex. In a simplified approach for the yield process we would have:

$$Y = A \frac{\mu}{\cos \theta} \int_0^\infty e^{-\frac{\mu x}{\cos \theta}} e^{-\frac{x}{L}} dx = A \frac{L}{L + \frac{\cos \theta}{\mu}} \quad \ldots \ldots \ldots \ldots (5)$$

In this expression A is the number of electrons per absorbed photon which are created in a depth L (escape depth) and are able to migrate to the surface. θ is the photon incidence angle. When the absorption length ($\lambda = 1/\mu$) is large as compared to L, the yield is approxiamately equal to AμL/cosθ. In this case the yield is directly proportional to the absorption coefficient μ. In case λ is comparable to L, then the expression (5) should be used for the estimation of the yield. This may lead to the saturation of the lines in the 3d-4f spectra of RE. Because of this saturation the observed FWHM is expected to become larger than the true width of the line.

Thole et al[9] examined the line saturation effect in the 3d-4f spectra of RE. Their approach was mainly based upon the $3d^9 4f^{n+1}$ multiplet calculations and the results were compared with the experimental yield spectra of RE. These calculations yield the absorption path lengths (λ) for each element at the maximum of the absorption line. The path lengths thus obtained varied from 29 Å for La to about 260 Å for Yb3+. In La and Gd the the electron escape depth estimated from thin films[19] was found to be of the same order of magnitude. In elements like La with very simple 3d absorption spectrum, a curve fitted to the wings of the peaks overestimated the observed peak intensity[9]. So appreciable saturation effects were suspected to occur in these elements. The observed 1.1 and 1.7 eV wide $^3D_1$ and $^1P_1$ lines of La besides their natural width are considered to contain appreciable saturation effect and the instrumental broadening. After taking account the saturation and the broadening the natural width of these lines were estimated to be 0.4 and 0.8 eV respectively.

The above mentioned observations emphasised on the importance and the implications of the saturation effects in the 3d-4f XAS obtained by the yield method. So interest naturally attaches to the precise knowledge and the quantative estimation of such effects. To this end, we performed some preliminary experiments on La by using various angles of photon incidence. We observed some variation in the relative intensity of the two main lines as a function of the photon incidence angle. This variation was some time found opposite to that expected for the increase or decrease of electron escape depth. This is most probably due to the non uniformity of the sample surface.

Some quantitative information on the line saturation effect can be obtained from the resonant photon scattering experiments. As opposed to the electron yield method, in these experiments the scattered photon is detected as a function of the incident photon energy. The spectra are scanned around the $3d_{5/2}$ and $3d_{3/2}$ resonance absorption lines. The process involved in the scattering is just complementary to the photoexcitation in which a reemission of photon due to the deexcitation of a 4f electron takes place. In this case the energy of the emitted photon is equal to that of the incident photon. This is demonstrated by the coincidence of the scattering and yield spectra peaks. For a resonance peak energy the photon penetration length is almost equal to the photon escape depth. This is equivalent to say that $\mu L = 1$ (one should remember that L can not be greater than the absorption path length).

If we suppose that the dominating Auger processes which govern the life time of the 3d hole are similar in resonant emission and the yield process, the saturation effect can be simulated by studying the change in the natural half width (Γ) of a lorentzian line as a function of μL. This is obtained by replacing μ in expression (5) by $\dfrac{\mu_0}{1+\left(\dfrac{E}{\Gamma}\right)^2}$ and we obtain

$$Y = A \dfrac{1}{1 + \dfrac{\cos\Theta}{\mu_0 L}\left[1 + \left(\dfrac{E}{\Gamma}\right)^2\right]} \quad \ldots\ldots\ldots\ldots (6)$$

where E is the photon energy and μ_0 is the absorption coefficient at the maximum of the peak.

For E=0 at the maximum, this expression reduces to

$$Y_{max} = A \dfrac{1}{1 + \dfrac{\cos\Theta}{\mu_0 L}} \quad \ldots\ldots\ldots\ldots\ldots (7)$$

so that its half width at half maximum Y/2 is obtained by using (6) and (7). Thus we obtain:

$$\dfrac{1}{2} A \dfrac{1}{1 + \dfrac{\cos\Theta}{\mu_0 L}} = A \dfrac{1}{1 + \dfrac{\cos\Theta}{\mu_0 L}\left[1 + \left(\dfrac{E}{\Gamma}\right)^2\right]}$$

From this the ratio E/Γ can easily be obtained. We thus have:

$$\left(\dfrac{E}{\Gamma}\right)^2 = 1 + \dfrac{\mu_0 L}{\cos\Theta} \quad \ldots\ldots\ldots\ldots\ldots (8)$$

The lorentzian natural width 2Γ is broadened to $2E$ by increasing the value of μL.

The typical values of E/Γ for the incidence angles of 45° and 60° are listed in table 2

TABLE 2

E/Γ	μL	
	$\theta=45°$	$\theta=60°$
1.01	10^{-2}	10^{-2}
1.1	0.15	0.105
1.5	0.875	0.625
1.55	1	
1.73		1
2	2.1	1.5

It is clear from (8) and table 2 that the E/Γ values are quite sensitive to the value of $\cos\theta$. This is illustrated in fig.2. In this figure the line saturation effect is demonstrated by simulating the broadening of a 1 eV wide lorentzian line. It should be noticed that for smaller values of $\mu L/\cos\theta$ the widths of the lines converge to the true width of the lorentzian line. As an example we present in fig.3 the $^3D_1$ line of La obtained by resonant scattering comparatively to that obtained by photo yield for 60° angle of incidence in both cases. The solid line in the photo yield spectrum is the result of a least square fit of a lorentzian line whose width was found to be 1 eV by considering $\mu L/\cos\theta$ 1. In this case a suitable slope was added to account for the rise in the absorption on the higher energy side. The solid line in the photon scattering spectrum is the result of a simple plot for 60° angle of incidence by using Γ =0.5 eV and $\mu L=1$. Considering the statistics of the experimental data and the precision in the angle of incidence the visual fit to the data seems satisfactory. The measured FWHM of the resonant scattering and the yield lines are respectively 1.7 eV and 1 eV. If we take the width of the scattering line as to be saturated to a maximum then the E/Γ value thus measured by scattering and yield indicate that the saturation in the photo yield peak is negligible.

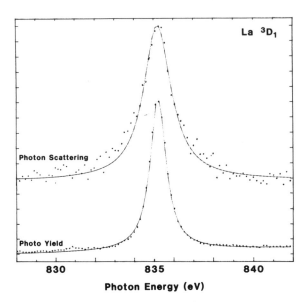

Fig.2. Simulation of line saturation effect.
Calculated broadening of 1 eV wide lorentzian line when the condition $\mu L \ll 1$ is not fulfilled.

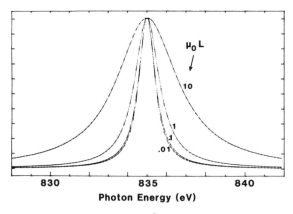

Fig.3. Observed broadening of the $^3D_1$ line of La obtained by photon scattering as compared to that by photoyield.

It is interesting to note that in the scattering experiments, the scattered photon intensity is very weak. So the results given in fig. 3 are the sum of 10 spectra in which each point was accumulated for 4 sec. Even then the statistics could not be improved. This leads to an error of about 10% in the width estimation.

In the preceeding paragraphs we discussed the influence of the resolution and the line saturation effects of the 3d-4f line widths. In our earlier work on the 3d-4f line widths, we performed the absorption measurements by transmission through 200-300 Å thick RE films. The line widths thus obtained by transmission method are found to be identical to those which we usually obtain by yield with the same resolution. So we emphasize here that for the RE the line saturation effects are negligible and the observed 3d-4f line widths generally correspond to the natural widths of the lines for the early RE. We estimate that the broadening due to crystal diffraction amounts to about 0.1 eV for Yb^{3+} at the end of RE series.

3d-4f line widths

At present the 3d-4f absorption spectra of all the RE metals, oxides and fluorides are available (excepting Pm). These measurements were performed under similar resolution and the 3d-4f spectra were found to be quite comparable. The over estimation of the line saturation effects and the monochromator resolution based on the rocking curve measurements lead automatically to the lower life time widths of the 3d-4f excitations. We have given in this paper a semi quantitative estimate of the monochromator resolution and the line saturation effect. The present work shows that the broadening introduced by these is less than 0.1 eV in the case of the 3d-4f lines and can be neglected. The line widths in the RE sequence shows a smooth plot as described in our earlier work[7]. This work suggests that in the 3d-4f excitation process, the excited 4f electron in some way is influencing the Auger decay rate of the 3d hole. Autoionization and the filling of the available short-lived 4f states may be the possible additional mechanism to explain the difference between the theoretical estimation and the experimental widths. The observation of the narrower 3d-4f line in RE fluorides suggests the influence of the surrounding fluorine cage on the available 4f states. In further paragraphs we will discuss such mechanism to explain the unusually broad 3d-4f lines observed in some RE dioxides.

New specific features in the 3d resonances of CeO_2, PrO_2 and TbO_2

The fluorite structure related CeO_2, PrO_2 and TbO_2 present special interest for the study of the valence problems in the RE. Their 3d spectra show some specific features which are not common to those of the normal trivalent RE compounds. The M_{IV-V} yield spectra of CeO_2, PrO_2 and TbO_2[13] are given in fig.4. The two 3d spin-orbit groups of lines and structures are found to be separated by 17.7, 19.5 and 31.5 eV respectively for CeO_2, PrO_2 and TbO_2. In these spectra additional weaker features (labelled Y in fig. 4) at about 3-5 eV on the higher energy side of the main line in each spin-orbit group are systematically observed.

One expects a diminution of a local f count at a given RE atom in a normal tri to tetra-valence change in RE oxides. So the 3d XAS transitions in RO_2 can be compared with the the corresponding transitions in Z-1 elements. In order to identify the main line features of these oxides, we compared them with the M_{IV-V} spectra of the trivalent compounds of the corresponding Z-1 elements. The M_{IV}/M_V intensity ratios observed in the present cases are similar to those for corresponding Z-1 element spectra. As expected we observe an increase in the 3d spin-orbit separation between the two main features of these spectra with respect to those obtained for Z-1 elements. In fig. 5(a) we have presented on a relative energy scale the M_V spectra of CeO_2 and PrO_2 comparatively to those of La_2O_3 and Ce oxalate. In fig. 5(b) the M_V spectrum of TbO_2 which is wider than the other oxide spectra is separately compared to that of Gd_2O_3. The correspondance between the various lines and the multiplet structures of RO_2 and the trivalent Z-1 compounds is indicated by connecting broken lines. The CeO_2 main lines are similar to the $3d^94f^1$ lines observed for La_2O_3. The PrO_2 broad main line features resemble closely the $3d^94f^2$ (trivalent Ce like) multiplet. Finally we find for TbO_2 a $3d^94f^8$ (Gd like) multiplets (fig. 5(b)) which are broad and the structures can easily be recognized by comparison with the Gd_2O_3 spectra. We, therefore, identify the main lines of CeO_2, PrO_2 and TbO_2 respectively as transitions from the atomic like f^0, f^1 and f^7 ground state configurations.

The most interesting fact in these oxide spectra is the observation of unusual broadening of the constituent main multiplet lines. The broad RO2 and the corresponding trivalent compounds spectra (fig. 5(a),(b)) were obtained with the same resolution. A comparison between the individual structures within the multiplets of RO_2 and those observed in the corresponding Z-1 elements reveals: (i) an increase in the widths of the $^3D_1$ and $^1P_1$ like lines in CeO_2, (ii) an increase in the separation between the structures within the multiplets of RO_2 and (iii) a correspondance between the relative intensity of RO_2 structures and those for Z-1 elements is observed. In table 3 we give the measured full width at half maximum (FWHM) of the 3d-4f lines or multiplets for RO_2 and those for the corresponding Z-1 elements in trivalent compounds[13]. The 40-50 % increase in the line or multiplet width in these oxides with respect to that observed in the corresponding Z-1 element is found to be too high to be explained by invoking the increase in the spin-orbit separation.

TABLE 3

RO_2	Z-1 element	Width (eV)	
		M_V	M_{IV}
CeO_2		1.7	2.6
	La (a)	1.2	1.8
PrO_2		4.1	5.0
	Ce (b)	2.4	3.2
TbO_2		4.8	6.8
	Gd (a)	2.6	4.6

(a) FWHM obtained from the spectra in ref.7
(b) FWHM obtained from the spectra in ref.20

These broadened multiplets roughly keep their usual forms as observed for the normal RE elements. A careful examination of the $^3D_1$ and $^1P_1$ like lines of CeO_2 show deviations from their usually expected lorentzian or Fano profiles. A detailed analysis of the line broadening requires a suitable function to take into account the interactions of the excited 4f electron with the complex conduction states in RO_2. Such interactions may also explain the further line splitting within the multiplets which leads to an unusual increase in the FWHM of the multiplets. For this reason a detailed analysis of the complex multiplets in PrO_2 and TbO_2 is presently difficult.

The structure labelled Y (fig.4) appearing at about 5 eV above the main lines of CeO_2 and PrO_2 have similar forms and the M_{IV} Y structures are somewhat higher in intensity than the M_V ones. They are asymmetic on the low energy side and appear to extend up to the bottom of the main lines. In TbO_2 these structures are found to be within the wide Gd like multiplets. The Y type structure at about 3 eV above the main M_V line can easily be recognized as a feature (fig.5(b)) contributing to the enhancement in the intensity between the main line and the high energy weaker structures of the Gd like multiplet. For the M_{IV} it is probably overlapping the high energy main line structure. In the M_V spectrum of TbO_2 we observe a weak shoulder at about 1.5 eV below the main line (fig. 5(b)). Its position corresponds approximately to the expected energy of the main M_V peak of the trivalent oxide. It may be due to the presence of some residual trivalent sites in TbO2 since it is rather difficult to reach an ideal oxygen stoichiometry of 2 even on prolonged oxidation.

Thus we see that the Y structures in CeO_2 and PrO_2 do not correspond to any known multiplet features of nearby RE elements. The systematic increased separation (3-5 eV) between RO_2 atomic main lines and the Y structures as compared to $RO_2-R_2O_3$ multiplet separation and the similar form of the Y structures indicate that the origin of these structures is closely related to the degree of the core hole screening by the excited f electron. These considerations suggest that the Y structures in RO_2 can easily be attributed to the transitions to the 4f admixture conduction states. The interaction of an excited 4f electron with a finite f admixture continuum may lead to further broadening of the 3d - 4f lines in these oxides.

CONCLUSION

The knowledge of the quantative estimate of instrumental and other broadenings is necessary to obtain the true widths and shapes of spectral lines. In the context of the 3d-4f transitions in RE we emphasise in the present paper that besides the spectral positions, the measurement of the line widths and the shapes are also important. The present high resolution line width measurements reveal the physics involved in the narrowing and the broadening of the 3d-4f lines. We further remark that deepening of such knowledge will certainly help us in the further understanding of the intermediate valence phenomenon in RE.

ACKNOWLEDGEMENT

We thank Professor J.P.CONNERADE for helpful discussions. Thanks are also due to Dr. P.E.CARO for introducing us to the fascinating studies of the RE oxides.

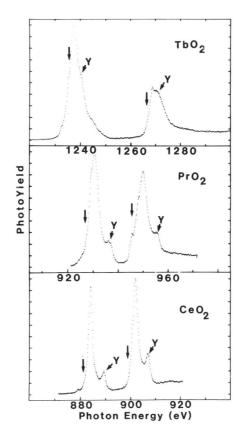

Fig.4. The M_{IV-V} spectra of CeO_2, PrO_2 and TbO_2. The vertical downward arows indicate the positions of the main lines from the corresponding trivalent oxides.

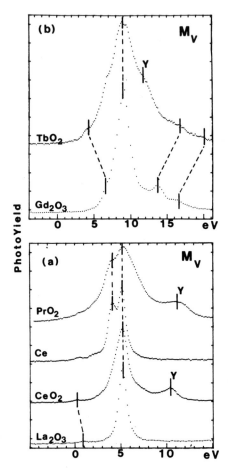

Fig.5. Comparison of $\dot{M}_V$ multiplet structures in RO_2 with those of the corresponding Z-1 elements in the trivalent oxides. The spectra are presnted on relative energy scales.

REFERENCES

1. J.P.Connerade, Contemp. Phys.19,415(1978);
   This volume and references therein
2. E.R.Radtke, J.Phys.B At. Mol. Phys.12,L73(1979);
   12,L81(1979).
3. C.Bonnelle,R.C.Karnatak,J.Sugar, Phys. Rev.A9,1920(1974).
4. V.F.Demekhin, Sov. Phys. Solid State 16,659(1974).
5. C.Bonnelle,R.C.Karnatak,N.Spector, J.Phys.B At.Mol.Phys.10,795(1977).
6. A.Belrhmi-Belhassan,R.C.Karnatak,N.Spector,C.Bonnelle,
   Phys. Lett.82A,174(1981).
7. R.C.Karnatak,J.M.Esteva,J.P.Connerade,
   J.Phys.B At.Mol.Phys.14,4747(1981).
8. J.Sugar,W.D.Brewer,G.Kalkowski,G.Kaindl,E.Paparazzo,
   Phys.Rev.A32,2242(1985).
9. B.T.Thole,G.van der Laan,J.C.Fuggle,G.A.Sawatzky,
   R.C.Karnatak,J.M.Esteva, Phys.Rev. B32,5107(1986).
10. E.J.McGuire, Phys.Rev.A5,1043(1972).
11. G.Crecelius,G.K.Wertheim,D.N.E.Buchanan,Phys.Rev.B18,6519(1978).
12. R.C.Karnatak,J.M.Esteva,H.Dexpert,M.Gasgnier,P.Caro,L.Albert,
    J.Mag.Mag.Mat. in press
13. R.C.Karnatak,J.M.Esteva,H.Dexpert,M.Gasgnier,P.Caro,L.Albert,
    Phys.Rev. B to be published
14. M.Lemonnier,O.Collet,C.Depautex,J.M.Esteva,D.Raoux,
    Nucl.Instrum.Meth. 152,109(1978).
15. Z.Hussain,E.Umbach,D.A.Shirley,J.Stöhr,J.Feldhans,
    Nucl.Instrum.Meth. 195,115(1982).
16. E.J.McGuire, Phys.Rev.185,1(1969);
    V.O.Kostroum,M.H.Chen,B.Crasemann, Phys.Rev.A3,533(1971);
    D.L.Walter,C.P.Bhalla, Phys.Rev.A3,1919(1971).
17. J.M.Esteva,B.Gauthé,P.Dhez,R.C.Karnatak,
    J.Phys.B At.Mol.Phys. 16,L263(1983).
18. M.H.Tuilier,D.Laporte,J.M.Esteva, Phys.Rev.A26,372(1982).
19. J.M.Esteva,R.C.Karnatak,J.P.Connerade
    J.Electr.Spectr.Rel.Phen. 31,1(1983)
20. J.C.Fuggle,F.U.Hillebrecht,J.M.Esteva,R.C.Karnatak,O.Gunnarsson,
    K.Schönhammer, Phys.Rev.B27,4637(1983).

THE EFFECTS OF 4f LEVEL OCCUPANCY, COULOMB INTERACTIONS, AND HYBRIDIZATION

ON CORE LEVEL SPECTRA OF LANTHANIDES

J.C. Fuggle

Research Institute of Materials, Faculty of Science
University of Nijmegen
Toernooiveld, 6525 ED Nijmegen, The Netherlands

ABSTRACT

One of the necessary conditions for giant resonances is the presence of empty valence states with compact wave functions. This characteristic, found typically in transition metals, lanthanides, and actinides, is also associated with formation of narrow bands, strong Coulomb interactions between the narrow band electrons and weak hybridization between the narrow valence states and the rest of the valence bands in solids. It leads to complex behaviour in various core level and other spectroscopies; but it also leads to many puzzles in the physical and chemical properties of these elements, and also their alloys and compounds. These properties are related to the influence of both band structure and atomic correlation effects. The theoretical treatment of these properties is terribly difficult, so that provision of any experimental guidance on the size of the parameters involved and the appropriate approximations, assumes great importance. This has led many workers in the last decade to try to deduce ground state properties from complex, high energy spectroscopies, rather than to investigate the complex, high energy spectroscopies for themselves. The results have been surprisingly good and some of the background will be discussed here, using the lanthanide 4f levels as an example.

We will consider screening of a core hole in a solid and the satellites and dynamics of core hole creation and destruction. It will be shown how these effects can be related to problems in 4f level occupancy, Coulomb interactions between valence electrons, and hybridization of the 4f levels with conduction states.

INTRODUCTION

Atomic Correlation Effects in Solids

The electrons in an atom or solid are in states characterized by a wave function and an energy level. These states are subdivided into the valence levels, whose wave functions overlap in a compound or solid and which control its physical and chemical properties, and the core levels, which are confined to the core regions of the atoms. High energy spectroscopy of these levels generally involves excitation of an electron from a core level to an empty valence level (X-ray absorption or XAS), removal of an electron from the atom or solid (photoemission or PS) or addition of an electron (inverse photoemission or Bremsstrahlung Isochromat spectroscopy BIS). However paradoxial it may appear, high energy spectroscopies often provide the best starting point to establish a framework within which the ground or near ground state, properties of a solid, such as magnetism, conductivity or semiconductivity, and optical properties can be understood. This situation has arisen as a result of development of experimental and above all theoretical methods to relate fine details of Auger, core level PS and XAS spectra to correlation and hybridization effects in the properties of the valence electrons[1-7].

The use of core level spectroscopic methods for studies of ground state properties is of most interest for narrow band materials. For simple metals and many insulators and semiconductors our understanding of the electronic structure is quite good. By contrast, transition metals, actinides and some rare earths present many puzzles. For these systems it has become apparent that atomic correlations play an extremely important role in determining the physical properties. To understand these a detailed knowledge is required of the electronic structure which is determined not only by either band structure or atomic correlations, but is strongly influenced by both. The theoretical treatment of this situation is associated with horrendous difficulties and is one of the central scientific problems of the last quarter of this century[8-15]. To come to a more complete understanding of this problem it is necessary to develop experimental techniques to answer questions like
(a) How strong are the atomic correlations when an atom is embedded in a solid?
(b) What is the ground state, local electronic configuration and term? What is the local magnetic moment and how is it oriented?
(c) How large is the hybridization of the correlated atomic-like states with the conduction band?
(d) How important are spin-orbit and crystal field splittings.

It is only in the last few years that high energy spectroscopies have been able to give information on these questions, but literature examples can now be given for almost all these questions. The purpose of this paper is to explain some of these developments in the simplest possible terms, using examples from studies of lanthanides.

The Lanthanides

In the lanthanides and their compounds and alloys it is the 4f electrons which are strongly correlated. It is instructive to recall the situation, not so long ago, when the 4f states were regarded as chemically totally inert. The 4f orbitals in solid lanthanides were known to be very small and they were supposed to be filled up one at a time as the atomic number was increased from 57 (La, $4f^0$) to 71 (Lu $4f^{14}$). Eu and Yb were exceptions and had 7 and 14 4f electrons because of the special stability of the half-filled and filled shells respectively and not all the textbooks got even that right. In the solid the entirely localized 4f electrons were thought to determine the magnetic moments of the rare earth ions, even in metals and alloys whilst the three 6s5d electrons were responsible for the chemical properties and metallic conductivity, etc. Variation in chemical valence was a little known phenomena in the lanthanides. For many lanthanides (e.g. La, Ph, Gd, Tb, Dy, Ho and Er) there is little need to modify this picture because of new chemical or physical observations. Even in the solid the spectroscopic behaviour can be taken as prototypical for systems with 4f levels where atomic correlations dominate.

Atomic correlation effects can be illustrated in a Hubbard-like[11] model where the effective correlation interaction between the electrons, U_{eff} is the difference of the first ionization potential (IP) and electron affinity

(EA) of the 4f levels, i.e.

$$U_{eff} = IP - EA \tag{1}$$

Care is necessary with the sign in this equation because the IP is defined as the energy required to <u>remove</u> an electron and the EA as the energy <u>gained</u> when one is added. The first IP and EA in the rare earths can be measured by photoelectron spectroscopy and inverse photoemission, or Bremsstrahlung Isochromat spectroscopy (BIS) respectively[3]. The measured value of U_{eff} varies between ~ 5 and ~ 11 eV[3], which is much larger than the 4f band width.

The consequences of the large value of U_{eff} can be illustrated in a very simplified model if we write the ionization potential of the 4f levels in the absence of repulsion between the 4f electrons as ε_m, where m is the number of the lanthanide (La = 0, Ce = 1, etc.) and the effective repulsion between the 4f electrons (with concomitant adjustment of the valence electrons) as U_{ff}. Then the contribution of n 4f electrons to the total energy is[4,5]:

$$E = n\varepsilon_m + n(n-1)U_{ff}/2 \tag{2}$$

Then for Ce (m = 1) the contribution of the 4f electrons, if $\varepsilon_1 = -2$ eV and $U_{ff} = +6$ eV are used as reasonable values, is

$$\begin{aligned}
E &= 0 & \text{for zero 4f electrons} &= 0 \text{ eV} \\
E_1 &= \varepsilon_1 & \text{for one 4f electron} &= -2 \text{ eV} \\
E_2 &= 2\varepsilon_1 + U_{ff} & \text{for two 4f electrons} &= +2 \text{ eV} \\
E_3 &= 3\varepsilon_1 + 3U_{ff} & \text{for three 4f electrons} &= +12 \text{ eV} \\
E_4 &= 4\varepsilon_1 + 6U_{ff} & \text{for four 4f electrons} &= +28 \text{ eV} \\
E_5 &= 5\varepsilon_1 + 10U_{ff} & \text{for five 4f electrons} &= +50 \text{ eV}
\end{aligned}$$

etc. (3)

From this it can be understood that a plot of energy as a function of 4f count will show a minimum, as in Fig. 1[16,17], for which ε_1 was set to -2 eV and U_{ff} was set to 6 eV. In general the position of the minimum will shift by one f electron for each unit increase in atomic number. Note also that if we consider two atoms with n = f electrons and convert them to a pair with n + 1 and n - 1 f electrons, then the energy cost will be U_{ff}, i.e. U_{ff} is identical to U_{eff} here. However, multiplet splittings due to interaction between the electrons give rise to many levels for each f-count, and together with chemical effects give rise to many exceptions to this simple rule. In particular multiplet effects may be such as to make configurations with different f-count nearly degenerate. This gives rise to a second group of lanthanides which exhibit variable chemical valence. Because of the 4f electrons were so strongly localized and could not take part in chemical bonds, those elements which exist in two valent (Pr, Nd, Eu, Tm or Yb) or four valent (Ce, Pr, Nd, Tb, Dy) compounds were assumed to do so via transfer of an f electron to or from the ligands. The possibility of fractional occupancy of an f-orbital was not considered until quite recently. To analyse this possibility one needs to consider first the concept of valence.

Valency was a chemical concept, formulated in the middle of the nineteenth century. The valency of an element was defined as the number of hydrogen atoms with which an atom of an element would combine, or one half of the number of oxygen atoms with which an atom of an element would combine. Later variable valency was admitted, first for transition metals, and an element would be described as three-valent or four-valent, etc. according to the compound in which it was found. Some lanthanides (and also transition metals) have compounds whose stoichiometry can only be explained in terms of atoms with more than one valency, e.g. Eu_3O_4 must contain Eu^{2+} and Eu^{3+}. In such a system one may speak of heterogeneous mixed valence, which does not present many major puzzles.

A phenomenon which does present more serious problems is that of inter configuration fluctuations or "homogeneous mixed valence". This phenomenon is hypothesized for some compounds of Ce, Eu, Sm, Tm and Yb[17], although as we shall see below it may be better to treat Ce separately. The phenomenon affects many physical properties but is easily explained in terms of lattice constants. If n represents the position of a rare earth atom (La = 0, Ce = 1, Pr = 2, Yb = 13, La = 14) then the $4f^n$ configuration is associated with a three-valent ion. There are then three "valence" electrons. A $4f^{n+1}$ or $4f^{n-1}$ configuration will be associated with 2 or 4 valent compounds respectively[18]. In metallic compounds an increase in the number of valence electrons results in a decrease in the lattice constant which may therefore be used as a monitor of the number of 4f electrons. In general the lattice constants of the rare earth metals follow a simple curve as a function of atomic number, with jumps at the elements where a $4f^{n+1}$ or $4f^{n-1}$ configuration occurs (see Fig. 2). However, some compounds exhibit lattice constants which can only be explained if there is effectively a non-integral number of 4f electrons on every site. It is here that one speaks of inter-configuration fluctuations or homogeneous mixed valence. Both terms are unsatisfactory: the term "interconfiguration fluctuation" encourages an oversimplistic picture of the phenomena and the term "mixed valence" ignores the chemical concepts behind the term "valency". However, the terms are now established.

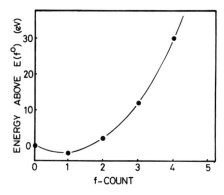

Fig. 1. Schematic diagram of the variation of total energy with f-count for Ce atoms.

The theoretical description of homogeneous mixed valence is problematic but usually involves some form of coupling or hybridization[19] of the f levels to the valence states. In the late rare earths the strength of this hybridization is not well known but is at present thought to be weak and not to contribute significantly to the chemical stability of the material. In the light rare earths, especially Ce, the hybridization is sufficiently large that it has direct influence on the chemical stability, the lattice constants, and other physical properties of Ce and its compounds. It is only in the last decade that one has become aware of the size of this hybridization[36] and its consequences and that we have begun to understand many of the properties of Ce and its compounds.

As many of our illustrations of the high energy spectroscopies will involve Ce it is worthwhile to give two illustrations of just how curious, even bizarre, its properties are and why so many hundreds of papers about Ce have been published in the last decade. One of the first important observations was the ~ 16.5% volume contraction of Ce which occurred at the phase transition when γ-Ce was cooled under mild pressure to produce α-Ce. This occurs without a change in crystal structure. The first theory was that in α-Ce an electron was promoted from the 4f level to the valence states[21]. Later there was a suggestion that the promotion was only partial (i.e. mixed valence), then that the transition was more similar to a Mott-transition, in which the f-electrons formed a band[22]. Most recently, as a result of Compton scattering and high energy spectroscopic measurements it has been concluded that f-counts in α-Ce and intermetallic compounds of Ce are never less than ~ 0.7 [6,23-29], and that the α-γ-Ce transition is best explained by a rather modest change in f-count and hybridization between f levels and the conduction states.

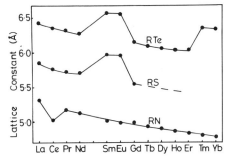

Fig. 2. Lattice constants for lanthanide nitrides, sulphides, and tellurides. Note the excursions of the curves to higher (two valent) values at Sm, Eu, Tm and Yb for the chalcogenides and to a lower (more four-valent like) value for CeN. Data taken from reference 57.

As a second puzzle in Ce compounds the very topical heavy Fermion problem is chosen. In most metallic materials the effective mass of an electron at the Fermi level is quite near one, and effective masses which differ by a factor of five are regarded as exceptional. In $CeCu_2Si_2$, $CeAl_3$, $CeCu_6$, and $CeH_{2.7}$ (and some U intermetallic compounds) effective masses of more than 1000 have been found[30-32], which is quite spectacular. There is, as yet, no generally accepted theoretical description of heavy Fermion materials and there is uncertainty over the ramifications of such observations. There is, however, already spectroscopic evidence that the number of 4f electrons in these materials is very close to one[32].

THE ROLE OF BIS AND VALENCE BAND PHOTOEMISSION

These techniques are not central to the theme of this school although there will be some discussion of their role in the chapters by Gunnarsson and by Oh. As mentioned above, they can give good measurements of U_{eff}^3 and of the f-count in the heavy lanthanides[33,34]. At the present time it is quite hard to identify with BIS and VB PS the strength of hybridization and the atomic configurations which play a role in the ground state. Further, without this information one cannot really use these techniques to determine the ground state f count[5-7,35] in the light rare earths where hybridization can be important. It is in these areas that core spectroscopies are most helpful[4-7,36,37].

THE CORE HOLE POTENTIAL

When a core hole is created the effective attractive interaction of the core hole with the small 4f orbitals is larger than with the more distended 5s6d valence states by of the order of 10 eV. If we term this difference U_{ac} then equation (2) can be modified to read (for Ce)[45]:

$$E = \Delta + n\varepsilon_1 + nU_{ac} + n(n-1)/2\, U_{ff} \qquad (4)$$

where Δ is the core electron binding energy in the absence of a core hole. We can illustrate the behaviour of equation (4) for Ce by choosing $\varepsilon_1 = -2$ eV and $U_{ff} = 6$ eV, as for Fig. 1 and $U_{ac} = -8$ eV. Then:

$$E_0 = \Delta \qquad \text{for zero f electrons} = \Delta$$
$$E_1 = \Delta + \varepsilon_1 + U_{ac} \qquad \text{for one f electron} = \Delta - 10 \text{ eV}$$
$$E_2 = \Delta + 2(\varepsilon_1 + U_{ac}) + U_{ff} \qquad \text{for two f electrons} = \Delta - 14 \text{ eV}$$
$$E_3 = \Delta + 3(\varepsilon_1 + U_{ac}) + 3U_{ff} \qquad \text{for three f electrons} = \Delta - 12 \text{ eV}$$

etc. (5)

From this it may readily be judged that the minimum in the plot of total energy against f-count is moved to higher f-counts in the presence of a core hole[34] as shown for Ce in Fig. 3. This distortion of the E vs f-count plot by the core hole potential is central to a large effort built up to measure f-counts using core spectroscopies, particularly X-ray absorption spectroscopy (XAS). If two lanthanide atomic configurations with different f-count have similar energies in the absence of a core hole then they may both contribute to the ground state. However, they will then have different energies in the presence of a core hole, i.e. in the final, excited states of a core level spectroscopy. There will thus be two peaks in the core level spectrum whose relative intensity can in principle be used to determine the relative weights of the two configurations in the ground state. However, mixing of the final state configurations does complicate this determination, as discussed below.

MULTIPLET INTERACTIONS IN CORE LEVEL SPECTROSCOPY

The interaction between a core hole and the localized 4f electrons does not in fact constitute just the spherical monopole term ($F^{(0)}$), but also there are non-spherical terms normally expressed as Slater $F^{(2)}$, $F^{(4)}$ and G integrals. The nearly spherical, monopole-like screening of the core holes in a solid strongly reduces $F^{(0)}$ with respect to its free-atom value, but has very little effect on the other integrals[38] so that atomic-like multiplet structure is observed in the core level spectra of lanthanides, and also of many transition metals. Here we treat the 3d core level spectra in XPS and XAS.

In 3d XAS the local atomic configuration is changed from $3d^{10}4f^n$ to $3d^9 4f^{n+1}$. The observed spectrum should thus reflect the $3d^9 4f^{n+1}$ multiplet structure and it has been clear since the pioneering work of Sugar[39] and Demekhin et al.[40] that this is the case (Other electrons are nearly passive spectators and thus not referred to in the configurations). The $3d^9 4f^{n+1}$ multiplet structure can be very well reproduced by modern atomic computations including relativistic effects[41]. The interaction of $3d^9$ and $4f^{n+1}$ electrons must be described in intermediate coupling. In XAS dipole selection rules apply so that the multiplet observed is a subset of the total multiplet.

In Fig. 4 the calculated $3d^9 4f^3$ multiplet of Pr^{3+}, is presented along with the measured XAS spectrum and the part of the total multiplet calculated to be observed in XAS under diple selection rules (see ref. 41 for further details) for a $3d^{10}4f^2 \rightarrow 3d^9 4f^3$ excitation. Also shown is the XAS spectrum expected for the 3d absorption in a $Pr^{4+}4f^1$ ion. The XAS spectral shapes in Fig. 4 are unlike normal "edge" structures in that they

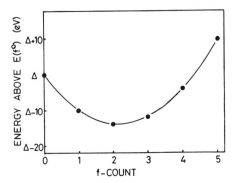

Fig. 3. Schematic diagram of the variation of total energy of Ce atoms with f-count for a Ce atom and for a Ce atom with a core hole. Note the shift of the energy minimum.

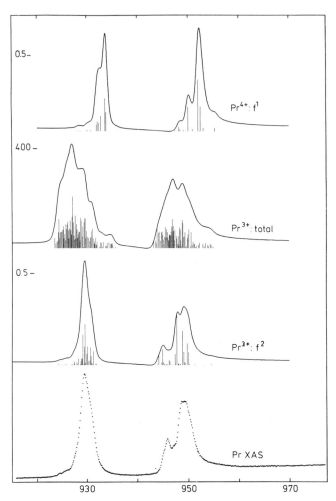

Fig. 4. Multiplet structure in Pr. The spectrum labelled XAS is the experimental 3d XAS spectrum. The label f^n denotes the theoretical dipole excitation spectrum from the Hund's rule ground level of the configuration f^n to the levels of $3d^9 4f^{n+1}$. The spectrum labelled "Pr^{3+} total" gives each level of the $3d^9 4f^3$ with weight $2J+1$. The horizontal axis gives the excitation energy in eV. The scale of the theoretical spectrum, at the left, gives the number of states per eV for the "total" curve and the absorption cross section in Å^2 for the dipole spectra. The theoretical spectra have been folded with an experiment and a Lorentzian contribution, as described in ref. 41.

have sharp peaks attributable to the 3d → 4f transitions which led early workers to refer to them as "line" absorptions. The same effect is seen for all the lanthanides except Lu[41]. The spectral features are in two groups whose separation corresponds roughly to the $3d_{3/2} - 3d_{5/2}$ spin-orbit splitting. Note, however, that the individual groups show considerable structure which is due to coupling of the spin and orbit moments of the 3d and 4f electrons. The spectra must be described in intermediate coupling and the designation of the groups as $3d_{3/2}$ (M_4) and $3d_{5/2}$ (M_5), whilst convenient, is only approximate. It is interesting to note that the $3d_{3/2}$ and $3d_{5/2}$ lineshapes in the full multiplet are rather similar, but the observed XAS lineshapes are quite different. This is because the dipole matrix elements give appreciable intensity to different parts of the total multiplet.

The integrated intensity of the M_4 absorption lines is comparable to that of the M_5 lines in the early rare earths, but its relative intensity rapidly decreases in the heavier rare earths (see for example the illustrations for Tm in Fig. 5). The origin of the effect is certainly in the selection rules because the ratio of numbers of levels in the $5d_{5/2}$ and $5d_{3/2}$ states is always 3:2. However, the basic physics is not easy to visualize in intermediate coupling. For La and Yb an explanation where spin-orbit coupling is the dominant interaction and $\Delta j = 1$ is the dominant channel is not completely unphysical. For La the "weight" of the $3d_{3/2} \to 4f_{5/2}$ excitation is $4 \times 6 = 24$ and that of the $3d_{5/2} \to 4f_{7/2} = 6 \times 8 = 48$. Other possibilities are forbidden so that the ratio of the $3d_{3/2} : 3d_{5/2}$ peaks is 1:2. The full calculation in intermediate coupling yields 1.09:1.70[42]. For Yb the only 4f hole has $4f_{7/2}$ character so that only the $3d_{5/2} \to 4f_{7/2}$ excitation is possible and only the $3d_{5/2}$ line should be observed. If spin-orbit coupling were the dominant interaction the $4f_{5/2}$ states would be filled first and only the $3d_{5/2}$ lines would be observed for all the late rare earths. This is not in agreement with experiment but the general trend that the $3d_{3/2} : 3d_{5/2}$ ratio decreases across the rare earth series is probably related to a tendency for the occupied 4f levels in the ground state to have more $4f_{5/2}$ character than the empty states[41].

When we simulated the lanthanide 3d XAS spectra from the calculated multiplets we included a Lorentzian broadening which ranged from $2\Gamma \sim 0.4$ eV for La $3d_{5/2}$ to $2\Gamma \sim 1.2$ eV for Tm $3d_{3/2}$ (A Fano asymmetry was introduced in the M_4 lines as discussed in ref. 41). This broadening is attributed to the core hole lifetime. The values are not quite accurate because we did not attempt to take full account of saturation effects, but they are a good guide to the lifetime effects. Even though the radiative transition rates we calculate for the 3d edges are enormous by normal standards (e.g. integrated cross sections over the $3d_{5/2}$ group range up to ~ 1.4 Å^2 eV) they still only account for a few percent of the lifetime broadening. Thus even here, Auger processes are the predominant 3d hole decay channel.

Before proceding to the uses of 3d XAS spectra from the lanthanides it is interesting to make a comparison with XPS and electron energy loss data, where different selection rules apply. As we will discuss later, it is possible to observe features in the XPS spectrum of the 3d levels corresponding to the $(3d^9 4f^{n+1} + e^-)$ final states, where an electron is taken from the valence band to increase the 4f occupation number by one. This extra electron may, in effect, have any j value, which means that the total $3d^9 4f^{n+1}$ multiplet can be observed, and not just those permitted under dipole selection rules. However, as illustrated in Figs. 4 and 5, there is a 2 - 4 eV difference in the energy of the centroids of the full multiplets and that part of the multiplet allowed to be observed in XAS under dipole selection rules. This leads to an apparent 2 - 4 eV discrepancy between peaks observed in XAS and XPS[42].

Similar breakdown of dipole selection rules has been observed in electron energy loss (EELS) by Netzer et al.[43,44]. When the primary beam energy in EELS is ~ 1000 eV higher than the threshold for core hole excitation a reflection electron energy-loss spectrum is equivalent to an XAS spectrum. However, near threshold the diple selection rules break down and the peaks of the full multiplet are observed. For the early rare earths

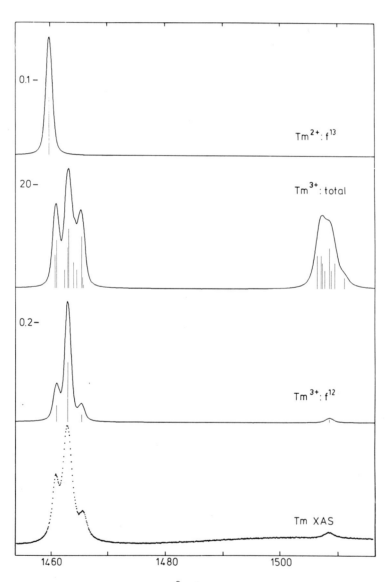

Fig. 5. Calculated and observed $3d^9 4f^n$ multiplet structure for Tm. Vertical scales and nomenclature are similar to Fig. 4.

this extra structure comes on the low-energy side of the dipole-allowed peaks, as our calculations indicated[41]. For the heavy rare earths we predicted that extra intensity should be observed on the high-energy side, as shown in Fig. 5 for Tm, and in Yb the $3d_{3/2}$ line should become visible[41].

Finally we note the shift of the lines and change of shape of the multiplet when the number of f electrons is changed. As will be illustrated below, these changes result in a potential use of XAS to determine weights of the different $4f^n$ contributions to the ground state, in both heterogeneous and homogeneous mixed valence materials.

MULTIPLET SPLITTING IN CORE LEVEL XPS

Much experimental work on multiplet splittings in XPS, reported until now, relates to lighter atoms where L-S coupling limits can be assumed[62-65]. In the rare earths the intermediate, or sometimes j-j limits are more appropriate. XPS multiplet effects are often physically less transparent in the intermediate coupling limits and we start with an example from the 3d transition metal elements, namely Mn^{2+}. Because full "spectator" subshells do not contribute to this ground state configuration of this may be written

$$(nl)^x 3d^5$$

where n represents the principle quantum number, l is the angular momentum quantum number, x is the number of electrons in the subshell to be ionized. The core ionized final state is then

$$(nl)^{x-1} 3d^5$$

To a first approximation for XPS the core and valence electrons are treated separately and $\Delta S = \pm 1/2$ and $\Delta L = 0, \pm 1 \ldots \pm L$ for the core electrons. Thus, for the Mn^{2+} the ground state $3d^5$ term is ^6S (2S + 1 = 6, L = 0) and this can be coupled with the ^3S (S = 1/2, L = 0) hole to give ^7S and ^5S final states. Consequently two 3s XPS peaks are expected at this level of approximation. In practise, the $3d^1 3d^5$ (^5S) configuration interactions with other (^5S) configurations of similar energy; (^5S) = $3s^2(^1S)3p^4(^3P)3d^6(^3P_1)$, ($^5S$) = $3s^2(^1S)3p^4(^3P)3d^6(^3P_2)$ and (^5S) = $3s^2(^1S)3p^4(^1D)3d^6(^5D)$. The ($^5S$) XPS intensity is thus distributed amongst four ^5S peaks, three of which have sufficient intensity to be found in the measured spectrum[64,65]. A general consequence of this configuration interaction (CI) in the final states is a reduction of the observed multiplet splittings with respect to calculated Hartree-Fock values without CI[62,63].

A study of the 4s and 5s XPS peaks of the rare earths revealed doublet structure[66] similar to that found for Mn^{2+} at low resolution. The 4f and 5s peaks have large lifetime broadening (~ 4 eV FWHM for 4s[67]) and low XPS cross sections so that it is hard to detect CI satellites directly. However, the 4s splittings are reduced to ~ 55% of the theoretical values[66] which is an indication that they are present.

Before proceding to discussion of XPS of core states with l > 0 we should note that hybridization of the rare earth 4f levels with the conduction states can lead to transitions to final states with a higher f-count than in the ground state (see next section). The extra f electron is taken from the conduction states and apparently acts as if it can have any s, l and j value. Thus all values of L, S and J are allowed and these spectral features exhibit the shape of the full local multiplet, not just those allowed under dipole selection rules[68] as applied to the local atomic configuration. Here we treat multiplet splittings in the peaks due to transitions to states with the same f-count as in the ground state.

When the core hole created in XPS (or XAS) has $l > 0$ the coupling of the core hole to the $4f^n$ configuration is more complex. Multiplet structure must be calculated in intermediate coupling for most levels of interest in XPS. We will illustrate a few features of such spectra with the 4d and 3d levels of La and Ce. The 4f levels of La in Fig. 6 show the 4d spin orbit splitting quite clearly because there are no 4f electrons. There are also small shoulders due to transitions to $4f^1$ final states and energy loss processes. The 4d XPS region of Ce is much more complex and shows multiple peaks due to the $4d^9-4f^1$ interactions, which are comparable in size to the spin-orbit interactions. In this case there is no direct way of tracing the spin-orbit coupling in the measured spectrum. For the Ce 3d XPS spectrum shown in Fig. 7[69] multiplet structure due to $3d^94f^1$ interaction again leads to complex lineshapes, but on a scale which is smaller by comparison with the spin-orbit splittings. The "$3d_{3/2}$" peak has a flat top and a shoulder on the high binding energy side, the $3d_{5/2}$ peaks is also not Lorentzian. Fig. 7 also illustrates another point, namely that in XPS too the observed multiplet structure can be used as a diagnostic test of the ground state configuration. Elsewhere[64] we have calculated the Ce 3d XAS spectrum using the assumption of $4f^1_{5/2}$ and $4f^1_{7/2}$ ground states. The experimental curve shows features which are more closely related to the $4f^1_{5/2}$ curve. In the width Δ of the 4f "band" were large by comparison with the $4f^1$ spin-orbit splitting, $\delta_{s.o.}$, then the experimental spectrum should more closely resemble a mixture of the two curves. For Ce itself $\Delta \ll \delta_{s.o.}$ but this is not always the case[37]. The potential of such XPS multiplet splittings for diagnostic purposes is not yet fully clear. In Ce 3d XPS spectra the differences between the curves calculated for $4f^1_{5/2}$ and $4f^1_{7/2}$ are not large. The effects may be larger and of more utility for other levels and other elements.

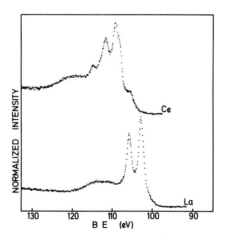

Fig. 6. The 4d XPS spectra of La and Ce.

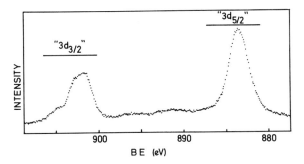

Fig. 7. The 3d XPS spectra of Ce.

SHAKE UP

Shake up is a general name for satellite features that occur in XPS, XAS, BIS and Auger spectroscopies. As we will discuss here, shake up phenomena can be used to make deductions about the ground state electronic structure. This will be described primarily for XPS, although the method can be used in XAS and BIS, as illustrated in Fig. 8 for Ce. In Ce, the ground

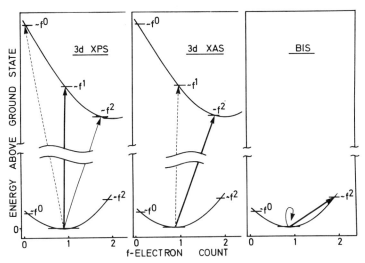

Fig. 8. Schematic representation of the transitions observed in XPS, 3d XAS and BIS of Ce and its intermetallic compounds. The multiplet structure has been neglected and the strengths of the arrows are indicative of the weight of the transition. The upper curves represent the energy as a function of f-electron count in the presence of a core hole. Note that the labels of the configurations are approximate, the true configurations are hybridized, with non-integer f-counts of approximately zero, one and two f-electrons. The BIS illustration is oversimplified.

state has predominantly $4f^1$ character but a little f^0 and f^1 is mixed in. The XPS transition removes a core hole and transitions occur to final states of predominantly f^0, f^1 and f^2 character respectively. As discussed in connection with Fig. 3, these states have quite large energy separation so that the spectral intensity is divided into three peaks. In 3d XAS the principle excitation is $3d \to 4f$, so that there is a strong tendency for the 4f-count in the final states to be increased by one and the transitions to Ce $4f^0$ states are not observed, only $4f^1$ and $4f^2$. The representation of the BIS process in diagrams such as Fig. 8 is problematic. Suffice it here to say that the peaks in the spectrum are associated to transitions to final states with predominantly $4f^1$ and $4f^2$ local character. A fuller treatment may be found in the chapters by Gunnarsson, by Schneider, and by Oh.

As the peaks in the various spectroscopies result from transitions to different final states of predominantly f^0, f^1 and f^2 character our task is to relate the intensities of the peaks to hybridization and the coefficients of the f^0, f^1 and f^2-like wave functions in the ground state. To illustrate this we will consider the case of XPS of La and Ce, but first, in Fig. 9, we

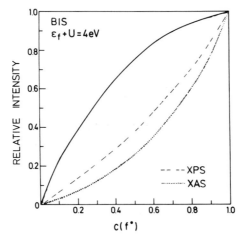

Fig. 9. Values relative to the total, of the f^1 peak intensity for BIS and 3d XAS, and the f^0 peak intensity for XPS, as a function of $C(f^0)$ (i.e. $1 - n_f$) for Ce compounds. The strength of hybridization between the 4f levels and the semi-elliptical valence band, $F(E)$ used was $\Delta = 120$ meV. $F(E) = 2V^2[B^2 - (\varepsilon - \varepsilon_0)^2]^{1/2}/B^2$ where $B = 2.785$ eV, $\varepsilon_0 = -1.215$ eV and $\Delta = 2V^2/B$. The value of ε_f was varied between -2.6 eV $[c(f^0) = 0.05]$ and -0.2 eV $[c(f^0) = 0.52]$. Details of the construction of these curves are given in ref. 6. Figure adapted from ref. 35.

illustrate the effects on the different spectroscopies. Here the relative intensities of the peaks are plotted as a function of the $4f^0$ weight (i.e. $\sim 1 - n_f$) for XPS, XAS and BIS of Ce. For XPS the curves represent the intensity of the f^0 peak relative to the total, for 3d XAS and BIS the curves respresent the f^1 intensity relative to the total, in each case with the 4f-valence band hybridization set at 120 meV. These results, taken from model calculations[6,35] indicate that one cannot, for Ce, simply assume a 1:1 correspondence between peak intensities and ground-state occupation coefficients. Indeed the distortion from a 1:1 correspondence between peak intensities and ground state weights is not only large, but also different for the different spectroscopies. To understand why we now consider XPS in detail.

The matrix elements in XPS can be written in the form

$$W_{fi} \propto |<\psi_f'(N)|t|\psi_i(N)>|^2 \rho(E_f) \tag{6}$$

Where W_{fi} is the transition rate and the $\psi_f'(N)$, $\psi_i(N) >|^2$ are the final and the initial state N-particle wave functions. t is the transition operator and $\rho(E_f)$ is the density of final states and is usually assumed to be independent of the energy E_f. The N-particle wave function can be written in terms of an antisymmetrized product of the single electron wave function, ϕ_j of the electron to be ionized and the wave function, $\psi_i(N-1)$, of the other N-1 particles. Equation (6) then reads[46]

$$W_{fi} \propto |<\psi_f'(N-1)|\psi_i(N-1)>|^2| <\phi_\varepsilon|t|\phi_j>|^2 \tag{7}$$

where ϕ_ε is a continuum state. The first term here introduces structure to each of the core level excitations in XPS because whilst the initial state, $\phi_j\psi_i(N-1)$ is a Eigenfunction of an N-particle Hamiltonian, $\psi_f'(N-1)$ is a Eigenfunction of an (N-1) particle Hamiltonian. In order to calculate the XPS spectrum it is necessary to expand the initial state wave function in terms of all the possible final state and the intensity of any one transition to the k-th final state with a j hole will be given by

$$P_{kj} = |<\psi_{fk}'(N-1)|\psi_i(N-1)>|^2 \tag{8}$$

It is common to refer to the observation of multiple XPS peaks associated with one given core level as "shake up". The idea behind this is that the higher binding energy XPS peaks correspond to states where a valence

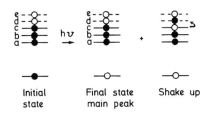

Fig. 10. A standard illustration of shake up.

electron has been excited to a higher unoccupied level, in addition to core level creation, as illustrated in Fig. 10. This picture of shake up contains an element which is not acceptable for our purposes. The picture neglects the fact that one-electron orbitals can be seen as mixtures from a basis set and that the coefficients of mixing can change dramatically in the presence of a core hole (i.e. it neglects the very fact that causes shake up). Indeed as described for Ce in connection with Figs. 1 and 3, the relative energies and order of the basis states changes in the presence of a core hole.

In the early lanthanides the dominant shake up takes a rather simple form because transitions to only a few final states dominate the spectra, and derivation of ground state mixing is quite simple. This will be illustrated with a hypothetical case of a lanthanum compound in which multiplet interaction is neglected. Consider the ground state, ψ_g of La atoms in this compound to have 0.2 4f electrons so that it can be written as the linear combination of the ortho normalized N-particle basis states, $\phi(f^0)$ and $\phi(f^1)$, with zero and one 4f electrons respectively, i.e.

$$\psi_g = a\phi(f^1) + b\phi(f^0) = 0.45\phi(f^1) + 0.89\phi(f^0)$$
$$\psi_g^2 = a^2\phi(f^1)^2 + b^2\phi(f^0) = 0.2\phi(f^1) + 0.8\phi(f^0)^2 \qquad (9)$$

For this system we will label the (N-1) particle XPS final states in the presence of a core hole in the level, $\psi_f(f^0)$ and $\psi_f(f^1)$. These states are linear combinations of the new, (N-1) particle core ionized basis states $\phi'(f^0)$ and $\phi'(f^1)$, with 0 and 14 f electron respectively, whereby $\phi'(f^1)$ is lower in energy than $\phi'(f^0)$ due to the core hole potential. Charge conservation in these basis states is achieved by adjustment of the charge in valence states.

In this simple situation the intensity of the $\psi_g \rightarrow \phi'(f^0)$ and $\psi_g \rightarrow \phi'(f^1)$ XPS transitions would give the weights of $\phi(f^0)$ and $\phi(f^1)$ in the ground state directly if there were no final state mixing. In practice final state mixing should be considered, not only core level XPS, but also for VBPS, XAS and BIS. As an example, consider that the final states for our La compound have 0.95 and 0.05 4f electrons respectively. The final states are then

$$\psi_f^1(f^1) = c\phi'(f^1) + d\phi'(f^0) = 0.97\phi'(f^1) + 0.23\phi'(f^0)$$
$$\psi_f^1(f^0) = d\phi'(f^1) - c\phi'(f^0) = 0.23\phi'(f^1) - 0.97\phi'(f^0) \qquad (10)$$

In the light rare earths the assumption that $\phi(f^n) \approx \phi'(f^n)$ seems to be reasonable so that the relative intensities of the transitions $\psi_g \rightarrow \psi_f(f^1)$ and $\psi_g \rightarrow \psi_f(f^0)$ are

$$I(f^1) = (ac + bd)^2 = (0.45 \times 0.95 + 0.89 \times 0.23)^2 = 0.41$$

$$I(f^0) = (ad - bc)^2 = (0.45 \times 0.23 - 0.89 \times 0.95)^2 = 0.59$$

i.e. the ratio of $f^1:f^0$ XPS peaks is 0.41:0.59 and not 0.2:0.8 as might naively be expected from the constitution of the ground state.

This illustrates how a small amount of final state configuration interaction (or mixing) leads to very significant transfer of intensity to the XPS peak nearer threshold. This is a general result and the similar transfer to the near-threshold peaks would be found in this simple model for BIS, XPS and XAS and was also found for the more sophisticated model, as shown in Fig. 9. The transfer of weight is stronger when the final state mixing is stronger. This final state mixing is increased when the energy separation of the final states is small, or when the hybridization potential mixing the states is large.

In the simple La system as described above there is one observable (the ratio of two XPS peaks) but two unknowns (the coefficients of mixing of f states in both initial and final states). In order to determine the ground state mixing coefficients and the strength of hybridization of the 4f states with the valence band it is necessary to use a model of the photoemission process. The model must include, for instance, degeneracy effects and must be tested with a variety of spectroscopies. Several authors have worked on such models[4-7,47-52]. Here we give a simple illustration of the Gunnarsson-Schönhammer (G-S) model and its implications. More details of the calculations are given in the chapter by Gunnarsson.

The G-S model utilizes the concept of a screening orbital. When a core hole is created all the wave functions of the other N-1 electrons of the systems change, or relax, to adjust to the new potential. In a metallic solid the creation or destruction of the impurity charge must be screened from the rest of metal, otherwise the potential would be changed throughout the metal. The valence electrons thus build up density around an atom where a core hole has been created, as is the case in XPS for example. The distribution of this extra charge may look rather like that in an atomic orbital, so that one can speak of "screening orbitals" and screening processes. The concept of a screening orbital is, however, not restricted to metals and yields good results for some insulators, such as La_2O_3[4] and for some adsorbates[53,54].

The G-S model is suitable for much solid state work because it incorporates partial occupation of the screening orbitals and hybridization effects as well as the dynamics of photoemission. This is done by giving the screening orbital a width and energy in both the initial and final states, as in Fig. 11, which illustrates the results of numerical calculations by Schönhammer and Gunnarsson (S-G model)[48].

In the initial state of photoemission the screening orbital is at an energy $\Delta+$ above the Fermi level and has a width, W, which represents the coupling to the other levels of the system. In the final state, in the presence of a core hole, the screening level is pulled down U_{ac} with respect to the Fermi level by the core hole potential. The level is now $\Delta-$ below E_F. In this model the probability that the screening orbital will be occupied after core level photoemission is related to the position of the orbital and its coupling to the other orbitals. Thus if the screening orbital is narrow and far above E_F in the initial state as shown in inset a, then its occupancy is low. There is then very little overlap between the initial state and the final state with the screening level occupied. This means that the probability of a transition to the "well-screened" final state is small. Most of the XPS intensity will be found in the "poorly-screened" peak lying approximately $\Delta-$ to higher BE and corresponding to the transition to a final state with the screening level almost unoccupied as shown in curve a. If, on the other hand, the width W of the screening level is of the same order as its position $\Delta+$, the level is strongly coupled to the system and its extensive tail below E_F can be interpreted as considerable occupancy, even in the ground state. Conversely in the final state W is also comparable to $\Delta-$ and there is considerable unoccupied character in the well-screened final state. The probability that the screening orbital becomes occupied in the final state of core-level photoemission, is then high and most of the XPS intensity is found in a so-called well-screened peak, as shown in curve d).

In all the final states of a core transition in a metal the change in core potential is screened from the rest of the metal within a very short distance (typically about the Wigner-Seitz radius). In the different final states this screening is done in different ways, often with accompanying plasmon excitations. In general the state with the smallest radius for the screening charge will have the lowest energy, although I suspect that this need not always be the case. In order to emphasize the common features of the screening processes in XPS from different systems we adopted the name "well-screened" to describe the peak due to transitions to the final state of the lowest energy and "poorly-screened" to describe the peaks at higher binding energy[4]. We debated with ourselves for a long time on what terms to use and eventually adopted this nomenclature to stress that the core-ionized final state of lowest energy in lanthanides or actinides was one where the screening cloud has a smaller radius and was <u>energetically more favourable</u>. Unfortunately some workers understood the term poorly-screened to imply a state where the core hole was incompletely screened from the metal and this is incorrect. It has since been suggested to us that "locally-screened" and "diffusely-screened" might have been better labels[55]. This is a good suggestion but to change the names now would only add to any confusion. I am also sure that some people would have chosen to be confused whatever names were used.

There are several examples now known in which the trend in core-level XPS lineshapes follows that predicted within impurity models of the SG type.

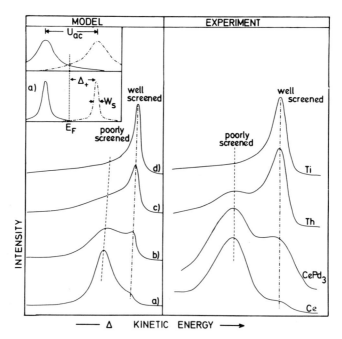

Fig. 11. Left: illustration of the Schönhammer-Gunnarsson model of screening and its implications for the core level line-shapes. The values of Δ_+/W were d = 0.94, c = 0.75, b = 0.56, a = 0.38 and U_{ac} was 1.5 Δ_+[48]. Right: Ce $3d_{5/2}$, Th $4f_{7/2}$, and Ti $2p_{3/2}$ XPS peaks from Ce, CePd$_3$, Th and Ti. The peak binding energies are 883, 333 and 454 eV for the Ce, Th and Ti levels respectively.

Fig. 10 illustrated the case of the Ti $2p_{3/2}$, Th $4f_{7/2}$ and Ce $3d_{5/2}$ XPS peaks from ref. 3. The screening levels in these cases are the Ti 3d, Th 5f and Ce 4f levels respectively. These levels are increasingly localized and decoupled from the other valence levels in the order Ti > Th 5f > Ce 4f so that the weight in the well-screened peak also decreases in that order. It should be recognized that the S-G model uses an impurity-like screening level. There have been attempts to study what happens when the screening level is part of a band but one is safest extrapolating from the S-G results when the direct interaction between screening orbitals on different sites is small. This is certainly the case for the Ce 4f and the Th 5f levels but must be less true for the Ti 3d levels.

The core level XPS lineshapes of about sixty La, Ce, Pr, Th and U intermetallic compounds have now been studied. A principle aim of many of these studies was to use the data to obtain trends in the degree of localization of the 4f and 5f states (see e.g. refs. 4-6,30,36,56). In those compounds whose magnetic properties were known, a good correlation was found between the occurence of "localized" magnetic behaviour and large XPS intensity in the poorly screened peaks; i.e. electrons which are poorly coupled to the system for XPS purposes are also poorly coupled for magnetic properties. Also Bremsstrahlung Isochromat (BIS) studies of U and Ce compounds usually show 5f and 4f levels near E_F when the core level XPS lineshapes indicate good coupling of the screening levels to the system. In contrast, in substances like Ce, where the poorly-screened peaks are relatively strong, the screening levels are found several eV above E_F by BIS. These observations thus support the use of the S-G model in such cases. Further support is given by the remarkable success of the S-G model in explaining the observed spectral functions in XAS[26], BIS[30,36] valence band photoemission[30,31,58,59] and ground state magnetic properties[59]. This success comes largely because the S-G scheme is applicable to many different spectroscopies and because it provides a suitable framework for inclusion of many effects of f-orbital degeneracy, multiplet interaction and the effects of the density of valence states.

THE ACHIEVEMENTS OF CORE LEVEL SPECTROSCOPY IN LANTHANIDE RESEARCH

The expansion during the last two decades of research into the lanthanides has been larger than in most scientific research. Our present understanding of the solid state properties of lanthanides is a result of studies using many techniques, not just core level spectroscopy and one must recognize that much of the information came from many sources simultaneously. The principle uses of core level spectroscopy in lantanide research have been to quantify the f-count and hybridization effects in materials exhibiting exotic phenomena. 4f hybridization values for early rare earths materials (defined as valence states) were found to be much higher than previously thought. Values for Ce range from ~ 25 to ~ 200 meV[36]. Because of this strong hybridization, f-count values cannot be taken directly from core level spectra but estimates must take hybridization and the dynamics of the spectroscopy into account. For Ce intermetallics core level spectroscopies indicate that the Ce 4f count ranges between ~ 0.7 and ~ 1.05, in contrast to the values of 0.00 to 1.00 current of few years ago.

For the later rare earths hybridization effects on the rare earths are much smaller and it appears that core level peak intensities are more or less proportional to the weights of the different f^n configurations in the ground state.

CONCLUDING REMARKS: THE FUTURE OF CONCEPTS GENERAL IN CORE LEVEL SPECTROSCOPY OF THE LANTHANIDES

It is safe to assume that the use of core level spectroscopy of lanthanide materials will continue and they will be used to determine f-counts under conditions which cannot be readily studied by more classical methods, e.g. for surface studies, under moderate pressures, and as a function of temperature. The use of multiplet structure as a diagnostic tool for the ground state properties is more recent (see e.g. ref. 38) but it shows great promise, in particular in combination with results from magnetic X-ray dichroism in XAS (see refs. 60,61).

In many ways the lanthanide materials provide a test case for core level spectroscopy because many effects are so clear cut. We hope and expect that the techniques developed for studies of 4f level occupancy, Coulomb interactions and hybridization by core level spectroscopies will prove applicable to the transition metal d electrons, especially in transition metal compounds. Increased applications of these concepts to surfaces, adsorbates, and actinides is also to be expected.

ACKNOWLEDGEMENTS

This paper contains elements from research that has been in progress for some years. The people involved in this cooperation included J.W. Allen, Y. Baer, M. Campagna, J.-M. Esteva, F.U. Hillebrecht, O. Gunnarsson, R. Karnatak, G. van der Laan, N. Mårtensson, D. Menzel, G. Sawatzky, K. Schönhammer, B.T. Thole and E. Umbach. I thank them for their various roles and for their very real contributions to this work. Thanks are also due to J.B. Goekdoop and J.F. van Acker for critical reading of the manuscript and D.D. van der Wey for typing it. I apologize to all the researchers I have not cited here, but it is no longer possible to cite all the papers on spectroscopy of lanthanides and if one considers all the administration I must do it is a miracle that I was able to cite anybody.

This work was supported in part by the dutch SON organization using ZWO funds. Thanks are also due to the LURE staff in Orsay, France, for their aid with the XAS measurements. This research would not have been possible without a NATO travel grant.

REFERENCES

1. M. Cini, Density of two interacting holes in a solid, Sol. State Comm. 20: 605 (1976).
2. G.A. Sawatzky, Quasiatomic Auger spectra in narrow-band metals, Phys. Rev. Lett. 39:504 (1977).
3. P.A. Cox, J.K. Lang and Y. Baer, J. Phys. F 11:113 (1981); J.K. Lang, P.A. Cox and Y. Baer, Study of the 4f and valence band density of states in rare-earth metals: II. Experiment and results, J. Phys. F 11:121 (1981).
4. J.C. Fuggle, M. Campagna, Z. Zolnierek, R. Lasser and A. Platau, Observation of a relationship between core-level lineshapes in photoelectron spectroscopy, Phys. Rev. Lett. 45:1597 (1980).
5. O. Gunnarsson and K. Schönhammer, Electron spectroscopies for Ce compounds in the impurity model, Phys. Rev. B 28:4315 (1983).
6. J.C. Fuggle, F.U. Hillebrecht, Z. Zolnierek, R. Lasser, Ch. Freiburg, O. Gunnarsson and K. Schönhammer, Electronic structure of Ce and its intermetallic compounds, Phys. Rev. B 27:7330 (1983).
7. O. Gunnarsson and K. Schönhammer, Double occupancy of the f orbital in the Anderson model for Ce compounds, Phys. Rev. B 31:4815 (1985).

8. N.F. Mott, The basis of the electron theory of metals, with special reference to the transition metals, Proc. Phys. Soc. London A 62:416 (1949).
9. N.F. Mott, "Metal Insulator Transitions", Taylor and Francis London (1974).
10. J.A. Wilson, Systematics of the breakdown of Mott insulation in binary transition metal compounds, Adv. in Phys. 21:143 (1972).
11. J.B. Goodenough, "Magnetism of the Chemical Bond" Interscience, New York (1963).
12. J. Hubbard, Electron correlation in narrow energy band. Parts I, II and III, Proc. Roy. Soc. London, 276:238 (1963); 277:237 (1964); 281:404 (1964).
13. C. Herring, "Magnetism", editors G.T. Rado and H. Suhl, Academic Press London (1966) Vol. IV.
14. P.W. Anderson, Localized magnetic states in metals, Phys. Rev. 124:41 (1961).
15. F. Kondo, "Theory of dilute magnetic alloys", in "Solid State Physics", editors F. Seitz and D. Turnbull, Academic Press, New York 23:183 (1969).
16. L.L. Hirst, Theory of magnetic impurities in metals, Phys. Kondens. Mat. 11:225 (1970).
17. See e.g. C.M. Varma, Mixed valence compounds, Rev. Mod. Phys. 48:219 (1976).
18. See. e.g. F.A. Cotton and G. Wilkinson, Adv. Inorg. Chem. J. Wiley, London (1962).
19. The term "hybridization" often has different meansing as used by physicists and chemists. Here it can usually be taken to mean mixing of orbitals on either the same or on different sites.
20. P.W. Bridgeman, Proc. Am. Acad. Arts Sci. 62:211 (1927).
21. W.H. Zachariasen, Quoted by A.W. Lawson and T.Y. Tang, Phys. Rev. 76:301 (1949); L. Pauling, Quoted by A.F. Schuck and J.H. Sturdivant, J. Chem. Phys. 18:145 (1950).
22. B. Johansson, The α-γ transition in cerium is a Mott transition, Philos. Mag. 30:469 (1974).
23. U. Kornstad, R. Lasser and B. Lengeler, Investigation of the γ-α phase transition in cericum by Compton scattering, Phys. Rev. B 21:1898 (1980); B. Lengeler, G. Materlik and J.E. Muller, L-edge X-ray absorption of γ- and α-Cerium, Phys. Rev. B 28:2276 (1983).
24. J.C. Fuggle, F.U. Hillebrecht, J.-M. Esteva, R.C. Karnatak, O. Gunnarsson and K. Schönhammer, f-count effects in x-ray absorption spectra of the 3d levels in Ce and its intermetallic compounds, Phys. Rev. B 27: 4637 (1983).
25. G. Krill, J.P. Kappler, A. Meyer, L. Abadli and M.F. Ravet, Surface and bulk properties of cerium atoms in several cerium compounds: XPS and x-ray absorption measurements, J. Phys. F: Metal Phys. 11:1713 (1981); D. Ravot, C. Godart, J.C. Achard and P. Lagarde, Rare earth valence evaluation by x-ray absoprtion spectroscopy in rare earth chalcogenides, "Valence Fluctuations in Solids", editors L.M. Falicov, W. Hanke, M.B. Maple, North Holland Publishing Company (1981)423.
26. K.R. Bauchspiess, W. Boksch, E. Holland-Moritz, H. Launois, R. Pott and D. Wohlleben, L_{III} absorption edges of Ce- and Yb-intermediate valence compounds: non-existence of tetravalent cerium compounds, "Valence Fluctuations in Solids", editors L.M. Falicov, W. Hanke, M.B. Maple, North Holland Publishing Company (1981) 417.
27. E. Wuilloud, H.R. Moser, W.-D. Schneider and Y. Baer, Electronic structure of γ- and α-Ce, Phys. Rev. B 28:7354 (1983).
28. J.W. Allen and R.M. Martin, Kondo volume collapse and the $\gamma \rightarrow \alpha$ transition in cerium, Phys. Rev. Lett. 49:1106 (1982).
29. J.C. Fuggle, J.W. Allen, O. Gunnarsson, B. Lengeler, N. Mårtensson, S.-J. Oh, K. Schönhammer, G. v.d. Laan, J.-M. Esteva and R.C. Karnatak, Spectroscopic studies of the α-γ Ce transition, in preparation.

30. K. Andres, J.E. Graebner and H.R. Ott, 4f virtual bound state formation in CeAl$_3$ at low temperatures, Phys. Rev. Lett. 35:799 (1975); F. Steglich, J. Aarts, C.D. Bredl, W. Lieke, D. Meschede, W. Franz and H. Schäfer, Superconductivity in the presence of strong Pauli paramagnetism: CeCu$_2$Si$_2$, Phys. Rev. Lett. 43:1892 (1979); G.R. Stewart, Heavy-fermion systems, Rev. Mod. Phys. 56:755 (1984) and references therein; Z. Fisk, H.R. Ott, T.M. Rice and J.L. Smith, Heavy-electron metals, Nature 320:124 (1986) and references therein.
31. P.A. Lee, T.M. Rice, J.W. Serene, L.J. Sham and J.W. Wilkins, Theories of heavy electron systems, Comments Cond. Mat. Phys. 12:99 (1986) and references therein.
32. L. Schlapbach, S. Hüfner and T. Riesterer, Core level spectroscopy of heavy-fermion Ce compounds, J. Phys. C: Solid State Phys. 19:L63 (1986).
33. J.F. Herbst, Replicate core levels and 4f excitation energies in fluctuating valence materials, in "Valence Fluctuations in Solids", editors L.M. Falicov, W. Hanke, M.B. Maple, North Holland Publishing Company (1981) 73.
34. See e.g. M. Campagna, G.K. Wertheim and Y. Baer, in "Photoemission in Solids II", editors L. Ley and M. Cardona, Springer Berlin (1979).
35. F.U. Hillebrecht, J.C. Fuggle, G.A. Sawatzky, M. Campagna, O. Gunnarsson and K. Schönhammer, Transition to nonmagnetic f states in Ce intermetallic compounds studied by bremsstrahlung isochromat spectroscopy, Phys. Rev. B 30:1777 (1984).
36. J.C. Fuggle, Spectroscopic studies of cerium compounds, Physica B 130:56 (1985) and references therein. This paper was intended as a summary of a large fraction of the high energy spectroscopic estimates of 4f hybridization in Ce compounds.
37. G. v.d. Laan, B.T. Thole, G.A. Sawatzky, J.C. Fuggle, R. Karnatak, J.-M. Esteva and B. Lengeler, Identification of the relative population of spin-orbit split states in the ground state of a solid, J. Phys. C: Solid State Phys. 19:817 (1986).
38. M. Ohno and G. Wendin, A many-body calculation of 3p XPS and Auger spectra for Zn, J. Phys. B 12:1305 (1979).
39. J. Sugar, Phys. Rev. A 6:1764 (1972); Potential barrier effects in photoabsorption. II. Interpretation of photoabsorption resonances at the 4d threshold, Phys. Rev. B 5:1785 (1972).
40. V.F. Demekhin, Fine structure of $M_{IV,V}$ and $N_{IV,V}$ absorption spectra of rare-earth elements, Sov. Phys. Solid State 16:659 (1974).
41. B.T. Thole, G. v.d. Laan, J.C. Fuggle, G.A. Sawatzky, R.C. Karnatak and J.-M. Esteva, 3d X-ray-absorption lines and the $3d^94f^{n+1}$ multiplets of the lanthanides, Phys. Rev. B 32:5107 (1985).
42. J.-M. Esteva, R.C. Karnatak, J.C. Fuggle and G.A. Sawatzky, Selection rules and multiplet effects in comparison of x-ray absorption and photoemission peak energies, Phys. Rev. Lett. 50:910 (1983).
43. F.P. Netzer, G. Strasser and J.A.D. Matthew, Selection rules in electron-excited 4f → 4f transitions at intermediate incident energies, Phys. Rev. Lett. 51:211 (1983).
44. J.A.D. Matthew, G. Strasser and F.P. Netzer, Multiplet effects and breakdown of dipole selection rules in the 3d → 4f core-electron-energy-loss spectra of La, Ce and Gd, Phys. Rev. B 27:5839 (1983).
45. Note that the precise definitions of the quantities we designate ε, U_{ff} and U_{ac} vary from one publication to another. The values and signs we use are (hopefully) self-consistent and realistic enough to illustrate the principles of 4f electron behaviour in elements like Ce, but should not be thoughtlessly compared with values given in other works.
46. R. Manne and T. Åberg, Koopman's theorem for inner shell ionization, Chem. Phys. Lett. 7:282 (1970).
47. A. Kotani and Y. Toyozawa, Optical spectra of core electron in metals with an incomplete shell. I. Analytic features, J. Phys. Soc. Japan 35:1073 (1973); A. Kotani and Y. Toyozawa, Optical spectra of core electrons in metals with an incomplete shell. II. Numerical calcula-

tions of overall lineshapes, J. Phys. Soc. Japan 35:1082 (1973).
48. K. Schönhammer and O. Gunnarsson, Shape of core level spectra in adsorbates, Solid State Comm. 28:691 (1977); see also Solid State Comm. 26:147,399 (1978); and Z. Phys. B 30:297 (1978).
49. A. Kotani, Japan. J. Phys. 46:488 (1979).
50. S.J. Oh and S. Doniach, Screening effects in the core-level spectra of mixed-valence compounds, Phys. Rev. B 26:2085 (1982).
51. A. Fujimori, 4f and core-level photoemission satellites in cerium compounds, Phys. Rev. B 27:3992 (1983); see also A. Fujimori, Phys. Rev. B 28:2881,4489 (1983).
52. A. Fujimori and J.H. Weaver, 4f-5d hybridization and the α-γ phase transition in cerium, Phys. Rev. B 32:3422 (1985).
53. J.C. Fuggle, E. Umbach, D. Menzel, K. Wandelt and C.R. Brundle, Adsorbate lineshapes and multiple lines in XPS; comparison of theory and experiment, Solid State Comm. 27:65 (1978).
54. E. Umbach, On the interpretation of XPS lineshapes of weakly chemisorbed N_2 on transition metals, Solid State Comm. 51:365 (1984).
55. G. Wendin, private communication.
56. W.-D. Schneider and C. Laubschat, Phys. Rev. Lett. 46:1023 (1981).
57. A. Jayaraman, "Handbook of the Physics and Chemistry of the Rare Earths", editors, K.A. Gschneidner and L.R. Eyring, North Holland Publishing Company Amsterdam (1979) p 730 in Vol. I and p. 578 in Vol. II.
58. O. Gunnarsson, this volume.
59. O. Gunnarsson, K. Schönhammer, J.C. Fuggle, F.U. Hillebrecht, J.-M. Esteva, R.C. Karnatak and B. Hillebrand, Occupancy and hybridization of the f level in Ce compounds, Phys. Rev. B 28:7330 (1983).
60. B.T. Thole, G. v.d. Laan and G.A. Sawatzky, Strong magnetic dichroism predicted in the $M_{4,5}$ x-ray absorption spectra of magnetic rare-earth materials, Phys. Rev. Lett. 55:2086 (1985).
61. G. v.d. Laan, B.T. Thole, G.A. Sawatzky, J.B. Goedkoop, J.C. Fuggle, J.-M. Esteva, R. Karnatak, J.P. Remeika and H.A. Dabkowska, Experimental proof of magnetic x-ray dichroism, Phys. Rev. B, submitted.
62. D.A. Shirley, "Photoemission in Solids I", editors M. Cardona and L. Ley, Springer Verlag Berlin (1978) 166 ff, and references therein.
63. C.S. Fadley, "Electron Spectroscopy, Theory, Techniques and Applications, Volume 2", editors C.R. Brundle and A.D. Baker, Academic Press London (1978) 98 ff and references therein.
64. S.P. Kowalczyk, L. Ley, R.A. Pollak, F.R. McFeely and D.A. Shirley, Phys. Rev. B 7:4009 (1973).
65. S.P. Kowalczyk, L. Ley, F.R. McFeely and D.A. Shirley, Phys. Rev. B 11:1721 (1975).
66. F.R. McFeely, S.P. Kowalczyk, L. Ley and D.A. Shirley, Phys. Lett. A 49:301 (1974).
67. J.C. Fuggle and S.F. Alvarado, Core-level lifetimes and determined by x-ray photoelectron spectroscopy measurements, Phys. Rev. A 22:1615 (1980).
68. J.-M. Esteva, R.C. Karnatak, J.C. Fuggle and G.A. Sawatzky, Selection rules and multiplet effects in comparison of x-ray absorption and photoemission peak energies, Phys. Rev. Lett. 50:910 (1983).
69. G.A. Sawatzky and J.C. Fuggle, to be published.

INTERMEDIATE VALENCE SPECTROSCOPY

O. Gunnarsson

Max-Planck Institut für Festkörperforschung
D-7000 Stuttgart 80, W. Germany

K. Schönhammer

Institut für Theoretische Physik
Universität Göttingen
D-3400 Göttingen, W. Germany

Spectroscopic properties of intermediate valence compounds are studied using the Anderson model. Due to the large orbital and spin degeneracy N_f of the 4f-level, $1/N_f$ can be treated as a small parameter. This approach provides exact $T = 0$ results for the Anderson impurity model in the limit $N_f \to \infty$, and by adding $1/N_f$ corrections some properties can be calculated accurately even for $N_f=1$ or 2. In particular valence photoemission and resonance photoemission spectroscopies are studied. A comparison of theoretical and experimental spectra provides an estimate of the parameters in the model. Core level photoemission spectra provide estimates of the coupling between the f-level and the conduction states and of the f-level occupancy. With these parameters the model gives a fair description of other electron spectroscopies. For typical parameters the model predicts two structures in the f-spectrum, namely one structure at the f-level and one at the Fermi energy. The resonance photoemission calculation gives a photon energy dependence for these two peaks in fair agreement with experiment. The peak at the Fermi energy is partly due to a narrow Kondo resonance, resulting from many-body effects and the presence of a continuous, partly filled conduction band. This resonance is related to a large density of low-lying excitations, which explains the large susceptibility and specific heat observed for these systems at low temperatures.

1. INTRODUCTION

The 4f-orbital in the rare earth series has a small spatial extent and is mainly located inside both the 5d, 6s conduction states and the 5s, 5p core states. The Coulomb interaction U between two 4f-electrons is therefore large and the hybridization strength Δ between the 4f orbital and the conduction states is small. For some systems two configurations f^n and f^{n+1} may, nevertheless, be almost degenerate on the energy scale Δ. The mixing between these two configurations can then become very important, leading to a socalled mixed valence behaviour. For such systems it is important to take into account the hopping between the f-state and the conduction states, although the hopping probability is rather small. The f-level couples to a continuum of conduction states and the f-electrons can show both itinerant and localized properties, depending on the experimental probe.

Here we focus on Ce mixed valence compounds, which have many interesting properties. The linear coefficient γ in the specific heat at low temperatures is usually large. While typical values of γ for simple metals and transition metals are in the range 1-10 mJ/mole K^2, mixed valence Ce compounds typically have γ values of the order 10-100 mJ/mole K^2 and some Ce compounds (heavy fermion systems) have a γ of the order 1000 mJ/mole K^2. [1]. Similarly, the static T = 0 susceptibility χ is usually very large, and for the heavy fermion systems it is one to two orders of magnitude larger than for a typical transition metal [1]. These results show that Ce compounds can have a very large density of low-lying excitations. The thermodynamic properties of Ce compounds have therefore often been interpreted in the so called promotional model. In this model the f^0 and f^1 configurations are assumed to be almost degenerate (to within 10^{-1} eV) and the coupling $\Delta \sim 10^{-2}$ eV is asumed to be very small. This can be modelled by noninteracting electrons with the f-level ε_f placed very close to the Fermi energy $\varepsilon_F = 0$ ($|\varepsilon_f| \sim 0.1$ eV) and with a weak coupling Δ between the conduction states and the f-level [2-5]. This leads to a narrow resonance at ε_f and a large density of states at ε_F, which could explain the large values of γ and χ.

5-10 year ago it was gradually realized that this model is hard to reconcile with spectroscopic properties. Photoemission spectroscopy (PES) shows an f-related feature about 2 eV below the Fermi energy for Ce [6,7] suggesting that $\varepsilon_f \sim -2$ eV. Furthermore, the 3d core level X-ray photoemission spectroscopy (XPS) measurements [8] are hard to understand unless Δ is assumed to be much larger than 10^{-2} eV. Later PES measurements showed that the f-related peak is rather broad suggesting that $\Delta \sim 1$ eV [9]. It was also pointed out that thermodynamic and spectroscopic data might be reconciled if Kondo effects are included [9]. Further interest in PES was generated by the observation that the PES spectrum has two f-related peaks [10], making the interpretation of the f-spectrum an intensively debated issue.

For many Ce compounds it is nontrivial to identify the f-features in the valence PES spectrum. The discovery that Ce compounds have a resonantly enhanced cross section at the photon energy $\omega \sim 122$ eV was therefore a great progress [7]. At this photon energy, in particular the emission of f-electrons is enhanced, which simplifies the identification of the f features in the spectrum. The two features mentioned above were, however, found to have different photon energy dependences [11], suggesting that the two features might have different characters. Later evidence showed that both have a substantial amount of f-character [12], but it has remained a puzzle why the photon energy dependencies are different.

The main reason for the discrepancy between the parameters deduced from thermodynamic and spectroscopic experiments, is not experimental uncertainties, but the interpretation of these experiments. It is therefore of great interest to study a microscopic model to investigate what information about the parameters is contained in the different experiments. Much of the theoretical work for Ce compounds has been based on the Anderson model. This model contains a localized level, extended conduction states, the hybridization between these states and the Coulomb interaction between the localized electrons. This may be considered as the simplest model which gives mixed valence behaviour. The model has been used both for the description of spectroscopic and thermodynamic properties. It is therefore well suited for attempts to relate thermodynamic and spectroscopic properties.

The large value of the Coulomb interaction U together with the weak coupling of the f-level to a continuum, makes the calculation of the properties of the Anderson model difficult. The Hartree-Fock approximation, for instance, leads to qualitatively incorrect results in the parameter range relevant for Ce. A great progress was therefore the realization of Anderson [13] and Ramakrishnan [14] that one can treat $1/N_f$ as a small parameter, where N_f is the degeneracy of the f-level. Ramakrishnan and Sur [15] have calculated some thermodynamic properties using this idea. We have used the $1/N_f$ idea to develop a method for calculating electron spectra [16]. This method can be used for T=0 and for a finite or infinite U. Alternative methods were later proposed for finite T and for $U=\infty$ [17].

In Sec. 2 we describe the model. In Sec. 3 the calculation of ground-state properties is described and a method for calculating spectra is developed in Sec. 4. Resonance photoemission is treated in Sec. 5. The results are compared with experiment in Sec. 6.

2. MODEL

We use the Anderson impurity model in the form

$$H = \sum_{\nu=1}^{N_f} \left\{ \int d\varepsilon\, \varepsilon \psi^+_{\varepsilon\nu} \psi_{\varepsilon\nu} + \varepsilon_f \psi^+_\nu \psi_\nu + \int d\varepsilon [V(\varepsilon) \psi^+_\nu \psi_{\varepsilon\nu} + H.C.] \right\} + U \sum_{\nu\mu} n_\nu n_\mu \tag{2.1}$$

where the first term describes the conduction states, the second the f-level, the third the hopping between these two types of states and the fourth the Coulomb interaction. The f-level has the degeneracy N_f and the index ν describes the spin- and orbital indices. The conduction states have been transformed to the same representation ν [16]. Eq. (2.1) is an impurity model, i.e. only the f-level on one Ce atom is treated explicitely. For the core spectroscopies we also include the terms [18]

$$H_c = \varepsilon_c \sum \psi^+_{c\nu} \psi_{c\nu} - U_{fc}(N_f - \sum \psi^+_{c\nu}\psi_{c\nu}) \sum n_\nu \tag{2.2}$$

where ε_c is the core level energy and U_{fc} the Coulomb interaction between a core hole and an f-electron. If all the N_f core states are filled the last term vanishes. This interaction is very important since it changes the ordering of the f-configurations when a core hole is created.

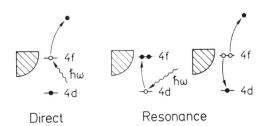

Fig. 1. Schematic description of the processes in resonance photoemission.

To be able to study resonance photoemission we must extend the model further. The processes [7] we have in mind are shown schematically in Fig. 1. To the left we show the direct process. This is described by the first term in the dipole operator

$$V_p = \int dE \tau(E) \psi^+_{E\nu} \psi_\nu + \sum_\nu \tau_c \psi^+_\nu \psi_{c\nu} \equiv V_d + V_r \qquad (2.3)$$

where $\psi^+_{E\nu}$ is the creation operator for a high-lying conduction state which is different from the states ε in Eq. (2.1). The z-axis has been placed so that the quantum number ν is conserved in the photoemission process. We also introduce the term

$$T = \sum_\nu \int dE \, E \, \psi^+_{E\nu} \psi_{E\nu} \qquad (2.4)$$

which is the kinetic energy of the emitted electrons. In the transition $4d^{10}4f^n \to 4d^9 4f^{n+1}$ the selection rules lead to a strong coupling to multiplets at high energies. In resonance photoemission, we take this into account by shifting the $4d^9 4f^{n+1}$ configurations by an energy ε_M^{n+1}. The right hand part of Fig. 1 shows the resonance process. First a 4d electron is excited to the 4f level in the photoemission process. This is described by the second term in (2.3). Then an f-electron falls back into the 4d hole and another f-electron is emitted. This leads to the same final state as the direct process and the amplitudes for these processes have to be added. The filling of the 4d hole is described by the term

$$V_A = \int dE \sum_{\nu\nu'\nu_1\nu_2} v_{\nu\nu'\nu_1\nu_2}(E) \psi^+_{E\nu} \psi^+_{c\nu'} \psi_{\nu_2} \psi_{\nu_1} + h.c. \qquad (2.5)$$

where v is a Coulomb matrix element. Since there is a transition from an f-level to a d-level, the leading Coulomb matrix elements occur for $|\nu'-\nu_2| \leqslant 1$ and $|\nu-\nu_1| \leqslant 1$. To simplify the problem, we assume the decay term

$$V_A = \int dE \, v(E) \sum_{\nu\nu'} \psi^+_{E\nu} \psi^+_{c\nu'} \psi_{\nu'} \psi_\nu + h.c. \qquad (2.6)$$

i.e., that $\nu'=\nu_2$ and $\nu=\nu_1$. The terms (2.1-4, 2.6) provide a simple model problem for resonance photoemission.

As mentioned in the introduction, we use a $1/N_f$ method. In particular we are interested in the limit $N_f \to \infty$. To define this limit, we must specify how $V(\varepsilon)$ depends on N_f. Since $V(\varepsilon)$ enters the calculation in the combinations $N_f V(\varepsilon)^2$ and $V(\varepsilon)^2$, it is useful to define

$$\tilde{V}(\varepsilon) = \sqrt{N_f} \, V(\varepsilon) \qquad (2.7)$$

and require that $\tilde{V}(\varepsilon)$ is independent of N_f. The leading terms, containing $N_f V(\varepsilon)^2$, are then independent of N_f, while the smaller terms, containing $V(\varepsilon)^2$, vanish for $N_f \to \infty$. The limit $N_f \to \infty$ is therefore particularly simple, but as shown later it is nevertheless closely related to the case of a finite but large N_f.

3. GROUND-STATE PROPERTIES IN THE $1/N_f$ METHOD

Below we illustrate how ground-state properties can be calculated and how the $1/N_f$ method can be applied. We use a variational approach [16]. The first step is the introduction of a many-electron basis set. As a starting point we introduce a "Fermi sea" state

$$|0\rangle = \prod_{\nu=1}^{N_f} \psi^+_{c\nu} \prod_{\varepsilon \leq \varepsilon_F} \psi^+_{\varepsilon\nu} |\text{vacuum}\rangle \qquad (3.1)$$

where all the conduction states below the Fermi energy $\varepsilon_F=0$ are filled and the f-level is empty. This state couples to

$$|\varepsilon\rangle = \frac{1}{\sqrt{N_f}} \sum_\nu \psi^+_\nu \psi_{\varepsilon\nu} |0\rangle \qquad (3.2)$$

These states were first introduced by Varma and Yafet [19] for $N_f=2$. The states $|\varepsilon\rangle$ couple to f^2-states

$$|\varepsilon'\rangle = \frac{1}{\sqrt{N_f(N_f-1)}} \sum \psi^+_\nu \psi_{\varepsilon\nu} \psi^+_{\nu'} \psi_{\varepsilon'\nu'} |0\rangle \qquad (3.3)$$

and to states with an electron-hole pair

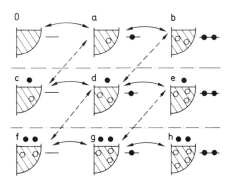

Fig. 2. Schematic representation of basis states. Solid circles show electrons and open circles holes. In each part of the figure the conduction electrons are shown to the left and the f-level to the right. The arrows show which states couple to each other. A solid line indicates a strength $\sim V$ and a dashed line a strength $\sim V/\sqrt{N_f}$.

$$|\tilde{\varepsilon}'\varepsilon\rangle = \frac{1}{\sqrt{N_f}} \sum_{\tilde{\varepsilon}'\nu} \psi^+_{\tilde{\varepsilon}'\nu} \psi_{\varepsilon\nu} |0\rangle \qquad (3.4)$$

where $\tilde{\varepsilon}' > \varepsilon_F$. These and further states are shown schematically in Fig. 2. A systematic discussion was given in Ref. 20. The ground-state $|E_0(N)\rangle$ is now written as

$$|E_0(N)\rangle = \sum_i c_i |i\rangle \qquad (3.5)$$

where the basis set $\{|i\rangle\}$, for instance, may include the states (3.1-3.4). For the calculation we need the matrix elements of the Hamiltonian (2.1). We obtain, for instance,

$$\langle \varepsilon | H | 0 \rangle = \tilde{V}(\varepsilon) \qquad (3.6)$$

and

$$\langle \tilde{\varepsilon}''\varepsilon | H | \varepsilon' \rangle = \frac{\tilde{V}(\tilde{\varepsilon}'')}{\sqrt{N_f}} \delta(\varepsilon-\varepsilon') \qquad (3.7)$$

In general, we find that the coupling between the states in the first row is of the order $\tilde{V}(\varepsilon)$, while the coupling between a state in the first row and a state in the second row is of the order $\tilde{V}(\varepsilon)/\sqrt{N_f}$. For $N_f \to \infty$, it is therefore sufficient to include the states in the first row, if U is so large that states with more than two f-electrons can be neglected. For a finite N_f the states in the second and third row give contributions of the order $(1/N_f)^1$ and $(1/N_f)^2$, respectively. Here we show explicitly a calculation for $U=\infty$ and to lowest order in $(1/N_f)$. The ground-state is then given by

$$|E_0(N)\rangle = A(|0\rangle + \int_{-B}^{0} d\varepsilon \, a(\varepsilon)|\varepsilon\rangle) \qquad (3.8)$$

We define the quantities ΔE and δ in terms of the ground-state energy $E_0(N)$

$$E_0(N)\rangle = \langle 0|H|0\rangle + \Delta E \equiv \langle 0|H|0\rangle + \varepsilon_f - \delta \qquad (3.9)$$

We now solve the Schrödinger equation

$$H|E_0(N)\rangle = E_0(N)|E_0(N)\rangle \qquad (3.10)$$

by applying $\langle 0|$ and $\langle\varepsilon|$ from the left. This gives the two equations

$$\Delta E = \int d\varepsilon \, \tilde{V}(\varepsilon) \, a(\varepsilon) \tag{3.11}$$

$$(\Delta E - \varepsilon_f + \varepsilon) \, a(\varepsilon) = \tilde{V}(\varepsilon) \tag{3.12}$$

Solving for ΔE we obtain

$$\Delta E = \int_{-B}^{0} \frac{\tilde{V}(\varepsilon)^2}{\Delta E - \varepsilon_f + \varepsilon} \, d\varepsilon = \tilde{V}^2 \ln \frac{\delta}{B+\delta} \tag{3.13}$$

In the last equality we have used Eq. (3.9) and asumed that $\tilde{V}(\varepsilon) \equiv \tilde{V}$ is a constant. If we further assume that $-\varepsilon_f \gg \tilde{V}^2$ we obtain

$$\delta = \exp(\varepsilon_f/\tilde{V}^2) \tag{3.14}$$

The quantity δ will be called the Kondo energy. It plays an important role both for the thermodynamic and some of the spectroscopic properties. Once ΔE or δ have been calculated it is easy to solve (3.12) for $a(\varepsilon)$, which gives the f-occupancy n_f.

An important issue is the accuracy which can be obtained in this approach. In Table 1 we show results for $N_f=1$. In this case double occupancy of the f-level is not possible and the problem therefore reduces to a one-electron problem, which can easily be solved exactly. The $N_f=1$ case is furthermore the most unfavourable case for a $1/N_f$ expansion and therefore a good test case. In the calculation we have used

$$\pi V(\varepsilon)^2 = \Delta \sqrt{1 - (\varepsilon/B)^2} \qquad |\varepsilon| \leq B \tag{3.15}$$

Table 1. The energy $\Delta E - \varepsilon_f$ and the f-occupancy n_f as a function of ε_f for $N_f = 1$, $\Delta = 0.75$ eV and $B = 6$ eV.

Order	$\varepsilon_f=0$		$\varepsilon_f=-1$		$\varepsilon_f=-2$	
	$\Delta E-\varepsilon_f$	n_f	$\Delta E-\varepsilon_f$	n_f	$\Delta E-\varepsilon_f$	n_f
$(1/N_f)^0$	-0.534	0.283	-0.054	0.813	-0.001	0.996
$(1/N_f)^1$	-0.714	0.487	-0.408	0.833	-0.291	0.919
$(1/N_f)^2$	-0.726	0.497	-0.416	0.822	-0.294	0.916
Exact	-0.726	0.500	-0.416	0.822	-0.294	0.916

The table illustrates that for $N_f=1$, the calculation to order $(1/N_f)^0$ can give rather unreliable results. However, already the $(1/N_f)^1$ results are quite good and the $(1/N_f)^2$ results are usually correct to three digits.

The static, $T=0$ susceptibility χ can also be calculated, for instance by using methods [21] similar to the ones described for the spectroscopies below. For $U=\infty$ and to order $(1/N_f)^0$ we find [16]

$$\chi = \frac{1}{3} j (j + 1) \frac{n_f}{\delta} \qquad (3.16)$$

where we have assumed that the localized level has a spin j ($N_f = 2j+1$). We note that χ is inversely proportional to δ. Since δ depends exponentially on ε_f and V^2, δ can become very small and χ very large. To illustrate the accuracy of the calculation of χ, we show in Fig. 3 results for $N_f= 2$ [22]. These results are compared with exact results obtained from the Bethe Ansatz [23].

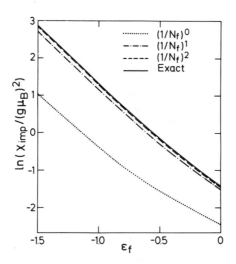

Fig. 3. The spin susceptibility as a function of ε_f for $N_f= 2$. The figure shows the exact Bethe Ansatz result (full curve) and the $1/N_f$ results to order $(1/N_f)^0$ (.....), $(1/N_f)^1$ (-.-.-) and $(1/N_f)^2$ (-----). The parameters are $\pi V(\varepsilon)^2 \equiv 0.5$ eV and $B = 10$ eV.

The figure illustrates that for N_f as small as 2, the $(1/N_f)^0$ results are quite inaccurate. The $(1/N_f)^1$ results are, however fairly accurate, and the $(1/N_f)^2$ results are very close to the exact Bethe Ansatz results.

4. SPECTROSCOPIC PROPERTIES

In the socalled sudden approximation [24] many of the spectroscopies can be related to a one-particle Green's function. The f-level PES is, for instance, related to the spectrum

$$\rho_f^<(\varepsilon) = \frac{1}{\pi} \sum_\nu \operatorname{Im} g_\nu^< (\varepsilon - io) \tag{4.1}$$

where

$$g_\nu^<(z) = \langle E_0(N) | \psi_\nu^+ \frac{1}{z - E_0(N) + H} \psi_\nu | E_0(N) \rangle \tag{4.2}$$

In the sudden approximation the main problem is therefore the calculation of Green's functions of the type (4.2). For this purpose we introduce a (N-1) electron basis set $\{|\mu\rangle\}$ and make the formal assumption that this basis set is complete

$$\sum |\mu\rangle\langle\mu| = 1 \tag{4.3}$$

Then the Green's function can be expressed as [16]

$$g_\nu^<(z) = \sum_{\mu\gamma} \langle E_0(N) | \psi_\nu^+ | \mu\rangle\langle\mu| \frac{1}{z-E_0(N)+H} |\gamma\rangle\langle\gamma| \psi_\nu | E_0(N)\rangle \tag{4.4}$$

where we have inserted the unit operator (4.3) on both sides of the resolvent operator in Eq. (4.2). The matrix elements $\langle\mu|z-E_0(N)+H|\gamma\rangle$ are calculated and the corresponding matrix is inverted. After the sums over μ and γ have been performed the Green's function (4.4) is obtained. The usefulness of this method depends on the size of the basis set $\{|\mu\rangle\}$. Therefore it is important to use the $1/N_f$ arguments to limt this size.

To calculate $g_\nu^<(z)$ for $U = \infty$ and to order $(1/N_f)^0$ it is sufficient to introduce the states [16]

$$|\varepsilon\nu\rangle = \psi_{\varepsilon\nu} |0\rangle \tag{4.5}$$

and

$$|\varepsilon\varepsilon'\nu\rangle = \frac{1}{\sqrt{N_f-1}} \sum_{\nu'(\neq\nu)} \psi_{\nu'}^+ \psi_{\varepsilon'\nu'} \psi_{\varepsilon\nu} |0\rangle \tag{4.6}$$

The state $|\varepsilon\nu\rangle$ results from the state $|\varepsilon\rangle$ when an f-electron with the quantum number ν is removed, and the state $|\varepsilon\varepsilon'\nu\rangle$ must be included since it couples to $|\varepsilon\nu\rangle$ with a strength of the order $\tilde{V}(\varepsilon)$. In this case the Hamiltonian matrix has such a simple form that the inversion can be performed analytically. It can be shown that close to the Fermi energy the f-spectrum has the form [16]

$$\rho_f^<(\varepsilon) = \frac{(1-n_f)^2 \, \tilde{V}(\varepsilon)^2}{(\delta-\varepsilon)^2} \qquad -\delta \leq \varepsilon \leq 0 \tag{4.7}$$

Thus the Kondo energy, which earlier was shown to enter the ground-state energy and the susceptibility, also enters the f-spectrum. Eq. (4.7) describes a steep rise in the spectrum as the Fermi energy is approached from below. This is the tail of the socalled Kondo peak, which appears in the Bremsstrahlung isochromat spectroscopy (BIS) at an energy δ above $\varepsilon_F = 0$. In Fig. 4 the f-spectrum is shown for both $N_f = 1$ and $N_f = \infty$. For $\varepsilon_f = -1.5$ eV there is a peak roughly at the f-level position. This peak corresponds to final states of primarily f^0 character. The $N_f = \infty$ results illustrate the

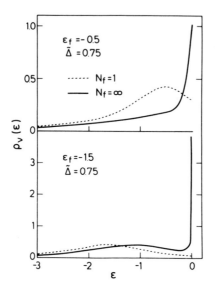

Fig. 4 The f-spectrum for $N_f = \infty$ (full curve) and for $N_f = 1$ (dashed curve). The parameters are $\pi V(\varepsilon)^2 \equiv 0.75$ eV and $B = 6$eV.

very steep rise (4.7). This is entirely a many-body effect. For the one-body case ($N_f = 1$) there is no sign of such a rise if ε_f 0. For $N_f = \infty$ the structure at $\varepsilon_F = 0$ is due to final states

$$\psi_{\varepsilon\nu} |E_0(N)\rangle \qquad (4.8)$$

which have the energy $E_0(N)+\varepsilon$ and therefore contribute weight to the spectrum at the energy ε. These states have the same f-occupancy as the ground-state, i.e. for typical parameters they have mainly f^1 character. States of the type (4.8) couple to the ground-state via ψ_ν with a strength $A^2 a(\varepsilon)$. The steep rise in the spectrum at ε_F therefore directly reflects the rapid growth in $a(\varepsilon)^2$ (see Eq. (3.12)) as ε_F is approached from below.

To test the accuracy of the f-spectrum, we have again peformed a calculation for $N_f = 1$, for which we can also easily obtain the exact result. In Fig. 5 we compare the exact results with the results for $N_f = \infty$ as well as

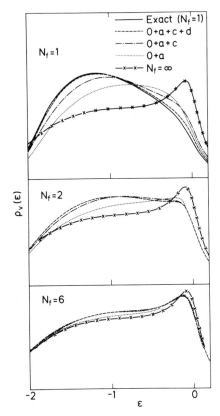

Fig. 5. The f-level PES spectrum for ε_f = -0.6 eV, $N_f\Delta$ = 0.8 eV, B = 2 eV and N_f = 1 (a), N_f = 2 (b) and N_f = 6 (c). The full line gives the exact result (only known for N_f = 1) and the -x- curve the N_f = ∞ results. The remaining curves show the results as the states 0, a, c and d in Fig. 2 are included. A Lorentzian broadening of 0.3 eV (FWHM) has been introduced.

with results based on calculations using the states 0 + a, 0 + a + c and 0 + a + c + d in Fig. 2 together with the corresponding (N-1)-electron basis states. The figure illustrates that the N_f = ∞ calculation is qualitatively wrong, while the calculations with a higher accuracy gradually converge towards the exact result. The best calculation, which is of order $(1/N_f)^1$ agrees very well with the exact result. Fig. 5 also shows approximate calculations for N_f = 2 and N_f = 6 for which the exact results are not known. For N_f = 2 and in particular for N_f = 6, the approximations using basis sets of different sizes differ from each other much less than for N_f = 1, as one would expect for a $1/N_f$ method. Since the $(1/N_f)^1$ results are quite accurate already for N_f = 1, we expect this to be even more true for N_f = 2 and N_f = 6.

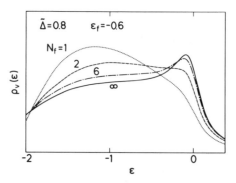

Fig. 6. The f-level spectrum calculated to order $(1/N_f)^1$ for $N_f = 1, 2, 6$ and ∞. The parameters are the same as in Fig. 5.

Fig. 6 shows the results for $N_f = 1, 2, 6$ and ∞, calculated to order $(1/N_f)^1$. The largest difference between the different curves is the Kondo peak at $\varepsilon_F = 0$ for $N_f \geq 2$. It is interesting that even such a small value of N_f as 6, leads to rather similar results as $N_f = \infty$. This illustrates that $N_f = \infty$ is a good starting point if N_f is not too small.

In a similar way the spectra of Bremsstrahlung isochromat spectroscopy (BIS) can be calculated. In BIS an electron is added to the system and we therefore study

$$g_\nu^>(z) = \langle E_0(N) | \psi_\nu \frac{1}{z + E_0(N) - H} \psi_\nu^+ | E_0(N) \rangle \qquad (4.9)$$

For $U = \infty$ we use the $(N+1)$-electron basis states

$$|\nu\rangle = \psi_\nu^+ |0\rangle \qquad (4.10)$$

$$|\tilde{\varepsilon}\nu\rangle = \psi_{\tilde{\varepsilon}\nu}^+ |0\rangle \qquad (4.11)$$

$$|\tilde{\varepsilon}\tilde{\varepsilon}\nu\rangle = \frac{1}{\sqrt{N_f - 1}} \sum_{\nu'(\neq\nu)} \psi_{\tilde{\varepsilon}\nu}^+ \psi_{\nu'}^+ \psi_{\varepsilon\nu'} |0\rangle \qquad (4.12)$$

For the BIS spectrum

$$\rho_f^>(\varepsilon) = \frac{1}{\pi} \sum_\nu \operatorname{Im} g_\nu^>(\varepsilon - io) \qquad (4.13)$$

we obtain in the limit $N_f = \infty$ [16]

$$\rho_f^>(\varepsilon) = \frac{(1-n_f)^2 \tilde{V}(\varepsilon)^2}{(\varepsilon-\delta)^2} \qquad \varepsilon < \delta \qquad (4.14)$$

which joins smoothly to the PES spectrum (4.7). For $\varepsilon = \delta$ the $N_f = \infty$ BIS spectrum has a peak with a half-width of the order $\pi n_f \delta / N_f$, i.e. a very narrow peak [16]. This peak corresponds to final states of mainly f^1 character. For a finite U there is a second peak at higher energies corresponding to f^2-like final states. Using these methods it is also possible to calculate spectra for core level X-ray photoemission spectroscopy (XPS) and X-ray absorption spectroscopy (XAS).

5. RESONANCE PHOTOEMISSION

Resonance photoemission has attracted much interest since this phenomenon was observed by Guillot et al. [25] for Ni and by several groups for rare earth compounds [26, 7]. Resonance photoemission for 3d elements has been studied theoretically by Penn [27], by Davis and Feldkamp [28] and by several other groups [29]. Resonance photoemission for Ce has been studied by Zangwill and Soven [30], who used the density functional formalism, and more recently by Sakuma et al. [31], who used a $1/N_f$ expansion. For systems with a large Kondo energy δ, Sakuma et al. obtained different photon energy dependences for the Fermi energy peak and the peak at -2 eV, as observed experimentally. We find, however, that their expression factorizes in the limit of $\delta \to 0$

$$P_f(\varepsilon, \omega) = \rho_f(\varepsilon) F(\omega) \qquad (5.1)$$

where $P_f(\varepsilon, \omega)$ is the f-spectrum for a given photon energy ω. Eq. (5.1) means that all structures in the f-spectrum has the same photon energy dependence, in contrast to experiment which shows different dependencies, even for systems with a very small δ [11]. Below we demonstrate that it is important to take into account that U is finite. This means that both f^1, f^2 and f^3 configurations are important in the presence of a core hole and that f^0, f^1 and f^2 configurations are important in the initial and final states. Using the quadratic response formalism [32], we have earlier developed a method for calculating Auger spectra [33]. This method is here adapted to resonance PES. Since the external electromagnetic field driving the PES process is normally weak, nonlinear effects can be neglected and the operator V_p (Eq. (2.3)) describing the PES process can be treated to lowest order only. It is, however, essential that the term V_A (Eq. (2.6)), describing the decay of the core hole, is treated to all orders. Our Hamiltonian is

$$H_T = H_A + T + V_p f(t) \qquad (5.2)$$

where

$$H_A = H + H_c + V_A \qquad (5.3)$$

and

$$f(t) = f\, e^{nt} (e^{i\omega t} + e^{-i\omega t}) \tag{5.4}$$

The terms H, H_c, V_A, T and V_P were defined in Eqs. (2.1 - 2.4, 2.6). The time dependence of $f(t)$ corresponds to an adiabatic switching on of an electromagnetic field with the frequency ω. In the absence of the term $V_P f(t)$, the ground-state satisfies

$$\psi_{E\nu}^+ \psi_{E\nu} |E_o(N)\rangle = 0 \tag{5.5}$$

where E refers to a high-lying continuum state. Using lowest order time-dependent perturbation theory in the operator $V_P f(t)$ we obtain the first order contribution to the wave function of the system

$$|\phi_1(t)\rangle = f\, e^{nt} \frac{1}{E_o(N)+\omega-H_A-T+i\eta} V_P |E_o(N)\rangle \tag{5.6}$$

where we have only considered processes annihilating a photon ($f \sim \exp(-i\omega t)$) and we have suppressed an irrelevant phase factor. From (5.5) it follows

$$P_f(E-\omega,\omega) = N_f \frac{d}{dt}\langle \phi(t)|\psi_{E\nu}^+ \psi_{E\nu}|\phi(t)\rangle =$$

$$= 2\eta f^2 N_f \langle E_o(N)|V_P^+ \frac{1}{E_o(N)+\omega-H_A-T-i\eta} \psi_{E\nu}^+ \psi_{E\nu} \frac{1}{E_o(N)+\omega-H_A-T+i\eta} V_P|E_o(N)\rangle \tag{5.7}$$

where $|\phi(t)\rangle$ is the total wave function. The resolvent operators in (5.7) could in principle be calculated using the methods in Sec. 4. In practice there is, however, a substantial difference. In Sec. 4 we always assumed that the emitted electron does not interact with the rest of the system (the sudden approximation). This electron did therefore not enter the calculation explicitly. When the term V_A is included, however, a state with an electron E can make a transition to a state without the electron E and with one core electron less but with two extra f-electrons. This state can then make a transition to a state with an electron E'. This follows from the necessity to treat V_A to all orders. It is therefore not sufficient to construct a basis set for

$$\psi_{E\nu}^+ |E_n(N-1)\rangle \tag{5.8}$$

with E fixed, as we did in Sec. 4, but we must allow E to take a continuum of values. This corresponds to an extension of the basis set, which makes the calculations more difficult and less transparent. We therefore prefer to use a "folding technique" [33] which effectively eliminates the states E from the problem. We introduce the notation

$$|\phi_{E\nu}\rangle \equiv \psi_{E\nu}^+ \psi_{E\nu} \frac{1}{E_o(N)+\omega-H_A-T+i\eta} V_P |E_o(N)\rangle \tag{5.9}$$

With $z = E_0(N)+\omega+i\eta$ we have the simple identity

$$(z-H_A-T)^{-1} = (z-H(n_c)-T)^{-1} + (z-H(n_c)-T)^{-1} V_A (z-H_A-T)^{-1} \quad (5.10)$$

where $H(n_c) = H+H_c(n_c)$ and n_c is the occupancy of the N_f core levels. We can then split the state (5.9) in a "direct term". (left hand side of Fig. 1) and a "resonance term" (right hand side of Fig. 1).

$$|\phi_{E\nu}\rangle_d = \frac{1}{z-H(N_f)-E} \tau(E)\psi_\nu |E_0(N)\rangle \quad (5.11)$$

Combining this term with its complex conjugate, Eq. (5.7) gives the "direct" contribution

$$P_f^d(E-\omega,\omega) = 2\pi f^2 \tau(E)^2 \langle E_0(N)|\psi_\nu^+ \delta(E_0(N)+\omega-H(N_f)-E)\psi_\nu)E_0(N)\rangle \quad (5.12)$$

where $P_f^d(\epsilon,\omega)$ depends on ω only via $\tau(E)$ and is proportional to $g_\nu^<(z)$ studied in Sec. 4. We now consider the "resonance term"

$$|\phi_{E\nu}\rangle_r = \frac{1}{z-H(N_f)-E} \psi_{E\nu}^+ \psi_{E\nu} V_A \frac{1}{z-H_A-T} V_P |E_0(N)\rangle \quad (5.12)$$

To perform the "folding" we introduce projection operators

$$P = 1 - \pi \sum_\nu \psi_{c\nu}^+ \psi_{c\nu} \quad (5.13)$$

$$Q = 1 - P \quad (5.14)$$

and define

$$\tilde{H} = H_A + T$$

Then we have

$$\psi_{E\nu}^+ \psi_{E\nu} V_A \frac{1}{z-\tilde{H}} V_P |E_0\rangle = \psi_{E\nu}^+ \psi_{E\nu} V_A P \frac{1}{z-\tilde{H}} (PV_r+QV_d)|E_0(N)\rangle \quad (5.15)$$

Using the folding technique we obtain [33]

$$g_P(z) \equiv P \frac{1}{z-\tilde{H}} P = \frac{1}{z-H(N_f-1)-T-\hat{\Gamma}} P \quad (5.16)$$

where

$$\Gamma = \int dE \sum_{\nu\nu'} v(E)^2 \psi_\nu^+ \psi_{\nu'} \psi_{c\nu'} \frac{1}{z-H(N_f)-E} \psi_{c\nu'}^+ \psi_{\nu'} \psi_\nu \qquad (5.17)$$

In the determination of Γ, we have used the fact that Γ only acts on a particular type of states. The operator (5.17) is further simplified by using the fact that the energy E is fairly large so that the relevant matrix elements can be assumed to be energy-independent. The resolvent operator in (5.17) then contributes an imaginary part while the real contribution vanishes

$$\Gamma \approx -i\Gamma \sum_{\nu\nu'\nu''} \psi_\nu^+ \psi_{\nu'}^+ \psi_{\nu''} \psi_\nu \psi_{c\nu'} \psi_{c\nu''}^+ \qquad (5.18)$$

where

$$\Gamma = \pi v(E)^2 \qquad (5.19)$$

Furthermore we obtain

$$P \frac{1}{z-\tilde{H}} Q = g_p(z) P \tilde{H} Q \frac{1}{z-H(N_f)-T} \qquad (5.20)$$

Using approximations of the same type as in (5.18) we find

$$P \tilde{H} Q \frac{1}{z-H(N_f)-T} V_d |E_o(N)\rangle \approx i\gamma \sum_\nu \psi_\nu^+ \psi_{c\nu} \sum_{\nu'} \psi_{\nu'}^+ \psi_{\nu'} |E_o(N)\rangle \qquad (5.21)$$

where

$$\gamma = -\pi v(E) \tau(E) \qquad (5.22)$$

Defining

$$\tilde{V}_r = \sum_\nu \psi_\nu^+ \psi_\nu [\tau_c + i\gamma \sum_{\nu'} \psi_{\nu'}^+ \psi_{\nu'}] \qquad (5.23)$$

we then obtain

$$P_f(E-\omega,\omega) = 2\pi f^2 N_f \langle \tilde{\phi}_{E\nu} | \delta(E_o(N)+\omega-H-E) | \tilde{\phi}_{E\nu} \rangle \qquad (5.24)$$

where

$$|\tilde{\phi}_{E\nu}\rangle = (\tau(E)\psi_\nu + v(E) \sum_{\nu'} \psi_{c\nu'}^+ \psi_{\nu'} g_p \tilde{V}_r)|E_o(N)\rangle \qquad (5.25)$$

The main problem is the calculation of (5.25), while (5.24) is a problem which was solved in Sec. 4.

[20] The ground-state $|E_o(N)\rangle$ is now calculated to lowest order in $(1/N_f)^0$

$$|E(N)\rangle = A(|0\rangle + \int_{-B}^{o} d\varepsilon\, a(\varepsilon)|\varepsilon\rangle + \int_{-B}^{o} d\varepsilon \int_{-B}^{\varepsilon} d\varepsilon'\, b(\varepsilon,\varepsilon')|\varepsilon\varepsilon'\rangle \qquad (5.26)$$

where $|\varepsilon\varepsilon'\rangle$ is a state with two f-electrons and two holes. To evaluate $\tilde V_r|E_o(N)\rangle$ we introduce the basis states

$$|1\rangle = \frac{1}{\sqrt{N_f}} \sum_\nu \psi_\nu^+ \psi_{c\nu} |0\rangle \qquad (5.27)$$

$$|\varepsilon 1\rangle = \frac{1}{\sqrt{N_f(N_f-1)}} \sum_{\nu\nu'} \psi_\nu^+ \psi_{c\nu} \psi_{\nu'}^+ \psi_{\varepsilon\nu'} |0\rangle \qquad (5.28)$$

$$|\varepsilon\varepsilon' 1\rangle = \frac{1}{\sqrt{N_f(N_f-1)(N_f-2)}} \sum_{\nu\nu'\nu''} \psi_\nu^+ \psi_{c\nu} \psi_{\nu'}^+ \psi_{\varepsilon\nu'} \psi_{\nu''}^+ \psi_{\varepsilon'\nu''} |0\rangle \qquad (5.29)$$

where $\varepsilon < \varepsilon'$ in (5,29) to avoid a linear dependence. It is now straightforward to express $\tilde V_r|E_o(N)\rangle$ in terms of the states (5.27 - 29). To lowest order in $1/N_f$ we have

$$\tilde V_r|E_o(N)\rangle = A\sqrt{N_f}\left(\tau_c|1\rangle + (\tau_c^o+i\gamma) \int_{-B}^{0} d\varepsilon\, a(\varepsilon) |\varepsilon 1\rangle \right. +$$

$$\left. (\tau_c^o + 2i\gamma) \int_{-B}^{0} d\varepsilon \int_{-B}^{\varepsilon} d\varepsilon'\, b(\varepsilon,\varepsilon') |\varepsilon\varepsilon' 1\rangle \right) \qquad (5.30)$$

To evaluate g_p we calculate matrix elements of $H(N_f-1)$ and Γ. In particular the matrix elements of Γ are of interest

$$\langle 1|\Gamma|1\rangle = 0 \qquad (5.31)$$

$$\langle \varepsilon 1|\Gamma|\varepsilon' 1\rangle = -i\Gamma(N_f-1)\delta(\varepsilon-\varepsilon') \qquad (5.32)$$

$$\langle \varepsilon\varepsilon' 1|\Gamma|\varepsilon_1\varepsilon_2 1\rangle = -2i\Gamma(N_f-2)\delta(\varepsilon-\varepsilon_1)\delta(\varepsilon'-\varepsilon_2) \qquad (5.33)$$

This clearly shows the operator character of Γ as the prefactors 0, 1 and 2 in (5.31 - 33) depend on the state considered. To calculate g_p we again use a folding technique, and fold the states $|\varepsilon\varepsilon' 1\rangle$. Then the operations in (5.25) are carried out and (5.24) is evaluated using methods descibed in Ref. 20.

In Fig. 7 we show $P_f(\varepsilon,\omega)$ as a function of ω and for different values of U. Instead of considering the kinetic energy E, we refer to the binding energy $\varepsilon_B = \omega-E = -\varepsilon$. Fig. 7 shows the ω-dependence for $\varepsilon_B = 0.2$ and $\varepsilon_B = 2.2$ eV, which corresponds to the two peaks in the spectrum. For each value of U the zero of the ω-axis is put at the maximum of the $\varepsilon_B = 0.2$ eV curve. The vertical scale has been normalized for each curve so that they have the same maximum value. For simplicity, we have assumed that $\gamma = 0$ (Eq. 5.23). The values of ratio $\tau_c v(E)/\tau(E)$ and the value of Γ were adjusted to give the experimentally observed Fano shape for the ω-dependence. In this

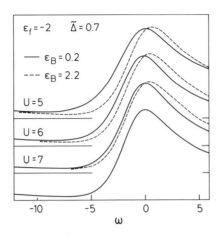

Fig. 7. The ω-dependence of $P_f(\varepsilon,\omega)$ (Eq. (5.24) for $\varepsilon = -\varepsilon_B = -0.2$ and -2.2 eV. The parameters are $\varepsilon_f = -2.0$ eV, $N_f\Delta = 0.7$ eV and $B = 6$ eV. The shape (3.15) was used for $V(\varepsilon)^2$. A Lorentzian broadening of 0.6 eV (FWHM) was introduced. For U=6 the intermediate f^2 and f^3 states are degenerate.

approach $\hat{\Gamma}$ also simulates the effects of the multiplet splitting of the intermediate f^2 and f^3 states. The multiplets in the intermediate states which have a strong coupling to the ground-state via V_r lie above the Fermi energy and become resonances. This additional broadening is also simulated by $\hat{\Gamma}$. For reasonable values of the f^2-f^3 splitting $\text{Re}(z_2-z_3)(|\text{Re}(z_2-z_3)| \sim 5\text{eV})$, here achieved by varying U, the theory then automatically gives a difference in the ω-dependence for the $\varepsilon_B = 0.2$ eV and the $\varepsilon_B = 2.2$ eV peaks of the type observed experimentally [11]. The unlabelled curve was obtained by suppressing the f^2 states in the initial and final states and the f^3 states in the evaluation of $g_p(z)$. The two curves for $\varepsilon_B = 0.2$ eV and $\varepsilon_B = 2.2$ eV are then identical within the plotting accuracy. Since the Kondo energy is small in Fig. 7, this corresponds to the result (5.1). Actually, we can recover the T = 0 limit of the results of Sakuma et al. [31], by suppressing the f^2 configurations in the initial and final states and the f^1 and f^3 configurations in the calculation of $g_p(z)$.

To simplify the discussion of the ω-dependencies for $\varepsilon_B = 0.2$ eV and $\varepsilon_B = 2.2$ eV we set $\tau(E) \equiv 0$. This changes the Fano line shape and shifts the peak positions as a function of ω, but the basic difference between $\varepsilon_B = 0.2$ eV and $\varepsilon_B = 2.2$ eV remains. We now notice that the $\varepsilon_B = 2.2$ eV peak corresponds to final f^0-like states, while the $\varepsilon_B = 0.2$ eV peak corresponds to f^1-like states. Thus the intermediate (in the presence of a core hole) f^2-states couple strongly to the $\varepsilon_B = 2.2$ eV peak, since two f-electrons are removed in the Auger decay. Therefore we define

$$f(\omega,\varepsilon) = \sum_i (g_p)_{\varepsilon,i}\, c(i) \tag{5.34}$$

where $c(i)$ are the coefficients in (5.30), e.g.

$$c(\varepsilon) = A\sqrt{N_f} \,(\tau_c + i\gamma)\, a(\varepsilon) \tag{5.35}$$

Since $a(\varepsilon)$ is large only for small values of $-\varepsilon$, $f(\omega,0)$ is representative for the ω-dependence of the ε_B = 2.2 eV peak. Since $f(\omega,0)$ primarily corresponds to the creation of intermediate f^2 states

$$f(\omega,0) \sim \frac{1}{\omega - z_2} \tag{5.36}$$

where $z_2 = 2\varepsilon_f - 2U_{fc} + U - \varepsilon_c - \Delta E - i\Gamma(N_f-1) + \varepsilon_M^2$, where ε_M^2 was defined below Eq. (2.4). The contributions from the initial f^0 and f^2 states in the sum (5.34) is more complicated. For the parameters considered here they are, however, numerically small. The final states contributing to the spectrum at $\varepsilon = 0$ are [20]

$$|\varepsilon\nu\rangle = \psi_{\varepsilon\nu}|E_o(N)\rangle = A(|\varepsilon\nu\rangle + \int_{-B}^{0} d\varepsilon' \, a(\varepsilon')|\varepsilon'\varepsilon\nu\rangle +$$

$$\int_{-B}^{0} d\varepsilon' \int_{-B}^{\varepsilon'} d\varepsilon'' \, b(\varepsilon',\varepsilon'')|\varepsilon'\varepsilon''\varepsilon\nu\rangle) \tag{5.37}$$

where $|\varepsilon\nu\rangle$ and $|\varepsilon'\varepsilon\nu\rangle$ were defined in Eqs. (4.5, 6) and $|\varepsilon'\varepsilon''\varepsilon\nu\rangle$ is an analogous f^2 state. From the folding technique it follos that the weight of the intermediate f^3 state $|\varepsilon\varepsilon'1\rangle$ is given by

$$\langle\varepsilon\varepsilon'1|g_pV_r|E_o(N)\rangle = A\frac{(\tau_c + 2i\gamma)b(\varepsilon,\varepsilon')}{\omega - z_3 + \varepsilon + \varepsilon'}(N_f-2) +$$

$$\frac{\tilde{V}(\varepsilon)f(\omega,\varepsilon') + \tilde{V}(\varepsilon')f(\omega,\varepsilon)}{\omega - z_3 + \varepsilon + \varepsilon'}(N_f-2) \tag{5.38}$$

where $z_3 = 3\varepsilon_f - 3U_{fc} + 3U - \varepsilon_c - \Delta E - 2i\Gamma(N_f-2) + \varepsilon_M^3$. Then the spectral weight at $\varepsilon = 0$ is proportional to

$$|\langle\varepsilon=0\nu|\tilde{\phi}_{E\nu}\rangle|^2 \sim A^2 \Big| f(0,\omega) + \int_{-B}^{0} d\varepsilon \frac{a(\varepsilon)[\tilde{V}(\varepsilon)f(0,\omega) + \tilde{V}(0)f(\varepsilon,\omega)]}{\omega - z_3 + \varepsilon} +$$

$$A\tau_c \int_{-B}^{0} d\varepsilon \frac{a(\varepsilon)b(\varepsilon,0)}{\omega - z_3 + \varepsilon} \Big|^2 \tag{5.39}$$

where for simplicity we have set $\gamma = 0$. We are now in the position to discuss the ω-dependence. Since the ε_B = 2.2 eV peak primarily is influenced by the intermediate f^2 states but the ε_B = 0.2 eV peak by both the f^2 and f^3 states, one might expect the relative energy of the f^2 and f^3 states to be decisive for which peak resonates at a lower photon energy. In Fig. 7 Re z_2 = Re z_3 for U = 6 eV and the f^2 and f^3 states resonate at the same energy. For U = 5 eV the f^3 states therefore resonate at a lower energy and for U = 7 eV they resonate at a higher energy. In all three cases, however, the ε_B = 0.2 eV peak resonates at a lower photon energy. To understand this we combine (5.36) and (5.39), and obtain the semi-quantitative approximation

$$|<\varepsilon=0\ \nu\ |\tilde{\phi}_{E\nu}>|^2 \sim |\frac{\alpha}{\omega-z_2} + \frac{\beta}{(\omega-z_2)(\omega-z_3)} + \frac{\kappa}{\omega-z_3}|^2 \quad (5.40)$$

where α, β and κ can be estimated from the quantities in (5.39). The relative signs are such that all three terms interfere constructively for $\omega\ll\mathrm{Re}\ z_2$ and $\omega\ll\mathrm{Re}\ z_3$. For the parameters considered here, the first and third terms have approximately the same phases also for larger values of ω. The phase of the second term, however, increases by 2π when ω grows from small values to large values, while the phases of the other two terms only increase by π. The result is a constructive interference for small photon energies and a partly destructive interference for larger photon energies. This displaces the maximum of Eq. (5.40) to lower photon energies than the maximum of Eq. (5.35), at least for the parameters we have considered. This results from the presence of both intermediate f^2 and f^3 states and from the interaction between these.

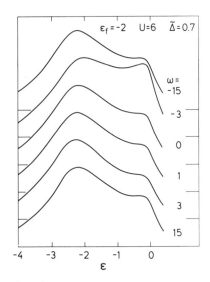

Fig. 8. The spectrum $P_f(\varepsilon,\omega)$ for different values of ω. The maximum of $P_f(-0.2, \omega)$ as a function of ω defines the zero of ω. The parameters are $\varepsilon_f = -2.\mathrm{eV}$, $U = 6$ eV, $N_f\Delta = 0.7$ eV and $B = 6$ eV. A Lorentzian broadening of 0.6 eV (FWHM) was introduced. The intermediate f^2 and f^3 configurations are degenerate.

In Fig. 8 we show spectra for different photon energies. The photon energy zero is the energy where the Fermi energy peak has its maximum. The spectra for $\omega = -15$ eV and $\omega = 15$ eV are weakly influenced by the resonance and these spectra are therefore very similar. For $\omega = -3$ eV, somewhat below the resonance, the Fermi energy peak is enhanced, as is also clear from Fig. 7. For the remaining photon energies the differences are not very large. The Fermi energy peak is slightly enhanced for $\omega = 0$ and somewhat suppressed for $\omega = 1$ eV and $\omega = 3$ eV.

This is in strong contrast to Ni, where the relative weight of the satellite is greatly enhanced at resonance [25]. To understand the difference we consider Eqs. (5.24 - 25). For Ni the ground-state is a mixture of $3p^6\ 3d^9\ 4s^1$ and $3p^6\ 3d^{10}$ configurations. Since there can be at most 10 d-electrons per atom the excitation of the p-electron leads to the intermediate configuration $3p^5\ 3d^{10}\ 4s^1$. Applying g_p and the Auger operator (second term in (5.25)) then leads primarily to states $|\tilde{\phi}_{E_\nu}\rangle$ of $3p^6\ 3d^8\ 4s^1$ character, which couple strongly to the satellite [25, 27]. The direct term (first term in Eq. (5.25)), on the other hand, contains both $3p^6\ 3d^8\ 4s^1$ and $3p^6\ 3d^9$ character, which leads to a coupling to both the main peak and the satellite. Thus it is clear that the satellite is enhanced at resonance. In Ce the ground-state is a mixture of f^0, f^1 and f^2 configurations. This leads to the intermediate configurations f^1, f^2 and f^3 since the maximum number of f-electrons is no limitation in this case. The Auger decay then leads to f^0 and f^1 configurations in $|\tilde{\phi}_{E_\nu}\rangle$. The same configurations are also generated by the direct term. These simple arguments are therefore consistent with the fairly weak photon energy dependence in the shape of the f-spectrum. This dependence is related to the interaction of the different f-configurations, as discussd below Eq. (5.40).

It is clear from Eq. (5.18) and (5.31 - 33) that $\hat{\Gamma}$ is an operator and not a c-number. To test the effect of this, we have assumed that the numerical factors are in (5.31 - 33) instead of 0, 1 and 2. The result is shown in the lower part of Fig. 9. Compared with the upper part of the figure, where Γ is treated as an operator, there is a substantial difference.

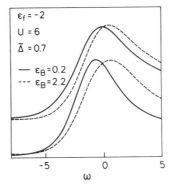

Fig. 9. The ω-dependence for the $\varepsilon_B = 0.2$ eV peak (———) and the $\varepsilon_B = 2.2$ eV peak (-----). In the upper part of the figure, $\hat{\Gamma}$ (Eq. (5.18) is treated as an operator and in the lower part as a c-number. The parameters are the same as in Fig. 8.

6. COMPARISON WITH EXPERIMENT

In this section we discuss how the parameters in the Anderson model can be determined from experiment and to what extent the same parameters can be used to describe different experiments for a given compound. In this section we do not consider resonance photoemission. The model (2.1 - 2) then has four parameters ε_f, $\Delta(\varepsilon) \equiv \pi V(\varepsilon)^2$, and U and U_{fc} (ε_c only enters as a displacement of the core spectra). U and U_{fc} have been calculated by Herbst et al. [34], who obtained values which are close to the ones we need to describe the relative peak positions. We therefore focus on ε_f and $\Delta(\varepsilon)$ as our main parameters. The conduction density of states $\rho(\varepsilon)$, is absorbed in $\Delta(\varepsilon)$. We therefore use conduction band PES as a crude estimate of the energy dependence of $\rho(\varepsilon)$ and $\Delta(\varepsilon)$. Having fixed the energy dependence of $\Delta(\varepsilon)$ we are left with the overall strength of $\Delta(\varepsilon)$, e.g. its average Δ_{av} over the occupied band, as an adjustable parameter. In the following we therefore refer to the parameters ε_f and Δ_{av} or equivalently n_f and Δ_{av}. As an illustration we show in Fig. 10 how n_f and Δ_{av} can be estimated from the 3d XPS spectrum. The spectrum has two components due to the 3d spin-orbit splitting. This is included in the theory by adding two calculated spectra with the relative weights 4 : 6 and the energy separation 18.4 eV.

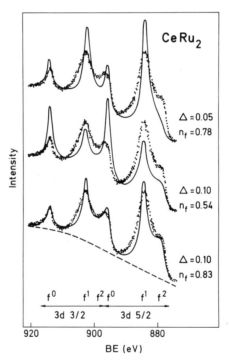

Fig. 10. The 3d core spectrum for $CeRu_2$ according to experiment (dots) and theory (full curve) [35]. We have added an inelastic background (broken curve) and used a Lorentzian broadening of 1.8 eV (FWHM).

For each of the two spin-orbit split components the spectrum has three features, corresponding to f^0, f^1 and f^2 like final states. The lowest panel in Fig. 10 illustrates that the experimental results can be rather well described if the appropriate values of n_f and Δ_{av} are used. The somewhat too small width of the calculated peaks is probably due to the neglect of multiplet effects. The weight of the f^0 peak is related to the amount of f^0 weight in the ground-state. A reduction of the n_f value (by raising the f-level) therefore leads to a much larger f^0 peak in disagreement with experiment (middle panel). The f^2 shoulder is sensitive to the value of Δ_{av}. A reduction of Δ_{av} by a factor of two therefore leads to strong reduction of the weight of the f^2 feature, which again disagrees with experiment (top panel). The core level spectrum therefore provides a convenient method for estimating Δ_{av} and n_f.

In Fig. 11 we show results for $CeNi_2$ [34]. The 3d XPS spectrum has been used to estimate the values of n_f and Δ_{av} as in Fig. 10. These parameters are then kept fixed in the calculation of the 3d→4f XAS, the BIS and the 4f PES spectra. For the XAS spectrum a background was added and for the BIS spectrum a background, multiplet effects and broadening effects were included as discussed in Ref. 36. The figure illustrates that one set of parameters, determined from the 3d XPS spectrum, gives a fairly good description of the four spectroscopies discussed here. For a more extensive discussion see Ref. 36 and 37.

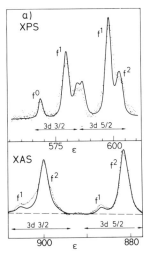
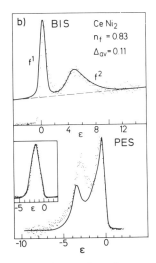

Fig. 11. The experimental (dots) and theoretical (full line) spectra of $CeNi_2$ [36, 38]. The dashed lines in the BIS and XAS calculations show the assumed backgrounds. In the BIS spectrum the background must go to zero below $\varepsilon = 0$, but it was simply assumed to be a straight line due to our lack of any precise knowledge about its shape. For the XPS (XAS) spectrum ε refers to the kinetic (binding) energy.

In the introduction we discussed the discrepancy between estimates of ε_f and Δ based on spectroscopic and thermodynamic data. These estimates may differ by one to two orders of magnitude. It is therefore interesting to calculate a thermodynamic quantity such as the static, $T = 0$ susceptibility χ, using the spectroscopic parameters. In the calculation of χ we have included the 4f spin-orbit splitting [21]. The $\Delta(\varepsilon)$ deduced from the core level XPS was used. The f-level position was then adjusted so that the experimental susceptibility was obtained. The corresponding value of n_f was calculated and used as an thermodynamic estimate of n_f. In Table 2 we compare spectroscopic and thermodynamic estimates of n_f [36]. The spectroscopic estimates show a spread, since we have adjusted n_f for each spectroscopy to obtain the optimum agreement with experiment. Our calculated susceptbility only includes the 4f contribution. To subtract the conduction contribution from the experimental susceptibility, the susceptibility of reference compounds (e.g. with La, Y, Lu) without 4f electrons was subtracted. Depending on which reference compound is used, somewhat different estimates of the experimental f-susceptibilities are obtained, which is reflected as a spread in the thermodynamic values of n_f. Table 2 shows, however, that the spread in the estimates of n_f is fairly small and that there is no systematic discrepancy between thermodynamic and spectroscopic estimates.

Table 2   Estimates of the f-occupancy n_f based on spectroscopic and susceptibility measurements. For details see Ref. 34.

System	Spectroscopic	Susceptibility
CeNi$_2$	0.78 - 0.83	0.76 - 0.83
CeIr$_2$	0.77 - 0.82	0.81 - 0.82
CeRu$_2$	0.79	0.84 - 0.89
CeAl	1.04	1.04

We therefore find that essentially the same parameters can be used to describe spectroscopic and thermodynamic properties, in strong contrast to earlier assumption that "spectroscopic" and "thermodynamic" parameters must differ by 1 - 2 orders of magnitude.

To understand these results we observe that there are two energy scales in the problem, a "large" and a "small" energy scale. The input parameters, ε_f and $N_f \Delta = \pi V^2$, are of the order 1 eV. Within the model these parameters produce a "small" energy scale δ, which for instance leads to the large values of $\chi \sim 1/\delta$ (Eq. (3.15)). The coupling of the localized level to a continuum of conduction states leads to the logarithm in Eq. (3.13), which leads to the exponential dependence on ε_f and $\bar{V}^2$. The value of δ can therefore become very small if $\bar{V}^2$ is small and/or ε_f is far below the Fermi energy. The many-body effects causing this phenomenon are often referred to as Kondo like effects. Characteristic is a large quasi-particle density of states at the Fermi energy and a narrow resonance in the f-spectral function (see e.g. Fig. 4).

References

1. J.M. Lawrence, P.S. Riseborough and R.D. Parks, Rep.Progr.Phys. 41: 1 (1981).
2. B. Coqblin and A. Blandin, Adv. Phys. 17, 281 (1968).
3. "Valence fluctuations in solids", L.M. Falicov, W. Hanke and M.P. Maple eds., North-Holland, Amsterdam (1981).
4. "Valence instabilities", P. Wachter and H. Boppart eds., North-Holland, Amsterdam (1982).
5. "Valence Fluctuations", E. Müller-Hartmann, B. Roden and D. Wohlleben eds., J. Magn.Magn.Mater, 47 & 48 (1985).
6. A. Platau and S.E. Karlsson, Phys.Rev. B18: 3820 (1978).
7. L.I. Johansson, J.W. Allen, T. Gustafsson, I. Lindau and S.B.M. Hagström, Solid State Commun. 28: 53 (1978).
8. J.C. Fuggle, F.U. Hillebrecht, Z. Zolnierek, R. Lässer and A. Platau, Phys.Rev.Lett. 45: 1597 (1980).
9. J.W. Allen, S.-J. Oh, I. Lindau, J.M. Lawrence, L.I. Johansson and S.B.M. Hagström, Phys.Rev.Lett 46: 1100 (1981); M. Croft, J.H. Weaver, D.J. Peterman and A. Franciosi, Phys.Rev.Lett. 46: 1104 (1981).
10. N. Mårtensson, B. Reihl and R.D. Parks, Solid State Commun. 41: 573 (1982).
11. J.M. Lawrence, J.W. Allen, S.-J. Oh and I. Lindau, Phys.Rev. B26: 2362 (1982).
12. R.D. Parks, N. Mårtensson, B. Reihl, in Ref. 4, p. 239; J.W. Allen, S.-J. Oh, M.B. Maple and M.S. Torikachvili, Phys.Rev. B28: 5347 (1983).
13. P.W. Anderson, Ref. 3, p. 451.
14. T.V. Ramakrishnan, Ref. 3, p. 13.
15. T.V. Ramakrishnan and K. Sur, Phys.Rev. B26: 1798 (1982).
16. O. Gunnarsson and K. Schönhammer, Phys.Rev.Lett. 50: 604 (1983); Phys.Rev. B28: 4315 (1983).
17. Y. Kuramoto, Z. Phys. B53: 37 (1983); Grewe, Z. Phys. 53: 271 (1983); H. Keiter and G. Czycholl, J.Magn.Magn.Mater, 31: 477 (1983); P. Coleman, Phys.Rev. B29: 3035 (1984); F.C. Zhang and T.K. Lee, Phys. Rev. B30: 1556 (1984); N.E. Bickers, D.L. Cox and J.W. Wilkins, Phys.Rev.Lett. 54: 230 (1985).
18. A. Kotani and Y. Toyozawa, J.Phys.Soc. Japan 37: 563 (1974); 37: 912 (1974).
19. C.M. Varma and Y. Yafet, Phys.Rev. B13: 2950 (1976).
20. O. Gunnarsson and K. Schönhammer, Phys.Rev. B31: 4815 (1985).
21. O. Gunnarsson and K. Schönhammer, J.Magn.Magn.Mater, 52: 227 (1985).
22. T.C. Li, O. Gunnarsson, K. Schönhammer and G. Zwicknagl, (to be published).
23. P. Schlottmann, Phys.Rev.Lett. 50: 1697 (1983); E. Ogievetski, A.M. Tsvelick and P.B. Wiegmann, J.Phys. C16: L797 (1983).
24. L. Hedin and S. Lundqvist, Solid State Phys. 23: 1 (1969).
25. C. Guillot, Y. Ballu, J. Paigne, J. Lecante, K.P. Jain, P. Thiry, R. Pinchaux, Y. Petroff and L.M. Falicov, Phys.Rev.Lett. 39: 1632 (1977).

26. W. Lenth, F. Lutz, J. Barth, G. Kalkoffen and C. Kunz, Phys.Rev.Lett. 41: 1185 (1978); J.W. Allen, L.I. Johansson, R.S. Bauer, I. Lindau and S.B.M. Hagström, Phys.Rev.Lett. 41: 1499 (1978); W. Gudat, S.F. Alvarado and M. Campagna, Solid State Commun. 28: 943 (1978).
27. D.R. Penn, Phys.Rev.Lett. 42: 921 (1979).
28. L.C. Davis and L.A. Feldkamp, Phys.Rev.Lett. 43:151 (1979); 44: 673 (1980); Phys.Rev. B23: 6239 (1981); L.C. Davis, Phys.Rev. B25: 2912 (1982).
29. Y. Yafet, Phys.Rev. B21: 5023 (1980); B23: 3558 (1981); S.M. Girvin and D.R. Penn, Phys.Rev. B22: 4081 (1980); S.-J. Oh and S. Doniach, Physics Letter, 81A: 483 (1981); Phys. Rev. B26: 1859 (1982); J.C. Parlebas, A. Kotani and J. Kanamori, Solid State Commun. 41: 439 (1982); G. v.d. Laan, Solid State Commun. 42: 165 (1982).
30. A. Zangwill and P. Soven, Phys.Rev.Lett.45: 204 (1980).
31. A. Sakuma, Y. Kuramoto, T. Watanabe and C. Horie, J.Magn.Magn.Mater. 52: 393 (1985).
32. W.L. Schaich and N.W. Ashcroft, Phys.Rev. B3: 2452 (1971).
33. O. Gunnarsson and K. Schönhammer, Phys.Rev. B22: 3710 (1980).
34. J.F. Herbst and J.W. Wilkins, Phys.Rev.Lett. 43:1760 (1979); J.F. Herbst, R.E. Watson and J.W. Wilkins, Phys.Rev. B17: 3089 (1978).
35. O. Gunnarsson, K. Schönhammer, J.C. Fuggle, F.U. Hillebrecht, J.-M. Esteva, R.C. Karnatak and B. Hillebrand, Phys.Rev. B28: 7330 (1983).
36. J.W. Allen, S.-J. Oh, O. Gunnarsson, K. Schönhammer, M.B. Maple, M. Torikachvili and I. Lindau, (to be published).
37. O. Gunnarsson and K. Schönhammer, in Handbook on the physics and chemistry of rare earths, K. Gschneider and L.R. Eyring eds., North-Holland, Amsterdam, Vol. 10 (1987).
38. J.C. Fuggle, F.U. Hillebrecht, J.-M. Esteva, R.C. Karnatak, O. Gunnarsson and K. Schönhammer, Phys.Rev. B27: 4637 (1983).

RESONANT PHOTOEMISSION USING THE d → f GIANT RESONANCES

Se-Jung Oh

Department of Physics
Seoul National University
Seoul 151, Korea

INTRODUCTION

Since the first observations[1,2] of resonant enhancements of the valence band and shallow core level photoemission from the rare-earth metals and compounds at the 4d → 4f edge, a lot of work has been done both to understand the nature of this resonance process itself and also to take advantage of this phenomena to study the electronic structure of solids. The research on the nature of these giant resonances has been discussed in other parts of this volume, so I will concentrate here mainly on the application of the resonance phenomena to study interesting electronic properties of solids. In particular, I will discuss the resonant photoemission study of the electronic structures of the mixed-valence or Kondo-like lanthanide compounds (especially cerium) using the 4d → 4f giant resonance, and the uranium based heavy-fermion systems using the 5d → 5f resonance. These studies illustrate how the resonant photoemission techuique can be used to understand electronic structures and hence help to solve the long-standing puzzle of the solid state physics, for example the cerium $\alpha \leftrightarrow \gamma$ phase transitions. And then the occurrence of the resonant photoemission phenomena at the deeper 3d → 4f absorption edges of rare earth compounds will be shown, with a discussion on the difference in the decay mechanisms involving deeper 3d core holes, in particular the importance of the Auger resonant Raman process.
I should mention at the outset that this article is not intended to be a comprehensive review — the field of the resonant photoemission has grown too big to be adequately covered in a one-hour talk. The choice of topics here is strictly out of my personal prejudice and involvement, and I apologize for those researchers whose important works are omitted here either due to my ignorance or lack of space.

MECHANISM OF RESONANT PHOTOEMISSION

As a simple example of the resonance phenomena in the photoemission spectra, let's look at Fig.1 (a). There are shown the photoemission spectra[3] of the outermost levels of Yb metal exposed to the 100 langmuir (1 L = 10^{-6} Torr sec) of oxygen, taken at different incident photon energies in the vicinity of the 4d absorption edge using tunable synchrotron light source. Yb metal is in the divalent state having a closed 4f-shell configuration $4f^{14}$, but becomes trivalent with the configuration $4f^{13}$ when oxidized to the Yb_2O_3 state by oxygen exposures. The peaks in the spectra at approximately 7-12 eV in binding energies originate from the $4f^{13} \to 4f^{12}$ photoionization transition, as confirmed by the agreement with the theoretical fit using the Cox's fractional parentage coefficients[4]. Fig.1 (a) clearly shows the resonant enhancement of these peaks near hν=181 eV, close to the 4d $\to$ 4f absorption of the Yb^{+3} ions. This enhanced emission can be thought of as arising from the two step process[5] where the incident photon makes a transition

$$4d^{10}4f^{13} + h\nu \to 4d^{9}4f^{14} \qquad (1)$$

and then the resulting intermediate state decays by the autoionization process involving super Coster-Kronig Auger transitions

$$4d^{9}4f^{14} \to 4d^{10}4f^{12} + \text{photoelectron} \qquad (2)$$

Since this final state is the same as the state resulting from the ordinary 4f photoemission process

$$4d^{10}4f^{13} + h\nu \to 4d^{10}4f^{12} + \text{photoelectron} \qquad (3)$$

it gives rise to the resonant enhancement of 4f emissions at the 4d $\to$ 4f threshold. In additon, this "resonance" process (egs. (1) and (2)) and the "direct" process (eg. (3)) can interfere quantum mechanically because both processes involve the same initial and final states, and the resulting resonant behavior as a function of photon energy has a characteristic line shape, as shown by U. Fano[6] in 1961. Fig.1 (b) shows that this characteristic Fano lineshape fits the cross section variation of Yb 4f emission in the vicinity of the resonance quite well. Fano lineshape is given by the expression

$$\sigma(h\nu) = \sigma_a \frac{(q+\varepsilon)^2}{1+\varepsilon^2} + \sigma_b \qquad (4)$$

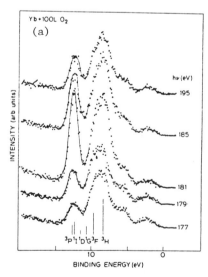
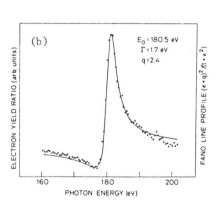

Fig. 1. (a) Photoemission spectra of the outermost levels of oxidized Yb excited with different photon energies near the 4d-absorption edge.
(b) Experimentally obtained line profile of the 4f emissions from Yb_2O_3 near the 4d threshold (dotted curve) and a best fit of a Fano line profile (solid curve) to the data.

where $\varepsilon = \frac{h\nu - E_o}{\Gamma}$, E_o is the resonance energy, Γ is the half-width of the line, q is the line profile index given by the ratio of matrix elements, and σ_a, σ_b correspond to transition cross sections to states of the continuum that do and do not interact with the discrete autoionizing state, respectively. A more detailed theoretical and experimental study on the resonant photoemission of oxidized Yb can be found in ref.7.

The giant enhancement of 4f emissions in the oxidized Yb due to the resonance process comes about because of the strong oscillator strength for the 4d → 4f absorption process in eq(1) and the strong super Coster-Kronig autoionization matrix element for the process (2). These two matrix elements tend to be large for the localized levels, and therefore we expect to see strong resonance effects for photoemissions from 3d, 4f and 5f electrons. This indeed is the case, and resonant phtoemission has been mostly utilized to study the electronic structures of transition metal and rare earth compounds. Due to their rather localized wavefunctions, these electrons exhibit relatively large electron correlation effect, and to interpret the photoemission data it is frequently necessary to take into account the many-body nature. This turned out to be the case for the resonant photoemission from cerium compounds, which I will discuss next.

APPLICATION OF RESONANT PHOTOEMISSION: STUDY ON THE CERIUM α ↔ γ PHASE TRANSITION

As discussed above, the resonant photoemission preferentially enhances emissions from localized electronic levels, such as 4f electrons of lanthanide compounds. This was utilized quite nicely to study the electronic structures of cerium compounds where 4f emission from conventional X-ray Photoelectron Spectroscopy (XPS) or Ultraviolet Photoelectron Spectroscopy (UPS) is relatively weak compared with the emissions from other underlying valence electrons and therefore difficult to locate its exact position relative to the Fermi level. By using resonant photoemission technique it was possible to identify 4f emissions unambiguousy, and this data completely changes our understanding on the origin of the cerium α ↔ γ phase transition.

Background and the Promotion Model

Cerium is the first element in the lanthanide series (atomic number 56) with the atomic ground state configuration $[Xe][4f]^2[5d]^0[6s]^2$. When it froms a metal, one or two 4f electrons are transferred into the 5d/6s shell to gain the cohesive energy by the hybridization with neighboring atoms[8]. This cerium metal has a very interesting phase transition, as illustrated in its equilibrium phase diagram of Fig.2. At room temperature in the atmospheric pressure, it is in the γ-phase. At low temperature (below ~ 100°K) or at high pressure (above ~8 kbar), it transforms into the α-phase. Both γ and α phases have f.c.c. crystal structures, but the volume of α-Ce is ~17% less than that of γ-Ce, orders of magnitude larger change than expected from the thermal expansion alone. In addition, the magnetic properties are completely different. The magnetic susceptibility of γ-Ce follows the Curie law, suggesting the existence of a localized 4f electron, whereas α-Ce seems to have lost the localized moment, showing only Pauli paramagnetism due to conduction electrons. For this reason the cerium phase transition is called magnetic-nonmagetic transition. Many compounds made of cerium also appear to mimic these differences in physical properties between α- and γ-phases. There are γ-like compounds which have larger lattice constants and show the Curie-Weiss law magnetic susceptibility (e.g. CeAl, CeBi, etc), and there are α-like compounds which show smaller lattice parameters and no magnetic moments (e.g. CeN, $CeRu_2$, etc). Some compounds such as $CeAl_2$ and CeP even show pressure-induced transition similar to the Ce-metal.

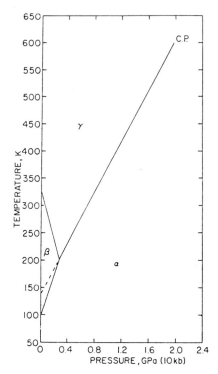

Fig. 2. Equilibrium phase diagram of cerium metal. The letter C.P. means a critical point.

There were several models to explain this interesting phase transition. One of the oldest, and the most widely-accepted for a long time, is the promotion model[9] proposed more than 30 years ago by Pauling and Zachariasen. They proposed that the change of the 4f electron energy level position is responsible for this transition. That is, in the γ-phase the Ce 4f level is below the Fermi level and therefore occupied, which gives the localized moment and the Curie-Weiss type magnetic susceptibility. In the α-phase, on the other hand, this Ce 4f level is promoted above the Fermi level and become unoccupied, hence the local magnetic moment is lost. The volume collapse in the α-phase can also be understood, at least qualitatively, due to the decreased shielding of the nuclear charge and increased hybridization. Since the energy needed to make the phase transition is small, the 4f level in the γ-phase must be very close to the Femi level, and it is estimated to be of the order of 0.1 eV. The 4f level width due to the hybridization must also be very small.

Resonant Photoemission and the Breakdown of the Promotion Model

Although a lot of experimental data (specific heat, thermopower, optical transition, etc) were interpreted using the promotion model, there was no direct determination of the Ce 4f level position experimentally. Photoemission spectroscopy is ideal for this purpose, and there had been early attempts using either UPS[10] or XPS[11] technique. However, it was difficult to identify the 4f emission from the underlying Ce 5d/6s conduction electron states, because there is only one 4f electron (so the emission is weak) and the 4f emission has no identifiable fingerprint.

By using the resonant photoemission technique, however, we can enhance weak 4f emissions preferentially, and thus identify the 4f level unambiguously. Fig.3 shows the resonant photoemission data on γ-Ce metal[2] using the tunable synchrotron radiation source. In this case the resonant enhancement occurs near the Ce 4d → 4f absorption edge (hν ≅ 120 eV) due to the following resonance process.

$$4d^{10}4f^1 + h\nu \rightarrow 4d^9 4f^2 \rightarrow 4d^{10}4f^0 + \text{photoelectron} \qquad (5)$$

The spectral feature that is most strongly enhanced on resonance in this data is the peak at ~2 eV below the Fermi level. This suggests that the 4f level is below ~2 eV from the Fermi level, which is far from the prediction of the promotion model. To understand whether this is the general behavior for all the cerium compounds, we have measured several cerium compounds[12], $CePd_3$ which is α-like, and $CeIn_3$, $CeAl_2$, CeBi which are all γ-like. As can be seen from the data in Fig.4, they all show 4f emissions on resonance at 1-3 eV below the Fermi level, in contradiction to the promotion model. Even the extremely α-like compounds $CeRu_2$ and $CeCo_2$, which should have no occupied 4f electrons according to the promotion model, show 4f emissions on resonant photoemission[13]. From these data, one can conclude that the promotion model is not correct, and that cerium and its compounds do not lose 4f electrons completely upon going from the γ-phase to the α-phase. The resonant photoemission data on α-Ce metal[14] shown in Fig.5 are not much different from the γ-Ce data, but rather there is a small shift of the weight from the peak at ~2 eV below the Fermi level to the structure near the Fermi level. This conclusion from resonant photoemission studies that α-like cerium compounds have substantial 4f occupation numbers is also consistent with the deep core level photoemission spectra, as discussed in this volume by Prof. J.C. Fuggle and Dr. O. Gunnarsson.

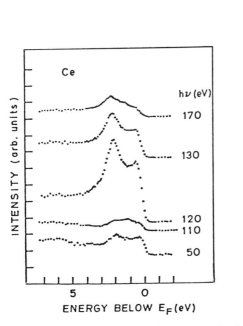
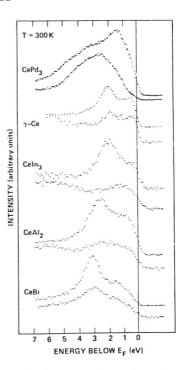

Fig. 3. (Left panel) Valence band photoemission spectra of γ-Ce film taken at different photon energies from the synchrotron radiation light source. We see the giant resonance at hν =120 eV.

Fig. 4. (Right panel) On-resonance (upper curve : hν =122 eV) and off-resonance (lower curve : hν =80 eV) pairs of valence band resonant photoemission spectra of several cerium compounds.

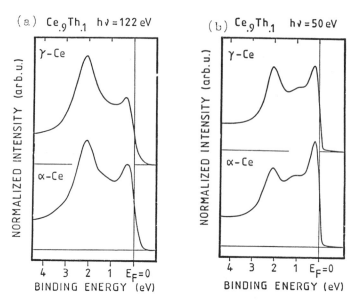

Fig. 5. Comparison of the photoemission spectra of α-Ce and γ-Ce (a) on resonance (hν =122 eV) and (b) off resonance (hν=50 eV).

Mott Transition Model and Kondo Volume Collapse Model

Since the long-believed promotion model is discredited on the basis of the resonant and core-level photoemission data, what are alternative explanations for the cerium α ↔ γ phase transition? One is the Mott transition model[15] proposed by Johansson in 1974. He used the cohesive energy analysis to estimate the binding energy of the Ce 4f level, and predicted that it is about 2.1 eV. At that time there was no resonant photoemission data to confirm his prediction yet, but he went on and proposed that the promotion model cannot work and that the cerium α↔γ phase transition comes from a mechanism similar to the Mott metal-insulator transition. That is, in the γ-phase 4f electrons are localized, but in the α-phase they become itinerant to form a band, resulting in the loss of the localized moment. For this model to work, however, the 4f bandwidth W in the α-phase must be comparable in magnitude to the 4f electron Coulomb correlation energy U. The correlation energy U for the γ-phase is known to be about 5-6 eV and the bandwidth W is at most 0.5-1.0 eV, and it is not obvious how these quantities make an order of magnitude change in the α-phase.

Another model proposed recently in 1982 by Allen and Martin is called the Kondo volume collapse model[16]. In this model, the order of magnitude for the values of U and W are kept about the same in both α- and γ-phases, and the loss of magnetism in the α-phase is explained on the basis of the Kondo effect[17]. That is, at low temperature (in the α-phase), because of the exchange interaction the localized 4f electron and the conduction electrons form a singlet (nonmagnetic) bound state, the binding energy of which is of the order $kT_k \cong \sqrt{J} \exp[-1/J]$, where J is the exchange coupling constant. Since J depends strongly on the volume, (empirically $J \sim 1/V^6$), the volume is collapsed in the α-phase to gain larger binding energy. However, as the temperature is raised, the entropy term TS in the Gibb's free energy G = E - TS + PV becomes important, and it is more profitable for the system to gain the spin entropy by breaking up the Kondo bound state. This results in the phase transition to the γ-phase, with the appearance of the localized magnetic moment.

The natural question is then "Is there a way to distinguish between these two models spectroscopically?". The key difference between these two models is that the Mott transition model requires $U \cong W$ in the α-phase, whereas the Kondo volume collapse model assumes $U \gg W$ even in the α-phase. Therefore the empty 4f energy level (i.e. the energy required to put an extra electron into the 4f orbital) should be close to the Fermi level in the Mott transition model since $U \cong W < \sim 1$ eV, whereas in the Kondo volume collapse model, the empty 4f level is separated from the occupied 4f level by the amount $U \cong 5 - 6$ eV. Hence the spectra on the empty 4f states in the α-phase should be able to distinguish between these two models.

Empty 4f States on Cerium Metal and Its Compounds

The most direct way to measure empty electronic states spectroscopically is the Inverse Photoemission (IPES) technique, which utilizes the time-reversed process of the photoemission as its name suggests. In this experiment, monoenergetic electron beam of kinetic energy E_K impinge

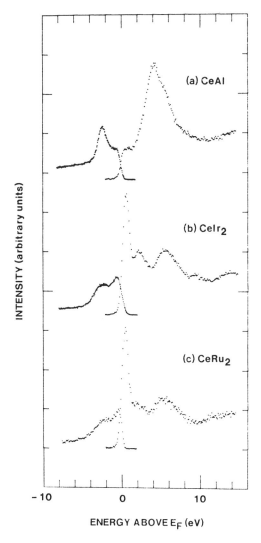

Fig. 6. Combined resonant-photoemission and IPES spectra for (a) CeAl, (b) $CeIr_2$, and (c) $CeRu_2$.

on the sample, where some of them make transitions to the empty states of the solid by a radiative decay. By analyzing the energy $h\nu$ of photons coming from this procss, we can determine the energy E_A of the empty electronic state from the energy conservation relation

$$E_K = E_A + h\nu \; ; \; E_A = E_K - h\nu$$

where energies are measured relative to the vacuum level. In practice, the energy of the incident electron beam is scanned and photons of fixed energy are detected, hence the name Bremsstrahlung Isochromat Spectroscopy (BIS) is often used. By modifying PES setup slightly[18] IPES experiment is also possible, and then the occupied and unoccupied states can be measured on the same sample in the same vacuum chamber.

In Fig.6, the experimental IPES data using X-ray ($h\nu$ =1486.6 eV) on CeAl$_2$ (γ-like compound), and CeIr$_2$, CeRu$_2$ (α-like compounds) are shown along with their resonant photoemission spectra[19]. Both spectra are dominated by the Ce 4f states, because of the resonant enhancement in the photoemission case (below the Fermi level E_F) and because of the cross section and the large orbital degeneracy in the IPES case (above E_F). The CeAl$_2$ data show the occupied 4f level at ~2 eV below E_F and the empty 4f level at ~4 eV above E_F, confirming U $\simeq$ 6 eV for the cerium 4f electrons in the γ-phase. Data on α-like compounds CeRu$_2$, CeIr$_2$ also show peaks at -2 eV and 4 eV from E_F, showing U $\simeq$ 6 eV in these α-like compounds as well, but in this case there are additional peaks at the Fermi level in both PES and IPES spectra. If we interpret these additional peaks near E_F as the Kondo resonance, whose existence can be inferred from Friedel sum rule for the Anderson model, these data are consistent with the spectra expected in the Kondo volume collapse model. Recent theories[20] based on the large degeneracy of the 4f orbital were able to fit core-level photoemission spectra, valence band resonant photoemission spectra, and the empty 4f inverse photoemission spectra with the same set of parameters, and the result is consistent with the Kondo picture. We have also measured IPES data[21] on Ce$_{.7}$Th$_{.3}$ as a function of temperature, where thorium was added to make a continuous second order phase transition, and the spectrum shows that the peak at ~4 eV above E_F hardly moves upon going from the γ-phase to the α-phase, suggesting little change in the U value, whereas the spectral weight at E_F grows as we approach the α-phase. Taken together with the PES data in Fig.5, these IPES data are also consistent with the Knodo volume collapse model.

RESONANT PHOTOEMISSION STUDY ON THE URANIUM HEAVY-FERMION SYSTEMS

The 5f electrons of actinides are believed to have properties intermediate between lanthanide 4f electrons and transition-metal 3d electrons. In particular, 5f electrons seem capable of being either localized or itinerant depending on the actinide-actinide internuclear distance judging from their magnetic properties[22]. As in the case of lanthanide mixed-valence or Kondo-like compounds, many actinide compounds show very interesting anomalous magnetic and electronic properties. Recently, actinide compounds in the category of "heavy-fermion" systems[23] with a large linear coefficient of the specific heat γ and effective mass m*, attracted much attention after it was discovered that some of them undergo superconducting transitions. Since most of heavy-fermion systems are cerium or uranium compounds, and their 4f or 5f electrons are believed to be responsible for the heavy-fermion properties, it is interesting to do resonant photoemission study of these compounds to learn their electronic structures and the role of f-electrons in the system.

The 5f emissions in actinide compounds are resonantly enhanced at the 5d → 5f absorption threshold due to the process

$$5d^{10}5f^n + h\nu \rightarrow 5d^95f^{n+1} \rightarrow 5d^{10}5f^{n-1} + \text{photoelectron} \qquad (6)$$

The resonance behavior as a function of photon energy is similar to the lanthanide 4d → 4f case, except that the spin-orbit split $5d_{3/2}$ and $5d_{5/2}$ edges are clearly resolved[24]. By tuning the incident photon energy to the maximum point of the resonance, we can effectively look at only the 5f-derived states in the photoemission spectra, whereas by using photon energy at the minimum (interference dip) of the Fano resonance (see fig. 1-(b)), we can suppress 5f emissions and thereby obtain underlying densities of states. An example of this technique is shown in fig.7, where the resonant photoemission data at hν=92 eV and hν=98 eV are shown along with the BIS spectrum taken with hν=1486.6 eV on two heavy-fermion uranium compounds UAl_2 and UPt_3 [Ref.25]. These compounds are believed to be typical spin fluctuating materials[23] and

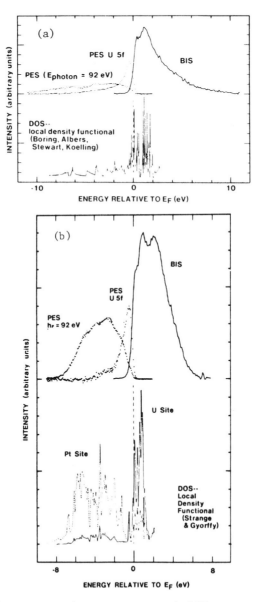

Fig. 7. Combined resonant-photoemission and BIS spectra for (a)UAl_2 and (b)UPt_3 along with theoretical one electron density of states.

have large γ-values (142 mJ/mole K² for UAl$_2$ and 450 mJ/mole K² for UPt$_3$); furthermore UPt$_3$ becomes a superconductor below T$_c$=0.54 K. The PES spectra for photon energy hν =92 eV have the U 5f emission resonantly suppressed and thereby reveal Al 3s-3p or Pt 5d emission, plus some U 5d emission which is only partly suppressed by the resonance. The PES spectra labeled U 5f result from subtraction of the hν =92 eV spectrum from that for hν =98 eV, a photon energy for which the U 5f emission is resonantly enhanced. By this technique we can separate out the partial densities of states of the particular orbital angular momentum character and local atomic site from the energetically overlapping valence band photoemission data[26].

Comparing these PES and BIS data with the theoretical one electron densities of states shown in the same figure, we immediately notice that although Pt 5d states of UPt$_3$ or Al 3s-3p states of UAl$_2$ show fair agreements between theory and experiment, the U 5f states near the Fermi level show much broader spectrum experimentally than the calculated curves in both compounds. This indicates appreciable values of electron correlation energy U_{ff} between U 5f electrons, which will increase the spreading of the 5f weight by displacing energetically unfavorable 5f valence states away from the Fermi level. From these data we can set a conservative lower limit on U_{ff} to be ~1.5 eV for these uranium compounds. This example shows that the resonant photoemission and BIS spectra can determine large energy scales of the system such as U_{ff}, 5f band width, and 5f electron occupation number, although the present resolution of the expermental setup at the resonance region is usually inadequate to observe sharp features in the density of states near the Fermi level expected from the large γ values of these heavy-fermion systems[27]. These values of the large energy scale parameters will set the frame for the proper microscopic theory to explain the interesting low temperature properties.

DETERMINATION OF THE VALENCE FROM THE RESONANCE BEHAVIORS

The usefulness of the resonant photoemission technique is not limited to enhancing or suppressing f-derived emissions discussed above. Since the resonance behavior at the 4d edge of the lanthanides shows characteristic multiplet structures of the absorption process $4d^{10}4f^n \rightarrow 4d^94f^{n+1}$ depending on the number of 4f electrons, this can serve as a spectroscopic "fingerprint" for the valence state in the photoemission spectra of lanthanide compounds. A nice example is illustrated for the mixed-valence TmSe resonant photoemission[28] shown in Figures 8 and 9. TmSe is a homogeneous mixed-valence compound with the ground state fluctuating between Tm^{2+} [$4f^{13}(5d6s)^2$] and Tm^{3+} [$4f^{12}(5d\,6s)^3$] configurations. The photoemission spectra, therefore, show two sets of 4f level structures corresponding to $4f^{13} \rightarrow 4f^{12}$ and $4f^{12} \rightarrow 4f^{11}$ transitions. The valence states of these 4f level structures can be identified by comparing with the expected final state multiplets calculated from the coefficient of the fractional parentage[4] as shown in the XPS spectrum of Fig.8, where peaks A and B are identified with the $4f^{13} \rightarrow 4f^{12}$ transition and peaks C,D, and E with the $4f^{12} \rightarrow 4f^{11}$ transition. However, these identifications can be made much more easily if we look at the resonance behaviors of these peaks near the Tm 4d edge (hν ≅ 170 eV) shown in Fig.9. These data are taken with the technique called Constant Initial-State Spectroscopy (CIS), where the intensity of the photoelectrons with a fixed <u>binding</u> energy is measured as the incident photon energy is varied. We can see that divalent (Tm^{2+}) peaks A and B show only one resonance peak, whereas trivalent peaks (Tm^{3+}) C,D, and E show three resonances as a function of photon energy. (The CIS spectrum of peak C is distorted because it rides on the shoulder of a strong divalent peak B.) This difference in the resonance behavior can be easily understood from the multiplet structures of the $4d^94f^{n+1}$

intermediate state of the resonance process. From the selection rule in the LS (Russell-Saunders) coupling scheme, the allowed photoabsorptions for the resonance process of the $4f^{12}$ and $4f^{13}$ initial configurations are,[29] respectively,

$$4d^{10}4f^{12}(^3H_6) + h\nu \to 4d^9 4f^{13}(^3H_6)$$
$$\to 4d^9 4f^{13}(^3H_5)$$
$$\to 4d^9 4f^{13}(^3G_5)$$

and

$$4d^{10}4f^{13}(^2F_{7/2}) + h\nu \to 4d^9 4f^{14}(^2D_{5/2}).$$

This accounts for the three peak resonances of the $4f^{12}$ initial state and a single resonance peak of the $4f^{13}$ initial-state configuration. By studying resonance behaviors, therefore, we can unambiguously identify the initial 4f valence of a particular peak in the photoemission, and this can be used as a spectroscopic "fingerprint."

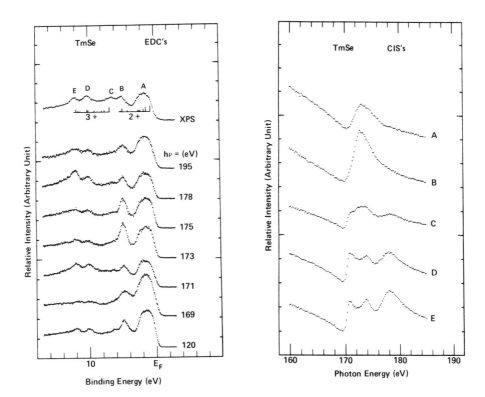

Fig. 8. (Left panel) Photoemission spectra of the TmSe valence band at various photon energies including XPS data (hν =1253.6 eV). Solid bars under the XPS curve represent positions and intensities of the final state multiplets for the divalent (2†) and trivalent (3†) Tm 4f emissions calculated from the fractional parentage coefficients.

Fig. 9. (Right panel) CIS spectra of the 4f peaks labeled in fig.8 near the Tm 4d threshold. Peaks A and B belong to the divalent part, while peaks C, D, and E belong to the trivalent structures.

RESONANT PHOTOEMISSION AT THE 3d ABSORPTION EDGES

Up to now we have discussed resonance phenomena of 4f and 5f photoemissions at the 4d and 5d absorption edges respectively, i.e., from the core level with the same principal quantum number. You might ask whether similar resonance phenomena occur at the deeper core level edges, for example 3d absorption threshold. The answer is yes, and we can see an example in Figs.10 and 11. Fig.10 shows valence band photoemission data of $TbCo_2$ at three different photon energies[30]: below Tb 3d edges ($h\nu=1229$ eV), at the Tb $3d_{5/2}$ edge ($h\nu=1236$ eV), and at the $3d_{3/2}$ edge ($h\nu=1268.5$ eV). We note that the 4f spectral intensity near 10 eV below E_F becomes nearly 72 times stronger at the Tb $3d_{5/2}$ edge than below the edge due to the resonance process

$$3d^{10}4f^n + h\nu \rightarrow 3d^9 4f^{n+1} \rightarrow 3d^{10}4f^{n-1} + \text{photoelectron} \qquad (7)$$

This giant resonance enhancement is more clearly seen in the CIS spectra of Fig.11. Two interesting features are worth mentioning for this 3d

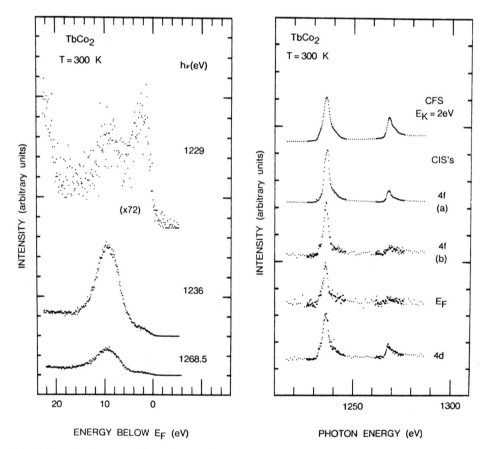

Fig.10. (Left panel) Valence-band photoemission spectra of $TbCo_2$ for several photon energies below and at the Tb 3d absorption edges. Curves have vertical zeros, defined by their portions above E_F, displaced upward from the figure boundary by 1, 3, and 8.5 units, but all have same vertical scale, for which one unit is 200 counts/sec, except as labeled.

Fig.11. (Right panel) Partial yield(CFS) and CIS spectra of $TbCo_2$ for photon energies near the Tb 3d absorption region.

resonance. The first point is that the ratio ξ of on-resonance to off-resonance intensity is very much greater for the 3d resonance ($\xi(3d) \cong 72$) than for the 4d resonance ($\xi(4d) \cong 7.4$ was obtained from the spectra near the 4d edge). However, detailed analysis[30] shows that the 3d resonance is intrinsically weaker than the 4d resonance, as might be expected from the reduced overlap of wavefunctions between 3d and 4f electrons (they belong to the different shell) compared with 4d and 4f electrons (the same shell). In other words, the 3d resonance is of the Coster-Kronig type which should be weaker than the super-Coster-Kronig type[31] 4d resonance. The second point is that although the constant-final-state (CFS) spectrum, which measures the emission intensity of the inelastically scattered electrons as a function of photon energy and is generally taken to be proportional to the absorption spectrum, shows the expected degeneracy ratio of 3:2 between resonance intensities at the $3d_{5/2}$ edge and $3d_{3/2}$ edge, CIS spectra show that the upper $3d_{3/2}$ edge is much less effective than the lower $3d_{5/2}$ edge in resonantly enhancing 4f photoemissions. This is due to the difference in the decay mechanism between the 3d hole and 4d hole, as will be discussed below.

A clue to the decay mechanism of the 3d hole created by the 3d → nf photoabsorption can be found in the resonant photoemission study of uranium 4d emissions of UO_2 at the U 3d edges[32]. Fig.12 shows photoemission spectra of the U 4d region at photon energies ranging from 10 eV below to 10 eV above the U $3d_{5/2}$ absorption peak ($h\nu = 3554$ eV). The spectra are plotted on a binding energy scale so that the $4d_{5/2}$ peaks, which show simple resonant enhancement, are aligned. For the U $4d_{3/2}$ region, however, we see very complicated spectral shape changes as the incident photon energy is varied. This turned out to be not due to the resonant enhancement of the U $4d_{3/2}$ peak or its loss structure, but comes from the following decay process of the 3d hole created by the 3d → 5f photoabsorption.

$$3d^{10}4f^{14}5f^2 + h\nu \rightarrow 3d^9 4f^{14} 5f^3 \rightarrow 3d^{10} 4f^{12} 5f^3 + \text{electron} \qquad (8)$$

That is, the additional 5f electron promoted from the 3d level in the photoabsorption process remains as a spectator in the decay process, where $MN_{6,7}N_{6,7}$ Auger process involving two 4f electrons takes place

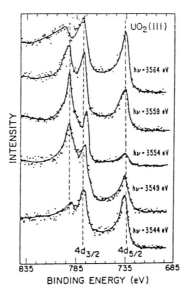

Fig.12. Photoemission spectra of the 4d levels of U near U $3d_{5/2}$ absorption maximum photon energy in UO_2. Relative vertical scale: 3544-1.0, 3549-1.2, 3554-5.3, 3559-1.6, and 3564-0.8.

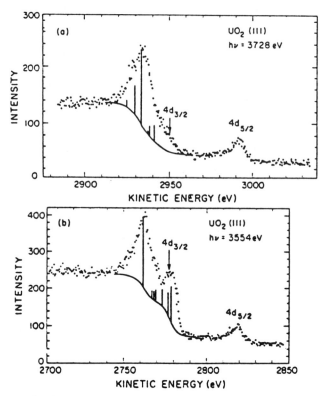

Fig.13. Photoemission spectra near the U 4d region at (a) $3d_{3/2}$ and (b) $3d_{5/2}$ resonant photon energies, along with calculated energies and transition probabilities of $MN_{6,7}N_{6,7}$ Auger decay final states.

instead of the autoionization involving 5f and 4d electrons. Experiment and theory[33] show that this Auger process is the dominant decay mechanism for U M_4 and M_5 holes, and indeed the calculated energies and lineshapes of the resulting two 4f hole final states give quite a good agreement with the spectra, as shown in Fig.13. Similar decay process was observed from the atomic Xe $2P_{3/2}$ hole[34], and was named Auger resonant Raman effect.

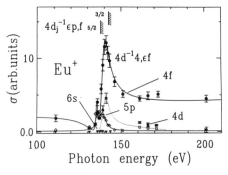

Fig.14. Partial cross sections of the 4d(squares), 4f(circles), 5p(diamonds), and 6s(triangles) hole final states in the vicinity of the 4d → 4f giant resonance of atomic Eu. The solid curves represent two-resonance fits with Fano profiles.

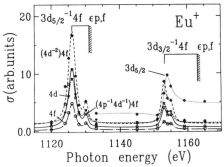

Fig.15. Partial cross sections of the 4d(triangles), 4f(circles), and $3d_{5/2}$(stars) hole final states, and the $4d^{-2}4f^8$ (diamonds) and $4p^{-1}4d^{-1}4f^8$ (squares) final states in the 3d → 4f excitation region.

The importance of this Auger resonant Raman process in the 3d resonant photoemission was confirmed by a recent experiment on atomic Eu[ref. 35]. In figures 14 and 15 are plotted the partial decay cross sections measured by the time-of-flight technique for several possible decay final states from the Eu 4d → 4f and 3d → 4f excitations, respectively. We see that in the case of the 4d → 4f resonance the autoionization to the 4f hole final state (circles) by the super-Coster-Kronig process, i.e. the process similar to eq.(2) leaving the ion in the same final state $4d^{10}4f^{n-1}$ as reached by direct 4f photoemission, is the dominant decay mechanism of the $4d^9 4f^{n+1}$ intermediate state. On the other hand, at the $3d \to 4f$ resonance the Auger resonant Raman process resulting in the $4d^8 4f^{n+1}$ final state (diamonds) has much larger cross section than the 4f hole final state $4d^{10}4f^{n-1}$ (circles). This is consistent with the UO_2 case discussed above. Furthermore, at the $3d_{3/2}$ edge the direct decay to the $3d_{5/2}$ hole state is possible and indeed this channel accounts for almost half of the total intensity of the $3d_{3/2}$ → 4f resonance. This gives the explanation for the ineffectiveness of the $3d_{3/2}$ resonance in enhancing Tb 4f emissions, as shown in the CIS spectra of Fig.11.

ACKNOWLEDGEMENTS

I wish to thank Dr. J.W. Allen, Prof. I. Lindau, Prof. S. Doniach and numerous other people who helped me understand the subject, perform the experiment and interpret the data during all these years.

REFERENCES

1. W. Lenth, F. Lutz, J. Barth, G. Kalkoffen and C. Kunz, Phys. Rev. Lett. 41, 1185 (1978).
2. L.I. Johansson, J.W. Allen, T. Gustafsson, I. Lindau and S.B.M. Hagström, Solid State Commun. 28, 53 (1978).

3. L.I. Johansson, J.W. Allen, I. Lindau, M.H. Hecht and S.B.M. Hagström, Phys. Rev. B21, 1408 (1980).
4. P.A. Cox, Struct. Bonding (Berlin) 24, 59 (1975).
5. This should not be considered as real two sequential processes. It should be taken as in the Feynman-diagram language.
6. U. Fano, Phys. Rev. 124, 1866 (1961).
7. J. Schmidt-May, F. Gerken, R. Nyholm, and L.C. Davis, Phys. Rev. B30, 5560 (1984).
8. D.C. Koskenmaki and K.A. Gschneidner, Jr., in Handbook on the Physics and Chemistry of Rare Earths, ed. by K.A. Gschneidner, Jr. and L. Eyring (North-Holland, Amsterdam, 1978) Vol. 1, p.337-377.
9. L. Pauling, 1950, and W.H. Zachariasen, 1949, unpublished.
10. C.R. Helms and W.E. Spicer, Appl. Phys. Lett. 21, 237 (1972).
11. Y. Baer and G. Busch, J. Electron Spectrosc. and Relat. Phenom. 5, 611 (1974).
12. J.W. Allen, S.-J. Oh, I. Lindau, J.M. Lawrence, L.I. Johansson and S.B.M. Hagström, Phys. Rev. Lett. 46, 1100 (1981).
13. J.W. Allen, S.-J. Oh, I. Lindau, M.B. Maple, J.F. Suassuna, and S.B. Hagstrom, Phys. Rev. B26, 445 (1982).
14. N. Mårtensson, B. Reihl, and R.D. Parks, Solid State Commun. 41, 573 (1982).
15. B. Johansson, Phil. Mag. 30, 469 (1974).
16. J.W. Allen and R.M. Martin, Phys. Rev. Lett. 49, 1106 (1982).
17. J. Kondo, Prog. Theo. Phys. (Kyoto) 32, 37 (1964).
18. J.K. Lang and Y. Baer, Rev. Sci. Instrum. 50, 221 (1979).
19. J.W. Allen, S.-J. Oh, M.B. Maple, and M.S. Torikachvili Phys. Rev. B28, 5347 (1983).
20. O. Gunnarsson and K. Schonhammer, Phys. Rev. Lett. 50, 604 (1983), and Phys. Rev. B28, 4315 (1983). See also the chapter by Dr. O. Gunnarsson in this volume.
21. J.W. Allen, S.-J. Oh and J.-S. Kang, to be published in J. Magn. Magn. Mat.
22. H.H. Hill, in Plutonium 1970 and other actinides, ed. by W.N. Miner (Metallurgical Society of AIME, New York, 1970) p.2.
23. For a review on heavy-fermion systems, see G.R. Stewart, Rev. Mod. Phys. 56, 755 (1984).
24. B. Reihl, N. Mårtensson, D.E. Eastman, A.J. Arko and O. Vogt, Phys. Rev. B26, 1842 (1982).
25. J.W. Allen, S.-J. Oh, L.E. Cox, W.P. Ellis, M.S. Wire, Z. Fisk, J.L. Smith, B.B. Pate, I. Lindau and A.J. Arko, Phys. Rev. Lett. 54, 2635 (1985).
26. G. Landgren, Y. Jugnet, J.F. Morar, A.J. Arko, Z. Fisk, J.L. Smith, H.R. Ott and B. Reihl, Phys. Rev. B29, 493 (1984).
27. A.J. Arko, C.G. Olson, D.M. Wieliczka, Z. Fisk and J.L. Smith, Phys. Rev. Lett. 53, 2050 (1984).
28. S.-J. Oh, J.W. Allen, and I. Lindau, Phys. Rev. B30, 1937 (1984).
29. J. Sugar, Phys. Rev. B5, 1785 (1972).
30. J.W. Allen, S.-J. Oh, I. Lindau and L.I. Johansson, Phys. Rev. B29, 5927 (1984).
31. E.J. McGuire, J. Phys. Chem. Solids 33, 577 (1972).
32. L.E. Cox, W.P. Ellis, R.D. Cowan, J.W. Allen and S.-J. Oh, Phys. Rev. B31, 2467 (1985).
33. R. Bastasz and T.E. Felter, Phys. Rev. B26, 3529 (1982) and E.J. McGuire, in Proc. of the Int. Conf. on Inner Shell Ionization Phenomena, ed. by R.W. Fink et al (Georgia Inst. of Tech. Atlanta, Georgia 1973). pp662-679.
34. G.S. Brown, M.H. Chen, B. Crasemann, and G.E. Ice, Phys. Rev. Lett. 45, 1937 (1980).
35. U. Becker, H.G. Kerkhoff, D.W. Lindle, P.H. Kobrin, T.A. Ferrett, P.A. Heimann, C.W. Truesdale and D.A. Shirley, to be published.

MIXED ELECTRONIC STATES AND MAGNETIC MOMENTS STUDIED WITH NEAR EDGE X-RAY ABSORPTION SPECTROSCOPY

Gerrit Van Der Laan

Materials Science Center, University of Groningen
Nijenborgh 16, 9747 AG Groningen, The Netherlands
Laboratoire pour l'Utilisation du Rayonnement Electromagnétique, Batiment 209D, Université Paris-Sud
91405 Orsay France

ABSTRACT

The line shapes of the near edge x-ray absorption spectra in 3d transition metal compounds, such as nickel dihalides and NiO show a direct relation with the amount of localization in the valence band states. Decrease in localization gives a reduction of the multiplet structure and an appearance of satellite structure due to core-hole potential effects. The observed multiplet structure can be used to diagnose the ground state. Another example is the near edge structure of magnetic materials using polarized synchrotron radiation. This can be used to determine the local magnetic moments in rare earth magnetic materials.

I - INTRODUCTION

The region in the vicinity of the x-ray absorption edge up to 10 eV above threshold, imaging the local electronic structure in the presence of a core hole, has received relatively little attention.[1-7] In the last decade much interest has been devoted to the energy region between 100 and 1000 eV above the absorption edge, the EXAFS region.[8] More recently the multiple scattering region between 10 and 100 eV has also been tackled.[8,9] Both regions give information about the perturbative scattering of free electron states by the nearby atoms.

The near edge fine structure upto 10 eV, or socalled Kossel structure, yields, particularly in the study of strongly correlated materials, useful information about the nature of the large-energy-scale dynamics that is in some respects complementary to what is found by other large-energy -scale probes as XPS, UPS, BIS and resonant photo emission. Usually one finds in these materials final state multiplet structure.[10,11] As was recently shown this can be used by virtue of the optical selection rules to determine the symmetry of the ground state including symmetry lowering by perturbations like crystal field[10] and spin-orbit coupling.[11]

The x-ray absorption spectra of transition metal compounds (TMC) are discussed in section II. They provide direct evidence that the one-electron approximation[5-7] is applicable for neither the ground state nor the final state. The creation of a core hole in x-ray absorption spectroscopy (XAS) changes the potential in the surrounding of the atom, and the final state eigenfunctions are different from the initial state wave functions.[12] This gives raise to many-body satellites. Because these satellites are normally small they have not been observed in well documented edges as the K-edge of 3d-TMC[6] or the $L_{2,3}$ edge of the 4d-TMC[5,7] due to the lifetime broadening of these levels.[13] The $L_{2,3}$ edges of 3d-TMC are ideal for studies of these effects because the lifetime broadening of the 2p level is only 0.3 eV. However, the x-ray region between 400 and 1000 eV has always been a difficult region, due to lack of high resolution monochromators, appropriate x-ray windows and due to severe vacuum requirements. Only recently the combination of high intensity sources (synchrotron radiation), high counting rates (electron yield method) and high resolution monochromators (double crystal monochromator)[14] has opened the way to the study of these phenomena.

Near edge absorption structure of rare earth magnetic materials using polarized synchrotron radiation is discussed in Section III. The polarization dependence of the 3d-4f XAS ($M_{4,5}$) spectra of rare earths is expected to be very sensitive to the magnetic structure of the ground state and to unequal population of the M_j sublevels as found in magnetic systems. This implies that XAS can be used to determine the local magnetic moment of a rare-earth atom in a magnetically ordered material, the orientation of the local moment relative to the total magnetization direction, as well as the temperature and field dependence thereof.

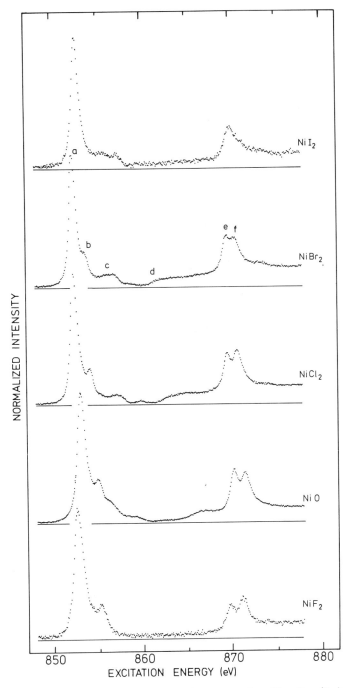

Fig. 1: $L_{2,3}$ absorption spectra of Ni compounds. The L_3 (L_2) edge is found at about 852 (870) eV.

II - LOCALIZED ELECTRONIC STATES IN NICKEL COMPOUNDS

Transition metal compounds have been an interesting and controversial topic of research since a long time, expecially regarding Mott-Hubbard transitions and electron correlation effects. As will be shown XAS can give valuable and supplementary information about these effects.

II.1 - Experimental and results

Ni $L_{2,3}$ x-ray absorption spectra were taken with the synchrotron radiation emitted by the Anneau de Collisions d'Orsay (ACO), using a double crystal monochromator equipped with beryl crystals (experimental resolution 0.3 eV).[14] The spectra were recorded in high vacuum by the electron yield method on -in situ- sublimed and on powdered samples of NiF_2, NiO, $NiCl_2$, $NiBr_2$ and NiI_2. The obtained $L_{2,3}$ spectra shown in Fig. 1 have a much higher resolution than earlier reported x-ray work on this edge.[15,16] As seen the near edge region is rich in structure which is strongly dependent on the compound. Notable is the second peak in the L_3 and L_2 edges, marked in Fig. 1 as respectively b and f, which gradually shifts towards the edge as the ligand becomes less electronegative. Also satellite structure is observed, especially in the L_3 (c in Fig. 1) and a continuum (d in Fig. 1) that shifts gradually towards lower excitation energy.

II.2 - Theory

$L_{2,3}$ multiplet structures of 3d-TMC have been calculated by Yamaguchi et al.[10], as in the case of Ni for the configuration $2p^5 t_{2g}^6 e_g^2$ ($^3A_{2g}$), taking the core hole exchange interaction and the crystal field interaction into account. A similar multiplet calculation is shown in Fig. 2 using Slater and exchange integrals from a relativistic Hartree-Fock calculation[17] and a 10 Dq value of 1.5 eV. The $2p_{3/2}$ and $2p_{1/2}$ edges show roughly a separation into two peaks with a splitting of about 2.5 eV, mainly determined by the exchange interaction $G_1(2p, 3d)$. This calculation for a transition from an octahedral ground state $3d^8$ to a final state $2p3d^9$ is in agreement with the experimental spectrum of NiF_2 (Fig. 2). In this compound the ground state is almost completely $3d^8$ due to the strong electronegativity of the fluoride anions.

In general the ground state of a TMC can be described as a mixture of configurations like $3d^n$, $3d^{n+1}\underline{k}$, $3d^{n+2}\underline{k}\underline{k}'$, etc. (where $\underline{k}$ denotes a

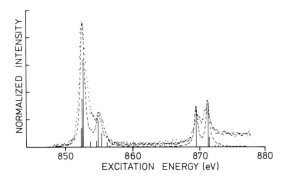

Fig. 2: Calculated multiplet structure for a dipole transition Ni $3d^8 \rightarrow 2p^5 3d^9$ (G_1 = 5.79, F_2 = 7.72, G_3 = 3.29, and 10Dq = 1.5 eV); the dashed line is a convolution by a 0.6 eV full width at half-height Lorentzian. For comparison, the experimental $L_{2,3}$ spectrum of NiF_2 is shown (dots).

hole in the ligand band).[18-20] The final states, in the presence of a core hole (denoted as $\underline{c}$) and an additional electron in the valence band, are linear combinations of the configurations $\underline{c}\ 3d^{n+1}$, $\underline{c}\ 3d^{n+2}\ \underline{k}$, etc. The x-ray absorption spectrum is given by the optical transition probability from the ground state to all accessible final states. Hybridization is caused by the transfer of an electron between a d-state and a k-state having the same irreducible representation. This transfer is taken into account by an effective mixing integral T.

In the specific case of the Ni compounds the ground state is a mixture of the configurations $3d^8$, $3d^9\ \underline{k}$ and $3d^{10}\ \underline{k}\ \underline{k}'$ with separations between these configurations equal to respectively Δ and $\Delta + U$, where Δ is the charge transfer energy and U is the effective 3d-3d Coulomb interaction. The states with a ligand hole are broadened into bands, as schematically illustrated in Fig. 3. The final states are a mixture of $\underline{2p}\ 3d^9$ and $\underline{2p}\ 3d^{10}\ \underline{k}$. The energy separation Δ' between the centroids of the multiplet split $\underline{2p}\ 3d^9$ states and the $\underline{2p}\ 3d^{10}\ \underline{k}$ band is equal to $\Delta + U - Q$, where Q is the effective 2p-3d coulomb interaction. For each irreducible representation in the symmetry group of the crystal field we have to calculate the configuration interaction. This has been done with

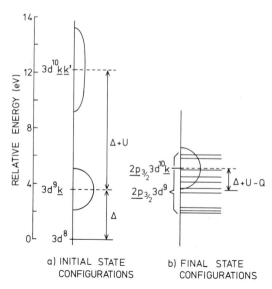

Fig. 3: Schematic representation of the energy for the different configurations: (a) initial state; (b) final states in the L_3 absorption for the case of $NiCl_2$. Configurations with a ligand hole ($\underline{k}$) have a bandwidth W. The $2p_{3/2}3d^9$ configuration is multiplet-split. Not drawn is the mixing between the configurations.

a cluster model[19-21] as well as with the Gunnarsson - Schönhammer impurity model[22]. An example of the results is given for NiI_2 in Fig.4. The satellite structure, which is due to the covalent mixing, is nicely reproduced in both models. The same type of calculation for the satellite structure can be used for XPS[21] and UPS[19] even though the spectral shapes in these methods are strongly different compared to XAS, due to the very different final states. The appearence of the satellite structure, with an energy separation from the main line equal, in first approximation, to $\Delta + U - Q$ is a clear proof for important correlation effects in the Ni band.

II.3 - Satellite Structure

In the more covalent compounds the mixed ground state also allows excitations to the $\underline{2p}\,3d^{10}\,\underline{k}$-like final state. This effect gives raise to

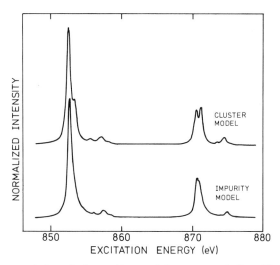

Fig. 4: Cluster model and impurity model compared for the case of NiI_2 ($\Delta = 1$ eV).

satellite structure in an analogous way as in XPS. The XAS of NiX_2 should be compared to the XPS of CuX_2 (see Fig. 5), where the same type of final state configurations appear.[21] These many body satellites in XAS, although small in intensity, are always expected due to the fact that the energy distance between the configurations is influenced by the presense of a core hole, so that $\Delta' \neq \Delta$. For this reason a frozen orbital approach, where the core hole potential is neglected can not predict satellite structure and would not be appropriate. The satellite intensity is a function of the difference in hybridization between the initial and final state, similar to the XPS case where this change in hybridization can be very large. Here, the core electron is completely removed and the screening of the core hole takes place by many-body processes like charge-transfer shake-up, polarization of the ligands, or plasmons. In XAS the core hole is already screened to a great extent by the additional electron in the d-band, but in correlated systems this screening is never complete. In XPS where the interaction between the core hole and a valance

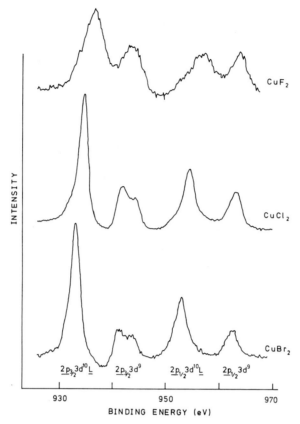

Fig. 5: Cu2p photoelectron spectra of Cu dihalides. The lines leading to a final state with a ligand hole ($\underline{L}$) show a chemical shift.

d-hole is equal to Q the energy levels in the final state are inverted and the dominant ground state component is in the satellite. In XPS this can give satellites with an intensity of the same order of magnitude as the main line, as is illustrated in Fig. 6. However, in XAS the interaction between the core hole and the remaining valence d-holes is reduced by the additional d-electron screening to a value of Q-U, which is approximately equal to 0.25 Q. The inversion of the energy levels does not occur so the lowest final state consists of states with the same phase in the wave function and also with the dominant ground state component. This results always in a much lower satellite structure normally smaller than 10% of the main peak intensity. This is in agreement with the experimental spectra of Fig. 1. For the Ni compounds these charge-transfer satellites are found at a distance between 3 and 5 eV from the edge as is clearly seen in the L_3 structure (Fig. 1, marked as c). The structure at

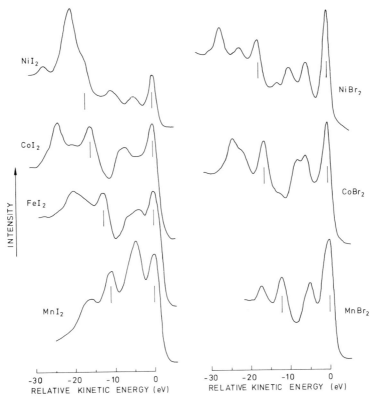

Fig. 6: Metal-2p photoelectron spectra of some 3d transition metal dihalides. The bars denote the positions of the $2p_{3/2}$ and $2p_{1/2}$ main lines. In NiI_2 the Ni $2p_{1/2}$ line is eclipsed by the I $3p_{3/2}$ line.

6.5 - 10 eV above the L_3 edge (Fig. 1, marked as d) comes from the onset of continuum states, where the core electron is excited into the Ni 4s state of the conduction band.

II.4 - Multiplet structure

Whereas the influence of the configuration interaction results in a small satellite intensity, it shows another effect that is not small, namely the reduction of the multiplet splitting in the main peak structure. When the mixing between the final state configurations $\underline{2p}3d^9$ and $\underline{2p}3d^{10}\underline{k}$ becomes stronger, the multiplet splitting of the $\underline{2p}3d^9$-like configuration is more reduced. In Fig. 1 is seen that the multiplet splitting of ~ 2.5 eV in NiF_2 is reduced to less than one eV in NiI_2. The gradual reduction of the main peak multiplet structure follows the decrease in Δ

(and hence in Δ') along the nephelauxetic series. This can be understood as follows: the width of the multiplet is roughly proportional to the exchange integral $G^1(2p, 3d)$, and a decrease in the localization of the 3d empty states will thus diminish the width of the multiplet structure. It is just the 3d final state character which is sampled by the 2p absorption spectrum. The main structures (a, b and e, f in Fig. 1) are $\underline{2p3d^9}$-like states, which become more and more delocalized in compounds toward NiI_2. This must be the reverse for the satellite structure (c in Fig. 1) which is a $\underline{2p3d^{10}}$ k-like band state. For more covalent compounds there will be more local 3d character present in this anti-bonding state.[23] This is seen for NiI_2, where structure in the satellite region is induced by localized states which are pushed out of the anion band.

As seen in Fig. 4, the cluster model greatly overestimates the peak splitting in the white line, whereas the impuity model gives good agreement for all measured Ni compounds. The reason for this is that in the impurity model the atomic exchange integral G^1 (2d,3d), which dominates the multiplet splitting is reduced by the hybridization with the valence band. In the cluster model this effect is underestimated.

II.5 - Determination of the ground state

The multiplet splitting in the main line can be used to obtain the relative population and the energy separation of the initial state configurations. Accurate values for these physical quantities can be obtained by comparison of the experimental spectra with the impurity model calculations (see Fig. 7). In Table I values are given for Δ, U and the $3d^8$, $3d^9$ and $3d^{10}$ characters in the ground state as obtained from a best fit. Details of these calculations, as well as extensive discussions can be found in references 24 and 25.

III - MAGNETIC X-RAY DICHROISM

The feasibility of using X-rays to determine the magnetic structure of magnetically ordered materials has recently been predicted theoretically[26]. Strong Magnetic X-ray Dichroism (MXD) is expected in the $M_{4,5}$ absorption edge structure of rare earth compounds. Polarized synchrotron radiation can therefore be used to reveal information on the local rare-earth magnetic moments in solids, thin films and surfaces.

Table I: The values for Δ and the $3d^8$, $3d^9$ and $3d^{10}$ characters in the ground state and for U as obtained from a best fit to an impurity model. Further constant for all compounds: Q = 7, T = 1.5 (1.75 in NiO), W = 3, 10 Dq = 1.5, G^1 = 5.79, F^2 = 7.72, G^3 = 3.29 eV, the effective spin-orbit splitting is 17 eV, the intrinsic line width is a 0.3 eV Lorentzian, the experimental resolution is a δ = 0.3 eV Gaussian.

Table I

	Δ(eV)	$3d^8$	$3d^9$	$3d^{10}$	U(eV)
NiI_2	1.5	0.47	0.44	0.09	4.5
$NiBr_2$	2.6	0.61	0.32	0.07	5
$NiCl_2$	3.6	0.71	0.23	0.06	5
NiO	4.6	0.73	0.21	0.06	5
NiF_2	7	~1.0	~0	~0	—

The $M_{4,5}$ absorption in rare earth compounds shows good agreement with the atomic Hartree-Fock calculations for the transitions from the $4f^n(J)$ Hund's rule ground state to the manifold of $3d^9 4f^{n+1}(J')$ final states [27-29]. Although in XAS hundreds of excited levels may be involved, one can distinguish between three different types of excitations, namely for J'-J = -1, 0, and 1.

In the presence of a magnetic field the (2J+1) degenerate ground state $4f^n(J)$ splits into sublevels M_J = -J, ..., +J. The relative population of these sublevels depends on the temperature. That the polarization vector of the X-rays can have a drastic effect on the spectrum can be seen from the simple case where only M = -J is populated (T = 0K). With the polarization direction parallel to the magnetization only the ΔM = 0 transitions are allowed. The transitions J'-J = -1 will then vanish, because the M' = -J sublevel is not present in the J' = J-1 state. An example is shown in Fig. 8. The calculated Dy M_5 absorption spectrum is given for the non-magnetic case and for the polarization direction of the X-rays perpendicular and parallel to the internal magnetic field at zero degrees Kelvin.[24]

A first experimental proof of this effect has been obtained for Terbium iron garnet (TbIG) at low temperature.[30] XAS spectra were measured using the electron yield method, and were obtained for different angles (α) between the polarization vector of the incident radiation and the [111] magnetization direction.

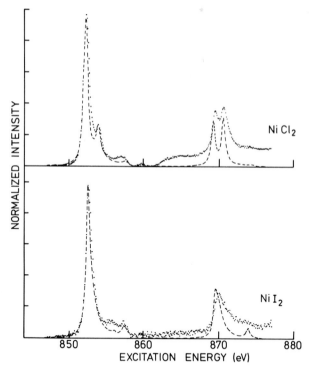

Figure 7: Impurity model calculation (dashed line) compared to the experimental $L_{2,3}$ x-ray absorption spectra of $NiCl_2$ and NiI_2 (dots). The values of the parameters are given in Table I and its caption.

The experimental M_5 spectra for various values of α are given in Fig. 9. The solid lines are theoretical curves. As seen, the intensities of the two major peaks are strongly polarization dependent. Parallel polarization in respect to the net magnetization enhances the left hand peak, whereas for perpendicular polarization the right hand peak increases in intensity. A multiplet calculation shows that the spectra agree with the magnetic structure of TbIG.[30]

MXD is complementary to other techniques, such as neutron diffraction and Mössbauer, because it can be applied to magnetically ordered thin films and surfaces, and because it is applicable for all magnetic rare earth compounds and probably can be extended to transition metal compounds.

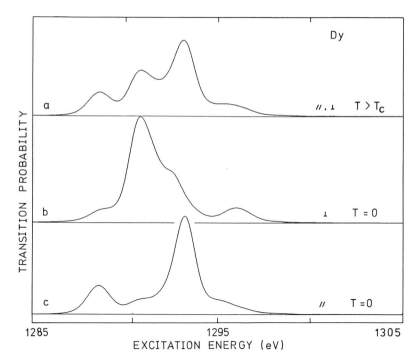

Fig. 8: The calculated XAS spectra of Dy (a) for $T > T_c$, and for $T = 0$ with polarization (b) perpendicular and (c) parallel to the internal magnetic field direction.

IV - CONCLUSIONS

The localization which influences the reduction of the exchange integral and the screening in the core ionization can be described by a formalism in which important parameters, as the charge-transfer integral can be estimated. The study of the near edge fine structure will then become an important tool in the determination of the ground state properties, for instance, the magnetic behavior.

ACKNOWLEDGEMENTS

I thank J.-M. Esteva, J.C. Fuggle, J.B. Goedkoop, C. Haas, R.C. Karnatak, G.A. Sawatzky, B.T. Thole, J. Zaanen and the LURE staff for valuable contributions.

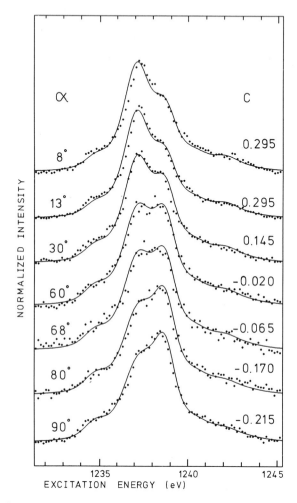

Fig. 9: Experimental M_5 absorption spectra of TbIG for various values of α, which is the angle between the polarization vector of the X-rays and the [111] magnetization direction. The solid lines are fits using Eq. (6) in Ref. 30. The optimum values of C are indicated.

This work was supported by the Netherlands Foundation for Chemical Research (SON) with financial aid from the Netherlands Organization for the Advancement of Pure Research (ZWO), and by the Centre National de la Recherche Scientifique (CNRS) in France.

REFERENCES

1. M. Brown, R. E. Peierls and E. A. Stern, Phys. Rev. B15, 738 (1977).
2. J.E. Muller, O. Jepsen, O.K. Andersen and J.W. Wilkins, Phys. Lett. 40, 720 (1978).
3. R.A. Bair and W.A. Goddard, Phys. Rev. B22, 2767 (1980).
4. E.A. Stern and J.J. Rehr, Phys. Rev. B27, 3351 (1983).
5. F.W. Lytle, P.S.P. Wei, R.B. Greegor, G.H. Via and J.H. Sinfelt, J. Chem. Phys. 70, 4849 (1979).
6. F.W. Kutzler, C.R. Natoli, D.K. Misemer, S. Doniach and K.O. Hodgson, J. Chem. Phys. 73, 3274 (1980).
7. J.A. Horsley, J. Chem. Phys. 76, 1451 (1982).
8. See for instance "EXAFS and Near Edge Structure" edited by A. Bianconi, L. Incoccia and S. Stipcich, Chemical Physics 27, Springer Series, Berlin (1983).
9. P.J. Durham, J.B. Pendry and C.H. Hodges, Solid States Commun. 38, 159 (1981).
10. T. Yamaguchi, S. Shibuya, S. Suga and S. Shin, J. Phys. C15, 2641 (1982).
11. J.-M. Esteva, R.C. Karnatak, J.C. Fuggle and G. A. Sawatzky, Phys. Rev. Lett. 50, 910 (1983); B.T. Thole R.D. Cowan, G.A. Sawatzky, J. Fink and J.C. Fuggle, Phys. Rev. B31, 6856 (1985).
12. U. von Barth and G. Grossmann, Solid State Commun. 32, 645 (1979).
13. J.C. Fuggle and S.F. Alvarado, Phys. Rev. A22, 1615 (1980).
14. M. Lemonnier, O. Collet, C. Depautex, J.-M. Esteva and D. Raoux, Nucl. Instrum. Methods 152, 109 (1978).
15. C. Bonnelle and C.K. Jørgensen, J. Chimie Physique 1964, 826 (1964).
16. C. Bonnelle, E. Belin and C. Sénémaud, Jap. J. Appl. Phys. 17, 125 (1978).
17. Robert D. Cowan, The Theory of Atomic Structure and Spectra (University of California Press, Berkeley, 1981), see Secs. 8-1 and 16-1.
18. J. Hubbard, Proc. R. Soc. London A276, 238 (1963).
19. G. van der Laan, Solid State Commun. 42, 165 (1982).
20. A. Fujimori and F. Minami, Phys. Rev. B30, 957 (1984).

21. G. van der Laan, C. Westra, C. Haas and G.A. Sawatzky, Phys. Rev. B$\underline{23}$, 4369 (1981).
22. O. Gunnarsson and K. Schönhammer, Phys. Rev. B$\underline{28}$, 4315 (1983); J. Zaanen, G.A. Sawatzky and J.W. Allen, Phys. Rev. Lett. $\underline{55}$, 418 (1985).
23. S. Antoci and L. Mihich, Phys. Rev. B$\underline{18}$, 5768 (1978); Phys. Rev. B$\underline{21}$, 3383 (1980); K. Terakura, A.R. Williams, T. Oguchi and J. Kubler, Phys. Rev. Lett. $\underline{52}$, 1830 (1984).
24. G. van der Laan, J. Zaanen, G.A. Sawatzky, R.C. Karnatak and J.M. Esteva, Phys. Rev. B$\underline{33}$, 4253 (1986).
25. G. van der Laan, J. Zaanen, G.A. Sawatzky, R.C. Karnatak and J.M. Esteva, Solid State Commun. $\underline{56}$, 673 (1985).
26. B.T. Thole, G. van der Laan and G.A. Sawatzky, Phys. Rev. Lett. $\underline{55}$, 2086 (1985).
27. J. Sugar, Phys. Rev. B$\underline{5}$, 1785 (1972), and Phys. Rev. A$\underline{6}$, 1764 (1972).
28. B.T. Thole, G. van der Laan, J.C. Fuggle, G.A. Sawatzky, R.C. Karnatak, and J.M. Esteva, Phys. Rev. B$\underline{32}$, 5107 (1985).
29. G. van der Laan, B.T. Thole, G.A. Sawatzky, J.C. Fuggle, R.C. Karnatak, J.M. Esteva, and B. Lengeler, J. Phys. C: Solid State Phys. $\underline{19}$, 817 (1986).
30. G. van der Laan, B.T. Thole, G.A. Sawatzky, J.B. Goedkoop, J.C. Fuggle, J.M. Esteva, R.C. Karnatak, J.P. Remeika, and H.A. Dabkowska, Phys. Rev. B, to be published.

PHOTOELECTRON SPECTROSCOPIC OBSERVATION

OF THE DENSITY OF LOW-LYING EXCITATIONS IN Ce-SOLIDS

W. D. Schneider, F. Patthey and Y. Baer

Institut de Physique, Université de Neuchâtel
CH-2000 Neuchâtel, Switzerland

B. Delley

RCA Laboratories ltd.
CH-8048 Zürich, Switzerland

INTRODUCTION

When atoms are brought close to one another the overlap of their outer charge densities will give rise to bond and band formation of condensed matter. A unique situation is encountered in the lanthanides where in the free atoms the centrifugal term $\ell(\ell + 1)/r^2$ combines with the atomic potential V_{at} to give an effective potential V_{eff} with a trough and a barrier for the high angular momentum ($\ell = 3$) electrons (see fig. 1(a))[1]. In the solid state (fig 1(b)) the peak of the barrier is higher than the zero of the atomic potential V_{at} leading to a trapping of the f-electrons in the trough. Their large kinetic energies lead to ionization energies lying within the bandwidth of the extended states. On the other hand, the atomiclike f-orbitals show a negligible direct f-f overlap between neighbouring atoms[2] but their hybridization with the band states of other symmetries can become significant for the physical and chemical properties of lanthanide materials[3,4,5]. It is the aim of the present paper to discuss within the Anderson single-impurity model the manifestations of this hybridization in the outer level excitations, observed with photoemission and inverse photoemission (BIS = Bremsstrahlungisochromat spectroscopy). Selected examples are presented (Ce-metal, CeN, CeCu$_6$) which map out the region from narrow f-band formation to heavy-electron behaviour.

DESCRIPTION OF HIGH-ENERGY EXCITATIONS WITHIN
THE ANDERSON SINGLE-IMPURITY MODEL

The hybridized states resulting from the coupling between states of completely different character can hardly be described in the pure atomic or in the pure band picture. A natural approach to such a situation is provided by the Anderson single-impurity model accounting equally for localized and extended states. If only outer level excitations are considered, the model Hamiltonian can be written as[5]

$$H = \sum_k \varepsilon_k n_k + \varepsilon_f \sum_m n_m + U_{ff} \sum_{m'>m} n_{m'} n_m + \sum_{km}(V_{mk} a_k^+ c_m + H.C.) \quad (1)$$

The first term describes the conduction band with energy dispersion $\varepsilon(\vec{k})$, the second term denotes the impurity atom level of degeneracy N_f characterized by the total energy difference ε_f between consecutive f-occupation numbers in the absence of hybridization and f-Coulomb correlation energy. The third term takes care of the Coulomb repulsion energy U_{ff} which is essentially an atomic quantity renormalized by the screening mechanisms in solids. Finally, the last term is responsible for the important and new aspects of the f-states encountered in the solid state. It contains the matrix elements V_{mk} that couple the f-level (index m) to the conduction states (index k); c_m and a_k^+ are the annihilation and creation operators for a f-electron in the state m and for a band electron in the state k, respectively.

The relevance of this Hamiltonian for high-energy spectroscopies like photoemission and inverse photoemission becomes immediately obvious if the coupling terms V_{mk} are neglected for a moment[6], which in fact corresponds to a pure atomiclike description of the f-states. For the discussion of excitation spectra it is convenient to choose as origin of the total energy scale the total electronic energy of the ground state configuration which for metals is at the Fermi energy[6]. The density of the outer level excitations induced in these spectroscopies is called "uncoupled $S(E_t)$". This quantity is composed of the density of the band states and of the discrete excitation spectrum of the uncoupled f-states. As shown in fig. 2(a), the uncoupled $S(E_t)$ can be split into electron-addition and electron-subtraction processes which must be represented on a common total-energy scale in order to be consistent with the Hamiltonian. It is more usual to represent these two types of excitations on a single-particle energy scale where the excitation energy is given as the energy required to modify the population of the considered level with one electron taken or added at E_F. This representation is simply obtained from the previous one by a 180^0 rotation of the electron-addition spectrum. Within the uncoupled scheme, for two atoms in a $4f^n$ ground state configuration one can verify in the single-particle or total-energy representation that U_{ff} corresponds to the common definition of the Coulomb correlation energy involved in the formation of the two non-interacting polar states $4f^{n-1}$ and $4f^{n+1}$. For an uncoupled $4f^1$ ground state the parameters lin-

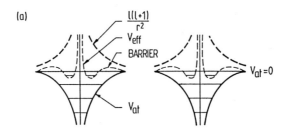

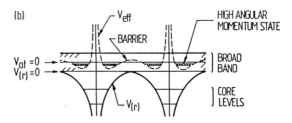

Fig. 1 Schematic illustration of 4f-states in (a) free atoms and in (b) condensed matter. (After Ref. 1)

king the different 4f populations are shown in the top of fig. 2(a). This atomiclike description of the f-states in solids has been very successful[7,8] but it must be considered only a useful approximation.

The situation changes dramatically if the hybridization between the f- and the band states, represented by the hopping terms V_{mk} in eq.(1), is taken into account. Now the calculation of excitation spectra is much more complicated since a degenerate f-state hybridizes with a continuum of extended states. Gunnarsson and Schönhammer[5] (GS) solved this problem for metallic Ce-systems using a limited number of many-body basis states with integral f-occupation and taking advantage of the degeneracy N_F of the f-states. The coefficients determining the weight of these basis states in the ground state are calculated in a variational procedure minimizing the total energy. Then, within the sudden approximation the excited eigenstates observed in high-energy spectroscopies are computed. The hopping terms V_{mK} are treated as constants yielding a hybridization strength $\Delta = \pi$ max $(V^2(\varepsilon))$ as an adjustable parameter. This GS scheme has proven to be extremely successful for the description of all kinds of excitation spectra[6,9-11]. The calculation yields in general ground states with fractional f-counts n_f. For example, the calculated f-contribution $S_F(E_t)$ to the outer level excitations of a typical α-Ce like material with a ground state f-occupation $n_F \approx 0.8$ is shown in fig. 2(b)[6,3]. The parameters used ($\varepsilon_f = -1.2$ eV, $U_{ff} = 6$ eV) correspond roughly to the uncoupled situation sketched in the top of fig. 2(a). The new aspects of the f-states in solids become apparent : in the presence of hybridization (here:$\Delta \approx 0.1$eV) the discrete atomiclike excitations transform into a continuum of f-excitations spread over the whole bandwidth of the extended states. Moreover, for the set of parameters chosen here, considerable spectral weight is found around E_F (Kondo resonance) which is absent in the atomiclike case.

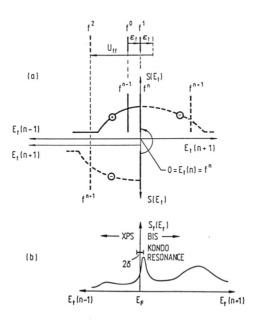

Fig. 2 (a) Screened single-particle excitations of the extended and localized outer levels without coupling as a function of the total electronic energy E_t of the final state (Ref. 6)
(b) f-contribution to the outer level excitations of a Ce-impurity in a metal in the presence of hybridization between f- and band states. (After Ref. 3)

THE KONDO RESONANCE IN EXCITATION SPECTRA

One of the fundamental motivations for spectroscopic studies is to deduce ground state properties of electronic systems from their excitation spectra. In principle, this task is accomplished if the parameters defined in the Anderson impurity model are determined. For example, the high-energy scale (≈ 1 eV) parameters ε_f, $N_f\Delta$ and the bandwidth B define a low-energy (<30 meV) scale $\delta = B \exp(\pi \varepsilon_f/N_f\Delta)$ [5] corresponding to the relevant energy scale of the quasi-ground state measurements (transport properties, specific heat etc.). This low-energy quantity δ is the lowering of the ground state resulting from the hybridization of the f- with the band states. When the f-population is close to 1, i.e. ε_f well below E_F and/or Δ small, δ/K_B is called the Kondo temperature T_K and the low-lying excitations within the energy δ form the Kondo peak[5] or Kondo resonance[12] (see fig. 2(b)). It is therefore a fundamental challenge to investigate more closely the relationship between the f-excitation spectrum and the nature of the many-body ground state.

A formal analysis of the structure of the N-electron ground state[13] and of the (N-1) electron final states involved in the photoemission process yields two types of final states [5,14,15]. At the Fermi level, corresponding to the ground state of the electronic system, the low-lying excitations $\rho_f^s(\varepsilon)$ have a maximum value proportional to $1/\Delta$ and fall off rapidly within δ. When $\Delta \to 0$ ($\delta \to 0$) the maximum diverges but its integral weight $n_f(1-n_f)$ vanishes. The intensity distribution of the split-off states $\rho_f^s(\varepsilon)$ maps out (proportional to $(1-n_f)$) the weight of the f-character n_f admixed to the band states at energy ε in the many-body ground state. Intuitively $\rho_f^s(\varepsilon)$ can be regarded as the adiabatic excitations of the f-projected part of the band. Towards higher energies ($\varepsilon > \delta$) another type of final states is formed, namely a continuum of f-excitations with increasing f-hole character having no close relationship to the ground state. Similar arguments hold for the (N + 1) electron final states encountered in BIS, where the relative intensity of the Kondo peak probed by electron addition is expected to be about 13 times stronger than in the photoemission spectrum (see fig. 2(b)). In order to observe directly the low-lying excitations $\rho_f^s(\varepsilon)$ in photoemission and inverse photoemission the instrumental resolution has to be of the order of δ which is always small (< 30 meV) in the Ce-systems studied so far.

CASE STUDIES

Ce-Metal - the γ-α Phase Transition

The isostructural γ-α phase transition in Ce-metal has always been associated with some major change of the electronic structure. During the last few years the accumulation of many experimental results [16,17,18] pointed to the fact that a change in hybridization between the f-states and the conduction band states is at the origin of the phase transition. This situation called for a high-resolution photoemission study of the outer level excitations in Ce-metal. With the HeI and HeII resonance lines produced in a commercial discharge lamp and an electron spectrometer carefully tuned to give a total instrumental resolution of 20meV the γ- and α-phases of Ce have been investigated at low temperature [15]. Fig. 3 shows, between 600 meV and $E_F = 0$ eV, for each phase the two photoemission spectra recorded with 40.8 eV and 21.2 eV photon energy. They have been weighted and subtracted from eachother in order to isolate as much as possible the f-contribution. The new feature observed at high resolution is the existence of two closely spaced peaks, one at E_F, the other one at 280 meV. This is precisely the energy separation of the $4f_{7/2}$ and $4f_{5/2}$ levels in atomic Ce. In the solid state a fractional j-population of the excited $4f_{7/2}$ state becomes possible by the hybridization mediated screening of a hole with predominantly $j = 5/2$. It is important to note that this type of many-body excitation probes the multiplet structure of the 4f ground state population of Ce. A first order GS model calculation has been performed including the spin-orbit interaction [15], the parameters were chosen in order to obtain the best possible agreement with many different spectroscopic data [6]. The essential result of this analysis is that a modest 30 % increase from Δ = 82 meV (n_f = 0.98) to 105 meV (n_f = 0.88) accounts correctly for all changes observed in the different spectra of the γ- and α-phases. The calculation yields for δ the values of 5 meV (γ-Ce) and 26 meV (α-Ce). The contribution $\rho_f^s(\varepsilon)$ of the low-lying excitations has been extracted from the calculation and convoluted with a Lorentzian in order to simulate the instrumental broadening and to tentatively account for the influence of the temperature. This is represented in fig. 3 by the curves defining the hatched areas which are 2 % and 12 % of the total f-spectral weight in γ-Ce and α-Ce, respectively. The good agreement between experimental and calculated f-contributions to the photoemission spectra shown in fig. 3 establishes the validity of the single-impurity approach and demonstrates that high-energy photoemission is well adapted to probe the crucial low-energy range of excitations closely related to the ground state.

Up to now the available instrumental resolution in BIS of ≈ 0.6 eV does not permit to observe directly the low-lying excitations of the (N+1) electron system. For example in α-Ce only a superposition of $\rho_f^s(\varepsilon)$ with the spin-orbit side band[12] of predominantly $j = 7/2$ character, convoluted with the instrumental broadening, is accessible to experiment[18].

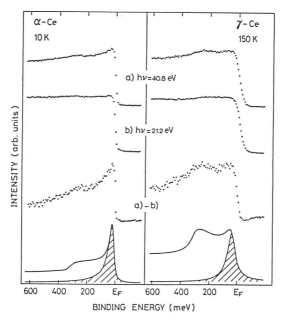

Fig. 3 Curves (a) and (b) : Photoemission spectra of α- and γ-Ce. Curve (a)-(b) : difference of weighted spectra (a) and (b) for the same phase. Full curve : GS-calculation. The curves defining the hatched areas correspond to $\rho_f^s(\varepsilon)$ convoluted with a Lorentzian. (Ref. 15)

CeN - a Narrow Band Material

The unusual properties of the compound CeN[19,20] were at the origin of many electron spectroscopic attempts[21,22,23] to clarify its electronic structure. The encouraging high-resolution photoemission results obtained for Ce-metal stimulated a similar study at ~30 K for CeN[24]. Fig. 4 a,b shows the photoemission spectra in the range between the top of the full p-band (mainly derived from N-states) and E_F. Fig. 4 c,d displays the result of a GS calculation including the 4f spin-orbit interaction. The conduction band contribution to the total emission spectrum is indicated by the partly dashed lines, their relative intensities have been estimated from the atomic cross section ratios for f- and p-states at the two photon energies. As in fig. 3, the hatched area corresponds to the energy region δ of the split-off states. The calculation yields δ = 15 meV and an f-population n_F = 0.82. The shoulder at 280 meV accounts for excited final states as discussed for Ce-metal. The striking similarity between the measured and calculated spectra yields a further proof that the single-impurity model contains the essential features for the description of the electronic structure of Ce-based materials. At low temperature ($T < \delta/K_B$) the f-states in CeN and α-Ce have no longer an atomiclike character, they form a metallic band of low-lying excitations of highly correlated states.

$CeCu_6$ - a Heavy-Fermion System

The compound $CeCu_6$ has one of the highest linear coefficients of the specific heat [25] (γ = 1600 $mJ/molK^2$) and shows no low-temperature magnetic phase. It crystallizes in an orthorhombic structure where each Ce atom is surrounded by 19 Cu atoms forming a cage[26]. The f-states can be visualized as embedded in a Fermi sea with the wave-function tails overlapping only weakly with the Cu sp-band states. The single-impurity model seems to be well adapted to account for this situation so that a spectroscopic study of this compound appeared to be promising [27]. Fig. 5 (a) shows the He II excited spectra of $CeCu_6$ in a range of 600 meV below E_F recorded with 20

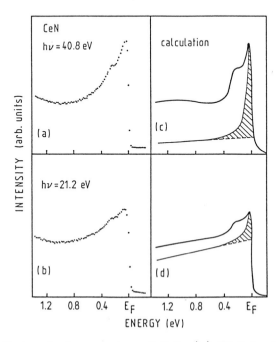

Fig. 4 (a) Photoemission spectra of CeN. (b) GS-Calculation (Ref. 24)

meV resolution at 15 K sample temperature. The only structure which is observed at 280 meV accounts for final states with mainly $j = 7/2$, as discussed previously for Ce-metal and CeN. These excitations are the only manifestations of the hybridization which can be detected in this spectrum, the Kondo resonance pinned at E_F cannot be discerned. This result is in perfect agreement with the GS-calculation ($\varepsilon_f = -1$ eV, $U_{ff} = 7$ eV, $\delta <$ 1 meV) including the 4f spin-orbit interaction, as shown in fig. 5(b). The dashed line denotes the assumed contribution of the Cu band states to the spectrum and the weak intensity above it acounts for the excitations of f-character. The inset of fig. 5 shows a narrow energy range around E_F recorded with a resolution of 12 meV. Under these conditions the Kondo peak should be discernable as a very weak intensity enhancement at E_F. The fact that this prediction is not confirmed by the experiment, performed at 15 K is probably due to the thermal broadening expected to wash out the Kondo peak for $T > T_K$ [28]. The message of this study is that in a system with extremely heavy electrons, the Kondo peak in photoemission and BIS spectra has a too small weight $((1-n_f) \approx 0.005)$ to be observed with the best resolution and low temperatures achieved until now.

CONCLUSIONS AND OUTLOOK

High-resolution photoemission spectroscopy of Ce-solids has proven to be well adapted to probe the low-lying excitations closely related to the ground state. The ultimate resolution of these methods is determined by the lifetimes of the states which are involved but is not affected by the high energy of the induced transitions. The Anderson single-impurity model, formulated for high-energy spectroscopies by GS, has demonstrated an undisputable success for describing these low-energy excitations in Ce-

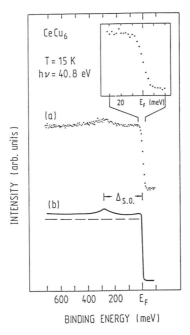

Fig. 5 (a) Photoemission spectrum of $CeCu_6$ measured with a resolution of 20 meV. (b) GS-calculation. Inset: Fermi edge of $CeCu_6$ measured with a resolution of 12 meV. (Ref. 27)

solids. The present fine-structure measurements, revealing the spin-orbit splitting of the 4f-ground state population, provide a decisive argument for the importance of the hybridization[29] taken explicitely into account by GS. Since in all Ce-systems studied so far the energy $\varepsilon_f \approx -1$ eV $> \Delta$, these systems have to be placed in the Kondo regime and not in the valence fluctuation regime, where $\varepsilon_f \approx \delta$ [4]. It must be emphasised that the hybridization strength Δ is the main adjustable parameter in the GS-formalism[3] so that methods to determine Δ from first principle band structure calculations are valuable[30]. For the description of heavy-Fermion systems the GS-calculation used so far appears to be only a first approach since it neglects the presence of the lattice and it implies $T = 0$. Coherence effects are expected to induce a splitting of the Kondo peak[28] at an energy scale $(1/N_f)\delta$ [31] and for $T > 0$ a weakening and/or broadening of the Kondo resonance is calculated[28,31]. A test of these predictions with electron spectroscopic techniques remains an experimental challenge for the future.

REFERENCES

1. D.J. Lam and A.T. Aldred, in : "The Actinides : Electronic Structure and Related Properties", A.J. Freeman and J.R. Darby Jr. eds, (Academic, New-York 1974), Vol. 1, Chap. 3
2. A.J. Freeman , Physica 102B, 3, (1980)
3. P.A. Lee, T.M. Rice, J.W. Serene, L.J. Sham and J.W. Wilkins, Comments Cond. Mat. Phys. 12, 99 (1986)
4. Y. Baer and W.-D. Schneider, in : "Handbook on the Physics and Chemistry of Rare Earths, K.A. Gschneidner, L. Eyring and S. Hüfner, eds. (North-Holland, Amsterdam 1987) Vol. 10, Chap. 1
5. O. Gunnarsson and K. Schönhammer, Phys. Rev. Lett. 50, 604 (1983); Phys. Rev. B 28, 4315 (1983)
6. W.-D. Schneider, B. Delley, E. Wuilloud, J.-M. Imer and Y. Baer, Phys. Rev. B 32, 6819 (1985)
7. J.F. Herbst, R.E. Watson and J.W. Wilkins, Phys. Rev. B 17, 3089 (1978) and refs. therein.
8. J.K. Lang, Y. Baer and P.A. Cox , J. Phys. F 11, 121 (1981)
9. J.C. Fuggle, F.U. Hillebrecht, J.-M. Esteva, R.E. Karnatak, O. Gunnarsson and K. Schönhammer, Phys. Rev. B 27, 4637 (1983)
10. J.C. Fuggle, F.U. Hillebrecht, Z. Zolnierek, R. Lässer, Ch. Freiburg, O. Gunnarsson and K. Schönhammer, Phys. Rev. B 27, 7330 (1983)
11. F.U. Hillebrecht, J.C. Fuggle, G.A. Sawatzky, M. Campagna, O. Gunnarsson and K. Schönhammer, Phys. Rev. B 30, 1977 (1984)
12. N.E. Bickers, D.L. Cox and J.W. Wilkins, Phys. Rev. Lett 54, 230 (1985)
13. C.M. Varma and Y. Yafet, Phys. Rev. B 13, 2950 (1976)
14. J.W. Allen, J. Magn. Magn. Mat. 52, 135 (1985)
15. F. Patthey, B. Delley, W.-D. Schneider and Y. Baer, Phys. Rev. Lett. 55, 1518 (1985)
16. N. Mårtensson, B. Reihl and R.D. Parks, Solid State Commun. 41, 573 (1982)
17. D.M. Wieliczka, J.H. Weaver, D.W. Lynch and C.G. Olson, Phys. Rev. B 26, 7056 (1982)
18. E. Wuilloud, H.R. Moser, W.-D. Schneider and Y. Baer, Phys. Rev. B 28, 7354 (1983)
19. R. Didchenko and F.P. Gortsema, J. Phys. Chem. Solids 24, 863 (1963)
20. J. Danan, C. de Novion and R. Lallement, Solid State Commun. 7, 1103 (1969)
21. Y. Baer and Ch. Zürcher, Phys. Rev. Lett. 39, 956 (1977)
22. W. Gudat, R. Rosei, J.H. Weaver, E. Kaldis and F. Hulliger, Solid State Commun. 41, 37 (1982)

23. E. Wuilloud, B. Delley, W.-D. Schneider and Y. Baer, J. Magn. Magn. Mater. 47&48, 197 (1985)
24. F. Patthey, S. Cattarinussi, W.-D. Schneider, Y. Baer and B. Delley, Europhys. Lett. 2, xxxx (1986)
25. G.R. Stewart, Rev. Mod. Phys. 56, 755 (1984)
26. D.T. Cromer, A.C. Larson and R.B. Roof, Acta Crystallogr. 13, 913 (1960)
27. F. Patthey, W.-D. Schneider, Y. Baer and B. Delley, Phys. Rev. B 34, xxxx (1986)
28. N. Grewe, Solid State Commun. 50, 19 (1984)
29. S.H. Liu, F. Patthey, B. Delley, W.-D. Schneider and Y. Baer, Phys. Rev. Lett. 57, 269 (1986)
30. R. Monnier, L. Degiorgi and D.D. Koelling, Phys. Rev. Lett. 56, 2744 (1986)
31. O. Gunnarsson and K. Schönhammer, in : "Handbook on the Physics and Chemistry of Rare Earths", K.A. Gschneidner, L. Eyring and S. Hüfner eds. (North-Holland, Amsterdam, 1987) Vol. 10, Chap.3

THE 4d → 4f GIANT RESONANCES FROM BARIUM THROUGH THE RARE-EARTHS

U. Becker

Institut für Strahlungs- und Kernphysik
Technische Universität Berlin
Hardenbergstr. 36, D-1000 Berlin 12, FRG

INTRODUCTION AND REVIEW OF GIANT RESONANCES AND ORBITAL COLLAPSE

The inner-shell photoabsorption spectra of atomic transition metals are characterized by giant resonances. These strong and broad absorption resonances are due to excitation of inner-shell electrons into partially occupied nd or nf subshells.[1,2] Prominent examples of this behavior are 3p→3d excitation in the iron group and 4d→4f excitation in the rare earths.[3] A giant resonancelike structure appears also in the photoabsorption spectra of the preceeding elements (Ar, Ca) and (Xe, Ba). The discussion of the new results regarding the deexcitation of giant resonances presented here requires a short summary of the theoretical interpretaion of this phenomenon which was observed in atoms as well as in solids. The nature of the giant resonances is closely related to the formation of transition elements due to the filling of the 3d and 4f shell in the iron group and rare earth region of the periodic table.[4] The properties of these atoms as well as the giant resonances depend on the specific features of the effective potential for d and f electrons.[1] This effective potential is determined by the atomic central potential $V(r)$ and the centrifugal term $l(l+1)/2r^2$. It consists of two wells separated by a centrifugal barrier. The specific character of this double well potential divides the elements which exhibit giant resonances roughly into two groups: The lighter elements of each transition period including the preceeding noble gases and the heavier elements throughout the rest of the period. The resonances in the first group are described mainly due to a delayed onset followed by a centrifugal-barrier shape resonance[5] whereas the resonances in the second group in particular the rare earths, are described mainly due to 4d→4f discrete transitions, which are raised into the f-continuum by the exchange interaction.[6] The essential physical difference between these two cases is the spatial extent of the major contributing final wave function i.e. uncollapsed f-wave function for $Z \leq 54$ versus collapsed f-wave functions for $Z \geq 55$. The changeover between the two cases is abrupt for the major component of the 4f wave function, hence the term collapse.

The appearance of giant resonances in the absorption spectra of Xe, Ba and La is very similar, strong and broad resonances in the 4d continuum as shown in Fig. 1. However they are assumed to be of different origin, Xe belongs to the first group whereas Ba and La belong to the second group of elements described above.

Following this interpretation the giant resonance in Xe results from a one step process: continuum enhancement due to a centrifugal-barrier shape resonance.

$$\text{Xe } 4d^{10} \xrightarrow{\text{photoabsorption}} \text{Xe } 4d^9 \varepsilon f$$

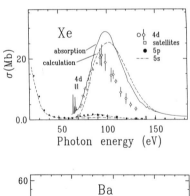

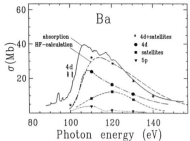

Fig. 1. Partial cross sections of Xe and Ba in the energy region of the giant resonances compared with corresponding absorption profiles[7]

In contrast the giant resonances in Ba and La result from a two-step process: photoexcitation of a 4d-subshell electron into the 4f-subshell

$$4d^{10}4f^N \xrightarrow{\text{photoexcitation}} 4d^9 4f^{N+1}$$

where N is the occupation number of the 4f-subshell, followed by autoionization of the $4d^9 4f^{n+1}$ configuration into the alternative continuum channels including the already open channel of direct 4d photoemission $4d^9 4f^N + \epsilon p,f$

$$4d^9 4f^{N+1} \xrightarrow{\text{autoionization}} 4d^9 4f^N + \epsilon p,f$$

where the continuum electron has predominantly an orbital momentum l=3. There was a theoretical discussion[8] about the possibility that the giant resonance in Ba could also be due to a one-step continuum enhancement similar to Xe:

$$Ba\,4d^{10} \xrightarrow{\text{photoabsorption}} Ba\,4d^9 \epsilon f\,(^1P)$$

It appeared during this controversy, that the final outcome of a theoretical calculation is strongly model dependent, therefore when one starts from a different basis set of eigenfunctions in a CI-calculation one may end with a different state associated with the giant resonance. A more unified approach developed recently shows the importance of shape resonances to the orbital collapse phenomenon.[9,10] It reveals the interaction between the inner well eigenstate (shape resonance) and the outer well nf, ϵf states. The calculations show that the inner well becomes deeper and deeper as nuclear charge increases. As a result the shape resonances gradually move towards the 4d threshold, interact with the nf Rydberg states and when the inner well is deep enough to support a bound state of its own, the 4f orbital starts to "collapse" into the inner well. This picture, which regards the orbital collapse basically as a resonance phenomenon, is supported by the appearance of a partial collapse along the isoelectronic sequence of Xe-like ions. The partial-collapse model explains the structure of the observed Ba^{2+} spectrum,[11] which is characterized by intense absorption lines in contrast to the giant resonance-like structure seen in the spectra of Ba and Ba^+. The calculations show in addition that the partial collapse is strongly term dependent corroborating the findings of earlier work. How does this partial collapse along the isoelectronic and isonuclear sequence fit to the sudden collapse observed along the periodic table? The main difference between the isoelectronic sequence and the period of transition elements is the variation in the height of the centrifugal barrier. This potential barrier disappears quickly along the isoelectronic sequence while it is reduced only slightly for neutral atoms

along the periodic table. Consequently the outer well of the
neutral atoms remains very shallow and the eigenstate of the
inner well (shape resonance) can sink below the nf series so
quickly that the whole process can be completed by stepping
from one element (Xe) to the other (Ba), hence the name
collapse. Along isoelectronic sequences however, the sinking
of the inner well eigenstate is competing with the lowering of
the outer well with increasing Z, and the process can take
place over several ions resulting in a gradual or partial
collapse. The key point of this review for our further
discussion is now that all the different theoretical models
show strong term dependence for any kind of orbital collapse.
Thus in quantitative discussions of the collapse phenomenon,
the term dependence of the 4f orbital should be taken into
account. The number of terms to be considered in this respect
increases rapidly with the number of 4f electrons in the
partially filled 4f shell of the rare earths. This introduces
the question of the term dependency of the orbital collapse
along the period of the transition elements although it seems
to be correct and in agreement with the predictions of all
existing models that the main component of the 4f orbital
(terms based on ^{3}L coupling) collapses abruptly around Z=56.
But what happens with the minor ^{1}P based components for which
the potential barrier may be higher or less exactly defined
due to stronger exchange forces than for the main component,
their wave functions may be uncollapsed or hybridized. This in
turn leads to the question about the strengths of the shape
resonance beyond Ba, because the remaining strengths of the
shape resonance after the collapse of the 4f main component is
closely related to the minor fraction of partially or still
uncollapsed multiplets of the 4f wave function as a result of
sum rules. The basic features of the orbital collapse as
discussed above are shown schematically in Fig. 2. This figure
shows the double well potential on a nonlinear scale with the
4f orbital before and after collapse.

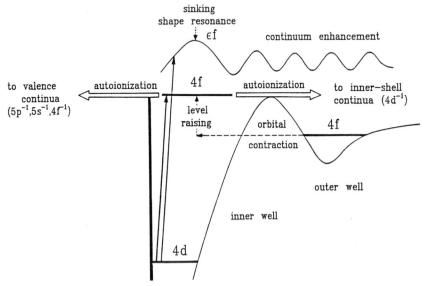

Fig. 2. Schematic representation of the collapse of the 4f
wave function in a double well potential together with
the decay channels of the associated 4d→e,εf
excitations

It shows further how the collapsed wave function is raised into the continuum close to the sinking shape resonance in the case when this continuum resonance contains still sufficient oscillator strengths. The decay pathways of the discrete 4f state, which are the subject of the next paragraph, are marked by arrows labelled with autoionization.

DECAY OF GIANT RESONANCES

Fig. 2 shows the different decay pathways of a giant resonance above the 4d-ionization threshold. Regarding the different final-state wave functions of the photoexcitation and photoionization process we distinguish between

1.) autoionization of <u>discrete states</u> into adjacent photoemission continua and

2.) <u>continuum enhancement</u> due to centrifugal-barrier shape resonances with delayed-onset character.

The decay channels of discrete states can be further distinguished with respect to the final ionic states of the autoionization process. There are basically two autoionization channels, which are of particular interest for the deexcitation of giant resonances:

a) autoionization into the inner-shell 4d continuum without core-hole relaxation as mentioned in the introduction and
b) autoionization into the outer-shell, also referred to as valence continua.

The second process which is associated with core-hole relaxation corresponds to regular autoionization of discrete resonances which are normally located below the ionization threshold of the subshell from where the excitation occurs (Feshbach resonances). Further differentiation allows one to distinguish between participator and spectator transitions with respect to the excited electron corresponding to decay into the photoemission "main line" and "satellite" channels. This differentiation is less important for the 4d→4f giant resonances, but it becomes helpful for the discussion of the related 3d→4f excitations. Summarizing the different decay modes gives the following:

1. Autoionization

$$4d^{10}4f^N \rightarrow 4d^9 4f^{N+1} \quad (a) \rightarrow 4d^9 4f^N \varepsilon f$$

$$(b) \quad \begin{array}{l} \rightarrow 4d^{10}4f^{N-1}\varepsilon g \\ 4d^{10}4f^N 5(s+p)^7 \varepsilon p,d \end{array}$$

2. Continuum enhancement

$$4d^{10}4f^N \rightarrow 4d^9 4f^N \varepsilon f$$

The different decay modes of giant resonances yield additional information on the character of these resonances. However, this information is unfortunately not unambigous with respect to the distinction between continuum enhancement and photoexcitation of discrete states. Autionization into the valence channels is a clear indication of deexcitation of collapsed 4f states because it reflects directly their overlap with the core-hole. The favoured decay mode of these states should be the super-Coster-Kronig transition[12]

$$4d^9 4f^{N+1} \rightarrow 4d^{10} 4f^{N-1} \varepsilon d,g$$

leaving the ion in the same final state as reached by direct 4f photoemission. However, autoionization into the inner-shell 4d continuum leaves the ion in the same final states as populated by the shape resonance, therefore there is no clear distinction between this decay mode of a discrete excited state and the continuum enhancement induced by a shape resonance. Even though there are different tests based on ion yield and partial cross measurements which should enable us to decide which of the two mechanism is more likely to contribute to the giant resonance intensity.

1. The energetic distribution of the various decay modes over the resonance region. Having in mind the sinking shape resonance along increasing Z a variation of the enhanced 4d partial cross section along the elements of the transition period may point to continuum enhancement rather than to autoionization into this continuum.

2. The analysis of the shape of the resonances should yield further insight into their character: Autoionizing resonances show Beutler-Fano profiles[13] while shape resonances are described by a sum of Breit-Wigner profiles.[14] The two profiles are most clearly distinguished with respect to their onset and the existence of possible minima.

3. The ratio between double and single ionization below and on resonances should vary significantly if autoionization into the inner-shell channel plays an important role because this channel decays via subsequent Auger decay into double charged ions.

Fig. 3 shows the partial photoemission cross sections of atomic Sm, Eu and Tm. These spectra clearly show the variation of the 4d cross section enhancement along the period.[15] This trend is strongly supported by additional measurements by the Sonntag group on Cs, Ce and Gd[16] and fits directly to the partial cross sections of the giant resonances in Ba and Xe as shown in Fig. 1.

The 4d cross section enhancement varies with respect to the excitation energy and the branching ratio between the 4d and 4f enhancement. The peak of the 4d enhancement moves closer to the autoionizing resonance structure and shrinks going along the periodic table. This behavior pointing to continuum enhancement rather than autoionization is complemented by the shape of the 4d cross section enhancement, which is compatible with profiles derived for shape resonances. Measurements of the double to single ionization ratio were performed by Schmidt and Zimmermann[17] for most of the 3d elements. Their results show that this ratio is nearly unaffected by most pronounced resonances below and above the 3p threshold indicating strong autoionization into the valence

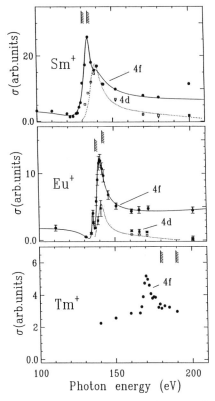

Fig. 3. 4f and 4d partial cross sections of Sm, Eu and Tm in the giant resonances region. Other partial cross sections have been omitted for clarity. The enhancement of the 4f-channel results from super-Coster-Kronig (sCK) decay of the collapsed 4f wave function, while the enhanced 4d cross section could be either due to autoionization of discrete 4f states into the 4d continuum or due to shape-resonant enhancement of this continuum channel. The solid lines represent a fit of the 4f partial cross section data with Beutler-Fano profiles, whereas the dotted lines show an approximate profile for shape-resonant enhancement of the 4d cross section.

channels with their associated shake-off continua.
Autoionization into the 3p continuum would show up in
overfractional enhancement of the double-ion yield due to
Auger recombination, which is seen to be a minor process in
the 3d elements, in particular in Sc and Ti, where many strong
resonances are above the 3p threshold. The double to single
ionization ratio of these elements mimics approximately the 3p
cross section.

In order to explore the giant resonance phenomenon
associated with 4d→4f excitation in more detail, theoretical
calculations of partial cross sections in the vicinity of
these resonances are required. First results by many body
theory for transition elements with a half filled 3d or 4f
subshell (Mn or Eu) show promising agreement with the
experiment.[18]

A last point worth mentioning here is the change in the
resonance profiles and partial cross sections if the
excitation of the collapsed 4f orbital occurs from the 3d
instead from the 4d subshell.[15] Because the 3d→4f resonances
are below their associated thresholds, 3d continuum
enhancement is not possible. Autoionization into the continua
of the outer shells including the 4d subshell is now competing
with strong spectator transitions reflecting the vanishing
overlap between the strongly localized 3d hole and the outer
subshells. Only the localized character of the collapsed 4f
orbital contributes to valence-shell enhancement via
participator transitions. From the other subshells the 4d
partial cross section seems also to receive appreciable
intensity from the deexcitation of the 3d→4f giant resonances.

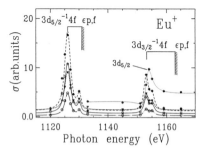

Fig. 4. Partial cross sections in the 3d→4f excitation region.
Participator transitions involving the 4f and 4d
subshell are shown by circles and triangles. Spectator
transition to final states $4d^{-2}4f^8$ and $4p^{-1}4d^{-1}4f^8$ are
given by diamonds and squares. The stars represent
autoionization of the $3d_{3/2}$→4f excitation into the
continuum of the $3d_{5/2}$ spin orbit component.

The large spin-orbit splitting between the 5/2 and 3/2 components of the 3d subshell allows autoionization of the $3d_{3/2} \to 4f$ resonance into the $3d_{5/2}$ continuum. This situation is similar to the Beutler-Fano resonances[19] in rare-gas atoms between the $^2P_{3/2}$ and $^2P_{1/2}$ thresholds, where the $np_{1/2} \to n's_{1/2}$, $n'd_{3/2}$ Rydberg series are imbedded in the $np_{3/2}$ continuum. The strengths of this decay channel without core-hole relaxation shows the small overlap between the 3d hole and the other electrons even the more localized ones. Fig. 4 shows the partial cross sections in the vicinity of the $3d \to 4f$ excitations. The positions of $3d \to 4f$ excitation below the 3d thresholds may be related to the weaker exchange forces in the 3d subshell compared to the 4d subshell, but could also be a result of the increase in effective nuclear charge seen by the excited electron if the core hole is a 3d hole instead of a 4d hole. The last interpretation would fit smoothly into the conceptual frame work outlined before. However, the presented results are only a first step to a deeper understanding of the intrinsic properties of giant resonances and their relation to other phenomena like orbital collapse.

CONCLUSIONS

Partial cross section measurements inside "giant resonances" of rare earths via deexcitation spectroscopy yield evidence that the orbital collapse of the excited 4f wave function is strongly term dependent and may be not completed before the 4f shell is more than half filled. In this case the ratio between autoionization and continuum-shape-resonance enhancement can be regarded as a direct measure of the ratio between collapsed discrete states which were pushed into the continuum by exchange forces and shape-resonantly enhanced continuum states carrying the oscillator strengths of the uncollapsed 4f components. This ratio therefore probes the collapse of the minor component of the 4f wave function which is equivalent to a measure of the partial collapse of these components. Corresponding measurements of $3d \to 4f$ excitations and ion yield measurements along the period of the 3d elements support this interpretation.

ACKNOWLEDGMENTS

The author is indebted to the research group of Prof. D.A. Shirley where part of this work was performed. He likes to thank further Prof. B. Sonntag and the members of his research group for their interest and help in the measurements at the Hamburger Synchrotronstrahlungslabor at Deutsches Elektronensynchrotron. He acknowledges funding by the Bundesminister für Forschung und Technologie and the Deutsche Forschungsgemeinschaft.

REFERENCES

1. J.P. Connerade, Contemp.Phys. 19, 415 (1978)
2. R.I. Karaziya, Usp.Fiz.Nauk. 135, 79 (1981) (Sov.Phys.Usp. 24, 775 (1981) and references therein
3. E.-R. Radtke, J.Phys.B 12, L71, L77 (1979)

4. M. Goeppert-Mayer, Phys.Rev. 60, 184 (1941)
5. J.W. Cooper, Phys.Rev.Lett. 13, 762 (1964)
6. J.L. Dehmer, A.F. Starace, U. Fano, J. Sugar, and J.W.Cooper, Phys.Rev.Lett. 26, 1521 (1971)
7. U. Becker, T. Prescher, E. Schmidt, B. Sonntag and H.-E. Wetzel, Phys.Rev. A 33, 3891 (1986); U. Becker, R. Hölzel, H.-G. Kerkhoff, B. Langer, D. Szostak, and R. Wehlitz, in Abstracts of Contributed papers, Fourteenth Intern. Conference on the Physics of Electronic and Atomic Collisions, Palo Alto 1985, ed. by M.J. Coggiola, D.L. Huestis, and R.P. Saxon (ICPEAC, Palo Alto, 1985), p.12 and references therein
8. G. Wending and A.F. Starace, J.Phys.B. 11, 4119 (1978)
9. K.T. Cheng and C. Froese-Fischer, Phys.Rev. A 28, 2811 (1983)
10. K.T. Cheng and W.R. Johnson, Phys.Rev. A 28, 2820 (1983)
11. T.B. Lucatorto, T.J. McIlrath, J. Sugar, and S.M. Younger, Phys.Rev.Lett 47, 1124 (1981)
12. E.J. McGuire, J.Phys.Chem Solids 33, 577 (1972); M.Y.Amusia, S.I. Sheftel, and L.V. Chernysheva, Sov.Phys.Tech.Phys. 26, 1444 (1982)
13. U. Fano, Phys.Rev. 124, 1866 (1961)
14. J.P. Connerade, J.Phys. B 17, L165 (1984); see also R.J. Peterson, Y.K. Bae, and D.L. Huestis, Phys.Rev.Lett. 55, 692 (1985)
15. U.Becker, H.G.Kerkhoff, T.A.Ferrett, P.A.Heimann, P.H.Kobrin, D.W.Lindle, C.M.Truesdale and D.A.Shirley, submitted to Phys.Rev. A (1986) and references therein
16. Th. Prescher, M. Richter, E. Schmidt, B. Sonntag, and H.E. Wetzel, J.Phys.B (1986) in press; M. Richter, private communication (1986)
17. M. Schmidt, Doctoral thesis, Technische Universität Berlin, 1986 (unpublished); P. Zimmermann, private communication (1986)
18. L.J.Garvin, E.R.Brown, S.L.Carter and H.P.Kelly, J.Phys.B. L269 (1983); H.P.Kelly, private communication (1986)
19. H. Beutler, Z.Phys. 93, 177 (1935); U. Fano, Nuovo Cimento 12, L125 (1935).

APPEARANCE-POTENTIAL STUDY OF RESONANT TRANSITIONS IN LANTHANIDES

D. R. Chopra

Department of Physics
East Texas State University
Commerce, Texas 75428

ABSTRACT

Soft x-ray appearance potential spectroscopy (SXAPS) is a high resolution core level technique for probing the local density of conduction band states of the solid surfaces. It exploits the abrupt changes in the total x-ray emission associated with the threshold for the excitation of core levels. The sharp increases in the total x-ray fluorescence are enhanced by potential modulation technique. SXAPS has been utilized to investigate the distribution of 4f states of the lanthanides. The spectra reflect the transitions of $N_{4,5}$ core electrons to unoccupied 4f states. These states split into a multiplet as a result of exchange interaction between the 4f electrons and 4d vacancy. Multiplicity of these configurations depends upon the number of 4f electrons. A systematic study of spectral features of the lanthanides does not reveal any regular or orderly changes unlike the spectra of 3d transition metals. These results confirm that the spectra of lanthanides are amenable to interpretation in terms of the atomic-like resonances involving exchange interaction rather than in terms of the continuous interband transitions. The experimental results are discussed in the light of the above theoretical considerations.

INTRODUCTION

SXAPS has been cited as a high resolution core level spectroscopy for the study of core level binding energies (BE) and the density of unoccupied states above the Fermi level (E_F) of atoms in the surface region.[1-3] In SXAPS, the total x-ray fluorescence of a sample bombarded by monoenergetic electrons of linearly increasing energy is measured as a function of primary electron energy. If the energy of the primary electron is at, or slightly above, the threshold of core level excitation, both the primary electron and the core electron enter into the unoccupied states above E_F. The de-excitation process following the excitation leads to the appearance of characteristic x-ray emission superimposed on a slowly varying Bremsstrahlung background. The sharp increases in the total x-ray fluorescence yield are enhanced by potential modulation as in Auger electron spectroscopy. Since for most metals the inelastic mean free path for incident electrons in the energy range 0-2000 eV is only a few atomic layers, the resultant radiation will represent the properties of the surface atoms. Information on the BE of core electrons of the surface atoms is gleaned from the energies of the incident electrons at which the sharp changes occur in the total x-ray fluorescence. The shape of the peaks is determined by the distribution of unfilled states of the valence band. SXAPS offers both experimental and conceptual simplicity in providing information on the electronic properties of the solid surfaces.

Various measurements have been reported on the SXAPS of lanthanides.[3-8] The purpose of this paper is to present a systematic study of the $N_{4,5}$-level SXAPS of the lanthanide series. The lanthanides are often referred to as the 4f metals because each new electron which is added as one proceeds from La to Lu enters into the 4f shell. Furthermore, since the 4f shell is located inside the 5d6s-conduction states, the nature of the latter changes little as a function of Z. The lanthanides exhibit similarities in many of their physical properties, and it is mainly only when the 4f electrons are directly involved that they do show distinctly different behavior. The position of the 4f states relative to E_F in lanthanides is important for understanding their properties. In the present paper we have attempted to compare their spectra in terms of the total x-ray fluorescence, the width and height of the peaks and the fine structure features and to correlate the findings with increasing Z.

EXPERIMENTAL

The appearance potential spectrometer, which has been described in detail elsewhere,[6] consists simply of a photoelectron collector which measures the soft x-ray fluorescence of an electron-bombarded sample. The tungsten filament cathode is directly heated by 20 V, 400 Hz current pulses with a duty cycle of 8% to minimize the energy spread of the incident electron beam. The cathode-target assembly is electrically isolated from the x-ray detector by a grid which is negatively biased with respect to the cathode. The potential on the target is linearly increased by a variable-slope ramp generator. The x-ray current is measured as the potential of the target is swept through the region of threshold excitation of a particular core level. The total fluorescence yield is the sum of the intense Bremsstrahlung continuum and the weaker characteristic x-rays due to radiative decay of the core hole. A derivative of the total fluorescence yield is obtained using the potential modulation technique in order to enhance the abrupt structure in the threshold region. This is accomplished by superimposing a 5.6 kHz frequency oscillation of 0.4 V_{pp} on the target potential and detecting the resulting synchronous variations in the photocurrent using a phase-lock amplifier.

In each case, the samples were high purity metal foils which were argon ion sputtered and annealed in situ. The K-level SXAPS of oxygen and carbon were barely detectable, thus ensuring clean samples for examination. The system pressure was on the order of 10^{-9} Torr. The resolution of the spectrometer is limited by the stability of the target potential, the amplitude of the modulating voltage and the voltage drop across the filament. The estimated instrumental resolution by comparison to the reported values was about 1.0 eV.

RESULTS AND DISCUSSION

We have recorded the $N_{4,5}$-level SXAPS of lanthanides (fig. 1). The metals selected range from light to heavy lanthanides (Pm excepted) in order to enlarge the scope of the investigation. The spectra are plotted on the same energy scale and aligned for purposes of making systematic comparison of their features. The arrows labeled N_4 and N_5 have been

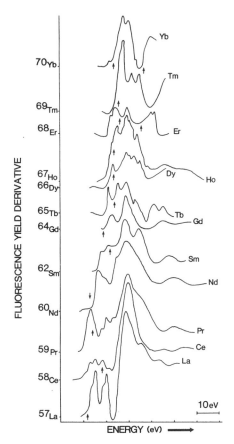

Fig. 1. $N_{4,5}$-level SXAPS of selected lanthanides. The x-ray fluorescence yield derivative is plotted vs. the energy of incident electrons. The spectra were recorded with an excitation current of 2 mA and with a modulation voltage of 0.4 V_{p-p}. The arrows mark the tabulated energies (Ref. 9) for exciting $4d_{3/2}$ and $4d_{5/2}$ electrons to the Fermi level. The spectra are aligned for making comparisons.

taken from standard tables[9] and locate E_F with regard to doubly degenerate levels $4d_{3/2}$ and $4d_{5/2}$. The $N_{4,5}$ spectral region is associated with the variable structure superimposed on a broad continuum. The width and the strength of this structure decreases with increasing Z. The complexity of the structure increases with Z initially, reaching a maximum for medium lanthanides and then begins to decrease for heavy metals towards the end of the series. The light lanthanides give evidence of the sharp lines below the threshold. We have recorded the N_2 and N_3-level SXAPS of the same elements and find these to be more than an order of magnitude weaker relative to the $N_{4,5}$ peaks. This indicates that the lanthanides exhibit intense SXAPS resonances only for those core levels which have proper symmetry to satisfy dipole selection rules for transition to the final states. The $N_{4,5}$-level SXAPS represent the excitation of 4d electrons to the final f states.

The reported soft x-ray absorption spectra (SXA) of trivalent lanthanides[10-12] show a group of weak narrow resonances near the 4d absorption edges, and for all except Er, Tm, and Yb a broad strong absorption feature at higher energy beyond the 4d edge. The typical intensity ratio between the prominent peak and weak discrete lines is about 100. The spectra do not exhibit a double peak due to spin-orbit splitting. The spectral features depend upon the degree of filling of the 4f shell with increasing Z. After the shell becomes half full, the width and separation of the prominent peak relative to the $N_{4,5}$ absorption edge decreases.

Sugar, Dehmer and Starace[13-15] have interpreted the SXA spectral features as arising from the $4d^{10}4f^N \to 4d^94f^{N+1}$ (N varies from 0 for La to 14 for Yb) transitions in the one-electron scheme. According to their interpretation, a centrifugal potential barrier of 10-20 eV prevents low energy f orbits from overlapping the 4d orbits for $Z < 57$. However, for $Z \geq 57$, 4f orbits contract and overlap the 4d orbits significantly. Accordingly $4d \to 4f$ SXA spectrum involves transitions from ground level of a $4d^{10}4f^N$ configuration to numerous levels of a $4d^94f^{N+1}$ configuration which extend over a broad range of 20 eV as a result of exchange interaction between the f electrons and d vacancy. Theoretical calculations indicate most of the oscillator strength to be associated with the transitions to the higher levels of $4d^94f^{N+1}$ configuration, when the 4f shell is not nearly filled. The interaction of the levels of $4d^94f^{N+1}$ configuration with the continuum channel $4d^94f^N\varepsilon f$ (where εf is a continuum state) shifts the energy position of $4d^94f^{N+1}$ levels and causes broadening of the prominent peak above the threshold due to autoionization. Analogous transitions to the lower levels of $4d^94f^{N+1}$ configuration give rise to numerous weak lines near the ionization threshold. The observed decrease of the 4d oscillator strength along the lanthanide series could be explained in terms of the $4d \to 4f$ transitions since the final state intensity is proportional to the number of vacancies in the 4f subshell.[16] Since the exchange interaction is stronger between subshells with the same principal quantum number, it affects $4d \to 4f$ more than $3d \to 4f$ or $4d \to 5f$ transitions.

For the less than half-filled 4f shell the intensities are controlled by the exchange interaction that preserve the $4d^94f$ ($^1P_1$) parentage. The position of prominent peak based on $^1P_1$ remains at about the same high energy $\sim$20 eV relative to the onset of absorption. The coupling of additional f electron to the $4d^94f$ system through the exchange interaction contributes to the decreasing intensity of the main absorption peak with increasing Z. Past the middle of the 4f shell the $^1P_1$ parentage model breaks down because the $^1P_1$ parentage cannot be traced to a single level.[13] In the latter half of the series the $4d^94f^{13}$ parentage is preserved. The oscillator strength is divided between the ^3H and ^3G terms in the ratio of 3:1. These terms are separated by ∼10 eV resulting in an abrupt drop of the interval between the onset of absorption and the prominent peak to ∼10 eV past the middle of the 4f shell.[13]

Following the filling of the 4f shell, the regularity in the spectra will again be interrupted. The decrease in exchange interaction is also expected because of the increase in the binding energy of the 4d electrons with increasing Z.[17] Sugar's calculations,[13] which are based on the assumption that lanthanides give up three electrons for bonding in the metal, provide a detailed agreement with the SXA data. For La^{+3} and Tm^{+3} only three lines are allowed and only three are observed. At the end of the period a single line is observed as expected in Yb_2O_3. Since both SXA and SXAPS probe, with different techniques, the same excited states of lanthanide ions, the two types of spectra should have related characteristics.

Wendin[18] has proposed three competing transition processes for explaining the SXAPS of lanthanides which have a localized excited level at energies near the threshold of a core level excitation. The normal Bremsstrahlung results from the scattering of the incident electron into the unfilled 4f state. In the second process both the incident electron and core electron are excited to the unfilled 4f state. The exchange interaction involving the 4d core hole and two 4f electrons splits the final state leading to a multiplicity of transitions to the initial state. The third process involves the inelastic scattering of the incident electron into the itinerant-electron state accompanied by the excitation of the 4d core electron into the localized 4f state. The exchange interaction involving the excited one 4f electron and core hole is expected to be weak relative to that of two 4f electrons and core hole. The first two processes represent resonant phenomena and are major contributors to the threshold splitting as measured by SXAPS. The third process of inelastic scattering is a nonresonant transition and plays a dominant role in the SXA.[19] SXAPS of lanthanides provides an excellent testing ground for determining the relative strength of the above process.

In understanding the relationship between the different spectra, it is logical to examine the simplest case of Lu which has completely filled 4f shell. We have not been able to observe any SXAPS associated with the excitation of 4d core levels in Lu. In Lu the possible transition underlying the SXAPS is $4d^{10}4f^{14}(5d^16s^2) + e \rightarrow 4d^94f^{14} + \varepsilon d + \varepsilon f$. Since εd and εf are delocalized states, no $4d \rightarrow 4f$ resonance is expected consistent with the present results. The divalent Yb with the ground state configuration of $4d^{10}4f^{14}(6s^2)$ is similarly not expected to yield any distinct 4d SXAPS and none is observed. However in Yb_2O_3 transitions of the type $4d^{10}4f^{13} + e \rightarrow 4d^94f^{14} + \varepsilon d$ are permissible. The N_5-level SXAPS of Yb_2O_3 exhibits a broad doublet peak with two components separated by an interval of 3.1 eV. No contribution to the spectrum lies in the region corresponding to the N_4 threshold. It is logical to assign the observed peak to the transition $^2D_{5/2} \rightarrow {}^3F_{7/2}$ which is allowed by the dipole selection rule $\Delta J = \pm 1, 0$. The broad spectrum of Yb_2O_3 is seen to be 20 times weaker relative to other lanthanides (La-Tm) suggesting that those transitions $(3d^{10}4f^N + h\nu \rightarrow 3d^94f^{N+1})$ responsible for SXA do not play a dominant role in the SXAPS of lanthanides, where there is more than a single 4f vacancy. The considerable reduction in the Yb_2O_3 spectral response compared to that of other lanthanides reflects the relative intensity of the two electron resonance and the single electron contribution to the SXAPS of lanthanides.

The next element to be considered is Tm where the transition of the type $4d^{10}4f^{12} + e \rightarrow 4d^94f^{14}$ is plausible. The resonant recombination is represented by the transition $^3H_6 \rightarrow {}^2D_{5/2}, {}^2D_{3/2}$. The dipole forbidden transition is allowed because of the interaction of an incident electron of 200 eV energy. The fast moving incident electron penetrates the s, p, d, or f subshells and the angular momentum, J is conserved. The term dependent transitions are referred to as giant resonances. The 4d-level SXAPS of Tm consists of an intense resonant peak above the 4d ionization edge and is followed by a couple of weak peaks. Because of the complete filling of 4f shell in the final state, the 4d-4f exchange interaction is nonexistent and no fine structure associated with the 4d-level SXAPS of Tm is observed as predicted. The present results correlate very well with the TmF_3 SXA measurements of Lynch, et al. and electron energy loss spectrometry (EELS) data of Strasser et al.[21] Strasser, et al., have assigned the main peak to $^3G_5$ and the weak peaks to $^3H_6$ and $^3H_5$ respectively.

At the end of the lanthanide series the SXAPS of La is represented by the resonant transition $4d^{10} + e \rightarrow 4d^9 4f^2$. The configuration of the ground state is $^1S_0$ and the multiplet of final states arising from the combination of ^2D with ^1S, ^3P, ^1D, ^3F, ^1G, ^3H, and ^1I states is complex. Dipole selection rules, however, narrow the transitions down to intense $^1P_1$, $^3P_1$, and $^3D_1$ multiplets. The 4d hole pulls the $^3P_1$ and $^3D_1$ terms to below the 4d ionization threshold resulting in sharp pre-threshold peaks in the 4d-level SXAPS of La. The prominent peak above the 4d edge is broadened as a result of the autoionization of $^1P_1$ term.

The spectra of the remaining lanthanides is more complex. Because the 4d-4f interaction is strong relative to the 4f-4f interaction, the configurations of the complex final states $4d^9 4f^{N+2}$ can be deduced from the $4d^9 4f^2$ parentage. For the first half of the lanthanide series the addition of a further 4f electron results in an increase in the exchange interaction and further splitting of the final states. Each of the final states of these configurations consists of a large number of terms and the number of possible transitions to these terms is very large. The experimentally observed maxima represent some groups of these transitions. It will be interesting to compare the $N_{4,5}$-level SXAPS results with the corresponding SXA data. We have attempted to identify the observed SXAPS main peaks with the prominent SXA peaks. In general the accuracy of agreement varies from 0.2 to 1.0 eV in the case of La, Pr, Er and Ho. Overall the two spectra are in agreement regarding the number and location of main peaks. The comparison is not strictly valid since SXAPS corresponds to the resonant scattering of two electrons into the unfilled 4f state in contrast with SXA where a single electron is involved. Since the selection rules are not rigorously obeyed in the SXAPS, as in the case of SXA; all symmetries in the SXA must be considered for correlating the spectral features of the SXAPS with SXA.

Certain general relationships can be observed in Fig. 1. The width of the $N_{4,5}$-level SXAPS decreases and the relative intensity of the maximum located near the 4d ionization threshold increases with the filling of the 4f shell. In the middle of the lanthanide series the pre-threshold peaks undergo marked changes in intensity and the oscillator strength is more widely spread. Due to the complexity of the spectra it is not possible to attribute any particular spectral features to one electron excitations. The spectra reported here are not amenable to reliable interpretation by the simple theory. The systematic trends in the spectral features can be understood in terms of the parentage arguments. In order to gain insight into the nature of 4d-4f interaction coupling of lanthanides, the calculations of lanthanide ions with the addition of a nearby electron but with the inclusion of multiplet structure are considered highly desirable.

ACKNOWLEDGEMENTS

The investigation was supported by the Robert A. Welch Foundation, American Chemical Society (PRF), and Texas Advanced Technology Research Program. The author wishes to thank Dr. A. R. Chourasia and Ms. Nancy Wallace for their valuable assistance in the final phases of this work.

REFERENCES

1. R. L. Park and J. E. Houston, J. Vac. Sci. Technol. 11:1 (1974).
2. R. L. Park, Surface Sci. 86:504 (1979).
3. W. E. Harte, P. S. Szczepanek and A. J. Leyendecker, J. Less Common Metals 93:189 (1983).
4. P. O. Nilsson, J. Kanski and G. Wendin, Solid State Commun. 15:287 (1974).
5. M. S. Murthy and P. A. Redhead, J. Vac. Sci. Technol. 11:837 (1974).
6. D. Chopra, H. Babb and R. Bhalla, Phys. Rev. B 14:5231 (1976).
7. R. J. Smith, M. Piacentini, J. L. Wolf and D. W. Lunch, Phys. Rev. B 14:3419 (1978).
8. D. Chopra and G. Martin, Jpn. J. Appl. Phys. Suppl. 17 (2):803 (1978).
9. J. A. Bearden and A. F. Burr, Rev. Mod. Phys. 39:125 (1968).
10. P. Trebbia and C. Colliex, Phys. Sat. Sol. (b) 58:523 (1973).
11. R. Haensel, P. Rabe and B. Sonntag, Solid State Commun. 8:1845 (1970).
12. T. M. Zimkina, V. A. Fomichev, S. A. Gribovskii and I. I. Zhukova, Sov. Phys. Solid State 9:1128 (1967).
13. J. Sugar, Phys. Rev. B5:1785 (1972).
14. J. H. Dehmer and A. F. Starace, Phys. Rev. B5:1792 (1972).
15. A. F. Starace, Phys. Rev. B5:1773 (1972).
16. J. H. Dehmer, A. F. Starace, U. Fano, J. Sugar and J. W. Cooper, Phys. Rev. Letts. 26:1521 (1971).
17. D. Chopra and G. Martin, in Rare Earths in Modern Science and Technology, Vol. 2, Eds. McCarthy, Rhyne and Silber, (Plenum, 1980) 489.
18. G. Wendin and K. Nuroh, Phys. Rev. Lett. 39:48 (1977).
19. W. E. Harte and P. S. Szcezepanek, Phys. Rev. Lett. 42:1172 (1979).
20. D. W. Lynch and C. G. Olson, Vacuum Ultraviolet Radiation Physics, edited by Koch, Haensel and Kunz, (Pergammon Vieweg, 1974) 254.
21. G. Strasser, G. Rosina, J. A. D. Matthew and F. P. Netzer, J. Phys. F. Met. Phys. 15:739 (1985).

ON THE EXCITATION AND RELAXATION OF 5p ELECTRONS NEAR THRESHOLD IN Eu and Yb METALS

Giorgio Rossi

Laboratoire pour l'Utilisation du rayonnement
Electromagnétique CNRS-CEA-MEN, Université de Paris-Sud
91405 Orsay, France

ABSTRACT

The optical absorption spectra of Yb and Eu near the 5p core level energy are understood by analyzing hν dependent photoemission and experimental partial photoionization cross sections of the valence and 4f subshells. Resonant photoemission of the top of the valence band, and L-edge X-ray absorption show that the highest partial density of states (DOS) in the Fermi level region is of d character, (occupied and unoccupied). An intense constant kinetic energy electron emission structure turns on at the $5p_{3/2}$ threshold of Yb and Eu, and is interpreted as the decay channel of a dynamical excitation process involving collapse of a localized 5d screening orbital when the 5p core hole is created ; this process is responsible of the sharp onset of the $5p_{3/2}$ optical absorption of the divalent rare earth metals.

Eu and Yb are the rare earth elements with the most stable 4f subshell configuration, respectively half full ($4f^7$) and full ($4f^{14}$), and with two valence electrons in the 6s subshell (1). The photoemission spectrum of atomic Yb is reproduced in Fig. 1 (after Svensson et al., 2) showing the two peaks due to the spin orbit splitting into j=5/2 and j=7/2 total angular momentum components of the 4f hole ($4f^{13}$ final state), and the $6s^1$ final state peak. Similar spectra exist for Eu (3). At the solid state both Yb and Eu maintain their 4f core configuration, assuming cubic crystal structure and forming delocalized band states with the two external electrons, likewise in the case of the alkaly earth elements. The two 6s valence electrons occupy in fact hybrid states with p and d character components to meet the solid state boundary conditions. The definition of the hybridization of the extended states in rare earths and particularly in the divalent rare earths is a difficult problem to solve quantitatively both theoretically and experimentally. Band structure calculations for Yb show the degree of sd band mixing to be totally dependent on the choice of the inner potential to the point that small changes of this parameter change the picture of the calculated Yb from a metal to a semimetal with a small gap between 6s and 5d states (1). The compression of the external charge in the solid state corresponds to an increased Coulomb interaction with the 4f localized electrons which raise in energy up to a few bolt binding

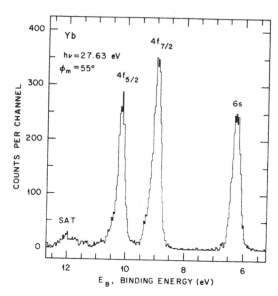

Fig. 1 Photoelectron spectrum of atomic Yb as obtained with hν=27.6 eV (From A. Svensson et al., reference 2).

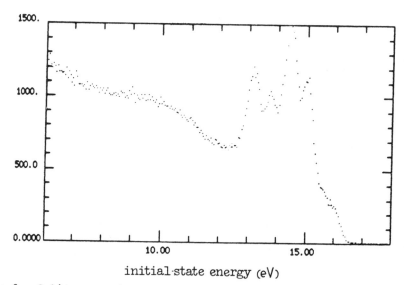

Fig. 2 Solid state photoelectron spectra of fcc Yb as obtained with hν=21 eV. The atmic 6s states are broadened in the sd band that extends up to E_F. The $4f^{13}$ final state structures are due to bulk and surface shifted $4f_{7/2}$ and $4f_{5/2}$ multiplets.

energy with respect to the Fermi level. Photoemission from solid state Yb is shown in Fig. 2. The DOS close to the Fermi level is due to the delocalized sd (see below) states ; between - 1.2 eV below E_F the and -3 eV the $4f^{13}$ j=7/3 and j=5/2 final state multiplets from bulk Yb atoms, and surface (0.6 eV shifted) low coordinated Yb atoms. The surface shifted peaks are prominent in the spectrum obtained with hν=21 eV, yielding final state kinetic energies of 15 eV, at the minium of the excape depth for Yb (4).

The energy distribution of the empty state DOS just above the Fermi level is observable by exciting the dipole transitions from deep core levels at threshold energy : $L_{2,3}$ edge resonances corresponding to $2p_{1/2} \rightarrow 5d_{3/2}$; $2p_{3/2} \rightarrow 5d_{3/2}, 5d_{5/2}$ and the L_1 edge absorption corresponding to 2s → 6p transitions. The hybrid nature of the empty DOS is evident from those absorption data(5).

The 5p core levels lie at 18.1 ± 0.3 eV and 23.4 ± 0.5 eV for Eu and at 24.1 ± 0.2 eV and 30.1 ± 0.2 eV for Yb, and lie externally with respect to the 4f electrons. A strong coupling between the 5p hole and the 5d empty states is expected due to the high overlap of localized states having the same principal quantum number (6).

The excitation of 5p core levels must induce relaxation of the valence states, but not of the 4f states (the process would only be possible in Eu which has the open $4f^7$ core, but requires a spin-flip transition). One may expect the excitation of the 5p electrons in the devalent rare earth to be exploitable for retriving informations on the valence states.

Optical absorption spectra in the 5p region have been measured with synchrotron radiation (SR) for Eu and Yb and by Miyahara et al. (7) and related data were obtained by means of electron energy loss spectroscopy by Netzer et al. (8), we have reproduced those data in the top of Figs. 3 and 4. The onset of $5p_{3/2}$ absorption is sharp in the case of Yb, and is followed by a second prominent peak at 3-4 eV above the edge, and then by the $5p_{1/2}$ edge, broader and followed by the intense resonance with the continua of both absorption edges, peaking above 35 eV photon energy.

The 5p absorption structures of Eu are broader, but qualitatively identical to the Yb. The spin orbit splitting of the 5p core is measured as 6 ± 0.1 eV for Yb, from the direct 5p core photoemisssion, (Fig. 9) and 5.5 ± 0.5 eV for Eu (not shown).

Selected photoelectron energy distribution curves as excited with hν values in the 5p excitation range are collected in Fig. 5 and Fig. 6 for Eu for Eu and Yb respectively. The curves obtained below the 5p threshold energy are explained by a narrow region of DOS emission (from E_F to - 0.8 eV for Yb and from E_F to - 1.3 eV for Eu), and by the $4f_{n-1}$ final state multiplets with their surface shifted components due to the low coordinated surface atoms (9).

As the photon energy reaches the $5p_{3/2}$ binding energy the photoelectron spectra undergo lineshape changes, and the optical absorption spectra display the onset of absorption. The expected electron transitions are : direct $5p_{3/2}$ photoemission, 5p-5d resonant photoemission $O_3N_{6,7}N_{6,7}$; $O_3N_{6,7}V$; O_3VV ; Auger decays, and O_3 radiative (x-ray emission) decays.

The direct photoemission for the 5p states is shown in Fig. 9 for Yb, as measured with hν well above threshold (hν=65 eV) so that the kinetic energy of the photoelectrons does not overlap the secondary emission peak, and the cross section is not influenced by the threshold phenomena. From these peaks, (showing the same surface shifted components as found in the 4f, the spin orbit splitting of 6 ± 0.1 eV is measured.

If the absolute and relative intensities of the photoemission features are plotted versus, photon energy, after normalization for the photon flux and measurement optical parameters (10,11), the orbital character of the related initial states can be recognized from the shape of the cross section curve, and the presence of edge singularities is pointed out. The experimental partial-subshell photoionization cross sections obtained are plotted on an arbitrary non normalized amplitude scale in figure 1 and 2.

Summarizing the observations : 1) the top of the occupied DOS undergoes photoemission intensity variations in correspondence of the 5p thresholds typical of resonant photoemission, through a mechanism described by the scheme 1 Fig. 10 (12). The maximum of the resonance is an indicator of the maximum of the empty 5d DOS, and results to be 3.5 eV above threshold, in agreement with the maximum of the empty 5d states as measured by photoabsorption on the 2p core levels, or by BIS (13). The results for Yb are directly seen in the top of the valence band cross sections, and are similar to the results on the 5d band of Lu (11) ; this is a direct measure of the occupation of states of 5d character in the ground state of fcc Yb. The result for Eu is better seen in Fig. 7 by plotting the VB and 4f experimental intensity ratio vs hν and the 5d/4f theoretical cross section ratio, from Yeh and Lindau (14) (assuming 20% of the 5d occupancy of Gd) apart a mismatch of the energy scale, due to the different threshold energies for the atomic (theory) core levels and the solid state measured core levels Fig. 7 confirms that the top valence states of Eu contain 5d character which undergoes resonances above the 5p edges.

2) the 4f final state peaks undergo strong amplitude modulations, due to the overlap of a constant kinetic energy electron emission structure (CKE1), which turns on abruptly at $5p_{3/2}$ edges, has an intensity comparable to that of the 4f direct photoemission, and remains fairly intense for several eV above threshold. The corresponding structures above the $5p_{1/2}$ edges are much less intense (CKE2).

3) ONN Auger transitions are recognized in the 7.5 eV (Yb $O_3N_{6,7} N_{6,7}$) and 12 eV (Yb $O_2N_{6,7} N_{6,7}$) kinetic energy peaks (Fig. 8), these structures are rather weak, and due to their low final state energy never overlap the photoemission nor the CKE peaks, threfore do not influence the partial corss section data. $O_3N_{6,7}V$; $O_3O_{4,5}O_{4,5}$ and $O_4O_{4,5}P_1$ decays are expected to yield weak and broad structures due to the low occupation of the O_3 (5d) states. Furthermore the line shape of such transitions should reflect the convolution of a broad and flat density of states. Such Auger decays are not apparent in the present data, but it should be kept in mind that OVV Auger emission may contribute unresolved intensity in the cross-section plots.

The CKE decay channel has been discussed in Ref. (10, 11). The data were explained by considering an Auger-like process as shown in the scheme, which involves a collapsed 5d atomic-like level as intermediate state. The high polarizability of the 5p-5d channel is known in atomic physics where $5p^5$ $(2P_{1/2})$ $6s^2$ $5d[_{3/2}]^1$ giant resonances occur (15). Our hypothesis is that the intermediate state is a 5d screening orbital that collapses under the attraction of the 5p core hole. Recombination of the intermediate state electron and the 5p core hole provides a fixed amount of energy for the ejection of a shallow electron (4f or valence). The transition is Auger-like, but the intermediate state is bonded only under the attraction of the 5p open core, and therefore depends on the availability of conveniently "collapsable" 5d states. Such process is resonant at threshold since the direct recombination of the excited intermediate state electron and the 5p hole is possible. Above threshold the collapsed screening charge can be provided by the conduction electrons, and the decay still exists.

The smaller feature in the 4f partial corss section that corresponds to the O_2 threshold is interpreted as the CKE2 decay of the $5p_{1/2}$ core level. The weak CKE2 structure is seen to determine an apparent branching ratio between bulk and the surface shifted 4f multiplets in the EDCs of Yb at 30.5 eV (Fig. 5) and of Eu at 25 eV (Fig. 6). The intensity difference between the CKE2 and CKE1 structures is much larger than explained by the dipole selection rules, this phenomenon can be compared to the branching ratio for the deexcitation of the $YbI(2p_{3/2})$ and $YbI(2P_{1/2})$ atomic configurations. In ejected electron spectroscopy of atomic Yb the photoexcitation of the $5p_{3/2}$ threshold is followed by autoionization decays of the type

$$YbI5p_5(2P_{3/2})\ 4f^{14}6s^25d^* \rightarrow YbII5p^64f^{14-x}6s^{2-y}$$

where $x \neq y$, $x,y = 0,1$ with ejection of a shallow electron (4f or 6s or 5d), in analogy with the CKE1 solid state process (15,16).

The CKE2 atomic analog would then be the autoionization of the excited configuration, $YbI5p_5(2P_{1/2})\ 4f^{14}6s^25d^*$, but this yields very little, or no intensity in atomic spectroscopy. This fact is accounted for by the presence of an intense nonradiative transition, $Yb15p^5(2P_{1/2})\ 6s^25d^* \rightarrow YbII(2P_{3/2})$, which represents a Coster-Kronig conversion of the $5p_{1/2}$ hole into a $5p_{3/2}$ core hole state of the Yb ion (15, 17).

If we compare the partial photoionization cross-section curves obtained with the optical absorption spectra and electron energy loss spectra as done in Figs. 3 and 4, we recognize that the onset of the absorption corresponds to the threshold resonant CKE transition onset, i.e. with the highly polarizable 5p-5d dipole involving locally collapsed 5d orbitals, and not to the empty 5d DOS above E_F. This is confirmed by two independent experiments on Yb (from references 10 and 11, not normalized with each other). The second absorption structure, 3 eV above threshold, does match with the peak of the resonant photoemission and therefore accounts for the transitions to the maximum of the 5d empty DOS.

It results therefore that optical absorption onsets in solids can be determined by dynamical processes involving localization of

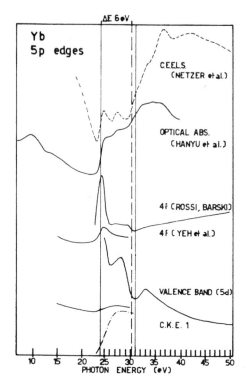

Fig.3 Optical absorption and partial cross sections for fcc Yb in the 5p excitation threshold region.

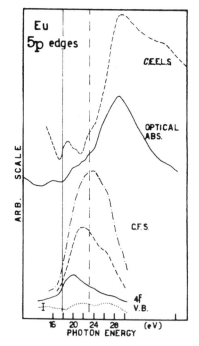

Fig.4 Optical absorption and partial photoionization cross sections for bcc Eu in the region of the 5p excitation threshold.

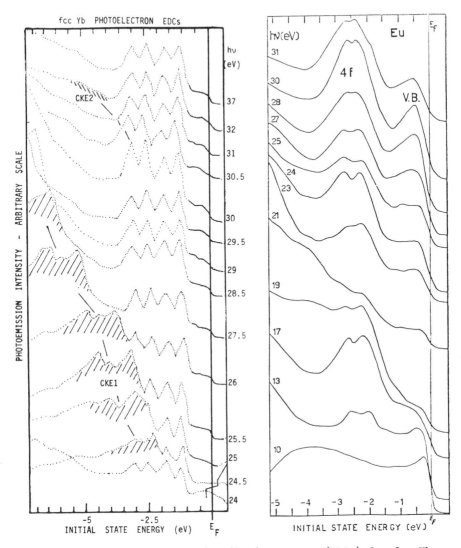

Fig.5 Photoelectron energy distribution curves (EDCs) for fcc Yb (film) as obtained with photon energy values in the 5p photoexcitation threshold regions ($5p_{3/2}$ edge = 24.1±0.2 eV; $5p_{1/2}$ edge = 30.1±0.2 eV). The CKE1 structure is visible for hν>24.5 eV. The weaker CKE2 structure determines the branching ratio of the 4f peaks at 30 eV.

Fig.6 EDCs for bcc Eu (film) in the 5p threshold region ($5p_{3/2}$ edge =18.1±0.3 eV; $5p_{1/2}$ edge = 23.4±0.5 eV). The CKE1 is seen in the lineshape change of the photoemission at hν=21 eV.

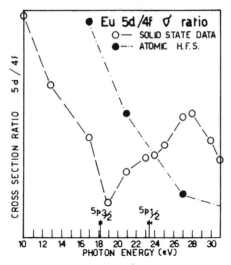

Fig.7 Photoemission intensity ratio for the valence band of Eu and the 4f (dashed). The data are compared with the ratio of the 5d versus 4f partial photoionization cross sections (14). The trend is interrupted by the resonant photoemission.

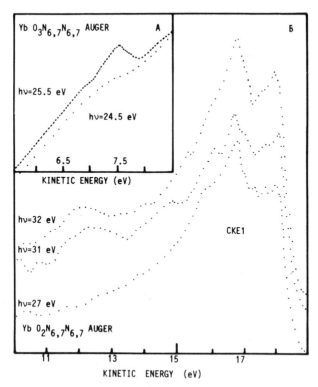

Fig.8 Yb ONN Auger transitions put in evidence by measuring the EDCs as excited with photon energies below and above the 5p thresholds.

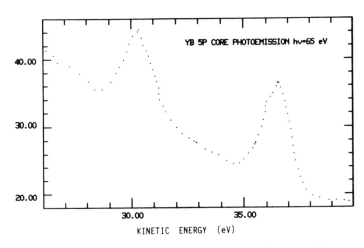

Fig.9　Yb 5p core photoemission as obtained with hν=65 eV. The lineshapes of the spin orbit splitted components are due to contributions both from bulk and surface atoms.

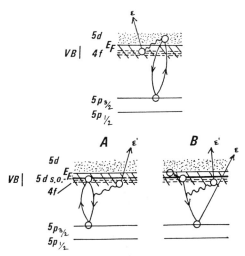

Fig.10　Schematic pictures of the decay processes of the 5p core hole;
-left: resonant photoemission involving the 5d DOS
-A: excitation of a localized screening orbital state (5d) and recombination with the 5p core hole with ejection of a CKE1 electron.
-B: collapse of a 5d screening orbital on the 5p core hole, and decay of one conduction electron in the core hole with ejection of a second CKE electron.

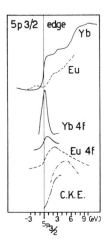

Fig.11　Detail of the $5p_{3/2}$ edge absorption and photoionization cross section of Yb and Eu solids. The delayed onset of Eu with respect to Yb is stressed. The top curves are the optical absorption data; below are the 4f partial cross sections for Yb (solid lines, non normalized) and Eu (dashel line), further down are the CKE1 intensity profiles for Eu (dot-dashed) and Yb (bottom curve).

atomic-like states, and not simply by interband, or core to band transitions. This raises important questions on the compatibility of dynamically localized intermediate states in the absorption process of solids, and the classical band picture of metals.

The broader onset of the $5p_{1/2}$ edges can be connected with the weakness of the CKE2 decay, and to overlap with the $5p_{3/2}$ photoionization continuum.

The difference in the width of the onsets is shown in Fig. 11. If the 5d screening orbital is localized by the 5p potential, as we propose, then, its interaction with the localized 4f charge should be large, with formation of a multiplet in interaction with the 5p hole. This effect, important for Eu which has an open 4f subshell but negligible for Yb ($4f^{14}$), would induce a broadening of the onset of excitation over the many terms of the multiplet. This idea was proposed first by Hanyu et al. and it was called "hole induced correlation" (7).

X-ray emission data for heavy rare earth metals show two lines corresponding to the spin-doublet of the $5p_{3/2}$ and $5p_{1/2}$ levels without a significant multiplet structure (18), so that some information on the band states can be retrieved. Both Eu and Yb did not display measurable 5p X-ray emission bands in the experiments of Shylakov et al. (18,19) due to the low population of the d-like states, but also possibly to bad quality of the samples (maybe oxydized).

In summary we have provided the evidence of electron emission processes playing a major role in the determination of the optical absorption spectra of divalent rare earth metals at the 5p edges. The occupation and large polarizability of the 5d band, have been shown to determine intense structures in the absorption and electron emission spectra. The role played by the 5d band in bonding other atoms in RE intermetallic compounds is reflected in the 5p threshold transitions. Eu and Yb interfaces with silicon have hv dependent photoemission spectra which do not show intense CKE structures as a consequence of the 5d band rehybridization, and smaller polarizability of the 5p-5d dipole transitions (or smaller "collapsability" of 5d screening orbitals) (20).

REFERENCES

1) S.H. Liu, in Handbook on the Physics and Chemistry of Rare Earths, edited by K.A. Gschneidner and L. Eyring (North Holland, 1978) p. 233

2) W. Agneta Svensson, M. Krause, T.A. Carlson, V. Radojevic and W.R. Johnson ; Phys. Rev. A 33, 10224 (1986)

3) S.T. Lee, S. Suzer, E. Matthias, R.A. Rosenberg and D.A. Shirley ; Journ. Chem. Phys. 66, 2496 (1977)

4) G. Rossi, thèse d'état, Paris 1985

5) G. Rossi et al. : unpublished results

6) G. Wendin, Comm. Modern Phys. F XVII, 115 (1986)

7) T. Ishii, Y. Sariteka, S. Yamaguchi, T. Hanyu and M. Ishii, J. Phys. Soc. Jan. 42, 876 (1977) ; T. Hanyu, T. Migahara, T. Kamada, K. Asada, H. Onkuma, M. Ishii, K. Naito, H. Naso and S. Yamaguchi, ibid 53, 3667 (1984)

8) F.P. Netzer, G. Strasser, G. Rosina and J.A.D. Mattew ; Surf. Sci. 152/153, 757 (1985)

9) Y. Tanakauwa, S. Suzuki, T. Yokotsuka and T. Sagawa, J. Phys. Soc. jpn. 53, 687 (1984) ; F. Gerken, A.S. Flodstrom, J. Bair, L.I. Johansson and C. Kunz ; Physica Scripta (1985)

10) G. Rossi, J.J. Yeh, I. Lindau and J. Nogami, Phys. Rev. B 29, 5989 (1984) ; Surf. Sci. 153/153

11) G. Rossi and A. Barski ; Phys. Rev. B 32, 5492 (1985) ; Solid State Commun. 57, 277 (1986)

12) For a recent review of photoemission and resonant photoemission see L.C. Davis, J. Appl. Phys. 59, R25 (1986)

13) J.H. Lang, Y. Baer and P.A. Cox, J. Phys. F 11, 121 (1981)

14) J.J. Yeh and I. Lindau ; At Data Nucl. Data Tables 32, 1 (1985)

15) S.M. Kazakov and O.V. Kristoforov, Zh. Eksp. Teor. Fiz. 84, 502 (1983) [Sov. Phys. JETP 57, 290 (1983)]

16) R.A. Rosenberg, S.T. Lee and D.A. Shirley, Phys. Rev. A 321, 132 (1980)

17) D.H. Tracy ; Proc. R Soc. London Ser. A 357, 485 (1977)

18) A.S. Shulakov, A.P. Braiko, T.M. Zimkina, V.A. Fomichev and L.P. Novikova ; Fiz. Tverd. Tela (Leningrad) 22, 442 (1980) [Sov. Phys. Solid State 22, 258 (1980)

19) A.S. Shulakov, A.P. Braiki and T.M. Zimkina ; Fiz Teverd Tela (Leningrad) 23, 1849 (1981) [Soc. Phys. Solid State 23 (1981)

20) G. Rossi in Proceedings of the 17th Int. Conf. on the Physics of semiconductors, San Francisco (1986) edited by J.D. Chadi and W.A. Harrison (North Holland, Amsterdam (1985), p. 149

IV. LASERS AND PLASMAS

THE SPECTRA OF LASER PRODUCED PLASMAS

WITH LANTHANIDE TARGETS

> Gerard O'Sullivan
>
> School of Physical Sciences
> National Institute for Higher Education
> Dublin 9, Ireland

ABSTRACT

The EUV spectra of laser generated rare earth plasmas have been studied in different regimes of optical thickness. When the rare earth ion concentration is close to 100% the spectra for elements with $Z \leq 60$ are found to consist of a recombination continuum overlaid with numerous strong emission lines. The most striking feature is an intense modulation of the continuum by $4p^6 4d^N - 4p^5 4d^{N+1} + 4p^6 4d^{N-1} 4f$ and $4d^{10} 5s^m 5p^k 4f^N - 4d^9 5s^m 5p^k 4f^{N+1}$ transitions which blend to form a UTA (unresolved transition array) some 30eV wide in the 100 Å region. For Z>62, the line emission disappears, the 4d-4f modulation decreases with Z and the targets yield a line free continuum which in some cases extends from 40 to 2,000 Å. With targets containing only ~1% concentration of rare earth in a low Z substrate the spectra obtained are very different: the recombination continuum is reduced and the spectra are dominated by a 4d-4f UTA about 10 to 15eV wide, the intensity of which decreases with Z.

This behaviour can be understood in terms of 4f wavefunction collapse. With increasing ionisation, the binding energy of the 4f electron increases rapidly. As a result 4f,5p and 4f,5s level crossing occurs and yields a range of low lying configurations containing open 4f, 5p and 5s subshells most of which will be thermally populated. Moreover the ground configurations for all the ion stages present for $Z \geq 62$ contain open 4f subshells. The resulting spectra are very complex, the oscillator strength is spread over a large number of lines and the transitions are so weak that they are submerged in the recombination continuum background. Moreover, with 4f collapse there is a gradual shift of oscillator strength from the continuum $4d-4,\epsilon f$ to the discrete 4d-4f transition. Thus in moderate to high ion stages 4d-4f UTA completely dominate the low density spectra while in the high density spectra their profile and intensity is altered by absorption.

1. INTRODUCTION

When the output from a 1.5J Q-switched ruby laser is focussed onto targets containing cesium, barium and the lower Z elements of the rare earth group the soft X-ray spectrum emitted by the plasma formed at the focal spot is dominated in the 50-120Å range by a region of intense emission which consists of a mixture of lines and continuum. The shape and extent of this

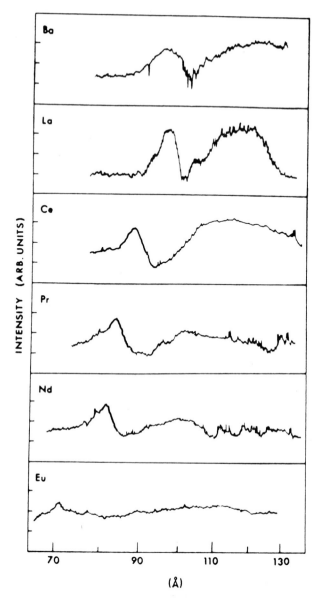

Fig.1 Densitometer traces of the spectra of the elements Ba,La,Ce,Pr,Nd and Eu obtained from compounds (after Carroll and O'Sullivan 1982)

feature was found to depend on the concentration of the element of interest present in the target and its atomic number (O'Sullivan and Carroll 1981, Carroll and O'Sullivan 1982). When the target used consisted of a salt (oxide or nitrate) compressed to form a pellet or mixed in an araldite base so that the metal of interest was only a minor constituent the features were found to be intense and resonance like, have a width of approximately 15eV and move towards shorter wavelength with increasing Z. In the lower Z elements the features contain a number of strong lines the intensity of which decrease

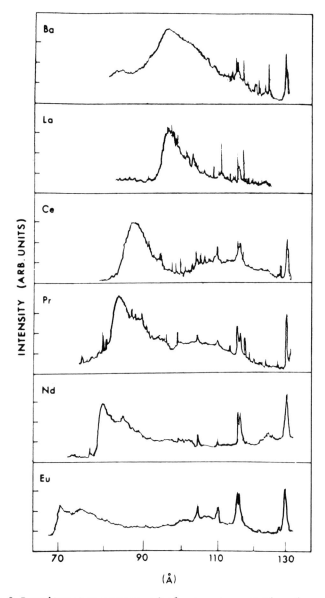

Fig.2 Densitometer traces of the spectra of the elements of fig.1 obtained from metals (after Carroll and O'Sullivan 1982)

along the sequence until the structure eventually becomes continuum like. With increasing Z the intensity of the underlying continuum also decreases so that the entire feature appears only weakly in the spectra of the higher Z elements of the rare earth sequence. Because of their profile and high intensity these features were originally called "emission resonances". Bauche-Arnoult et al (1978) have referred to similar features which often appear in the X-UV spectra of highly ionised atoms as unresolved transition arrays (UTA). In general, however, these features are considerably narrower and weaker and lie to much shorter wavelength than those observed for the rare earth elements in the present case.

Densitometer traces of some of the spectra obtained from salt targets are presented in fig.1. It is seen that the UTA are the strongest features in each case. Indeed away from these features a few weak lines were observed for the low Z elements only and the background continuum is extremely weak in every case.

In contrast, if pure metallic targets are used the soft X-ray spectra are completely different. For the elements from samarium to ytterbium the spectra consist of essentially a line free continuum from 35-350Å. Indeed the spectrum of samarium was found to be line free up to 2,000Å. Because of the absence of line emission laser produced plasmas of these elements have been proposed and used as VUV sources for a variety of applications such as absorption spectroscopy, lithography and UV radiometric transfer standards (Carroll et al 1978, 1980, O'Sullivan et al 1981, 1982, Nagel et al 1984). In these spectra some modulation of the continuum was observed in the vicinity of the UTA the profile and extent of which was very different to the preceding case. For the elements from cesium to neodymium the spectra contained numerous strong strong absorption and emission lines throughout the 35-200Å region and the modulation of the continuum was much more pronounced. Densitometer traces of some of these spectra are presented in fig.2. It is apparent from these that the width, profile and line density differ markedly from the features observed in the salt spectra.

More recently, studies on spectra from vacuum spark and Tokamak plasmas containing rare earths again showed the spectra to be dominated by intense bands of emission, which closely resemble the salt UTA of fig.1. (Finkenthal et al 1986a, Mandelbaum et al 1986). Again the brightness of these features was found to exceed that of any individual lines. Moreover the spectra of plasmas generated at very high laser powers where the dominant stage is ZnI like or CuI like were found by Finkenthal et al (1986b) to contain strong bands of quasicontinuum at the same position, even though the ion stages and plasma temperature are much higher than for the Q-switched ruby plasmas.

Two distinct phenomena must thus be explained. Firstly the appearance of the "emission resonance"/UTA, its brightness and its sensitivity to plasma parameters will be dealt with and secondly the complete absence of emission lines in the spectra of laser produced plasmas of the higher Z rare earths, which in some cases is observed over a 300eV energy range will be explained. In what follows the gross features of the salt spectra are dealt with in sections 2 to 8 and then the differences between salt and metal spectra are accounted for in section 9.

2. ION STAGE IDENTIFICATION

Before any of the above topics can be addressed it is important to know what ion stages are present in the plasmas, and of these which contribute to the emission spectra in the X-UV region. In the Q-switched laser experiments it was inferred that the plasma temperature lies in the range $50 \lesssim T_e \lesssim 60eV$. This value was deduced by studying the plasma emission from high Z targets such as molybdenum, niobium and zirconium whose spectra are well known under identical experimental conditions to the rare earth targets. The ion stages present in these plasmas were identified and this information was used in the collisional radiative model of laser-plasma interaction developed by Colombant and Tonon (1973). This model, which is particularly appropriate in the laser power density range $10^{11}-10^{13} W/cm^2$ (the power density in these experiments was $2 \times 10^{11} W/cm^2$) is intermediate between the high density ($n_e > 10^{22}$ electrons cm^{-3})

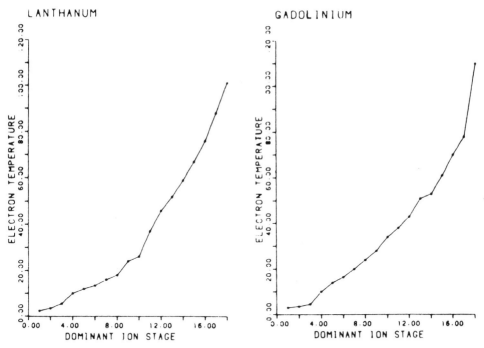

Fig. 3. Dominant ion stage as a function of electron temperature for lanthanum and gadolinium according to the plasma model of Colombant and Tonon (1973). The ordinate scale is in electron volts (After O'Sullivan and Carroll 1981)

local thermodynamic equilibrium model where collisions dominate and the low density ($n_e < 10^{19} cm^{-3}$) coronal model where collisional excitation is balanced by radiative decay. The theory enables the electron temperature to be predicted if the ion stage distribution is known or conversely may be used to predict the ion stage distribution for any element provided T_e, n_e and the ionisation potentials are known. For the rare earth plasmas the electron density was taken to be $n_e = 10^{21} cm^{-3}$ which is close to the cutoff density for ruby light and the ionisation potentials were a mixture of experimental and theoretical values. Application of the model predicted that ion stages up to and including XVI should be present in the rare earth plasmas while the dominant stage is XIII or XIV (fig.3).

In the vacuum spark and Tokamak spectra the electron temperature is considerably higher, while n_e is much less; $n_e \sim 10^{14} cm^{-3}$ in most inertial confinement devices. Considerably higher ion stages are attained under these conditions and in fact ZnI and CuI like lines of PrXXX and PrXXXI have been observed by Finkenthal *et al* (1986a).

3. RELATIONSHIP BETWEEN UTA AND $4d-\overline{4,}\varepsilon f$ GIANT RESONANCES

As noted in the introduction, the positions of the UTA shift towards shorter wavelength with increasing Z. For caesium, barium and the rare earth elements the neutral absorption spectra are dominated in the X-UV by $4d-4,\varepsilon f$ giant resonances. (Petersen *et al* 1976, Connerade and Mansfield 1976, Connerade 1978, Radke 1979a,b). Each UTA in fact lies to the short λ side of the position of the $4d-\overline{4,}\varepsilon f$ resonance in the corresponding element. Since each feature consists of a great number of lines it is difficult to assign any characteristic wavelength to the emission as a whole but the maximum wavelength of emission of each UTA was determined from densitometer traces. A plot of $\sqrt{E_{max}}$, the square root of maximum emission, vs Z is presented in fig.4. Also included in this plot is the position of the maxima of the 4d-4,εf resonances. The data are plotted only for those spectra which contain a single, well defined maximum. The resulting plot has a slope only slightly different from that shown by the UTA, which indicates that the two phenomena must be related in some way.

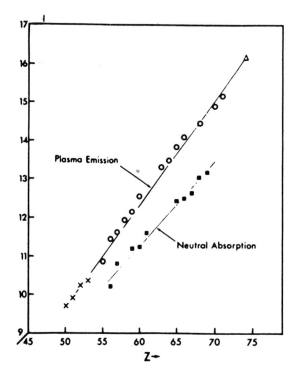

Fig.4. Dependence of 4d-4f transition energies on atomic number Z. (After O'Sullivan and Carroll 1981).

The behaviour of f electron wavefunctions with increasing ionisation has been discussed for BaI,II and the XeI sequence by Cheng and Froese-Fischer (1983). In BaI, the effective potential for f electrons consists of two wells separated by a centrifugal barrier. The outer well is essentially hydrogenic and contains the nf series of bound states. The first eigenstate of the inner well lies above threshold and has a large spatial overlap with the 4d subshell. The absorption spectrum of BaI is consequently dominated by a 4d-$\overline{4}$,εf giant

510

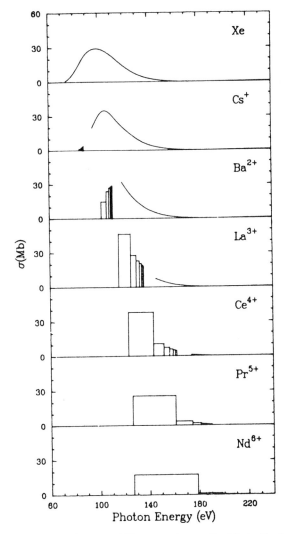

Fig.5. Effective oscillator strength distributions for the 4d-nf ^1P transitions (n=4-9) of Xe-like ions. (After Cheng and Froese Fischer 1983)

resonance. In low stages of ionisation where the potential barrier begins to disappear, the 4f wavefunction undergoes a gradual contraction as it changes from an "outer well" to an "inner well" state. The inner well state descends below threshold and mixes with the levels of the nf series to give them an appreciable amplitude in the inner well. This increase in overlap between 4d and nf functions causes a shift of oscillator strength from the continuum to the discrete spectrum. Thus in the Ba^{++} spectrum Lucatorto and McIlrath (1981) found the lowest members of the 4d-nf series to have comparable intensity. Furthermore, the amplitude in the inner well region is strongly term dependent, in particular, the contraction of the nf ^3D and ^3P terms is almost complete at Ba^{++} while the nf ^1P contraction takes place much more gradually. Younger (1980) performed a detailed analysis of the behaviour of the $4d^9 4f$ configuration along the Pd I sequence and found that the 4f wavefunction exhibits pronounced term dependence. In particular, he showed that the configuration averaged Hartree-Fock 4f ($4f_{av}$) orbital, which closely resembles the 4f ^3P and ^3D contracted with increasing Z and ζ much more rapidly than the 4f ^1P. Since the 4d orbital is localised in the core, calculations for the $4d^94f\ ^1P$ performed in a HF_{av} basis overestimate the values of the Slater Condon parameters F^k (4d,4f) and G^k (4d,4f). Since the $4d^94f\ ^1P$ energy in pure LS coupling may be written as

$$E(^1P) = E_{av} - \frac{24}{105} F^2(4d,4f) - \frac{66}{693} F^4(4d,4f) + \frac{137}{70} G^1(4d,4f)$$
$$- \frac{2}{105} G^3(4d,4f) - \frac{55}{2541} G^5(4d,4f)$$

it will be significantly overestimated because of the magnitude of the coefficient of the G^1 term. In fact in neutral barium and lanthanum this term dependent behaviour is so pronounced that the $4f\ ^1P$ function is an eigenstate of the outer well and lies below threshold while the $4f_{av}$ function lies in the inner well and the $4f_{av}\ ^1P$ state lies in the continuum (Sugar 1972, Hansen 1972, Hansen et al 1975, Wendin and Starace 1978, Connerade and Mansfield 1982). Since the εf function lies in the inner well and there is strong $4f_{av}\ \varepsilon f$ mixing, the giant resonances have been labelled $\overline{4,\varepsilon f}$ by Connerade (1978)

In high ion stages the inner well state descends below the entire nf series of the outer well and becomes the 4f wavefunction; hence the $4f\ ^1P$ state collapses from the outer to the inner well. Once collapse has occured, Cheng and Froese-Fischer found that strong transitions are observed only to 4f states as the higher nf states once more have little amplitude in the inner well (see table I and fig.5).

Table I. Absorption oscillator strengths of the $4d^{10}\ ^1S-4d^9nf\ ^1P$ transitions for Xe-like ions (Cheng and Froese Fischer 1983)

n	Xe	Cs^+	Ba^{2+}	La^{3+}	Ce^{4+}	Pr^{5+}	Nd^{6+}
4	0.00009	0.020	0.586	4.315	7.167	8.090	8.049
5	0.00007	0.018	0.503	1.297	0.879	0.440	0.164
6	0.00005	0.014	0.328	0.591	0.327	0.114	0.018
7	0.00004	0.010	0.213	0.321	0.155	0.043	0.002
8			0.143	0.194	0.086	0.020	0.000
9			0.100	0.127	0.052	0.011	0.000

In a somewhat similar calculation on the behaviour of 4d-4f transitions in Xe with increasing ionisation (O'Sullivan 1982) it was observed that the $4f\ ^1P$ wavefunction underwent a gradual contraction in proceeding from XeI to XeIX (fig.6). Since the 4d wavefunction is localised in the ionic core this contraction is accompanied by a gradual transfer of oscillator strength from the continuum to the $4d^{10}\ ^1S-4d^94f\ ^1P$ transition as the <4d/4f> increases. The summed oscillator strength for $4d^{10}nln^1l^1 \rightarrow 4d^94fnln^1l^1$ transitions are presented in table II and the increase in oscillator strength for 4d-4f transitions with increasing ionisation is clearly evident.

Since the UTA are extremely intense and since their position with increasing Z scales in the same way as the $4d-\overline{4,\varepsilon f}$ resonances, they must arise for the most part from 4d-4f transitions in ion stages where the 4d,4f overlap is large.

Table II. Total oscillator strength and mean wavelength for transitions of the type $4d^{10}(^1S_0)n\ell n'\ell'-4d^94f(^1P_1^0)n\ell n'\ell'$ in XeI to XeIX. The figures in brackets are the powers of ten by which the result is multiplied.

Ion	$\langle r \rangle_{4f}$ a.u	$\langle 4d/4f \rangle$	Σgf
XeI	17.9	-1.766(-3)	8.94(-5)
XeII	8.58	-2.867(-2)	2.08(-2)
XeIII	5.10	-1.222(-1)	3.5(-1)
XeIV	3.48	-2.543(-1)	1.44
XeV	2.72	-3.717(-1)	2.88
XeVI	2.28	-4.667(-1)	4.35
XeVII	2.0	-5.437(-1)	5.73
XeVIII	1.77	-8.472(-1)	7.05
XeIX	1.60	-8543(-1)	11.04

4. LINE IDENTIFICATION IN THE UTA

Sugar and Kaufman (1980, 1981, 1982) have made a detailed analysis of transitions along the AgI and PdI isoelectronic sequences. They identified $4d^{10}4f-4d^94f^2$ transitions up to HoXXI and $4d^{10}-4d^94f^1P,^3D$ lines up to HoXXII; these lines appear in the ruby plasmas of the elements up to and including neodymium (consistent with the ion stage predictions of section 2) and in the vacuum spark and Tokamak plasmas. In each case they correspond to the strongest lines observed in the UTA since for these particular sequences the entire 4d-4f oscillator strength is contained in a few lines. For cesium, barium and lanthanum the ion stages giving rise to the spectra of figs 1 and 2 have ground configurations with either a full or partially filled 4d subshell.

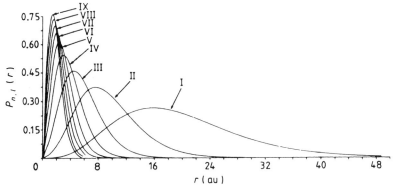

Fig 6. *The collapse in Xe with increasing ionization of the 4f wavefunction of the $4d^94f^1P$ configuration. (After O'Sullivan 1982).*

For an open 4d subshell the number of lines can be considerable and transitions of the type $4d^N-4d^{N-1}4f$ can generate a large number of lines. For example if N=5 or 6 we get almost 5,700 LS allowed transitions. As the principle quantum number doesn't change (i.e. $\Delta n=0$), transitions from adjacent ion stages will tend to overlap to yield a complex array while at the same time the intensity of individual lines will be reduced because of the sharing of oscillator strength. Since hundreds of lines are thus overlapped in a narrow spectral region the UTA thus takes on the appearance of a continuum upon which the strongest lines are superimposed. For transitions of the type $4d^{10}nln'l'-4d^94fnln'l'$ the complexity depends on the $nln'l'$ configurations involved. Since it turns out that different phenomena influence the spectra in this case we must distinguish clearly between situations where the 4d subshell is open or closed. Both of these situations are dealt with separately in what follows.

5. LINE DISTRIBUTIONS AND INTENSITY OF SIMPLER $4d^N-4d^{N-1}4f$ ARRAYS

Because of the enormous line density occurring in the vacinity of the UTA no other individual 4d-4f transitions along other isoelectronic sequences have been identified though detailed calculations for the RhI like ($4d^9$ ground state) RuI like ($4d^8$ ground state) SrI like ($4d^2$ ground state) and RbI like (4d ground state) have been performed recently by Mandelbaum et al (1986) for $4d^N-4d^{N-1}4f$ transitions in praeseodymium. These calculations gave some extremely interesting results. Firstly they yielded the energies and gf values of the expected transitions and secondly they showed the influence of a hitherto unobserved phenomenon, namely the effect of configuration interaction between $4p^64d^{N-1}4f$ and $4p^54d^{N+1}$ configurations. To estimate the configuration mixing, full jj coupling calculations were performed for an entire array with the Relac code of Klapisch et al (1977) which computed the radial integrals, and the code of Grant et al (1980) which gave the angular coefficients.

Cowan (1980) discussed the relative strength of $4p^64d^N-4p^64d^{N-1}4f$ and $4p^54d^{N+1}$ transitions. For N≥5, 4d-4f are the most intense, for N<5, 4p-4d should dominate. In the Q-switched laser experiments the only ion with a $4d^4$ ground configuration which should be encountered is CsXVI so the contribution to the UTA intensity by 4p-4d transitions will be negligible. However for the vacuum spark, Tokamak and high power laser experiments where the 4p subshell as well as the 4d becomes open in many of the ions produced, 4p-4d transitions will contribute to the UTA intensity. Indeed Mandelbaum's single configuration calculations show that strongest lines of the 4p-4d transitions in SrI and RbI like praeseodymium are significantly more intense than any of the 4d-4f lines.

The calculations also show that if single configuration 4d-4f and 4p-4d transitions are considered they predict the strongest lines to lie within a 5Å band between 80 and 85 Å in PrXV and PrXVI, and the 4d-4f lines shift towards longer wavelenths with increasing ionization.

Experimentally the centre of the UTA is found to lie near 84Å, regardless of experimental conditions but its width in the vacuum spark spectrum is only about 2Å. Moreover in PrXXII and PrXXIII which are present in the latter the $4d^2-4d4f$ and 4d-4f groups are predicted to lie in the 90-100Å region and when the stronger 4p-4d transitions are included the entire line group should extend from 80 to 100Å - a result which is not in accord with experiment. However, if allowance is made for $4p^54d^{N+1}+4p^64d^{N-1}$ 4f (N=9,8,2,1) configuration interaction the extent of the line groups shrink in every case to a width of about 3Å and lie in the 80-85Å region. These groups will then superimpose to yield a strong emission feature 2 to 3Å wide. Any discrepancy between the calculated and observed position of this feature may be attributed to the term dependence of the $G^1(4d,4f)$ integral and the fact that the Relac code used for these calculations essentially uses configuration average 4f wavefunctions. The reduction is width of the predicted line group is particularly dramatic for SrI like PrXXII, where the array is completely altered and which shows that $4p^64d^{N-1}4f$ and $4p^54d^{N+1}$ can no longer be considered separately in any calculation. The UTA in the vacuum spark spectra of the lanthanum, cerium and neodymium obtained by Mandelbaum *et al* and in the Tokamak spectrum of dysprosium obtained by Finkenthal *et al* also have a FWHM of 5Å or less which indicates that the $4p^64d^N4f, 4p^54d^{N+2}$ mixing is not confined to praeseodymium but is an important feature encountered throughout the rare earth group. Thus in any ion stages where the 4d subshell is opened the effect of configuration interaction must be included.

6. COMPLEX $4d^N - 4d^{N-1}4f$ ARRAYS; THE UTA MODEL

Because of the large number of lines that will result from excitation of $4d^N$, $3 \leq N \leq 7$, configurations it is generally impossible to identify individual lines or even to resolve them, especially since it has been observed that because of $4p^64d^{N-1}4f$, $4p^54d^{N+1}$ configuration interaction the emission from each ion stage overlaps in a narrow energy range. To deal with this situation, calculations using the statistical UTA model developed by Bauche-Arnoult, Bauche and Klapisch (1978,1979,1983) and which has been extremely successful in interpreting 3d-4l spectra should be most suitable. In this model, the UTA is described as a statistical distribution of transition energies rather than a superposition of many transitions computed individually. The moments of the distribution are given by

$$\mu_n = \sum_{a,b} E_{ab}^n w_{ab}/W_{ab}$$

where E_{ab} is the transition energy between states a and b of configurations A and B respectively; w_{ab} is a weighting factor related to the transition probably and $W_{ab} = \Sigma w_{ab}$. The intensity weighted mean energy of the array corresponds to the first moment and can be written as

$$\mu_1 = E_{av}(A) - E_{av}(B) + \delta E(A,B)$$

where δE is a shift originating in the weighting factor W_{ab}. δE vanishes for transitions of the type $l^N l' - l^N l''$ where the mean energy just corresponds to the difference of the configuration average energies of the two configurations. However for transitions of the type $l^{N+1} - l^N l'$

$$\delta E = \frac{N(2l+1)(2l'+1)}{4l+1} \left[\sum_{k \neq 0} f_k F^k(l,l') + \sum_k g_k G^k(l,l') \right]$$

where $f_k = \begin{pmatrix} l & k & l \\ 0 & 0 & 0 \end{pmatrix} \begin{pmatrix} l' & k & l' \\ 0 & 0 & 0 \end{pmatrix} \begin{Bmatrix} l & k & l \\ l' & 1 & l' \end{Bmatrix}$

$$g_k = \begin{pmatrix} l & k & l' \\ 0 & 0 & 0 \end{pmatrix} [2/3 \delta(k,1) - \frac{1}{2(2l+1)(2l'+1)}]$$

so that the shift is proportional to N. In the case of 4d-4f transitions, for PdI like $4d^{10} - 4d^94f$ the strongest transition is $^1S_0 - ^1P_1$ which exceeds the $^1S_0 - ^3D_1$ by about two orders of magnitude. We have already seen that the energy of the ^1P term may be written as $E \sim E_{av} + 2G^1(4d,4f)$ so that in this case the mean transition energy

$$\mu_1 \sim E_{av}(4d^94f) + 2G^1(4d,4f) - E_{av}(4d^{10})$$

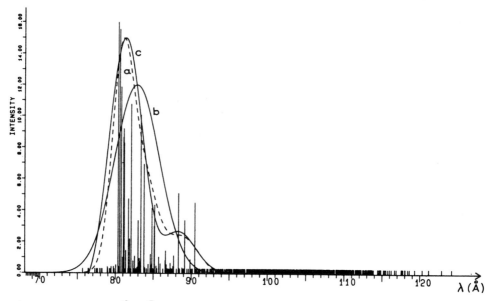

Fig. 7. Pr XVI $4d^8-4d^74f$ transition array. Each line is represented with a height proportional to its oscillator strength except those with <3% of the highest, which have all been given this value (a), envelope calculated by adding the lines after assigning each a FWHM of 0.5Å. (b) Gaussian curve, (c) skewed Gaussian curve (after Bauche-Arnoult et al 1984).

and $\delta E \sim 2G^1(4d,4f)$. If the mean transition energy was taken to be the difference between the average energies of both configurations the result would be at too low an energy, typically by a factor of about 20eV! In the case of the simple RbI like $4d(^2D)-4f(^2F)$ transitions the coefficients of the $F^k(4d,4f)$ and $G^k(4d,4f)$ integrals are much smaller and in this case.

$$\mu_1 \sim E_{av}(4f) - E_{av}(4d)$$

i.e. $\delta E \sim Q$ which explains the shift in the mean position of the 4d-4f array from $\sim$80Å in RhI like PdXV to $\sim$100Å in RbI like PrXXIII. Because δE varies linearly with N, the occupation number of the 4d subshell, the mean position of the UTA is predicted to shift to longer wavelengths with decreasing N. As we have seen $4p^64d^{N-1}4f$, $4p^54d^{N+1}$ configuration interaction destroys this trend so the simple UTA model has only limited applicability in interpreting the spectra of highly ionised rare earths. In the Q switched laser produced plasmas, where in general the 4d subshell does not become stripped, the UTA model should be ideal for interpreting the spectra.

The second moment of the distribution, μ_2 is related to the width of the array; in fact the variance σ is given by $\sigma^2 = \mu_2 - (\mu_1)^2$ and the FWHM of the array is then related to the variance by a simple expression, for example $\Delta E = 2\sqrt{2\ln 2}\sigma$ for a Gaussian distribution. Since in general μ_n can be written

$$\mu_n = \frac{1}{W_{ab}} \sum_{a,b} [\langle a|H|a \rangle - \langle b|H|b \rangle]^n |\langle a|D|b \rangle|^2$$

where H is the non relativistic Hamiltonian and D is the electric dipole operator, any moment of the distribution can in principle be calculated once its energy parameters are known. Bauche-Arnoult, Bauche and Klapisch have given simple analytical expressions for these moments in terms of Slater integrals, spin orbit parameters and combinations of 3j, 6j and 9j symbols. For example the variance can then be written as:

$$\sigma^2 = \sum_i N_i \left[\sum_{kk'} x_i(k,k',l,l'...)(X^k Y^{k'})_i \right]$$

where N_i are numerical factors containing the dependence on the subshell occupation numbers, X_i are combinations of 3j, 6j and 9j symbols and $(X^k Y^k)_i$ are products of Slater integrals. They also noted that many contributions to σ^2 are negative so that frequently the width of the transition array is smaller than the widths of the configurations involved. This is especially true in the case of $l^N l' - l^N l''$ transitions. Moreover to deal with transitions such as $4d^{10} 5p^N - 4d^9 4f 5p^N$ where most of the oscillator strength lies in the $4d^{10} 5p^N - 4d^9 4f(^1P) 5p^N$ transitions Bauche et al introduced the concepts of "emissive zones" of any array from which most of the strong emission lines originate. To allow for more complex distributions other than Gaussian, (see for example fig. 7.) Bauche-Arnoult et al (1984) used skewed Gaussian distributions originally introduced by Croxton et al (1967) and which could be formulated in terms of an assymmetry parameter defined in terms of the first, second and third moments of the distribution. Klapisch et al (1983) applied the UTA model to the spectrum of cesium and obtained a good fit with experimental data. (Fig. 8) In this figure it is seen that superpositions of UTA for two stages CsXV and CsXVI with ground states $4d^5$ and $4d^4$ reproduce the overall shape of the observed spectrum. Theoretical UTA for other stages are displaced towards shorter wavelengths. Since the energy parameters in these calculations are essentially computed in an E_{av} basis, F^k and G^k parameters will be overestimated, and the discrepancy is greatest for lower ion stages where the shape of the 4f wavefunction is strongly term dependent. Since δE is proportional to N, the 4d occupation number, (which increases as the ion stage is reduced) multiplied by combinations of Slater integrals, the

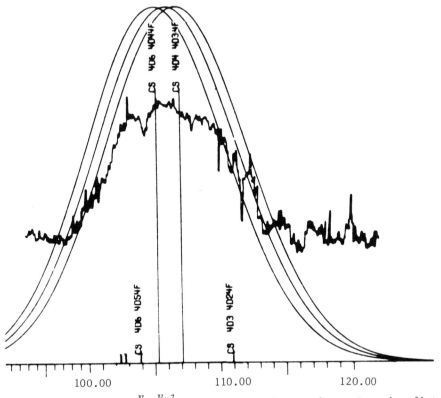

Fig. 8. Calculated $4d^N 4d^{N-1} 4f$ UTA for CsXVI assuming a Gaussian distribution (After Klapisch et al 1983).

shift of the theoretical UTA to shorter wavelengths is most likely an artifact of the calculation. If term dependent values of F^k and G^k parameters were included, the UTA in several ion stages should overlap. However Mandelbaum found that the UTA model did not reproduce the praeseodymium emission. Even of the F^k and G^k parameters were scaled to allow for term dependence there was still a shift towards longer wavelength with increasing ionisation, which in reality is removed by configuration interaction. Thus in the case of cesium, it is not clear if the UTA model is by itself entirely adequate to explain the spectrum or if configuration interaction between $4p^54d^{N+1}$ and $4p^64d^{N-1}4f$ states must also be included to account fully for the wavelength shift. Further calculations are needed to clarify this point.

7. EFFECT OF 4f COLLAPSE ON GROUND STATE CONFIGURATIONS

In the Q switched laser produced plasmas the 4d subshell only becomes opened in ground states of elements up to and including neodymium. In the higher Z elements the degree of ionisation is such that the ground configurations always contain a closed $4d^{10}$ subshell with variable numbers of outer electrons. The same is true for low to moderate ion stages of the elements from cesium to neodymium. The transitions then giving rise to the emission bands must be of the form $4d^{10}nln'l' - 4d^94fnln'l'$ i.e. pure 4d - 4f transitions, since 4p - 4d transitions cannot occur and hence configuration interaction is not a problem for these spectra. At first sight it might appear that $nln'l'$ would correspond to 5s and 5p, since the 4f electrons only become bound at cesium and would be removed easily in low stages of ionisation. However, this is not the case; with increasing ionisation as 4f contraction proceeds the binding energy of the 4f electron increases more rapidly than either the 5s or 5p and as a result, 4f,5p and 4f,5s level crossing takes place. In the absence of level crossing effects many of the transitions would be based on simpler ground configurations such as $4d^{10}5s^25p$, $4d^{10}5s^2$ and $4d^{10}5s$, so that individual lines would be expected to appear strongly. Hence the 4d-4f band should consist of a background of weak unresolved structure upon which a group of strong, well defined lines is superimposed. This is certainly true for cesium, barium and lanthanum but is not the case for the higher Z elements where from these considerations the only open subshells would be 5s and 5p and the spectra should contain many strong lines. In fact the opposite is observed, the spectra become so complex that the 4d-4f emission has a continuum appearance.

A range of direct evidence for 4f,5p level crossing exists in spectroscopic data. In the XeI isoelectronic sequence Reader and Ekberg (1972) observed the resonance transitions from the $5p^6{}^1S_0$ ground state up to CeV, some of these lines are observed for PrVI, but in NdVII the spectrum consists of hundreds of weak lines and the resonance transitions could not be identified. Edlen (1973) attributed this result to the 4f electron becoming more tightly bound than the 5p so that the ground configuration changes from $5p^6$ to $4f^6$. In the case of Pr, the ground configurations are known experimentally up to PrV ($5p^64f$). In PrVI because only the strongest lines associated with a $5p^6$ ground state are observed, the 4f and 5p orbitals must be very close in energy. In going to PrVII, because the core charge has increased it is reasonable to assume that 4f,5p crossing will again occur. At the high Z end of the rare earth sequence some evidence for the 4f and 5p energy trends can be derived from the behaviour of the $5p^55d$ and $4f^{13}5d$ excited configurations of the YbIII isoelectronic sequence which were shown by Kaufman and Sugar (1976) to cross at WVII. They also remarked that $4f^{13}5p^6$ rather than $4f^{14}5p^5$ is the most likely ground configuration for WVIII. Hence it would appear that 4f,5p level crossing occurs near the sixth ion stage in each of the rare earth elements past cerium. That it should occur for the same ion stage in each element is not unexpected since adjacent elements differ from each other by the presence of an extra 4f electron which is added to the core. This electron effectively

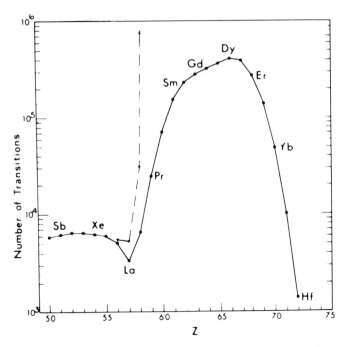

Fig. 9 Approximate number of lines in LS coupling predicted for 4d-4f transitions summed over ion stages VIII-XVI. Full curve: predictions without inclusion of 4f,5p and 4f,5s level crossing, broken curve: number of lines predicted with inclusion of low lying configurations (After Carroll and O'Sullivan 1982).

shields the outer electrons from the increased nuclear charge so that the coulomb potential for 5s and 5p electrons is approximately the same for a given ion stage in any rare earth element.

With increasing ionisation the 4f electron eventually becomes more tightly bound than the 5s. Experimental evidence for 4f,5s level crossing is provided by the analysis of Sugar (1977) of spectra along the AgI sequence. He showed that the ground state changes from $4d^{10}5s\ ^2S$ to $4^{10}4f\ ^2F$ between NdXIV and SmXVI. Cheng and Kim (1979) subsequently calculated that the crossing should take place at PmXV. Again this leads to the conclusion that 4f,5s crossing occurs near the fourteenth ion stage of each rare earth element.

To study the effect of 4f,5p/5s crossing on the ground and low excited configurations, the separation of the centres of gravity of the different configurations derived from combinations of 5p,5s and 4f electrons from the lowest configuration for the XeI sequence were calculated using the Froese-Fischer (1978) MCHF computer code (O'Sullivan 1983). In this work the effects of configuration interaction were ignored. The results of Bauche et al (1983) show this to be a valid approach for these configurations since in general the off diagonal Slater R^k integrals which describe the configuration interaction are much smaller than diagonal terms. The results of these calculations are

TABLE III: Hartree Fock Energies (eV.) of Configurations Containing Variable Numbers of 4f, 5p and 5s Electrons Along the Xe I Isoelectronic Sequence.

Ion	Configuration								
	$5s^25p^6$	$5s^25p^54f$	$5s^25p^44f^2$	$5s^25p^34f^3$	$5s^25p^24f^4$	$5s^25p4f^5$	$5s^24f^6$	$5s4f^7$	$4f^8$
Ba III	0.0	–	43.05	65.86	89.03	112.46	136.11	–	270.02
La IV	0.0	17.03	36.57	57.99	80.68	104.22	128.31	173.20	197.55
Ce V	0.0	11.78	26.97	45.00	65.28	87.33	110.60	149.95	181.06
Pr VI	0.0	5.49	15.10	28.34	44.68	63.61	84.63	120.23	154.33
Nd VII	1.73	0.0	3.03	10.36	21.53	36.03	53.40	82.18	118.55
Pm VIII	14.24	4.46	0.0	0.45	1.30	14.32	26.83	56.08	88.17
Sm IX	41.01	22.36	9.58	2.26	0.0	2.38	8.97	34.16	62.84
Eu X	81.81	53.53	31.63	15.71	5.40	0.30	0.0	20.14	44.40
Gd XI	136.44	97.80	66.01	40.71	21.53	8.09	0 0	14.17	33.06
Tb XII	195.74	146.00	103.59	68.14	39.30	16.71	0.0	7.34	19.98
Dy XIII	259.85	198.34	144.57	98.25	59.00	26.48	0.33	0.0	5.53
Ho XIV	338.86	264.86	199.04	141.11	90.71	47.60	11.16	2.37	0.0
Er XV	429.24	342.06	263.51	193.26	130.99	76.37	29.07	11.07	0.0
Tm XVI	525.33	424.30	332.32	249.06	174.21	107.46	48.48	20.52	0.0

shown in table III. It was found that the ground configuration changes from $5s^25p^6$ to $5s^25p^54f$ at NdVII and the occupation number of the 5p subshell decreases along the sequence until it becomes $5s^24f^6$ at EuX. At DyXIII $5s4f^7$ becomes the ground configuration and at HoXIV the 4f,5s crossing is complete and the ground state becomes $4f^8$. In addition to the configurations listed in table III all configurations of the type $5s^m5p^k4f^N$ (m=0,1,2, m+k+N=8) are permitted and at high ion stages where the 4f and 5s energies are nearly degenerate, configurations such as $5s4f^7, 5s5p4f^6$ and $5p4f^7$ will for example lie close to the $5s^24f^6$. Thus the origin of the exact ground state term is not clear and is in fact meaningless, since the spectra are so complex that no individual lines can ever be identified. Moreover, since these configurations occur in rare earth ions which require electron temperatures of tens to hundreds of eV to produce, all of the low lying configurations which overlap to produce a near continuum of states will be populated thermally. Hence the ground state is best described by $(4f5s5p)^q$ where q is the total number of electrons shared amongst 4f,5s and 5p orbitals.

Excitation of a 4d electron generates a family of excited $4d^95s^m5p^k4f^{N+1}$ configurations which again lie close in energy. Hence the 4d-4f transition effectively occurs between two continua of levels to produce a continuum like structure. Moreover cross transitions of the type $4d^{10}5s^m5p^k4f^N - 4d^95s^m5p^{k+1}4f^N$ are also expected through with less intensity than the 4d-4f, and which, because of the 4f, 5p degeneracy will lie close to the 4d-4f emission band in a number of stages. That many low lying configurations must be included in any interpretation of the spectra can be inferred from the fact that no 4d-4f lines are observed in the higher rare earths where the spectra should simplify as the 4f shell occupancy nears completion (fig.9). However a search for $^2D-^2F$ lines of the transition $4d^{10}4f^{13}-4d^94f^{14}$ proved fruitless which indicates that the large oscillator strength normally associated with such a simple transition is in fact distributed over a number of complex configurations of the type $4d^{10}5s^m5p^k4f^{13-(k+m)}-4d^95s^m5p^k4f^{14-(k+m)}$

8. SHAPE OF THE $4d^{10}5s^m5p^k4f^N - 4d^95s^m5p^k4f^{N+1}$ ARRAYS

It has been demonstrated above that all low lying configurations of the type $4d^{10}5s^m5p^k4f^N$ which contain variable numbers of 5s, 5p and 4f electrons must be included in any interpretation of the emission arrays observed in the laser produced plasmas of elements with Z>62. In the case of GdXII, Bauche et al (1983) have computed the mean energies and widths of the configurations involved and the mean wavelengths and widths of the transition arrays within the UTA formalism. The overall trends predicted by this calculation, though performed specifically for GdXII, will be true for other ion stages of gadolinium and also for other rare earths ions though the actual numerical results will of course be different. The results of their calculation are shown in fig. 10. Here each broad rectangle represents the position and width of a configuration; the average ordinate is the mean wavenumber of the configuration state distribution while its width corresponds to the FWHM, assuming a Gaussian distribution and is given by $\Delta E = 2(2\ln2)^{\frac{1}{2}}\sigma = 2.355\sigma$. For the lower configurations the spin orbit parameter ζ_{5p} essentially dictates the width while for the upper configurations the (4d,4f) Slater Integrals determine the size of σ and in fact $\sigma \alpha [N(14-N)]^{\frac{1}{2}}$ where N is the number of 4f electrons. In fig. 10 the narrow shaded rectangles correspond to "emissive zones", or zones of preferential transition probability, which are located, on average at δE above the centre of gravity of the configuration where, as before, $\delta E = \Sigma_{a,b} E_{ab} w_{ab}/W_{ab}$. δE is different for different types of transitions; for transitions of the type $l^{N+1}-l^N l'$ (l=1,3)

$$\delta E(d-p) = N/4l+1[-1/5 F^2(d,p) + \frac{19}{15} G^1(d,p) - 3/70 G^3(d,p)]$$

$$\delta E(d-f) = N/4l+1[-8/35 F^2(d,f) - 2/21 F^4(d,f) + \frac{137}{70} G^1(d,f) - 2/105 G^3(d,f) - 5/231\ G^5(d,f)]$$

and the location of the "emissive zone" depends on the type of transition. The emissive zones lie above the bulk of the distribution since $G^1(4d,4f) > G^1(4d,5p)$ and the coefficient of the G^1 parameter is larger in the 4d-4f case.

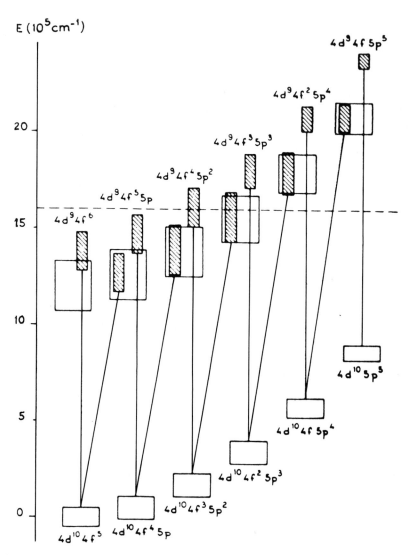

Fig. 10. *Low lying configurations and transition arrays in the spectrum of GdXII (After Bauche et al 1983).*

Bauche *et al* also considered the behaviour of the emissive zones in relation to the J-file sum rule of Condon and Shortley (1935). A J-file sum is simply the sum of the strengths of all the transitions connecting a given J level of one configuration with the J-1, J and J+1 states of the other configuration. They found that the J file sum for transitions of the type $4d^{10}4f^{N-1}-4d^94f^N$ could be written as

$$k[(4d^94f^N)\alpha J-4d^{10}4f^{N-1}] = (2J+1)[\tfrac{3N}{7} + C(G^1;\alpha J)]I^2{}_{4d,4f}$$

where $I_{4d,4f} = e\int_0^\infty R_{4d}(r)rR_{4f}(r)dr$ is the dipole radial integral and C is the coefficient of the G^1 (4d,4f) Slater Integral appearing in the expression for the electrostatic energy of state αJ. Therefore the J file sum and consequently the strength of a transition from a given J level of the upper configuration depends on the magnitude of the coefficient $C(G^1;\alpha J)$ which places the bulk of the oscillator strength well away from the centre of the configuration, since the most extreme cases, $\delta E \sim E_{av}+C(G^1;\alpha J).G^1(4d,4f)$ and $C(G^1;\alpha J)=2$.

In this treatment Bauche *et al* ignored the effects of 5s subshell stripping which gives rise to sets of arrays of the type $4d^{10}5s4f^N5p^k-4d^95s4f^{N+1}5p^k$ etc. These configurations will in fact lie in the same energy range as those presented in fig. 10, and will contain emissive zones identical to the cases already discussed. Hence any simplification of the spectrum arising from the absence of configurations which lie above the ionisation potential of GdXII is more than offset by the inclusion of these configurations.

Bauche *et al* also computed the mean wavelength and FWHM of all the possible arrays included in fig. 10. For 4d-4f transitions the arrays are all centred between 66.2 and 78.6Å and have widths ranging from 5.2 to 12.8Å. The observed 4d-4f feature in the salt spectrum of gadolinium peaks at 68Å and extends from 66 to 75Å, so the theory is in excellent agreement with experiment. The configurations in other ion stages of gadolinium will behave similarly since they differ only in the number of outer $(5s5p4f)^q$ electrons and the $4d^{10}5s^m5p^k4f^N-4d^95s^m5p^k4f^{N+1}$ arrays will all superimpose in the 65 to 76Å region. Thus the overall shape of the array in gadolinium and the other rare earths with Z>62 is determined by the position and width of the emissive zones, which in general lie to shorter wavelengths than the separation of the average energies of the upper and lower configurations involved and are considerably narrower than expected from a simple computation of the total widths of these configurations, a result which is identical to that predicted for the $4p^64d^N-4p^64d^{N-1}4f+4p^54d^{N+1}$ transitions in the lighter elements of the lanthanide sequence.

One further aspect of the features should be noted at this point. It was observed that their intensity tended to decrease past Z=62. Past this point the number of 4f electrons in the configurations of interest is increasing by one for identical ion stages of adjacent elements. Starace (1972) has shown that the summed oscillator strength for 4d-4f transitions depends in the number N of 4f electrons in the upper state according to $\Sigma gf \alpha (14-N)I^2{}_{4d,4f}$. Hence, as $I_{4d,4f}$ is almost constant for a given ion stage for any rare earth element, the oscillator strength decreases rapidly as N increases.

9. COMPARISON OF SALT AND METAL SPECTRA; OPTICAL DEPTH EFFECTS

Up to now we have concentrated on the emission from laser produced plasmas when the rare earth ions were only a minor constituent, or from Tokamak or vacuum spark plasmas where the ion and electron densities are orders of magnitude lower than in the laser plasma case. As is evident from fig. 2., the spectra from targets of the pure metal are considerably different. This behaviour has been dealt with at some length by Carroll and O'Sullivan and their interpretation is summarised in what follows. In fig.11 a direct comparison is made between the spectra from lanthanum metal (a) and lanthanum oxide in araldite (b) which had a lanthanum concentration of approximately 1%. It is seen that the overall extent of the continuum like structure is much greater in the metal than the salt spectra and consists of

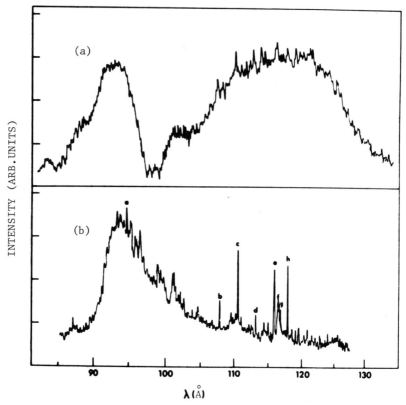

Fig. 11. Spectrum of lanthanum from laser produced plasmas (a) Target of lanthanum metal (b) Target of lanthanum oxide in araldite. Line identifications: LaXII $4d^{10}-4d^94f$ (a) $^1S-^1P$, (d) $^1S-^3D; 4d^{10}-4d^95p$ (b) $^1S-^3D$, (c) $^1S-^1P$ due to Sugar (1977). Impurity lines: (e) (f) and (h). (After Carroll and O'Sullivan 1982)

two regions of intense emission separated by dip which coincides with the falling edge of the salt emissions UTA. For the elements from cesium to neodymium the metal spectra contain both absorption and emission lines; as samarium is approached the intensity of individual lines decreases and the spectra become more continuum-like. Past samarium, no individual lines appear and the intensity of the continuum-like feature decreases until at thulium it has merged completely with the underlying recombination continuum. The difference in behaviour between the salt and the metal spectra can only arise from the fact that in going from the salt to the metal plasmas there is typically a hundredfold increase in the number density of the atoms of interest and at the same time because the average charge state obtained in the metal plasma (12 or 13) is higher than in the salt, where the presence of lines of OV, OVI and CV puts the average charge state at four or five times ionized, there is an increase in electron density.

The behaviour of the $4d^{10}-4d^94f$ lines of the PdI sequence provide an important clue to interpreting these differences in the spectra. The f-value of the $^1S-^1P$ line is between 10 and 100 times larger than the $^1S-^3D$, as the coupling in this configuration is almost pure LS. These lines have been observed in the spectra from CsX to NdXV. Whereas the $^1S-^1P$ line appears strongly in emission in the salt spectra for the corresponding metal spectra it appears in adsorption up to CeXIII, in PrXIV it is completely self absorbed and in NdXV it appears in emission. The $^1S-^3D$ line appears in emission in all cases and lies in the region of the longer wavelength intensity maximum. From this behaviour it can be inferred that transitions in lower ion stages with a large gf value will appear in absorption in the metal spectra while in the salt spectra the ion density is insufficient to cause appreciable absorption. Moreover, in the salt spectra the absolute emission rate is greatly reduced and consequently the low f-value transitions which are predicted to lie to the long wavelegth end of the 4d,4f array (see fig. 10) do not appear except in cases where all of the oscillator strength is concentrated in a few lines. These transitions originate from parts of the upper configuration away from the emissive zones or J file sums from terms with low values of $C(G^1;\alpha J)$. On the other hand radiation from the higher energy emissive zones is transmitted through the plasma with only slight attenuation. The metal plasmas however, contain a sufficient ion density to permit the appearance of weaker lines which overlap to form the long wavelength emission band. For the high f-value lines the plasma becomes optically thick and the intensity of these transitions is reduced, or in the most extreme cases, in lower ion stages actually appear in absorption and cause the pronounced dip in the centre of the array. In lanthanum the centre of the dip coincides approximately with the centre of the $4d-\overline{4,\epsilon}f$ feature of LaI; however in the higher rare earths the positions of the absorption dip and the $4d-\overline{4,\epsilon}f$ features do not correspond which indicates that in the laser plasmas significant absorption only begins at moderate ion stages. As already mentioned the average electron density in the metal plasmas will be larger by a factor of 2 to 3 than in the salt plasmas. Recently, Krasnitz (1985) has shown that at electron densities close to $10^{18}cm^{-3}$ the types of transitions occurring in the 4d-4f UTA are extremely density sensitive. If $n_e < 10^{18}cm^{-3}$ transitions from singly excited configurations of the type $4d^9 4f^N 5s^m 5p^k - 4d^{10} 4f^{N-1} 5s^m 5p^k$ dominate, while for $n_e > 10^{18}cm^{-3}$ transitions from doubly excited configurations of the type $4d^8 4f^{N+1} 5s^m sp^k - 4d^9 4f^N 5s^m 5p^k$ become important. In ruby laser produced plasmas, the cutoff electron density is $2 \times 10^{21} cm^{-3}$. Hence it is reasonable to assume an average density of $\sim 10^{18} cm^{-3}$ in the region of strongest emission. If this is so then the emission in both types of plasma could arise from different types of transition. The lower density salt spectrum could be dominated by transitions from singly excited configurations while in the metal spectra transitions from doubly excited configurations could be most important. This would naturally cause some differences in the emission profiles of the corresponding transition arrays.

10. ABSENCE OF LINES IN SPECTRA OF LANTHANIDES WITH X>62

The most striking feature of the spectra of the higher rare earths is the complete absence of line emission over large wavelength ranges (Carroll *et al* 1980). From the foregoing discussion it follows that once 4f,5p or 4f,5s level crossing begins to occur and excited configurations containing variable numbers of 4f,5p and 5s electrons are populated, no strong lines are emitted because the oscillator strength for resonance excitation is now shared amongst a number of complex configurations of the type $4d^{10}(4f5s5p)^q$. Only in cases where q=1 or 21, so that the range of available states is severely restricted, will lines be observed. In practice this means that for the elements from samarium to ytterbium no strong lines are emitted between the fifth and fifteenth ion stages except in the case of SmXVI which has q=1 and for which the resonance 4f-5d transitions identified by Kaufman and Sugar (1981) were observed. For the elements up to samarium the 4d subshell is opened to some extent in the ion stages produced and also, although 4f,5p crossing effects occur 4f,5s crossing does not take place (since the 4f and 5s subshells are stripped well before the thirteenth or fourteenth spectra and line emission is observed throughout the 30-2,000Å region. It was noted in the introduction that for samarium, for example, no lines were observed between 200 and 2,000Å. With increasing Z a narrow line group which first appears in the europium spectrum gradually increases in width and intensity until at erbium it extends from 600 to 1500Å. The work of Martin et al (1979) shows that this group is due to $4f^N$-5d transitions in the fifth spectrum and $4f^{N-1}5d - 4f^{N-1}6p$ transitions in the fourth and fifth spectra. Because of the low stages involved, these lines are emitted from the cooler outer plasma region and can be removed entirely from the spectra if a suitable optical system is used to focus the plasma core onto the spectrograph slit. Because the $4f^{N+k}5p^{6-k}$ manifold is relatively far from the ground configurations when compared to the average electron temperature, only transitions from the ground configurations need be considered and in some cases strong lines can result because of the distribution of terms with increasing energy (see Cowan 1980, pg. 601). For example, at GdIV, N=7 and the $4f^7$ $^8S_{7/2}$ level lies about 5eV lower than other levels of the configuration because of Hund's rule, and this value compares with the value of T_e (=5eV) required for observation of this stage according to the CR model. (fig. 3).

Past the fifth spectrum the resonance transitions anticipated at $\lambda > 100$Å are 4f-5d, 4f-5g, 5p-5d, 5p-6s and 5s-5p. The intrashell $5l-5l'$ transitions are predicted to overlap in different stages and lie in the 200-400Å region for each element and shift very slowly towards higher energy with increasing Z since the presence of an extra 4f electron shields them from the increased core charge for a given ion stage. The dominant transitions are expected to be 4f-5d and 4f-5g. The positions of these were calculated for different ion stages of samarium in fig. 12 (O'Sullivan 1983). Because of the complexity of the configurations involved these transitions should yield narrow UTA ~1eV wide as (4f,5d) and (4f,5g) Slater Integrals are small. However the oscillator strength is so reduced that no such features appear upon the background continuum which therefore must arise primarily from recombination and bremsstrahling.

11. ORIGIN OF THE XUV CONTINUUM

Recently the relative importance of the different continuous radiation mechanisms in laser produced rare earth plasmas has been considered in some detail (O'Sullivan 1983). In making a comparison between the emission from salt and metal plasmas it must be remembered that in the former, the rare earth ions essentially constitute a 1% impurity by number density. The primary effect of the presence of impurities of any kind, in any concentration, is to reduce the number of emitting ions of the element under consideration. In the core region practically every ion emits and hence the integrated radiation intensity of a particular element is proportional to the number of ions of that element present there. Consequently, if the population of a given species is reduced by a given factor, the absolute line intensity is reduced by the same factor. However the greatest effect of changing the number density is on the opacity where, to a first approximation, absorption in the cooler regions depends on $\exp(-n_\zeta \sigma r)/r^2$ where r is the distance from the plasma core, σ is the absorption coefficient and n_ζ is the number of ions of the species of interest in charge state ζ. However n_ζ depends both on distance from the core and the electron temperature. The highest stages are most abundant in the core but their populations decrease rapidly because of recombination and ion streaming. In a comparison of the ion distribution in metal and salt plasmas a complete treatment must allow for the different recombination rates in each due to the difference in electron density and average state of charge. If these effects are neglected, a decrease in the rare earth ion population by a factor of 100 effectively reduces the absorption by e^{100}. Hence the major difference between the two types of plasma arises because of their different optical thickness.

Theoretical expressions for the different emission mechanisms in an optically thin plasma in coronal equilibrium have been derived by Spitzer

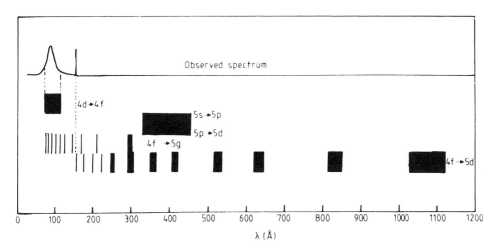

Fig.12. Schematic diagram showing the positions of predicted transition arrays in Sm and the observed spectrum (After O'Sullivan 1983)

(1962). The total power radiated by bremsstrahlung is

$$P_B = 1.5 \times 10^{-32} n_e T_e^{\frac{1}{2}} \sum_\zeta n_\zeta \zeta^2 \quad \text{Wcm}^{-3}$$

where n_e is the electron density and n_ζ is the density of ions of charge ζ. The total emitted power from recombination of electrons and ions of charge $\zeta+1$ is

$$P_R = 1.5 \times 10^{-32} n_e T_e^{-\frac{1}{2}} \sum_{\zeta,n} \zeta^2 \chi_{\zeta,n} n_\zeta \quad \text{Wcm}^{-3}.$$

Here n is the principal quantum number of the shell into which the electrons recombine and $\chi_{\zeta,n}$ represents the ionisation potential for electron removal from that shell. If recombination with the ground state predominates then $P_R/P_B = \chi_\zeta/T_e$, or $P_R \simeq 6 P_B$ in the present case. The equivalent expression for line ζ radiation is given by

$$P_L = 7 \times 10^{-25} n_e n_{\zeta,g} T_e^{\frac{1}{2}} \sum_n f_{n\zeta} \exp(-\varepsilon_n/T_e) \quad \text{Wcm}^{-3}$$

where $n_{\zeta,g}$ is the density of ions in the ground state, $f_{n\zeta}$ is the oscillator strength for excitation from level n to the ground state and ε_n corresponds to this transition energy. For the Q-switched laser plasmas, $\overline{\zeta} \simeq 13$, $T_e \sim 55$ eV so $P_L/P_B = 2 \times 10^3 \sum f_{n\zeta} \exp(-\varepsilon_n/t_e)$ which gives a typical ratio of $P_L/P_B \sim 10^3$ for 4d-4f transitions, ie. $P_L/(P_R+P_B) \sim 100$. For transitions involving 4f or 5p electrons such as 4f-5d etc, $P_L/P_R+P_B \sim 10$. In the spectra where the rare earth elements are present in $\sim 1\%$ concentrations the strongest features are indeed the 4d-4f UTA which are more than an order of magnitude more intense than the background which is itself reduced in intensity compared to the case of pure metals because the average charge state is quite low and $P_R \alpha \zeta^4$. Moreover the electron density in these plasmas is at least a factor of three lower than in the pure metal case so the salt plasmas are closer to a coronal type of equilibrium than the metals. However at the densities encountered in a laser produced plasma collisional deexcitation cannot be ignored. Weisheit (1975) has studied the effects of collisional interruption of satellite line intensity and found reductions of >50% in the observed intensity of He-I like satellites of CV. Similar considerations for rare earth plasmas will reduce the ratio of P_L/P_R+P_B by at least the same amount.

For the spectra from pure metal targets, the background continuum is enhanced because of the $P_R \alpha \overline{\zeta}^4$ dependence. Simultaneously, because of the increased ion density, the plasma becomes optically thick to high f-value transitions and the UTA show evidence of absorption. Hence the overall power radiated as lines is decreased; in fact Galanti and Peacock (1975) measured a reduction of 5×10^{-4} for the resonance line of CV in a ruby laser produced plasma. A similar reduction is the present case would imply that $P_R + P_B/P_L$ should exceed 100 which agrees with observation.

It should be noted that the vacuum spard and Tokamak plasmas have average ion and electron densities at least five orders of magnitude lower than the laser plasmas and so approximate to a coronal equilibrium. Thus they correspond more closely to the spectra of optically thin laser plasmas where the rare earth ion density is low and optical opacity effects are not important. In spectra produced by high power lasers, it was noted in the introduction that a 4d-4f UTA was observed by Finkenthal et al (1986b) even though the dominant ion stages have $3d^N$ and $3d^{10} 4s^m$ configurations. The density of ions with $4p^N$ or $4p^6 4d^N$ ground states is extremely low because of the high value of the plasma temperature. That these features are observed at all is a function of their intensity and the fact that the plasma is to them, optically thin.

12. CONCLUSION

The EUV spectra of laser generated rare earth plasmas have been studied in different regimes of optical thickness. In each case the most striking feature observed was an intense UTA, some 30eV wide in the case of optically thick plasmas and approximately 15eV wide in the optically thin case. The feature arises from transitions of the type $4p^6 4d^N - 4p^6 4d^{N-1} 4f + 4p^5 4d^{N+1}$ and $4d^{10} 5s^m 5p^k 4f^N - 4d^9 5s^m 5p^k 4f^{N+1}$. In the former, configuration interaction in the upper state causes the arrays to shrink in extreme cases, from some 20eV to less than 3eV, the shrinkage being directed towards the high energy end of the array. *Ab initio* calculations predict the $4d^N - 4d^{N-1} 4f$ transitions to move towards longer wavelengths as N decreases but inclusion of configuration interaction removes this trend and so transition arrays in different ion stages overlap in energy.

For transitions of the type $4d^{10} 5s^m 5p^k 4f^N - 4d^9 5s^m 5p^k 4f^{N+1}$ 4f,5s and 4f,5p level crossing causes complex configurations with different values of m,k and N to lie sufficiently close in energy that they are populated thermally in the plasma and the 4d-4f transition effectively occurs between two continua of levels. Again transitions in adjacent ion stages tend to overlap, this time because the bulk of the oscillator strength lies about $2G^1(4d,4f)$ above the centres of gravity of the upper and lower configurations, and the quantities $E_{av}(4d^9 4f) - E_{av}(4d^{10})$ and $G^1(4d,4f)$ are relatively insensitive to the ionic charge past $\zeta = 7$ or 8. Hence the feature has a continuum like appearence. It's high intensity is a result of the large 4d-4f oscillator strength which arises because the gf value associated with the giant $4d^{10} - 4d^9$ $4,\varepsilon f$ resonance in the neutral is transferred to 4d-4f transitions in the ion. The difference in appearance of the feature in salt and metal spectra is essentially a number density effect. In the salt plasmas, the rare earth density is very low, there is little absorption in strong transitions and weak lines do not appear. In metal plasmas absorption damps the intensity of the high f value end of the array while the weaker components appear because of the density enhancement.

For elements with Z>62 it was noted that the spectra are line free over large wavelength regions, which in the most extreme case ranged from 30 to 2,000Å. This again is due to the fact that all transitions occur essentially between continua of levels and the f value of individual lines is so low that they are completely submerged in the background continuum which is itself due primarily to recombination.

Finally it should be noted that although this paper concentrated on 4d-4f

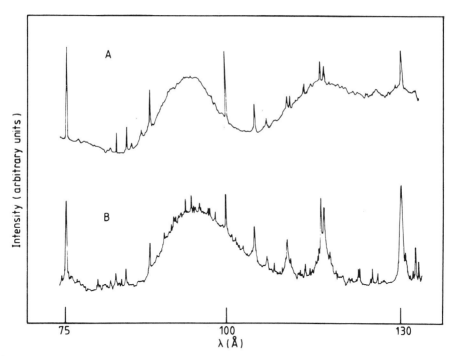

Fig. 13. Spectrum of uranium from laser produced plasmas. (A) Target of uranium metal, purity ~ 100% (B) Target of uranium oxide in Araldite, concentration ~1%. (After Carroll et al 1984.)

UTA, similar spectra are observed for thorium and uranium plasmas which also emit intense UTA centred at about 105 and 95Å respectively. These features have been shown by Carroll *et al* (1984, 1986) to arise from 5d-5f transitions in ion stages similar to those encountered in the lanthanides. The spectrum of uranium is presented in fig. 13. It can be seen from this figure that the 5d-5f UTA show the same sensitivity to optical depth effects as 4d-4f UTA and behave very similarly. However the width of the 5d-5f emission feature is broader than the 4d-4f. The effects of 5f,5p and 5f,5s level crossing as well as the $5p^6 5d^{N-1} 5f + 5p^5 5d^{N+1}$ configuration interaction are being investigated at the moment.

ACKNOWLEDGEMENT

I wish to express my gratitude to Prof. P.K. Carroll of University College Dublin and Prof. Eugene T. Kennedy and Dr. John Costello of NIHE Dublin for their continued interest in this work and their critical reading of the manuscript, to Prof. Marcel Klapisch of the Hebrew University in Jerusalem for many stimulating discussions on the UTA approach and especially to Drs. Michael Finkenthal and Pinchas Mandelbaum (Hebrew University) who communicated the full details of their results prior to publication. Work at U.C.D. was supported by N.B.S.T. grant URG-111-78.

REFERENCES

Bauche-Arnoult, C., Bauche, J., and Klapisch, M. "Mean wavelength and spectral width of transition arrays in X-UV atomic spectra" 1978. J. Opt. Soc. Am. 68, 1136.

Bauche-Arnoult, C., Bauche, J., and Klapisch, M. "Variance of the distributions of energy levels and of the transition arrays in atomic spectra." 1979 Phys. Rev. A20, 2424.

Bauche-Arnoult, C., Bauche, J. and Klapisch, M. "Variance of the distribution of energy levels and of the transition arrays in atomic spectra II. Configurations with more than two open subshells. 1982 Phys. Rev. A25, 2641.

Bauche, J., Bauche-Arnoult, C., Luc-Koenig, E., Wyart, J.F., and Klapisch, M. "Emissive zones of complex atomic configurations in highly ionised atoms". 1983 Phys. Rev. A28, 829.

Bauche-Arnoult, C., Bauche, J. and Klapisch, M. "Asymmetry of $l^{N+1}-l^N l'$ transition array patterns in ionic spectra." 1984. Phys. Rev. A30, 3026.

Carroll, P.K., Kennedy E.T. and O'Sullivan, G. "New continua for absorption spectroscopy from 40 to 3000 Å". 1978 Opt. Lett, 2, 72.

Carroll, P.K., Kennedy, E.T. and O'Sullivan, G. "Laser-produced continua for absorption spectroscopy in the vacuum ultraviolet and XUV" 1980. Appl Opt. 19, 1454.

Carroll, P.K. and O'Sullivan, G. "Ground-state configurations of ionic species I through XVI for Z=57-74 and the interpretation of 4d-4f emission resonances in laser produced plasmas" 1982 Phys. Rev. A25, 275.

Carroll, P.K., Costello, J.T., Kennedy, E.T. and O'Sullivan, G. "X-UV emission from uranium plasmas; the identification of UXIII and UXV". 1984. J.Phys B:Atom. Molec. Phys. 17, 2169.

Carroll, P.K., Costello, J.T., Kennedy, E.T. and O'Sullivan, G. "X-UV emission from thorium plasmas" 1986. J. Phys B: Atom.Molec. Phys. in press.

Cheng, K.T. and Froese-Fischer, C., "Collapse of the 4f orbital for Xe-like ions" 1983 Phys. Rev. A28, 2811.

Cheng, K.T. and Kim, Y-K., "Excitation energies and oscillator strengths in the silver isoelectronic sequence". 1979 J. Opt. Soc. Am. 69, 125.

Colombant, D. and Tonon, G.F., "X-ray emission in laser produced plasmas". 1973 J. Appl. Phys. 44, 3524.

Condon, E.U. and Shortley, G.H. "The Theory of Atomic Spectra " Cambridge University Press. 1935.

Connerade, J.P. and Mansfield, M.W.D. "Observation of an "atomic plasmon" resonance in the BaI absorption spectrum between 10 and 200Å" 1974. Proc. Royal Soc. Lond. A341,267.

Connerade J.P. "The Non-Rydberg Spectroscopy of Atoms" 1978 Contemp. Phys. 10,415.

Connerade, J.P. and Mansfield, M.W.D. "Term dependent hybridisation of the 5f wave functions of Ba and Ba^{++}." 1982, Phys. Rev. Lett 48, 131.

Cowan, R.D. "The Theory of Atomic Structure and Spectra" 1981 publ. University of Califoria. Press, Berkley.

Croxton, F.E., Cowden, D.I. and Klein, S. 1967. "Applied General Statistics" 3rd ed. publ. Prentice-Hall, Englewood Cliffs.

Edlen, B. "The term analysis of atomic spectra". 1973 Phys. Scr. 7,93.

Finkenthal, M., Lippmann, A.S., Huang, L.K., Yu, T.L., Stratton, B.C., Moos, H.W., Klapisch, M., Mandelbaum, P., Bar-Shalom, A., Hodge, W.L., Phillips, P.E., Price, T.R., Porter, J.C., Richards, B. and Rowan, W.L. 1986a. "The spectrum of highly ionised praeseodymium and dysprosium from the Texas tokamak plasma in the 50-250A range". J. Appl. Phys. (in press).

Finkenthal, M., Mandelbaum, P, Spector, N., Ziegler, A., Klapisch, M. 1986b. Private communication.

Froese-Fischer, C. "A multiconfiguration HF program for atomic calculations". 1978 Comput. Phys Commun. 14, 145.

Galanti, M. and Peacock, N.J. "Quantitative x-ray spectroscopy of the light-absorption region at the surface of laser irradiated polyethylene". 1975. J. Phys. B: Atom. Molec. Phys. 8,2427.

Grant, I.P., McKenzie, B.J. Norrington, P.H., Mayers, D.F. and Pyper, N.C., "An atomic multiconfiguration Dirac-Fock package." 1980. Conput. Phys. Commun. 21, 207.

Hansen J.E. "Hartree-Fock calculations for the $4d^9 5s^2 5p^6 4f$ configuration in LaIV". 1972 J. Phys. B: Atom Molec. Phys. 5, 1096.

Hansen, J.E, Flifet, A.W., and Kelly, H.P. "The position of the $4d^9 4f$ ^1P level relative to the ionisation limit in BaI" 1975 J. Phys. B: Atom. Molec. Phys. 8, L127.

Kaufman, V. and Sugar, J., "Wavelengths, classifications and ionisation energies in the isoelectronic sequences from YbII and YbIII through BiXV and BiXVI" 1976 J. Opt. Soc. Am 66, 1019.

Kaufman, V. and Sugar J. "AgI isoelectronic sequence: wavelengths and energy levels for IVII through LaXI.". 1981. Phys. Scr 24 738.

Klapisch, M., Schwob, J.L., Fraenkel, B.S., and Oreg J., "The 1s-3p K_β like x-ray spectrum of highly ionised iron." 1977 J. Opt. Soc. Am. 67, 148.

Krasnitz, A., M.Sc. thesis, Hebrew University of Jerusalem 1985.

Lucatorto, T.B., McIlrath, T.J., Sugar, J., and Younger, S.M. "Radical redistribution of the 4d oscillator strength observed in the photoabsorption of the Ba, Ba+ and Ba++ sequence." 1981 Phys. Rev. Lett. 47 1124.

Mandelbaum, P., Finkenthal, M., Schwob, J.L., and Klapisch, M., 1986 to be published in Phys. Rev. A.

Martin, W.C., Zalubas, R., and Hagan, L. "Atomic Energy Levels - The Rare Earth Elements". 1978 NSRDS - NBS 60.

Nagel, D.J., Brown, C.M., Peckerar, M.C., Ginter, M.L., Robinson, J.A., McIlrath, T.J., Carroll, P.K., "Repetitively pulsed-plasma soft x-ray source" 1984 Appl. Opt. 24 1428.

O'Sullivan, G. and Carroll, P.K., "4d-4f emission resonances in laser-produced plasmas". 1981 J. Opt. Soc. Am 71, 227.

O'Sullivan, G., Carroll, P.K., McIlrath, T.J. and Ginter, M.L. "Rare-earth plasma light source for VUV applications". 1982 Appl. Opt. 20, 3043.

O'Sullivan G., Roberts J.R., Ott, W.R., Bridges, J.M., Pittman, T.L. and

Ginter, M. "Spectral-irradiance calibration of continuum emitted from rare earth plasmas" 1982 Opt. Lett 7, 31.

O'Sullivan, G., "Charge dependent wavefunction collapse in ionised xenon" 1982. J. Phys. B: Atom. Molec. Phys. 15, L765.

O'Sullivan, G. "The origin of line-free XUV continuum emission from laser -produced plasmas of the elements $62<Z<74$" 1983. J. Phys. B: Atom. Molec. Phys. 16, 3291.

Petersen, H., Radler, K., Sonntag, B. and Haensel, R. "Photoabsorption of atomic Cs in the XUV." 1975 J. Phys. B: Atom. Molec. Phys. 8,

Radtke. E.R. "On the character of the intense 4d-4f resonances in atomic La and Tm". 1979a J. Phys. B.: Atom. Molec. Phys. 12 L71.

Radtke, E.R. "Systematic Comparison between the 4d spectra of lanthanide atoms and solids." 1979 b. J. Phys. B: Atom. Molec. 12. L77.

Reader, J. and Ekberg, J.O. "Resonance lines of CeV and CeVI" 1972 J.Opt. Soc. Am. 62, 464.

Spitzer, L. Jr. 1962 Physics of Fully Ionised Gases (New York: Interscience)

Starace, A.F. "Potential-Barrier Effects in Photoabsorption I. General Theory" Phys Rev B5, 1773.

Sugar, J., "Potential-Barrier Effects in Photoabsorption. II. Interpretation of Photabsorption Resonances in Lanthanide Metals at the 4d-Electron Threshold" 1972 Phys. Rev. B. 5. 1972.

Sugar, J. "Resonance lines in the Ag I and PdI isoelectronic sequences: CsIX through SmXVI and CsX through NdXV" 1977. J. Opt. Soc. Am. 67, 1518.

Sugar, J. and Kaufman, V. "Tokamak-generated tungsten radiation identified in AgI isoelectronic sequence (WXXVIII)" 1980 Phys Rev A21, 2096.

Sugar, J. and Kaufman, V. "Resonance lines in the PdI isoelectronic sequence: IVIII to HoXXII". 1982 Phys. Scr. 26, 419.

Sugar, J. and Kaufman, V. "AgI isoelectronic sequence: wavelengths and energy levels for CeXII through HoXXI and for WxxVIII" 1982 Phys. Scr. 24. 742.

Weisheit, J.C. "Recombination in dense plasmas". 1975, J. Phys. B: Atom Molec. Phys. 8, 2556

Wendin, G. and Starace, A.F. "Perturbation theory in a strong-interaction regime with applications to 4d-subshell spectra of Ba and La." 1978. J. Phys B: Atom. Molec. Phys. 11, 4119.

Younger S.M. "Theoretical line strengths for the $4d^{10}\,{}^1S - 4d^9 4f^1 P$ resonance transitions in the palladium isoelectronic sequence." 1980. Phys. Rev. A22, 2682.

ORDERED MANY-ELECTRON MOTIONS IN ATOMS AND X-RAY LASERS

Charles K. Rhodes

Department of Physics
University of Illinois at Chicago
P.O. Box 4348, Chicago, Illinois 60680

ABSTRACT

Subpicosecond ultraviolet laser technology is enabling the exploration of nonlinear atomic interactions with electric field strengths considerably in excess of an atomic unit. As this regime is approached, experiments studying multiple ionization, photoelectron energy spectra, and harmonically produced radiation all exhibit strong nonlinear coupling. Peak total energy transfer rates on the order of $\sim 2 \times 10^{-4}$ W/atom have been observed at an intensity of $\sim 10^{16}$ W/cm^2, and it is expected that energy transfer rates approaching $\sim 0.1 - 1$ W/atom will occur under more extreme conditions for which the ultraviolet electric field E is significantly greater than e/a_o^2. In this high intensity regime, a wide range of new nonlinear phenomena will be open to study. These will include the possibility of ordered driven motions in atoms, molecules, and plasmas, mechanisms involving collisions, and relativistic processes such as electron-positron pair production. An understanding of these physical interactions may provide a basis for the generation of stimulated emission in the x-ray range.

I. INTRODUCTION

Recently, a new experimental regime, in which the optical field strength E is considerably greater than an atomic unit (e/a_o^2), has become open for systematic study.[1] Indeed, electric field strengths on the order of 100 (e/a_o^2) should be attainable. This extraordinary range of field strengths has become accessible principally because of the availability of an ultraviolet laser technology capable of producing subpicosecond pulses[2,3] with energies approaching the joule level in low divergence beams at high repetition rates. Clearly, a technology of this genre will make possible the creation of physical conditions unachievable with any other known means. It appears possible that an understanding of the physical processes that will occur under these extreme conditions of irradiation will enable the generation of stimulated emission in the x-ray range to be produced by the direct, highly nonlinear coupling of ultraviolet radiation to atoms and molecules.

Figure (1) illustrates the parameters of the physical regime, in terms of pulse intensity (I) and pulse time scale (τ) that characterize our experimental studies. It is seen that the ultraviolet laser technology

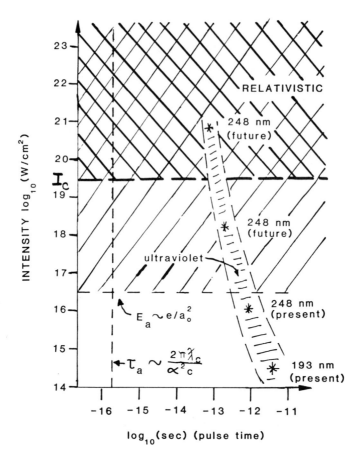

Fig. (1): Conditions of irradiation achievable with high brightness subpicosecond ultraviolet sources. Relativistic motions will be prominent above the Compton intensity I_c. The anticipated range of conditions that can be studied is shown as well as its relation to an atomic field strength (E_a) and an atomic time (τ_a).

will take us a considerable distance into the unexplored area, a zone associated with field strengths far greater than an atomic unit (e/a_0^2) and time scales that are advancing toward an atomic time τ_a. In addition, at the wavelength of 248 nm, intensities above $I_c \sim 5 \times 10^{19}$ W/cm^2 will cause strongly relativistic motions to occur.[4]

It has been conjectured that fundamentally different atomic motions may be driven under such extreme conditions of irradiation. Indeed, it has been suggested[5-8] that ordered many-electron motions of outer-shell electrons could lead to enhanced rates of coupling from the radiation field to an atom. In order to study the conditions necessary for this desired atomic response to occur and to determine the influence of possible damping mechanisms, the properties of ion charge state distributions,[9] electron energy spectra,[10] and harmonic radiation[11] produced by irradiation of atoms with ultraviolet radiation with different pulse lengths, wavelengths, and intensities have been investigated.

II. EXPERIMENTAL RESULTS

Recent experiments[12] have begun to explore the atomic response to intense 248 nm irradiation ($\sim 10^{16}$ W/cm^2) with subpicosecond pulses. Previous work, as cited in References (1), (9), and (10), had been conducted at a few specific wavelengths in the range between 193 nm and 10.6 µm with pulse lengths of $\geq$ 5 ps duration. Since the dynamics of electron ejection from an atom can[13-16] involve a time scale significantly shorter than $\sim$ 5 ps, it is expected[10] that the nonlinear coupling with intense radiation will be significantly modified if sufficiently short pulses in the subpicosecond domain are used.

For these measurements, two ultraviolet laser systems were available. One was a recently developed KrF* (248 nm) laser system[2] which produces pulses having a maximal energy of $\sim$ 23 mJ with a pulse duration of $\sim$ 0.5 ps. The second source used was an ArF* (193 nm) laser system[17] capable of producing a $\sim$ 5 ps pulse with a maximal energy of $\sim$ 40 mJ. The focusing system used in the studies of electron spectra[10,18] produced a maximum intensity of $\sim 10^{16}$ W/cm^2 at 248 nm and of $\sim 10^{15}$ W/cm^2 at 193 nm, respectively, in the experimental volume. The focusing lens was not corrected for spherical aberration and had, therefore, a focal diameter[10] of $\sim$ 20 µm which limited the achievable laser intensity, but made the experiments relatively insensitive to the detailed spatial properties of the laser beam. The apparatus used for the measurement of the ion states[9] and the electron spectra[10,18] have been described previously. Both are time-of-flight type spectrometers; for the electron measurements a magnetic mirror collimated the electrons while the ions were extracted with a static electric field. The harmonic radiation was generated in a medium produced by a pulsed value and subsequently detected with a grazing incidence spectrometer and a multichannel detector.[19]

Overall, the results of the subpicosecond studies at 248 nm (1) provide the first observation in ion spectra of the removal of an inner-electron in a direct multiquantum collision-free interaction, namely, an electron whose principal quantum number (n) is less than that characterizing the outer-most shell of the neutral atom, (2) demonstrate that the pulse width (τ) is an important parameter in the coupling, (3) reveal the characteristics of the spectra of the energetic electrons up to $\sim$ 250 eV formed by the interaction, and (4) exhibit the production of coherent radiation at 22.6 nm, the eleventh harmonic of 248 nm and the shortest wavelength produced by that means.

A. Ion Charge State Spectra

The basic process under study is

$$N\gamma + X \rightarrow X^{q+} + qe^-. \tag{1}$$

In comparison with the data obtained with the $\sim$ 5 ps, (193 nm) source,[9] marked differences are seen in the ion charge state spectra obtained with the subpicosecond 248 nm irradiation.

Figure (2) illustrates the results for krypton at an intensity of $\sim 10^{16}$ W/cm^2. In addition to the observation of Kr^{6+} which was seen in the earlier studies[9] with 5 ps 193 nm radiation, Kr^{7+} and Kr^{8+} are now visible, indicating the complete removal of both the 4p^6 and 4s^2 subshells.

Considerably different behavior was also observed in xenon. The experiments[9] conducted at 193 nm with pulses having a duration of $\sim$ 5 ps and an intensity of $\sim 10^{16}$ W/cm^2, gave explicit evidence for the production of Xe^{8+}, a feature indicating complete removal of the outer-shell of the atom comprised of both the 5s and 5p subshells. A careful search failed to reveal the presence of Xe^{9+}, a species which would involve the additional

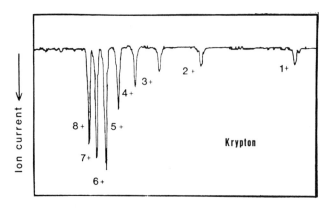

Fig. (2): Time-of-flight ion spectrum of krypton irradiated at $\sim 10^{16}$ W/cm^2 (20 cm lens) at 248 nm with sub-picosecond pulses. The gas pressure for this trace was 2×10^{-5} Torr, a value sufficiently high for detector saturation to cause some distortion in the recording of the lower charge states.

ionization of an electron from the 4d atomic inner-shell. This charge state is now seen and Figure (3) illustrates the detail of the Xe^{9+} signal which exhibits the expected isotopic signature,[9] an important feature confirming its identification. The presence of Xe^{9+} indicates ionization of the 4d-shell and a minimum total energy transfer[20] of ~ 630 eV. In addition, the abundances of charge states with $q > 5$ are substantially augmented. Of course, since the pulse width at 248 nm is considerably shorter, the energy transfer <u>rate</u> has sharply increased, in this case by a factor of ~ 20. The xenon ion spectrum gives an average energy[20] transfer of ~ 130 eV, a value corresponding to an effective average cross section for the nonlinear interaction on the order of $\langle \sigma_{NY} \rangle_{av} \simeq 4 \times 10^{-21}$ cm^2. The observation of Xe^{9+} implies the presence of a peak energy transfer rate on the order of $\sim 2 \times 10^{-4}$ W/atom.

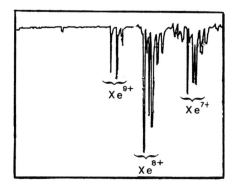

Fig. (3): Region of the ionic time-of-flight signal recorded at an intensity of $\sim 10^{16}$ W/cm^2 in xenon with the 248 nm (450 $\pm$ 150 fs) laser pulse which specifically illustrates the presence of Xe^{9+}. The peaks in the signals correspond to the distribution of known xenon isotopes.

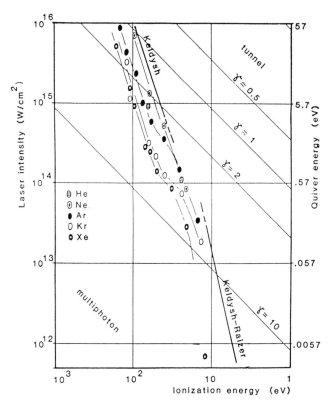

Fig. (4): Threshold laser intensity for ion charge state production with a 248 nm ∿ 500 fs laser pulse. A transition probability of T ≃ 10^{-3} has been assumed for the calculation (estimated detection limit). The quiver energy corresponding to a free electron in the radiation field is also shown on the ordinate.

The ion time-of-flight technique has been used to measure the threshold laser intensities for all charge states of the rare gases which appear below 10^{16} W/cm^2 with subpicosecond irradiation at 248 nm. The experimental values are plotted in Figure (4) as functions of the ionization energies for each charge state. They are compared to calculated threshold laser intensities for a transition probability of about 10^{-3}, which corresponds to the estimated detection limit of the apparatus, using Keldysh's formula[21] (including coulomb correction) in the tunneling approximation ($\gamma \ll 1$) and the Keldysh-Raizer formula[22] in the multiphoton approximation ($\gamma \gg 1$). The observed thresholds for neon, which range from ∿ 8 × 10^{13} W/cm^2 (Ne$^+$) to ∿ 7 × 10^{15} W/cm^2 (Ne^{4+}), agree remarkably well with the calculations, whereas the corresponding thresholds are consistently lower for heavier ions, apparently scaling with the atomic number and the complexity of the atom, a possible hint to many-electron effects.

A previously observed[9] property of the ion spectra of the rare earths is now recalled. On the basis of the patterns of ionization observed as a function of atomic number, the hypothesis was advanced that it was mainly the number of electrons in the outer subshells that governed the coupling. This picture fits with the expected effects of electronic screening[16,23] of the external field. A measurement of the response of elements in the lanthanide region, with the use of a method involving laser-induced evaporation to provide the material, enabled this view to be checked. As one moves from La (Z = 57) to Yb (Z = 70) in the lanthanide sequence, aside from slight rearrangements involving the 5d shell for Gd (Z = 64), 4f electrons are being added to interior regions of the atoms. The results obtained for $_{63}$Eu ($4f^76s^2$) and $_{70}$Yb ($4f^{14}6s^2$), which differ by seven 4f electrons, indicated that these inner electrons play a small role in the <u>direct</u> radiative coupling, a fact that is in rapport with the observed[9] dependence on the outer-shell structure and the expectations due to screening. This feature of the coupling could lead to some unusual characteristics of ion formation in the lanthanide and actinide regions, an aspect discussed below in Section III.A.

B. <u>Photoelectron Energy Spectra</u>

Photoelectron energy spectra[10,18] for process (1), recorded under collision-free conditions, provide information complementing that given by the measurement of ion charge state distributions. Commonly, the observed electron spectra consist of a characteristic pattern of interwoven above threshold ionization (ATI) ladder line series. As an example, the electron time-of-flight spectra from argon irradiated with a 248 nm, ∿ 500 fs laser pulse, at different peak intensities, are shown in Figure (5). At laser intensities too low for the production of higher charge states (e.g. ∿ 10^{14} W/cm^2), the principal ATI-ladders leading to the production of Ar$^+$ in the $^2P^o_{1/2}$ and $^2P^o_{3/2}$ ground multiplet states (which are resolved in the spectra at higher resolution) are readily observed up to the fourth ATI order. At higher laser intensities, new lines appear at ∿ 3.5 eV and ∿ 4.5 eV together with corresponding higher order features which can be related to the formation of multiply ionized species. However, in contrast to the results[10] with

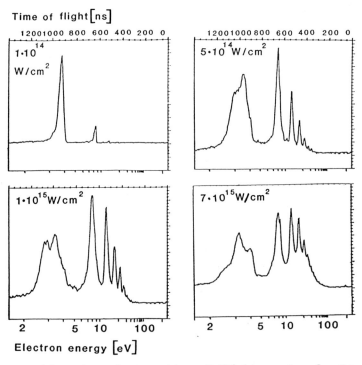

Fig. (5): Photoelectron time-of-flight spectra for Ar produced at different peak intensities with a 248 nm, ∿ 500 fs laser pulse.

the 193 nm, ∼ 5 ps pulse, the ATI-ladders associated with the higher ion charge states seem to be shifted in energy up to a value of about 1.5 eV, a finding which is confirmed in comparable spectra from helium, neon, and xenon at 248 nm. Again, similar to observations[10] at 193 nm, the low energy and shifted five photon line for Ne at 2 eV and the three photon absorption line for krypton at 0.97 eV are suppressed[24] in strength relative to the next higher order lines. This phenomenon has been attributed to the influence of the ponderomotive potential.[5,25-31] Another general feature of all electron spectra at 248 nm is the appearance of relatively high energy electrons with energies up to a maximum of ∼ 250 eV in neon, ∼ 200 eV for xenon, and ∼ 120 eV for argon. These were observed in the laser intensity range spanning from 10^{15} W/cm^2 to 10^{16} W/cm^2.

C. Generation of Harmonic Radiation

The nonlinear interaction, in addition to producing ions and electrons, leads to the generation of harmonic radiation according to the parametric process

$$n\gamma(\omega) + X \rightarrow \gamma(n\omega) + X \ . \tag{2}$$

In the electric dipole approximation in a spatially homogeneous medium, it is well known that n is restricted to odd integers. Previously, with radiation at 248 nm in pulses of ∼ 15 ps duration, 35.5 nm radiation was produced as the seventh harmonic.[32] In our experiment, conducted at a maximum intensity of ∼ 10^{16} W/cm^2, the eleventh harmonic at 22.6 nm, the shortest wavelength produced by such means was readily observed in neon, along with the lower harmonics at 27.6 nm, 35.5 nm, and 49.7 nm in neon and helium as shown in Figure (6). It was also found, as illustrated in the inset of Figure (6), that the signal strength fell rather slowly as the harmonic order increased. In our experiments with the 248 nm subpicosecond pulses, the intensity of the ninth harmonic was less than a factor of thirty below that of the fifth, a finding that contrasts with the earlier work[32] which showed a decrement of several hundred between adjacent orders.

Naturally, the potential role of multiply excited states is of considerable interest. Interestingly, in our preliminary measurements in neon, the eleventh harmonic (22.6 nm) of 248 nm was produced with what appears to be anomalous strength. In addition, it is known that a certain neon doubly

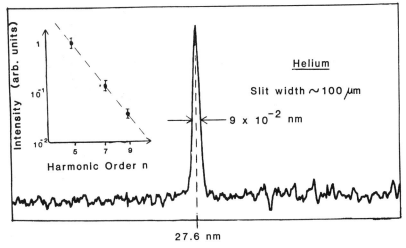

Fig. (6): Ninth harmonic (27.6 nm) signal produced by 248 nm in helium. The inset shows the relative strengths of the fifth, seventh, and ninth harmonics.

excited state [$2p^4(^3P) 3d(^2P) 4p^1P^o_1$, 442,700±100 cm^{-1}], connected with a dipole allowed transition to the neutral neon ground state, exists[33] exactly at the excitation energy (~ 442,800±500 cm^{-1}) corresponding to eleven 248 nm quanta. Such a state could produce a resonant enhancement in the generation of 22.6 nm radiation. This spectral region in neon has been recently examined in the study of resonant enhancement of satellites.[34]

Finally, it is noted for comparison that extremely high harmonic production (n ≤ 46) has been observed[35] in studies of plasmas irradiated at 10.6 μm with intensities greater than 10^{15} W/cm^2. A parameter found relevant in that case was the product of the intensity of irradiation I(W/cm^2) and the square of the wavelength λ(μm). High order harmonic generation occurred for $I\lambda^2 \gtrsim 10^{16}$ (W/cm^2)(μm)^2, a value not far from that applying in the experiments involving the production of the extreme ultraviolet radiation described above.

D. Experimental Summary

Therefore, all three classes of experiments, involving (a) the measurement of ion charge state distributions, (b) photoelectron energy spectra, and (c) harmonically generated radiation, exhibit substantially stronger nonlinear coupling in the subpicosecond regime. A common physical basis for these phenomena is strongly indicated by the confluence of these findings.

III. DISCUSSION

In this section we discuss several speculative aspects of atomic and molecular behavior produced by the influence of strong oscillatory fields.

A. Ion Production in Rare Earth Atoms

Ion production in the rare earths, and also the actinide sequence, may involve some anomalous characteristics. As presented in Section II.A above, the direct radiative coupling of rare earth atoms was found experimentally[9] to be insensitive to the electrons occupying the 4f-shell. As an example of a typical rare earth system, Figure (7) gives an approximate representation[36]

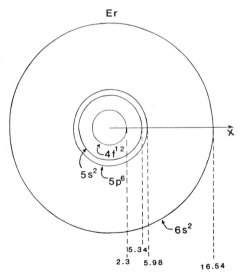

Fig. (7): Plot of the positions of the radial maxima of the charge distributions for the $4f^{12}$, $5s^2$, $5p^6$, and $6s^2$ shells of erbium. The radial scale parameter X corresponds to the data given in Ref. (36).

Table I: Total One-Electron Subshell Binding Energies of Er. Energies are in atomic units (27.2 eV). Data are taken from Ref. (37) with permission.

ERBIUM - 4F12
($Z = 68$, $A = 166$)

	E(TOT)
4S1/2	-16.809636
4P1/2	-13.622807
4P3/2	-11.989120
4D3/2	-6.807922
4D5/2	-6.419415
4F5/2	-0.216231
4F7/2	-0.152437
5S1/2	-2.203297
5P1/2	-1.255914
5P3/2	-1.050573
6S1/2	-0.176476

of radial distributions of orbital electron charge for Er($Z = 68$) for all shells outside the 4d orbital. The corresponding relaxed-orbital electron binding energies[37] of these shells are given in Table I. The ground state configuration[38] of Er is $(Pd)5s^2 5p^6 4f^{12} 6s^2\ ^3H_6$.

From these data,[36,37] we see that the 4f electrons are both weakly bound and situated in the interior of the atom. As a consequence of this structure and the shielding[16,23] provided by the outer-electrons, if in the process of multiple ionization by a nonlinear process, the electrons are removed progressively in the order of decreasing <u>radius</u>, the ionization will not occur in order of monotonically increasing binding energy. For example, Table II illustrates the respective ionization energies[20] that would apply for the energetically sequential removal of atomic electrons from Er.

Table II: Calculated Ionization Potentials For Multiply Charged Er Ions. Data taken from Ref. (20) and reproduced with permission.

ER   Z = 68

ION	CONFIG	IP, EV
NEUTRAL	(6S+) 2	5.227E 00
1+	(6S+) 1	1.152E 01
2+	(4F+) 6	2.503E 01
3+	(4F+) 5	5.889E 01
4+	(5P+) 4	7.876E 01
5+	(5P+) 3	9.629E 01
6+	(5P+) 2	1.138E 02
7+	(5P+) 1	1.314E 02
8+	(5P-) 2	1.576E 02
9+	(5P-) 1	1.765E 02
10+	(4F+) 4	2.007E 02
11+	(4F+) 3	2.364E 02
12+	(4F+) 2	2.684E 02
13+	(5S+) 2	2.900E 02
14+	(5S+) 1	3.112E 02
15+	(4F+) 1	3.448E 02

According to Table II, the sequential removal of electrons for the production of Er^{4+} corresponding to the minimum energy for ionization would yield the ground state ionic configuration

$$Er^{4+} (Pd) 5s^2 5p^6 4f^{10}. \qquad (3)$$

However, removal of electrons in order of decreasing radius to produce the same Er^{4+} charge state, if accomplished suddenly without internal rearrangements, as readily seen from Figure (7), would produce the ionic configuration

$$Er^{4+} (Pd) 5s^2 5p^4 4f^{12}. \qquad (4)$$

Configuration (4) is transformed into configuration (3) by the interchange $4f^2 \to 5p^2$. If we use the calculated orbital energies pertaining to the neutral atom contained in Table I as a very coarse estimate of the excitation energy of configuration (4) with respect to the Er^{4+} ground state, a value of ~ 49 eV is obtained. Since this energy appears to be considerably below that necessary for autoionization to produce Er^{5+}, we would expect the excited Er^{4+} configuration (4) to decay, depending upon the nature of its local environment, by either radiative or collisional processes. The electric dipole two electron transition

$$5p^4 4f^{12} \to 5p^5 5d 4f^{10} + \gamma, \qquad (5)$$

which is allowed by configuration interaction,[39] is estimated from the data in Table I and known[38] $4f \to 5d$ transitions in Er^{3+} to have a quantum energy of ~ 20 eV. The $4f \to 5p$ radiative channel could also occur[40-42] by M1 and E2 amplitudes in transitions with an energy estimated at ~ 24 eV. The detection of these radiative emissions could provide interesting information on the mechanism of electron removal in nonlinear processes of multiple ionization.

B. Many-Electron Motions

1. Excitation. Theoretical estimates[7,8] have shown that driven ordered many-electron motions of outer-shell electrons could lead to enhanced rates of coupling from the radiation field to an atom, and possibly, to inner-shell excitation. Collective responses of atomic shells have been discussed in relation to the mechanism of single-photon photoionization.[43-52] At this point, we simply indicate the possibility of a radiatively _driven_ nonlinear analog of this basic electronic mechanism. Of course, such a picture, which would be represented by a multiply excited configuration, could only be valid if the damping rate, presumably by electron emission, is sufficiently low.

A mechanism which may enhance the probability for this kind of _direct_ intra-atomic process in high intensity fields is suggested by studies[53-59] of photoelectron spectra generated by multiquantum ionization. These experiments and analyses have shown the importance of the ponderomotive potential in the suppression of photoelectron channels for intensities above certain specific values. The ponderomotive potential may, therefore, through the elevation of the effective ionization potential, tend to retard the total electron emission rate by reducing the number of channels available for the escape of electrons from the vicinity of the atomic potential.

In order to effectively excite multi-electron motions of this type, it is generally required that energy be invested in the electronic motion at a rate greater than that lost by damping of those motions by electron emission. The hypothesis has been advanced[10] that a necessary, although possibly not sufficient,[60] condition is that the ponderomotive potential increases by an amount equal to or exceeding the quantum energy ($\hbar\omega$) in the time scale

τ_e characterizing electron emission. In this situation, the atomic electrons experience a sequence of closing channels for emission at successively higher energies as radiation is absorbed by the atom. It follows, for a triangular pulse with maximum intensity (I_o) that rises linearly in time over a period (τ), that the requirement can be written[10] as

$$\frac{I_o}{\tau} \geq \frac{m\omega^3}{2\pi\alpha\tau_e} \tag{6}$$

in which α denotes the fine structure constant, m represents the electron mass, and ω is the angular frequency. For $\tau_e \sim 10^{-15}$ s, a value supported by experimental studies of electron spectra[10,18] and discussed in a related context below, it is found that so far no experiments have been conducted in the regime specified by Eq. (1). At a wavelength of 248 nm, Eq. (6) would be satisfied for $I_o \simeq 10^{18}$ W/cm^2 and $\tau \simeq 1.2$ ps, values certainly within the achievable zone shown in Figure (1).

2. <u>Simple Model</u>. In connection with the generation of coherent x-ray quanta from highly excited atomic states produced by multiphoton processes, it is natural to consider the following questions. Do particular classes of excited atomic states possess properties that cause them to be favored configurations for efficient radiation in the x-ray region? And can we form an orbital representation of such states? Finally, what requirements must be satisfied in order to produce such states at the densities necessary for stimulated emission, and how do these conditions for excitation depend upon the pulse width τ? Simply stated, is there an optimal class of atomic states and what are the collateral implications concerning the pulse width τ?

Ideally, excited atomic configurations are desired which have (1) a resistance against autoionization, (2) a sufficiently strong radiative rate so that the system has a high radiative yield, and (3) a lifetime matched to the time scale of excitation τ. These features would favor efficient energy flow in the production of the x-ray quanta. In order to have a high radiative yield, the radiative width Γ_r should be comparable to or greater than the width for autoionization Γ_a, namely, the condition characteristic of the K-shell fluorescence yield of heavy atoms.[61,62] However, K-shell energies for high Z materials are on the scale of 100 KeV, while in the present case of interest we desire states with $\Gamma_r/\Gamma_a > 1$ that radiate in the 0.1 - 10 KeV range. K-shell fluorescence yields in that lower energy range typically are small, since that region corresponds to relatively low Z systems.[61,63] From this we infer that neutral materials exhibiting a high radiative yield in the 0.1 - 10 KeV range will presumably involve unusual excited state configurations.

Certain multiply excited atoms in which the electrons are coherently excited, that is, occupy excited orbits with definite phase relationships[64-68] may possess these desired properties. Such states could be envisaged as planetary atoms[69] with appropriately phased and spatially oriented electronic motions whose nature favors the radiative rate and retards the autoionization rate. Clearly, it seems natural to excite such states with coherent energy.

Although very little is known about such states and their properties, so that this discussion must be regarded as extremely speculative, appeal to a simple classical picture suggests possible cases, provides a hint at the Z-scaling, and places emphasis on the desirability of producing pulses of radiation for excitation of subpicosecond duration, since pulses of this kind appear most applicable to the excitation of such unorthodox configurations.

A few general comments concerning multiply excited atomic electronic states, which comprise a large and extremely interesting class of highly excited systems, are now made. The simplest prototype of these multiply excited configurations is doubly excited helium, which has an observed sequence of double excitations[70-72] 2snℓ beginning at approximately 60 eV above the (1s)^2 ground state. More complex multiply excited systems, known as planetary atoms,[69] have also been considered, as noted above.

The possibility of direct radiative excitation of multiply excited states was initially raised by Heisenberg[73] and Condon.[74] Subsequently, a more detailed analysis, including the derivation of selection rules, was performed by Goudsmit and Gropper.[39] Physically, as mentioned above, violation of the normal single-electron rate for transitions arises from configuration interaction which is generated, for example, by the electrostatic interaction of the atomic electrons. This term in the atomic Hamiltonian, represented by the double sum $e^2 \sum_{i<j} (r_{ij})^{-1}$ over all pairs of electrons, causes mixing of the initial and/or final states with other configurations of the same symmetry. For example, the ground state of mercury would not be described solely by the $5d^{10}6s^2$ state, but also includes an admixture of the $5d^{10}6p^2$ level, so that the corrected wavefunction reads

$$\Psi_o = \alpha(5d^{10}6s^2) + \beta(5d^{10}6p^2) , \tag{7}$$

with appropriate amplitudes α and β. Effects of this general nature have been found to play a significant role in a variety of systems, including Zn, Cd, Hg, Ca, Sr, Ba, Sm, Eu, Yb, Pb, and Bi, in studies of both photoabsorption and photoelectron spectra.[75-78]

These intra-atomic electronic interactions, which give rise to a nonvanishing probability of multiple excitation, result in a variety of atomic rearrangements. The possibilities for the atomic response are described variously[79] as "shake-up," "conjugate shake-up," and "shake-off" processes. Auger processes, of course, can contribute to double ionization, as Holland and Codling pointed out[80] in the case of Pb, and Chang and Poe[81] in the case of Ne. Further notice should be taken of emission originating from two-electron one-photon transitions into doubly ionized K-shells[82] as well as recent work dealing with autoionizing levels,[83,84] particularly those of sodium,[85] lithium,[86] barium,[87] and strontium.[88] A common feature of these studies is the general problem of a discrete state coupled to an adjacent continuum.[89]

Two-electron processes have been intensively investigated. Double ionization measurements are summed up by Holland et al.[90] for photon energies in the 40- to 300-eV region. Whereas double ionization of He, Ne, and Ar in this region is due to pair excitation of electrons in the valence shell, that of Kr and Xe is mainly a result of Auger processes following one-electron ionization of the core. Shake-up experiments in the rare gases have also been performed. Spears et al.[91] have used x-ray sources to eject an inner-shell electron and observe the satellite lines in the photoelectron spectrum, which are due to valence-shell excitation. Finally, two-electron excitation in the photoabsorption of the rare gases was observed by Codling, Madden, and Ederer[33,70,92-95] in the 8- to 60-nm region.

Imagine an atomic excited state of the type illustrated in Figure (8). Although the figure is drawn for Z = 3, denoting the number of excited electrons, the extension to an arbitrary number of electrons is clear. For this discussion, we consider a spherical system with the electrons traveling on great circle routes through P_1 and P_2. If we assume that the electrons all oscillate at the same frequency and are phased so that they all reside at P_1 at a given time, a symmetric oscillation of charge in the direction of the x-axis, and evenly distributed about that direction, can be envisaged.

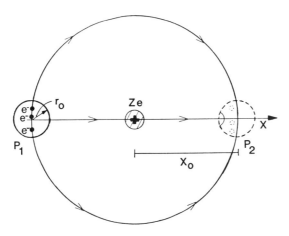

Fig. (8): Schematic representation of an atom in a coherently multiply excited configuration. The picture corresponds to Z = 3, the number of excited electrons. Consider a spherical system with the electrons traveling on great circle routes through points P_1 and P_2.

With this orbital picture, a large oscillating dipole, with magnitude $\mu \sim Z e X_0$ validly viewed[96] as a classical phenomenon, is established along the x-axis. Furthermore, for a symmetric distribution of great circle orbits, the particles are at large relative separations, approximately X_0, during the major fraction of the orbital period, although they can be regarded as confined temporarily to a somewhat smaller relative distance, shown as r_0, at the pole points P_1 and P_2.

The motion described above has two immediate consequences. The large dipole naturally leads to an enhanced radiation rate, since it acts like an atomic plasmon[97] which bears a resemblance to the nuclear giant dipole resonances.[98,99] The choice of orbital paths, which provides for substantial inter-particle separation during most of the orbital period, is expected to result in a tendency to suppress the autoionization rate. Therefore, states of this type, if they can be excited, are expected to exhibit favorable scaling properties for the ratio of Γ_r/Γ_a. We note that in the presence of configuration interaction many-electron radiative transitions become allowed, although with a rate that depends upon the strength of the configuration interaction. The special case, limiting the number of electrons simultaneously involved in a transition to three,[39] only applies to estimates taken in first order perturbation theory. It is important to realize, that although the ratio Γ_r/Γ_a can be optimized, it cannot be raised to indefinitely high values, since <u>both</u> Γ_r and Γ_a depend upon the presence of configuration interaction. Therefore, the presence of one physical mechanism allowing radiation implies the presence of the other process constituting a loss channel. In an intense radiative field, photoionization is also a loss mechanism, but this can be reduced by using appropriate configurations and frequencies.[83] Interestingly, the most stable orbits from the standpoint of autoionization alone are planar configurations of the excited electrons resembling the rings of Saturn. This problem was originally treated by Maxwell[100] and more recent studies involving two electrons in Ba have also been made.[101]

Although more complex theoretical procedures are available,[102] we now derive an approximate scaling relationship for Γ_r/Γ_a based upon the simple model described above and pictured in Figure (8). For a uniform distribution of Z electrons confined to a spherical volume of scale length r, the coulomb energy E can be estimated as

$$E_c \simeq \frac{Z(Z-1)e^2}{2r} \qquad (8)$$

by simply adding the electrostatic energies pairwise and assuming a mean particle separation equal to the scale length associated with the spherical volume. This result has essentially the same form as the correlation energy associated with collective motions of an electron gas.[103] An estimate of the magnitude of the force F felt by a single electron in this ensemble can then be represented by

$$F \simeq -\frac{1}{Z}\frac{dE_c}{dr} = \frac{(Z-1)e^2}{2r^2} \qquad (9)$$

which leads to an acceleration a_e given by

$$a_e \simeq \frac{(Z-1)e^2}{2mr^2} . \qquad (10)$$

If we assume a constant acceleration and wish to appraise the time τ for an electron to escape from the ensemble, we can write

$$\frac{1}{2} a_e \tau^2 \simeq r \qquad (11)$$

giving

$$\tau \simeq \sqrt{\frac{2r}{a_e}} = \sqrt{\frac{4mr^3}{(Z-1)e^2}} . \qquad (12)$$

Finally, if we take this time as a measure of the interval required for autoionization to occur, and assume that $r \sim X_o$ for the major fraction of the orbital period, we can represent the autoionizing width Γ_a as

$$\Gamma_a = \frac{1}{\tau} \simeq \sqrt{\frac{(Z-1)e^2}{4mX_o^3}} . \qquad (13)$$

We now appraise the corresponding value for the radiative width Γ_r. The dipole matrix element μ associated with the motion shown in Figure (8) is given approximately by

$$\mu \sim ZeX_o . \qquad (14)$$

If customary values of dipole matrix elements are taken as $\mu_o \sim ea_o$, with a_o denoting the Bohr radius, the conventionally derived radiative rate $1/\tau_o$ is enhanced by a factor γ of approximately

$$\gamma \sim \left(\frac{ZX_o}{a_o}\right)^2 \qquad (15)$$

so that the width Γ_r can be written as

$$\Gamma_r = \frac{1}{\tau_r} \simeq \frac{1}{\tau_o} \left(\frac{ZX_o}{a_o}\right)^2. \tag{16}$$

Combining expressions (13) and (16), and assuming Z >> 1, we obtain for the ratio

$$\rho \equiv \frac{\Gamma_r}{\Gamma_a} \simeq \left(\frac{2\sqrt{m}}{ea_o^2\tau_o}\right) Z^{3/2} X_o^{7/2}. \tag{17}$$

If $\tau_o \simeq 10^{-12}$ sec, a value in the range corresponding to an atomic transition with a quantum energy of a few hundred electron volts, $X_o = 5 \times 10^{-8}$ cm, and Z = 5, expression (17) evaluates to $\rho \simeq 1.4$. The radiative component dominates.

An excited atom behaving in the predicted fashion would exhibit the interesting property of having as the favored response a physical process which is essentially an electronic implosion accompanied with the emission of an x-ray quantum rather than the customary mechanism of de-excitation involving the emission of electrons. The super-excited atom "explodes" by emitting radiation in preference to particles. Furthermore, since equation (17) clearly favors the simultaneous excitation of several electrons in systems of relatively large scale size, heavy atoms are strongly suggested. We recall[9,104-106] that it is just this class of atomic materials which exhibits the anomalously strong coupling previously discussed. An example of a candidate material satisfying the conditions described above is U^{13+} with coherent excitation of electrons from the 5d-shell.[20] In this case, we note the observation of U^{10+} which was reported earlier[9,104] at the relatively modest intensity of $\sim 10^{14}$ W/cm^2. Naturally, there is a considerable range of possible alternatives.

Finally, we examine the pulse width τ necessary to couple with reasonable efficiency to such excited states. This is simply given by evaluating expression (13), the relation which gives a measure of the relevant loss rate. For the range of parameters being considered, it is found that 10^{-15} s $\leq \tau \leq 10^{-14}$ s, a value not far from the value of τ_e used in Eq. (6). Clearly, pulses in the femtosecond regime indicated in Figure (1) are near the range desirable for excitation of levels of this genre.

The speculative nature of this discussion should be emphasized. Its purpose is simply to indicate that certain classes of excited states may exist which have anomalous radiative properties and that these configurations could possibly be excited by high order nonlinear mechanisms.

C. Radiatively Excited Collisional Processes

Radiatively driven collisional processes may represent an important mode of absorption under appropriate circumstances. In addition to a possible intra-atomic process arising from driven many-electron motions discussed in Section III.B.1, a corresponding inter-atomic coupling may occur when the intensity (I) and medium density (ρ) are sufficiently high.[11] In the regime for which the electric field is greater than an atomic unit (e/a_o^2), nominally $I \gtrsim 10^{17}$ W/cm^2, the more loosely bound outer ionic electrons can be approximately modeled as free particles. Therefore, for those electrons, we can describe their motion as that of free electrons accelerating in intense coherent fields.[107,108] At $I \simeq 10^{17}$ W/cm^2 with λ = 248 nm, the maximum excursion[109] of an electron is $x_o \simeq 2.5$ nm, a parameter that scales nonrelativistically as

$$x_o \simeq 1.3 \times 10^{-20} \lambda^2 I^{1/2} \tag{18}$$

with x_0 in (cm), λ in (nm), and I in (W/cm^2). With motions of this type, appreciable inter-atomic coupling could be expected for densities greater than a critical value of $\rho_c \sim x_0^{-3}$, a relationship which yields a scaling of

$$\rho_c \simeq 4.6 \times 10^{59} \lambda^{-6} I^{-3/2} \qquad (19)$$

with ρ_c in (cm^{-3}), λ in (nm), and I in (W/cm^2).

Figure (9) illustrates the boundary separating the physical regimes in which inter-atomic and intra-atomic processes are expected to occur in the coupling. The Compton intensity (I_c) for 248 nm radiation is shown, the value for which the electrons acquire strongly relativistic velocities.[4] The absorption rate has a different density dependence for the intra-atomic and inter-atomic processes. The former depends linearly upon the density while the latter, on account of the collision induced nature of the interaction, would be expected to scale quadratically in the density. The quadratic density dependence is also a property of inverse bremsstrahlung, a process which has similarities to the inter-atomic mechanism under discussion, but differs in the intensity dependence.[110,111] In comparison to normal plasma processes, it should be noted that with 248 nm radiation at an intensity of $\sim 10^{18}$ W/cm^2, $\rho_c \simeq 1.8 \times 10^{18}$ cm^{-3}, a value nearly a full four orders of magnitude below the critical plasma density[109,112] $n_c \sim 1.6 \times 10^{22}$ cm^{-3} for that wavelength. Also, with the strong λ^{-6} scaling appearing in Eq. (9), if $\lambda \simeq 100$ nm at I $\sim 10^{18}$ W/cm^2, the critical density is substantially increased to $\rho_c \sim 4.2 \times 10^{20}$ cm^{-3}, a density of approximately 16 amagat. Furthermore, if we assume an average ionization per atom of Z_{av}, the electron density $Z_{av}\rho_c$ will equal the critical density[110] n_c when,

$$\frac{\lambda^4 I^{3/2}}{Z_{av}} \simeq 4.6 \times 10^{32}. \qquad (20)$$

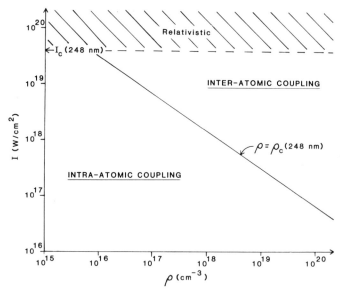

Fig. (9): Diagram showing the regions of intra-atomic and inter-atomic coupling as a function of laser intensity I and atomic density ρ. The Compton intensity I_c is indicated, above which strong relativistic motion occurs. $\rho = \rho_c$ corresponds to Eq. (19). The intensity corresponding to an electric field $E = e/a_0^2$ is also shown.

We note that a problem rather similar to that discussed above has been examined by Pert and coworkers.[113]

D. Molecular Coulomb Explosion

Sufficiently rapid processes of molecular ionization may produce anomalous production of ionic fragments, since inter-atomic couplings, as discussed above, could occur between neighboring atoms in molecules, independent of the medium density, provided that the radiative field is applied in a sufficiently short time. In order to estimate the conditions necessary for this phenomenon to occur, we examine the time τ_1 for two atoms, with a reduced mass of M initially situated at a typical bond distance of $\sim 2\, a_o$ and ionized suddenly by the radiation to charge states Z_1 and Z_2, to undergo a coulomb explosion[114,115] and develop a separation of x_o, the parameter given in Eq. (18). For $x_o/a_o \gg 1$, it can be shown that

$$\tau_1 \simeq \frac{\lambdabar_c}{c} \frac{1}{\alpha^2} \sqrt{\frac{M}{m_e} \frac{1}{Z_1 Z_2}} \left[\frac{x_o}{a_o} + \ln \frac{2 x_o}{a_o} \right] \tag{21}$$

in which α is the fine structure constant, m_e the electron mass, $\lambdabar_c$ the electron Compton wavelength, and c the speed of light. At an intensity of $\sim 2.5 \times 10^{18}$ W/cm² for 248 nm, with M equal to twenty atomic units, $Z_1 = 10$, and $Z_2 = 20$, the time scale evaluates as $\tau_1 \simeq 100$ fs. These values, $I \simeq 2.5 \times 10^{18}$ W/cm² and $\tau_1 \simeq 100$ fs, are well within the experimental capability represented in Fig. (1). Presumably, information on electron ionization rates, a critical question raised in Section III.B.1, could be obtained through study of the energy widths of the ionic spectra produced. Nearly spherical molecular systems composed of a central heavy atom surrounded by somewhat lighter constituents would be promising candidates to explore the possible presence of these effects. As an example, the class of hexafluorides, which includes MoF_6, and UF_6, has exactly this property.

E. Electron Positron Pair Production

At intensities significantly above the Compton intensity I_c shown in both Figure (1) and Figure (9), the electronic motions in a plasma will become highly relativistic[4] and processes involving the production of electron-positron pairs will occur.[7,116] If a fully ionized plasma is assumed with ions of charge Z, the relativistic electrons will lose energy mainly by two processes, Bremsstrahlung and pair production. The lowest order diagrams for these two processes are illustrated in Figure (10a) and Figure (10b), respectively. In the extreme relativistic case ($\gamma \gg 1$), the Bremsstrahlung cross section[117] is given by

$$\sigma_b \simeq 4 \alpha^3 \lambdabar_c^2 \, Z^2 \ln(2\gamma) \tag{22}$$

and the pair production cross section is given by

$$\sigma_p \simeq \left(\frac{28}{27}\right)\left(\frac{\alpha}{\pi}\right) \alpha^3 \lambdabar_c^2 \, Z^2 (\ln\gamma)^3 \tag{23}$$

in which α is the fine structure constant, $\lambdabar_c$ is the Compton wavelength, and γ is the customary Lorentz factor. For $\gamma = 10$, the ratio

$$\frac{\sigma_p}{\sigma_b} \simeq \left(\frac{28\alpha}{108\pi}\right) \frac{(\ln\gamma)^3}{\ln(2\gamma)} \simeq 2.4 \times 10^{-3}. \tag{24}$$

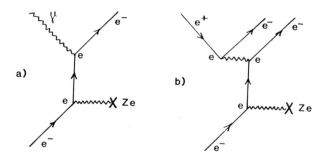

Fig. (10): Lowest order diagrams for (a) Bremsstrahlung and (b) electron-positron pair production. By inspection from the diagrams, the Bremsstrahlung cross section $\sigma_b \sim Z^2 \alpha^3$ and the pair production cross section $\sigma_p \sim Z^2 \alpha^4$ so that $\sigma_p/\sigma_b \sim O(\alpha)$.

Since it should be possible to arrange plasma conditions so that a fraction of at least $\sim 10^{-2}$ of the incident radiative energy is channeled into energetic Bremsstrahlung,[118,119] the pair production would then correspond to $\sim 10^{-5}$ of the total. Therefore, for an incident energy of ~ 1 J, the number of pairs N_p produced would be expected to be in the range $10^6 \leq N_p \leq 10^7$, a value that should be readily observable with standard coincidence instrumentation.[120]

IV. CONCLUSIONS

The availability of extraordinarily bright femtosecond ultraviolet sources is enabling the study of a wide range of new nonlinear phenomena. These will include the possibility of ordered driven motions in atoms, molecules, and plasmas as well as processes involving collisions, even those of relativistic electrons such as electron-positron pair production. Although the exploration of these phenomena is just beginning at relatively low intensities, on the scale of $\sim 10^{16}$ W/cm^2, it has been shown that the rate of energy transfer of the atomic coupling can be extremely high (~ 0.2 mW/atom) and exhibits a strong dependence on the time scale of irradiation. Furthermore, the removal of an electron from an inner principal quantum shell has been demonstrated. Under more extreme conditions, it is expected that energy transfer rates approaching $\sim 0.1 - 1$ W/atom will occur. There is the expectation that an understanding of these physical processes will enable the generation of stimulated emission in the x-ray range.

V. ACKNOWLEDGEMENTS

The author wishes to acknowledge stimulating conversations with H. Jara, U. Johann, T. S. Luk, I. A. McIntyre, A. McPherson, A. P. Schwarzenbach, K. Boyer, H. M. R. Hutchinson, A. Szöke, A. Hauer, and G. Schappert and the technical assistance of R. Slagle, J. Wright, T. Pack, and R. Bernico. This work was supported by the U.S. ONR, the U.S. AFOSR, the SDIO(ISTO), the U.S. DOE, the LLNL, the U.S. NSF, the DARPA, and the LANL.

VI. REFERENCES

1. C. K. Rhodes, Multiphoton Ionization of Atoms, Science 229:1345 (1985).
2. A. P. Schwarzenbach, T. S. Luk, I. A. McIntyre, U. Johann, A. McPherson, K. Boyer, and C. K. Rhodes, Subpicosecond KrF* Excimer Laser Source, Opt. Lett. in press.
3. J. H. Glownia, G. Arjavalingham, P. P. Sorokin, and J. E. Rothenburg, Amplification of 350 fsec pulses in XeCl excimer gain modules, Opt. Lett. 11:79 (1986).
4. F. V. Bunkin et A. M. Prokhorov, Interaction des électrons avec un champ intense de rayonnement optique, in: "Polarisation, Matiére et Rayonnement," édité par La Société Française de Physique, Presses Universitaires de France, Paris (1969).
5. A. Szöke, Interpretation of electron spectra obtained from multiphoton ionisation of atoms in strong fields, J. Phys. B 18:L427 (1985).
6. C. K. Rhodes, Studies of collision-free nonlinear processes in the ultraviolet range, in: "Multiphoton Processes," P. Lambropoulos and S. J. Smith, eds., Springer-Verlag, Berlin (1984).
7. K. Boyer and C. K. Rhodes, Atomic inner-shell excitation induced by coherent motion of outer-shell electrons, Phys. Rev. Lett. 54:1490 (1985).
8. A. Szöke and C. K. Rhodes, A theoretical model of inner-shell excitation by outer-shell electrons, Phys. Rev. Lett. 56:720 (1986).
9. T. S. Luk, U. Johann, H. Egger, H. Pummer, and C. K. Rhodes, Collision-free multiple photon ionization of atoms and molecules at 193 nm, Phys. Rev. A 32:214 (1985).
10. U. Johann, T. S. Luk, H. Egger, and C. K. Rhodes, Rare gas electron energy spectra produced by collision-free multiquantum processes, Phys. Rev. A in press.
11. U. Johann, T. S. Luk, I. A. McIntyre, A. McPherson, A. P. Schwarzenbach, K. Boyer, and C. K. Rhodes, Multiquantum processes at high field strengths, in: "Proceedings of the Topical Meeting on Short Wavelength Coherent Radiation," J. Bokor and D. Attwood, eds., AIP, New York (to be published).
12. U. Johann, T. S. Luk, I. A. McIntyre, A. P. Schwarzenbach, K. Boyer, and C. K. Rhodes, Subpicosecond studies of collision-free multiple ionization of atoms at 248 nm, Phys. Rev. Lett. (submitted).
13. M. Crance and M. Aymar, Dynamics of multiphoton ionisation to multiple continua, J. Phys. B 13:L421 (1980).
14. Y. Gontier and M. Trahin, Spatio-temporal effects in resonant multiphoton ionisation of the caesium atom, J. Phys. B 13:259 (1980).
15. P. Lambropoulos, Mechanisms for multiple ionization of atoms by strong pulsed lasers, Phys. Rev. Lett. 55:2141 (1985).
16. G. Wendin, L. Jönsson, and A. L'Huillier, Screening effects in multielectron ionization of heavy atoms in intense laser fields, Phys. Rev. Lett. 56:1241 (1986).
17. H. Egger, T. S. Luk, K. Boyer, D. F. Muller, H. Pummer, T. Srinivasan, and C. K. Rhodes, Picosecond, tunable ArF* excimer laser source, Appl. Phys. Lett. 41:1032 (1982).
18. U. Johann, T. S. Luk, I. A. McIntyre, A. McPherson, A. P. Schwarzenbach, K. Boyer, and C. K. Rhodes, Multiphoton ionization in intense ultraviolet laser fields, in: "Proceedings of the Topical Meeting on Short Wavelength Coherent Radiation," J. Bokor and D. Attwood, eds., AIP, New York (to be published).
19. A. McPherson, private communication.
20. T. A. Carlson, C. W. Nestor, Jr., N. Wasserman, and J. C. McDowell, Calculated ionization potentials for multiply charged ions, Atomic Data 2:63 (1970).
21. L. V. Keldysh, Ionization in the field of a strong electromagnetic wave, Sov. Phys. -JETP 20:1307 (1965).

22. Yu. P. Raizer, Breakdown and heating of gases under the influence of a laser beam, Sov. Phys.-USP 8:650 (1966).
23. L. I. Schiff, Measurability of electric dipole moments, Phys. Rev. 132:2194 (1963).
24. P. Kruit, J. Kimman, H. G. Muller, and M. J. van der Wiel, Electron spectra from multiphoton ionisation of xenon at 1064, 532, and 355 nm, Phys. Rev. A 28:248 (1983).
25. S.-I. Chu and J. Cooper, Threshold shift and above-threshold multiphoton ionization of atomic hydrogen in intense laser fields, Phys. Rev. A 32:2769 (1985).
26. H. G. Muller, A. Tip, and M. J. van der Wiel, Ponderomotive force and AC stark shift in multiphoton ionisation, J. Phys. B 16:L679 (1983).
27. H. G. Muller and A. Tip, Multiphoton ionization in strong fields, Phys. Rev. A 30:3039 (1984).
28. M. Edwards, L. Pan, and L. Armstrong, Jr., Model study of multiphoton ionization in strong fields, J. Phys. B 17:L515 (1984).
29. Z. Bialynicka-Birula, Strong-field effects in electron spectra from multiphoton ionisation, J. Phys. B 17:2097 (1984).
30. M. H. Mittleman, Kinematics of multiphoton ionization in a steady laser beam, Phys. Rev. A 29:2245 (1984).
31. M. H. Mittleman, Intensity dependence of the ionisation potential of an atom in a resonant laser field, J. Phys. B 17:L351 (1984).
32. J. Bokor, P. H. Bucksbaum, and R. R. Freeman, Generation of 35.5-nm coherent radiation, Opt. Lett. 8:217 (1983).
33. K. Codling, R. P. Madden, and D. L. Ederer, Resonances in the photoionization continuum of Ne I (20-150 eV), Phys. Rev. 155:26 (1967).
34. U. Becker, R. Hölzel, H. G. Kerkhoff, B. Larger, D. Szostak, and R. Wehlitz, Near-threshold resonance enhancement of neon valence satellites studied with synchrotron radiation, Phys. Rev. Lett. 56:1120 (1986).
35. R. L. Carman, C. K. Rhodes, and R. F. Benjamin, Observation of harmonics in the visible and ultraviolet created in CO_2-laser-produced plasmas, Phys. Rev. A 24:2649 (1981).
36. F. Herman and S. Skillman, "Atomic Structure Calculations," Prentice-Hall, Inc., Englewood Cliffs (1963).
37. K.-N. Huang, M. Aoyagi, M. H. Chen, B. Crasemann, and H. Mark, Neutral-atom electron binding energies from relaxed orbital relativistic Hartree-Fock calculations, Atomic Data and Nuclear Data Tables 18:243 (1976).
38. W. C. Martin, R. Zalubas, and L. Hagan, "Atomic Energy Levels - The Rare-Earth Elements," NSRDS - NBS 60, USGPO, WDC (1978).
39. S. Goudsmit and L. Gropper, Many-electron selection rules, Phys. Rev. 38:225 (1931).
40. S. Pasternak, Transition probabilities of forbidden lines, Astrophys. J. 92:129 (1940).
41. D. Layzer and R. H. Garstang, Theoretical atomic transition probabilities, An. Rev. Astron. Astrophys. 6:449 (1968).
42. R. H. Garstang, Forbidden transitions, in: "Atomic and Molecular Processes," D. R. Bates, ed., Academic Press, New York (1962).
43. M. J. van der Wiel and T. N. Chang, Intershell correlation in double-electron ejection from the outermost shell of Xe, J. Phys. B 11:L125 (1978).
44. M. Ya. Amusia, Collective effects in photoionization of atoms, in: "Advances in Atomic and Molecular Physics," D. R. Bates and B. Bederson, eds., Academic Press, New York (1981), Vol. 17.
45. M. Ya. Amusia and N. A. Cherepkov, Many-electron correlations in scattering processes, in: "Case Studies in Atomic Physics 5," E. W. McDaniel and M. R. McDowell, eds., North-Holland, Amsterdam (1975).
46. A. Zangwill and P. Soven, Density functional approach to local field effects in finite systems: Photoabsorption in the rare gases, Phys. Rev. A 21:156 (1980); ibid., Resonant photoemission in barium and cerium, Phys. Rev. Lett. 45:204 (1980); W. Ekhardt and D. B. Tran

Thoai, Collective excitations in atomic shells, Phys. Scr. 26:194 (1982); A. Zangwill, Ph.D. thesis, University of Pennsylvania (1981).

47. T. Åberg, Theory of atomic decay following inner-shell ionization, in: "Photoionization and Other Probes of Many-Electron Interactions," F. J. Wuilleumier, ed., Plenum, New York (1976).

48. G. Wendin, "Breakdown of the One-Electron Pictures in Photoelectron Spectra," Vol. 45 of "Structure and Bonding," Springer-Verlag, Berlin (1981); S. Lundquist and G. Wendin, Theoretical aspects on high-energy excitations, J. Electron. Spectrosc. Relat. Phenom. 5:513 (1974).

49. G. Wendin, in: "Vacuum Ultraviolet Radiation Physics," E. E. Koch, R. Haensel, and C. Kunz, eds., Pergamon, Braunschweig (1974).

50. G. Wendin, The random phase approximation with exchange, in: "Photoionization and Other Probes of Many-Electron Interactions," F. J. Wuilleumier, ed., Plenum, New York (1976); H. P. Kelly and S. L. Carter, Many body perturbation calculations of the interaction of atoms with electromagnetic radiation, Phys. Scr. 21:448 (1980).

51. J. A. R. Samson and G. N. Haddad, Multiple photoionization of the rare gases, Phys. Rev. Lett. 33:875 (1974); J. A. R. Samson, Future experimental problems in photoionization, in: "Photoionization and Other Probes of Many-Electron Interactions," F. J. Wuilleumier, ed., Plenum, New York (1976).

52. J. P. Connerade, On double photoionisation, J. Phys. B. 10:L239 (1977).

53. M. Hollis, Multiphoton ionization and EM field gradient forces, Opt. Commun. 25:395 (1978).

54. P. Agostini, F. Fabré, G. Mainfray, G. Petite, and N. K. Rahman, Free-free transitions following six-photon ionization of xenon atoms, Phys. Rev. Lett. 42:1127 (1979).

55. B. W. Boreham and B. Luther-Davis, High energy electron acceleration by ponderomotive forces in tenuous plasmas, J. Appl. Phys. 50:2533 (1979).

56. P. Kruit, J. Kimman, and M. J. van der Wiel, Absorption of additional photons in the multiphoton ionisation continuum of xenon at 1064, 532, and 440 nm, J. Phys. B. 14:L597 (1981).

57. P. Kruit, J. Kimman, and M. J. van der Wiel, Electron spectra for multiphoton ionisation of xenon at 1064, 532, and 355 nm, Phys. Rev. A. 28:248 (1983).

58. K. G. H. Baldwin and B. W. Boreham, Investigation of tunneling processes in laser-induced ionization of argon, J. Appl. Phys. 52:2627 (1981).

59. L. A. Lompré, A. L'Huillier, G. Mainfray, and C. Manus, Laser-intensity effects in the energy distributions of electrons produced in multiphoton ionization of rare gases, J. Opt. Soc. Am. B 2:1906 (1985).

60. T. J. McIlrath, M. Bashkansky, P. Bucksbaum, and R. R. Freeman, Suppression of multiphoton ionization with circularly polarized light, in: "Proceedings of the Topical Meeting on Short Wavelength Coherent Radiation," J. Bokor and D. Attwook, eds., AIP, New York (to be published).

61. P. Venugopala Rao, Inner-shell transition measurements with radioactive atoms, in: "Atomic Inner-Shell Processes II," B. Crasemann, ed., Academic Press, New York (1975).

62. W. Bambynek, K-shell flourescence yields, in: "Proceedings of the International Conference on Inner-Shell Ionization Phenomena and Future Applications," Vol. 1, CONF-720404 (1973).

63. A. Moljk, K-shell fluorescence yields of low-Z elements determined from gaseous samples, in: "Proceedings of the International Conference on Inner-Shell Ionization Phenomena and Future Applications, Vol. 1, CONF-720404 (1973).

64. G. S. Ezra and R. S. Berry, Collective and independent particle motion in doubly excited two-electron atoms, Phys. Rev. A 28:1974 (1983).

65. G. S. Ezra and R. S. Berry, Quantum states of two particles on concentric spheres, Phys. Rev. A 28:1989 (1983).

66. U. Fano, Correlation of excited electrons: progress in the alkaline earth and other spectra, Phys. Scr. 24:656 (1981).

67. M. E. Kellman and D. R. Herrick, Ro-vibrational collective interpreta-

tions of supermultiplet classifications of intrashell levels of two-electron atoms, Phys. Rev. A 22:1536 (1980).
68. D. R. Herrick, New symmetry properties of atoms and molecules, Adv. Chem. Phys. 52:1 (1982).
69. I. C. Percival, Planetary atoms, Proc. Roy. Soc. London A353:289 (1977).
70. R. P. Madden and K. Codling, Two-electron excitation states in helium, Astrophys. J. 141:364 (1965).
71. U. Fano, Doubly excited states of atoms, in: "Atomic Physics," B. Bederson, V. W. Cohen, and F. M. J. Pichanik, eds., Plenum, New York (1969).
72. The following volumes also contain considerable information on multiply excited atomic states: "Beam-Foil Spectroscopy," S. Bashkin, ed., Springer-Verlag, Berlin (1976); "Beam-Foil Spectroscopy," Vol. 1 & 2, I. A. Sellin and D. J. Pegg, eds., Plenum Press, New York (1976).
73. W. Heisenberg, Zur quantentheorie der multiplettstruktur und der anomalen zeemaneffekte, Z. Phys. 32:841 (1925).
74. E. U. Condon, The theory of complex spectra, Phys. Rev. 36:1121 (1930).
75. J. Berkowitz, J. L. Dehmer, Y.-K. Kim, and J. P. Desclaux, Valence shell excitation accompanying photoionization in mercury, J. Chem. Phys. 61:2556 (1974).
76. S. Süzer and D. A. Shirley, Initial-state configuration-interaction satellites in the photoemission spectrum of Cd, J. Chem. Phys. 61:2481 (1974); S. Süzer, S. T. Lee, and D. A. Shirley, Correlation satellites in the atomic photoelectron spectra of group-IIA and -IIB elements, Phys. Rev. A 13:1842 (1976); S. T. Lee, S. Süzer, E. Mathias, R. A. Rosenberg, and D. A. Shirley, Configuration interaction effects in the atomic photoelectron spectra of Ba, Sm, Eu, and Yb, J. Chem. Phys. 66:2496 (1977).
77. S. Süzer, M. S. Banna, and D. A. Shirley, Relativistic and correlation effects in the 21.2-eV photoemission spectrum of atomic lead, J. Chem. Phys. 63:3473 (1975).
78. S. Süzer, S. T. Lee, and D. A. Shirley, PES of atomic and molecular bismuth, J. Chem. Phys. 65:412 (1976).
79. M. W. D. Mansfield and J. P. Connerade, On the simultaneous excitation of two electrons in neutral atomic zinc, Proc. Roy. Soc. London A359:389 (1978); M. W. D. Mansfield and G. H. Newsom, The Ca I absorption spectrum in the vacuum ultraviolet: excitation of the 3p-subshell, Proc. Roy. Soc. London A357:77 (1977); M. W. D. Mansfield, The simultaneous excitation of two electrons in atomic cadmium, Proc. Roy. Soc. London A362:129 (1978); J. Stöhr, R. Jaeger, and J. J. Rehr, Transition from adiabatic to sudden core-electron excitation: N_2 on Ni(100), Phys. Rev. Lett. 51:821 (1983); G. B. Armen, T. Åberg, K. R. Karim, J. C. Levin, B. Crasemann, G. S. Brown, M. H. Chen, and G. E. Ice, Threshold double photoexcitation of argon with synchrotron radiation, Phys. Rev. Lett. 54:182 (1985).
80. D. M. Holland and K. Codling, Double photoionisation in Ti and Pb in the region of the 5d transition, J. Phys. B 13:L745 (1980).
81. T. N. Chang and R. T. Poe, Double photoionization of neon, Phys. Rev. A 12:1432 (1975).
82. C. Stroller, W. Wölfli, G. Bonani, M. Stöckli, and M. Suter, Two-electron one-photon transitions into the doubly ionized K-shell, Phys. Rev. A. 15:990 (1977).
83. C. K. Rhodes and R. M. Hill, "Laser Excitation of Inner-Shell Atomic States by Multiquantum Amplitudes," Molecular Physics Laboratory Memorandum MP 77-80a, SRI International (May, 1977).
84. P. Feldman and R. Novick, Auto-ionizing states in the alkali atoms with microsecond lifetimes, Phys. Rev. 160:143 (1967).
85. J. Sugar, T. B. Lucatorto, T. J. McIlrath, and A. W. Weiss, Even-parity autoionizing states in neutral sodium (350-400 Å), Opt. Lett. 4:109 (1979).
86. J. R. Willison, R. W. Falcone, J. C. Wang, J. F. Young, and S. E. Harris, Emission spectra of core excited even-parity ^{2}p states of neutral lithium, Phys. Rev. Lett. 44:1125 (1980).

87. J. J. Wynne and J. P. Hermann, Spectroscopy of even-parity autoionizing levels in Ba, Opt. Lett. 4:106 (1979); D. L. Ederer, T. B. Lucatorto, E. B. Saloman, R. P. Madden, M. Manalis, and J. Sugar, Photoabsorption of the 4d electrons in xenon and barium: a comparison, in: "Electron and Photon Interactions with Atoms," H. Kleinpoppen and M. R. C. McDowell, eds., Plenum Press, New York (1976).
88. W. E. Cooke and T. F. Gallagher, Calculation of autoionization rates for high-angular-momentum Rydberg states, Phys. Rev. A 19:2151 (1979).
89. "Etats Atomiques et Moléculaires Couplés a un Continuum Atoms et Molécule Hautement Excites," Colloques Internatoinaux de C.N.R.S. no. 273, Paris (1977).
90. D. M. P. Holland, K. Codling, J. B. West, and G. V. Marr, Multiple photoionisation in the rare gases from threshold to 280 eV, J. Phys. B 12:2465 (1979).
91. D. P. Spears, H. J. Fischbeck, and T. A. Carlson, Satellite structure in the x-ray photoelectron spectra of rare gases and alkali-metal halides, Phys. Rev. A 9:1603 (1974).
92. R. P. Madden, D. L. Ederer, and K. Codling, Resonances in the photoionization continuum of Ar I (20 - 150 eV), Phys. Rev. 177:136 (1969).
93. K. Codling and R. P. Madden, Newly observed structure in the photoionization continua of Kr and Xe below 160 Å, Appl. Opt. 4:1431 (1965).
94. K. Codling and R. P. Madden, Resonances in the photoionization continua of Kr and Xe, Phys. Rev. A 4:2261 (1971).
95. K. Codling and R. P. Madden, The absorption spectra of krypton and xenon in the wavelength range 330 - 600 Å, J. Res. Nat. Bur. Stand. 76a:1 (1972).
96. J. D. Jackson, "Classical Electrodynamics," Second Edition, John Wiley and Sons, New York (1975).
97. F. Bloch, Bremsvermögen von atomen mit mehreren elektronen, Z. Phys. 81:363 (1933); E. T. Verkhovtseva, E. V. Gnatchenko, and P. S. Pogrebnjak, Investigation of the connection between 'giant' resonances and 'atomic' bremsstrahlung, J. Phys. B 16:L613 (1983).
98. M. Goldhaber and E. Teller, On nuclear dipole vibrations, Phys. Rev. 74:1046 (1948).
99. "Giant Multipole Resonances," F. E. Bertrand, ed., Harwood Academic, London (1980); F. Cannata and H. Überall, "Giant Resonance Phenomena in Intermediate-Energy Nuclear Reactions," Springer-Verlag, Berlin (1980).
100. On the stability of the motion of Saturn's rings, "The Scientific Papers of James Clerk Maxwell," W. D. Niven, ed., Dover, New York (originally published in 1890). This reference was first indicated to the author by William E. Cooke.
101. R. M. Jopson, R. R. Freeman, W. E. Cooke, and J. Bokor, Electron shake-up in two-photon excitation of core electrons to Rydberg autoionizing states, Phys. Rev. Lett. 51:1640 (1983).
102. D. Chattarji, "The Theory of Auger Transitions," Academic Press, New York (1976).
103. D. Pines, "Elementary Excitation in Solids," W. A. Benjamin, Inc., New York, (1964).
104. T. S. Luk, H. Pummer, K. Boyer, M. Shahidi, H. Egger, and C. K. Rhodes, Anomalous collision-free multiple ionization of atoms with intense picosecond ultraviolet radiation, Phys. Rev. Lett. 51:110 (1983).
105. A. L'Huillier, L.-A. Lompré, G. Mainfray, and C. Manus, Multiply charged ions formed by multiphoton absorption processes in the continuum, Phys. Rev. Lett. 48:1814 (1982).
106. A. L'Huillier, L.-A. Lompré, G. Mainfray, and C. Manus, Multiply charged ions produced by multiphoton absorption in rare gas atoms, in: "Proceedings of the Conference on Laser Techniques in the Extreme Ultraviolet," Vol. 119, S. E. Harris and T. B. Lucatorto, eds., AIP, New York (1984).
107. M. J. Feldman and R. Y. Chiao, Single-cycle electron acceleration in focused laser fields, Phys. Rev. A 4:352 (1971).

108. E. S. Sarachik and G. T. Schappert, Classical theory of the scattering of intense laser radiation by free electrons, Phys. Rev. D 1:2738 (1970).
109. T. P. Hughes, "Plasmas and Laser Light," John Wiley and Sons, New York (1975).
110. P. Mora, Theoretical model of absorption of laser light by a plasma, Phys. Fluids 25:1051 (1982).
111. P. D. Gupta, R. Popil, R. Fedosejevs, A. A. Offenberger, D. Salzmann, and C. E. Capjack, Temperature and x-ray intensity scaling in KrF laser plasma interaction, Appl. Phys. Lett. 48:103 (1986).
112. P. Kaw and J. Dawson, Relativistic nonlinear propagation of laser beams in cold overdense plasmas, Phys. Fluids 13:472 (1970).
113. G. J. Pert, Inverse bremsstrahlung absorption in large radiation fields during binary collisions in the Born approximation, J. Phys. B 11:1105 (1978); R. J. Dewhurst, G. J. Pert, and S. A. Ramsden, Laser-induced breakdown in the rare gases using picosecond Nd:glass laser pulses, J. Phys. B. 7:2281 (1974).
114. T. A. Carlson and M. O. Krause, Relative abundances and recoil energies of fragment ions formed from the x-ray photoionization of the N_2, O_2, CO, NO, CO_2, and CF_4, J. Chem. Phys. 56:3206 (1972).
115. W. Eberhardt, J. Stöhr, J. Feldhaus, E. W. Plummer, and F. Sette, Correlation between electron emission and fragmentation into ions following soft x-ray excitation of the N_2 molecule, Phys. Rev. Lett. 51:2370 (1983).
116. J. W. Shearer, J. Garrison, J. Wong, and J. E. Swain, Pair production by relativistic electrons from an intense laser focus, Phys. Rev. A. 8:1582 (1973).
117. A. I. Akhiezer and V. B. Berestetskii, "Quantum Electrodynamics," Wiley-Interscience, New York (1965).
118. D. R. Bach, D. E. Casperson, D. W. Forslund, S. J. Gitomer, P. D. Goldstone, A. Hauer, J. F. Kephart, J. M. Kindel, R. Kristal, G. H. Kyrda, K. B. Michell, D. B. Hulsteyn, and A. A. Williams, Intensity-dependent absorption in 10.6 μm laser-illuminated spheres, Phys. Rev. Lett. 50:2082 (1983); A. Hauer et al., Super thermal electron generation, transport and deposition in CO_2 laser irradiated targets, in: "Laser Interaction and Related Plasma Phenomena," Vol. 6, H. Hora and G. Miley, eds., Plenum, New York (1984).
119. R. A. Granden, The status of laser fusion, in: "Strongly Coupled Plasmas," G. Kalman and P. Corini, eds., Plenum Press, New York (1978).
120. E. L. Chapp, "Gamma Ray Astronomy," Reidel, Dordrecht (1976).

USEFUL ACRONYMS

APS
Appearance potential spectroscopy

BE
Binding energy

BIS
Bremsstrahlung isochromat spectroscopy

CD
Circular dichroism

CI
Configuration interaction

CIC
Configuration interaction in the continuum

EA
Electron affinity

EELS
Electron energy loss spectroscopy

ELS
Energy loss spectroscopy

ESD
Electron stimulated desorption

EXAFS
Extended X-ray absorption fine structure

FWHM
Full width at half maximum

g-H
g-Hartree method

g-TF
g-Thomas Fermi method

GDR
Giant dipole resonances

HF
Hartree Fock calculations

IP
Ionization potential

LDA
local density approximation

LDRPA
Local density random phase approximation

LEED
Low energy electron diffraction

LSHF
LS-dependent Hartree-Fock calculations
(L-total angular mementum, S=total spin)

MBPT
Many body perturbation theory

MCD
Magnetic circular dichroism

MCHF
Multi configurational Hartree-Fock

MPI
Multi-photon ionization

MXD
Magnetic X-ray dichroism

NATO
North Atlantic Treaty Organisation

OTAN
Organisation du Traité Atlantique Nord

PES
Photoemission spectra

PEY
Total photoelectron yield

PSD
Photon stimulated desorption

QED
Quantum electro-dynamics

QFT
Quantum field theory

RBIS
Resonant bremsstrahlung isochromat spectroscopy

RLDA
Relativistic local density approximation

RLDI
Resonant laser driven ionization

RMPI
Resonant multi-photon ionization

RPA
Random phase approximation

RPES
Resonant photoemission spectra

RRPA
Relativistic random phase approximation

RRPAE
Relativistic random phase approximation with exchange

SAC
Symmetry adapted configuration

SEXAFS
Surface extented X-ray absorption fine structure

SXA
Soft X-ray absorption

SXAPS
Soft X-ray appearance potential spectroscopy

TDLDA
Time dependant local density approximation

UPS
Ultra violet photoelectron spectroscopy

UTA
Unresolved transition arrays

VUV
Vacuum ultra violet

XANES
X-ray absorption near edge structure

XAS
X-ray absorption spectroscopy

XPS
X-ray photoelectron spectroscopy

PARTICIPANTS

ADAM Maryvonne
Laboratoire LURE
Batiment 209 D,
Université de Paris-Sud,
91405 Orsay Cedex
FRANCE

BECKER Uwe
Institut fur Strahlungs-
und Kernphysik
Technisches Universitat Berlin,
Hardenberg strasse 36,
1000 Berlin 33,
WEST GERMANY

BRIGGS John S
Facultat fur Physik
Universitat Freiburg,
Hermann Herder strasse 3,
7800 Freiburg,
WEST GERMANY

CANKURTARAN M.E.
Faculty of Engineering,
Department of Physics,
Hacettepe University,
Beytepe Ankara,
TURKEY

CHOPRA D.R.
Department of Physics,
East Texas Station,
Commerce,
Texas 75428,
USA

CLARK Charles
US Department of Commerce,
National Bureau of Standards,
Gaithersburg,
Maryland 20899,
USA

CONNERADE Jean-Patrick
The Blackett Laboratory,
Imperial College,
London SW7 2AZ,
UNITED KINGDOM

DECLEVA Piero
Dipartimento di Scienze Chimiche,
Universita degli studi di Trieste
Piazzala Europa 1
34127 Trieste,
ITALY

DIETZ Klaus
Physikalisches Institut
Universitat Bonn,
53 Bonn 1,
Nussallee 12,
WEST GERMANY

ESTEVA Jean-Marc
Labo de Spectroscopie
Atomique et Ionique
Batiment 350,
91405 Orsay Cedex
FRANCE

FAUCHER Michèle
Laboratoire des Eléments de Transition dans les Solides (ER N° 210)
1 place A. Briand,
92195 Meudon
FRANCE

FUGGLE John C.
Laboratorium voor Fysische Chemie,
University Nijmegen,
Toernooive
Nijmegen,
THE NETHERLANDS

COMBET FARNOUX Françoise
Labo de Photophysique Moléculaire
Batiment 213,
Université de Paris-Sud,
91405 Orsay Cedex,
FRANCE

GARCIA Denis
Laboratoire des Eléments de Transition dans les Solides (ER No210)
1 place A. Briand,
92195 Meudon
FRANCE

GARTON W.R.S.
Blackett Laboratory,
Imperial College,
London SW7 2AZ,
UNITED KINGDOM

GRIESMANN Ulf
Physikalisches Institut
Universitat Bonn,
53 Bonn 1,
Nussallee 12,
WEST GERMANY

GRIFFIN Donald C.
Department of Physics,
Rolling College, Campus Box 2743,
1000 Holt Avenue,
Winter Park, Florida 32789
USA

GUNNARSSON D.R.
Max Planck Institut fur
Festkoperforschung
Heisenbergstrasse 1,
7000 Stuttgart 80,
WEST GERMANY

HITCHCOCK Adam P.
Institut for Materials Research,
McMaster University,
1280 Main Street West, Hamilton,
Ontario L8S 4MI,
USA

HORMES Josef
Physikalisches Institut,
Universitat Bonn,
53 Bonn 1,
Nussallee 12,
WEST GERMANY

FRYE Daniel D.
Physics Department,
University of Virginia,
Charlottesville,
Virginia 22901,
USA

KARNATAK Ramesh
Labo de Spectroscopie
Atomique et Ionique,
Batiment 350,
Université de Paris-Sud,
91405 Orsay Cedex,
FRANCE

KELLY Hugh P.
Department of Physics,
University of Virginia,
Charlottesville,
Virginia 22901,
USA

KERHOFF Hans Georg
Institut fur Strahlungs-
und Kernphysik
Sekr PN 3 2
1000 Berlin 12,
WEST GERMANY

KOMNINOS Y.
Theoretical and Physical Chemistry
Institute, NHRF,
48 Vassileos Constantinou avenue,
Athens 501/1,
GREECE

KORKMAZ Mustafa
Hacettepe University,
Faculty of Engineering,
Department of Physics,
Beytepe, Ankara,
TURKEY

LEFEBVRE BRION Helene
Labo de Photophysique Moléculaire,
Batiment 213,
Université de Paris-Sud,
91405 Orsay Cedex,
FRANCE

LISINI Adriana
Dipartimento di Scienze Chimiche,
Universita degli studi di Trieste,
Piazzale Europa 1,
34127 Trieste,
ITALY

LOUBRIEL Guillermo
Division 1152,
Sandia National Laboratories,
Alburquerque,
New Mexico 87185,
USA

MANSFIELD Mike W.D.
Physics Department,
University College,
Cork,
IRELAND

MARTIN Philippe
Laboratoire de Chimie Physique,
11 Rue Pierre et Marie Curie,
75231 Paris Cedex,
FRANCE

MATTHEW James
Department of Physics,
University of York,
Heslington,
York YO1 5DD,
UNITED KINGDOM

MILLACK Thomas
Physikalisches Institut,
Universitat Bonn,
53 Bonn 1,
Nussallee 12,
WEST GERMANY

MAYHEW Chris
The Blackett Laboratory,
Imperial College,
LONDON SW7 2AZ,
UNITED KINGDOM

MORIN Paul
Laboratoire LURE,
Batiment 209 D
Université de Paris-Sud,
91405 Orsay Cedex,
FRANCE

MOSER Hans R.
Institute for Theoretical Physics,
Chalmers University,
41296 Goterborgs,
SWEDEN

NENNER Irène
Laboratoire LURE
Batiment 209 D,
Université de Paris-Sud,
91405 Orsay Cedex
FRANCE

NICOLAIDES Cleanthis A.
Theoretical and Physical Chemistry
Institute, NHRF,
48 Vassileos Constantinou Avenue,
Athens 501/1,
GREECE

NUROH Kofi
Department of Mathematics,
University of Benin,
Benin City,
NIGERIA

OH Se Jung
Department of Physics,
Seoul National Universty
Seoul 151,
KOREA

O'SULLIVAN Gerard
School of Physical Sciences,
National Institute for Higher
Education
Dublin 9,
IRELAND

PETROFF Yves
Laboratoire LURE,
Batiment 209 D,
Université de Paris-Sud,
91405 Orsay Cedex
FRANCE

QUINEY Harry M.
Department of Theoretical Chemistry,
1 South Parks Road,
Oxford OX1 3TG,
UNITED KINGDOM

RHODES Charles K.
Department of Physics,
University of Illinois at Chicago,
P.O. Box 4348,
Chicago, Illinois 60680,
USA

RICHTER M.
Institut fur Experimentalphysik,
Universitat Hamburg,
Luruperchaussee 149,
2000 Hamburg 50,
WEST GERMANY

ROSSI Georgio
Laboratoire LURE,
Batiment 209 D,
Université de Paris-Sud,
91405 Orsay Cedex
FRANCE

SCHNEIDER W.D.
Institut de Physique,
Université de Neuchatel,
Rue A.L. Bréguet 1,
2000 Neuchatel,
SWITZERLAND

SMITH James
Los Alamos National Laboratory,
Los Alamos
New Mexico 87545,
USA

SUSKIN Mark
Physics Department,
John Hopkins University,
Baltimore,
Maryland 21218,
USA

Van der LAAN G.
Laboratory of Physical Chemistry,
Nijenborgh 16,
9747 AG Groningen,
THE NETHERLANDS

WENDIN Goran
Institute of Theorical Physics,
Chalmers University of Technology,
412 96 Goterborg,
SWEDEN

YOUNGER Stephen M.
Lawrence Livermore Natl Lab,
L 17 Livermore,
California 94550,
USA

ZANDY Hassan F.
Chairman Sciences
Maths Depaqrtments
Milford Academy
Milford, Connecticut 06460,
USA

ZANGWILL Andrew
School of Physics,
College of Sciences & Liberal
Studies,
Georgia Institute of Technology,
Atlanta, Georgia 30332,
USA

ZIMMERMANN Peter
Institut fur Strahlungs-
und Kernphysik
Sekr PN3 2,
Hardenbergstrasse 36,
1000 Berlin 12,
WEST GERMANY

AUTHOR INDEX

ALTUN Zikiri	71		SCOTT M.J.	115
BAER Y.	463		SMITH J.L.	311
BAIG M.A.	463		SOMMER K.	225
BECKER U.	473		SONNTAG B.	251
BORING A.M.	311		VAN DER LAAN G.	447
CHOPRA D.R.	483		WENDIN G.	171
CLARK	137		WETZEL H.E.	251
COMBET FARNOUX FRANCOISE	153		WEYMANS G.	243
CONNERADE J.P.	3,225		YOUNGER S.M.	237
COWAN ROBERT D.	25		ZANGWILL ANDREW	321
DELLEY B.	463		ZAREMBA E.	115
DIETZ K.	49		ZIMMERMANN P.	247
ESTEVA J.M.	361			
FISHER P.	281			
FUGGLE J.C.	381			
GRIFFIN D.C.	25			
GUNNARSSON O.	405			
HITCHCOCK A.P.	281			
HORMES J.	225			
KARNATAK R.C.	361			
KELLY HUGH P.	71			
LEFEBVRE-BRION H.	301			
LOUBRIEL G.M.	339			
LUCATORTO T.	137			
MANSFIELD M.W.D.	91			
MATTHEW J.A.D.	353			
McLAREN R.	281			
MILLACK Th.	243			
MORIN P.	291			
NENNER I.	259			
NICOLAIDES C.A.	213			
NUROH K.	115			
O'SULLIVAN G.	505			
OH S.J.	431			
PATTHEY F.	463			
PINDZOLA M.S.	25			
RHODES C.K.	533			
ROSSI G.	491			
SCHMIDT E.	251			
SCHMIDT M.	247			
SCHNEIDER W.D.	463			
SCHONHAMMER K.	405			

INDEX

Accordion resonance, 14, 15
Anderson model, 405, 407, 426, 463-465
Appearance potential spectroscopy (APS), 557
Atomic correlation in the lanthanides 382-386
Atomic dipole-polarizability, 382-386
Atomic response to intense radiation 535-540
Auger decay, 163, 187, 191, 248, 253, 259, 267, 278, 370, 422, 444, 478, 492, 493, 544
Auger electron process, 298
Auger resonance, 431, 444
Auger spectrum, 271, 279, 417
Autoionizing resonances, 3, 9, 232

Beutler-Fano resonance 10, 11
Born-Approximation, 353
Born-Coulomb approximation, 238, 241
Born-Oppenheimer approximation, 265
Brillouin's Theorem, 35
Bruecker Orbitals, 75

Clebsch Gordon coefficients, 121
Cluster-model, 452
Collective effects
  in electron impact interactions, 192-200
  and giant resonances, 174-176, 200-207, 219, 322
  and linear response, 172-174, 176-184, 203
Collective model, 5
Collisonal processes
  radiatively excited, 547-549
Configuration interaction, 316
Continuum resonance in molecules, 273, 277-278

Core level spectroscopy in lanthanides, 399-400
Curie-Weiss magnetic behaviour, 317

Density correlations, 57
Density quasiparticle, 57
Desorption in rare earths
  electron stimulated, 347-350
  photon stimulated, 339-347
Dipole collision strength, 38
Dipole exchange integral, 38
Dirac-Fock, 19
Dirac-Hartree-Fock, 62
Dirac sea, 49, 52, 54, 69
Dirac spinors, 60, 61
Direct-ionization cross-section, 41-45
Discrete resonances
  in molecules, 260, 266-272
  in molecular ions, 275-276
Distorted-wave calculations, 41
Double excitation, 94
Double well potential, 12, 13

EELS in rare earths, 353-360, 390
Effective quantum number, 26, 28-31, 138
Electron impact interactions, 36, 35-41, 43-45, 238-242
Electron positron pair production in plasma, 549-550
Electrons in atoms, 312-313
Electrons in metals, 313-317
Electrostatic interactions
  direct, 33
  exchange, 33
Energy density functional, 58
EXAFS, 11, 284, 447
Fermi-Dirac statistics, 51
Fermi energy, 57, 312, 323, 406, 409, 424
Fermi distribution, 53, 69

Fermi gas, 58
Fermi level, 435, 483
Fermi liquid theory, 318
Fermi sea, 216, 218, 409, 468
Feynman gauge, 59
Feshbach rope resonance, 14, 15
Fluorine cage, 281

g-Hartree energy functional, 66
g-Hartree mean-field theories, 49, 54, 56, 58-69, 243, 246
g-Hartree negative energy states, 68
g-Hartree orbitals, 55
g-Hartree total energies, 66
g-Thomas Fermi equation, 224-246
Giant resonance, 3, 4, 7-23, 34, 74, 91, 95-98, 111, 116, 127, 172, 181, 186-187, 190, 199, 215, 292, 304, 306, 321
  in actinides, 101-102
  in closed shell atoms, 75-82
  conditions for, 381
  decay of, 477-480
  discrete decaying, 162-164
  double, 329
  in heavy elements, 153
  in lanthanides, 91-101, 251-253
  in open shell atoms, 82-87
  in organic molecules, 283-289
  in photoabsorption spectra, 91, 140
  in transition elements, 102-110, 137-150, 247-249, 253-254, 333-336, 353
Goldstone diagrams, 73
Grand canonical potential, 54, 56, 58

Hartree-Fock, 6, 8, 15, 18, 25, 33-35, 45, 77-78, 80, 82, 95, 102, 106, 137, 141-143, 145-148, 150, 157, 158, 160, 165, 167, 171, 174, 184, 204, 215, 218, 233, 315, 323, 356, 407, 450
  energies, 84
  multiconfiguration (MCHF), 146, 149
Hartree-Fock-Slater, 25, 27, 147, 206

Hartree plus statistical exchange, 25, 27, 32
Herman-Skillman potential, 76, 79, 157, 160
Hubbard-Stratonovitch transformation, 51, 54
Hund's rule, 312, 318

Independent particle model, 92, 157
Interacting resonance method, 84
Interchannel interactions, 6, 16-18, 160, 161-168, 228
Ion production in rare earths, 540-542
Ionization cross-section, 42, 44
Isoelectronic sequences, 33, 36, 44-45

Jahn-Teller effect, 265

Keldysh-Raizer formula, 537
Kondo energy, 411, 414, 422
Kondo resonance, 315-316, 318, 405, 466
Kondo volume collapse model, 436-437

Levinson's Theorem, 13
Lifetime broadening, 9
Linear response, 116-118
  non-, 201, 202
Linear saturation, 368-373
Local density approximation (LDA), 117, 118, 123, 204, 323
Local density functional approx., 315
Local-moment magnetism in transition elements, 311, 318

Magnetic X-ray dichroism, 456-458
Many body interactions, 11, 17, 19
Many body perturbation theory, 71-72, 75, 82-86
Many electron system
  relativistic, 50
  non-relativistic, 50
Mayer-Fermi Theory, see Thomas-Fermi
Molecular coulomb explosion, 549
Monochromator resolution, 346-368
Mott transition model, 436-

Mott transition model (continued), 437, 450
MQDT, see QDT
Multichannel scattering theory, 162, 214
Multichannel quantum defect theory, 227, 228, 235
Multiphoton absorption, 139
Multiplet interacions, 387-391
Multiplet splitting in XPS, 391-392

Orbital collapse, see Wave function collapse
Optimal basis, 56, 58
Optical depth effects, 523-525

Partial cross-section, 97, 124-125, 185, 190
Partition function, 50-54, 59, 61, 243
Photoabsorption
  by atoms, 91, 323
  by fluorine cage molecules, 91
  in molecules, 269, 272, 291
  spectra of alkaline atoms, 231-232
  spectra of rare gases, 225, 227-231
  spectra in transition elements, 93, 97-98, 104-105, 107, 109, 114, 125-137, 154, 158, 162, 164
  spectra of uranium, 186, 192, 199
  spectra of xenon, 92, 93
  transitions, 34, 41

Photo electron spectra, 97
Photo-emission cross-section, 81
Photoionisation cross-section, 72, 77, 79, 92, 140, 146, 179, 183
  of Barium, 34
  with excitation, 84
Photon stimulated desorption, 20
Promotion model, 433-435

Quantum scattering model of 'giant-resonance', 6, 8, 11, 13, 18
Quantum defect theory (QDT), 13, 17
Quantum defect, 226, 230, 233, 235

R-matrix theory, 84
Random phase approximation (RPA), 115, 132, 134, 168, 173, 324
  with exchange (RPAE), 71, 72, 75-76, 79, 82-83, 85, 105, 165, 179, 181
Resonance photoemission, 417-425, 431-445
Resonance in rare earths, 374-376
Resonant photoemission in solids, 330-333
Resonant satellites, 188, 191
  inner shell, 259
Resonant transitions in lanthanides, 484-488
Response function, 118-123, 124, 127, 133
Rydberg series, 148-149
Rydberg transitions, 285, 291

Satellite structure, 452-455
Schwinger's theorem, 6, 8-10
Seaton's theorem, 13
Second quantization, 57
Self-consistent mean field, 54
Shake up, 544
Shape resonance, 155-161, 181, 191, 215, 282, 291, 304-305, 323, 478
  giant, 301
  in molecules, 259, 263, 301-304
Slater parameters, 33
Spin orbit interaction, 167-169
Stark broadening, 207
Super-Coster-Kronig transition, 478
Super-conductivity in transition elements, 311, 313, 318

Time dependent local density approximation TDLDA, 115, 118, 128, 129, 132, 134, 147, 325-329
Thomas Fermi, 55, 58, 245
  dirac model, 25, 245
  potential, 30
Two electron ionization ladder, 220-222

Unrelaxed orbitals, 78
Unresolved transition array, 515-518
  line identification in, 513-514

Wannier theory, 222
Wavefunction collapse, 12, 18, 25-26, 29, 31, 33, 45, 94, 96, 102-103, 108-109, 131, 137, 141-142, 144, 147, 154, 157, 160-163, 174, 199, 232, 361, 473-477, 480, 494
Woods-Saxon potential, 323

Xanes, 11, 558
XUV continuum
  origins of, 526-528